HOMOLOGICAL ALGEBRA

PRINCETON MATHEMATICAL SERIES

Edited by PHILLIP A. GRIFFITHS, MARSTON MORSE, and ELIAS M. STEIN

1. The Classical Groups, by HERMANN WEYL
3. An Introduction to Differential Geometry, by LUTHER PFAHLER EISENHART
4. Dimension Theory, by W. HUREWICZ and H. WALLMAN
6. The Laplace Transform, by D. V. WIDDER
7. Integration, by EDWARD J. MCSHANE
8. Theory of Lie Groups: 1, by C. CHEVALLEY
9. Mathematical Methods of Statistics, by HARALD CRAMÉR
10. Several Complex Variables, by S. BOCHNER and W. T. MARTIN
11. Introduction to Topology, by S. LEFSCHETZ
12. Algebraic Geometry and Topology, edited by R. H. FOX, D. C. SPENCER, and A. W. TUCKER
14. The Topology of Fibre Bundles, by NORMAN STEENROD
15. Foundations of Algebraic Topology, by SAMUEL EILENBERG and NORMAN STEENROD
16. Functionals of Finite Riemann Surfaces, by MENAHEM SCHIFFER and DONALD C. SPENCER
17. Introduction to Mathematical Logic, Vol. 1, by ALONZO CHURCH
19. Homological Algebra, by H. CARTAN and S. EILENBERG
20. The Convolution Transform, by I. I. HIRSCHMAN and D. V. WIDDER
21. Geometric Integration Theory, by H. WHITNEY
22. Qualitative Theory of Differential Equations, by V. V. NEMYTSKII and V. V. STEPANOV
23. Topological Analysis, by GORDON T. WHYBURN (revised 1964)
24. Analytic Functions, by AHLFORS, BEHNKE and GRAUERT, BERS, et al.
25. Continuous Geometry, by JOHN VON NEUMANN
26. Riemann Surfaces, by L. AHLFORS and L. SARIO
27. Differential and Combinatorial Topology, edited by S. S. CAIRNS
28. Convex Analysis, by R. T. ROCKAFELLAR
29. Global Analysis, edited by D. C. SPENCER and S. IYANAGA
30. Singular Integrals and Differentiability Properties of Functions, by E. M. STEIN
31. Problems in Analysis, edited by R. C. GUNNING

HOMOLOGICAL ALGEBRA

By

HENRI CARTAN

and

SAMUEL EILENBERG

LONDON: GEOFFREY CUMBERLEGE
OXFORD UNIVERSITY PRESS
PRINCETON:
PRINCETON UNIVERSITY PRESS
1956

Published, 1956, by Princeton University Press
L.C. Card No. 53-10148
I.S.B.N. 0-691-07977-3

Second Printing 1960
Third Printing 1961
Fourth Printing 1965
Fifth Printing 1966
Sixth Printing 1970

COMPOSITION BY THE PITMAN PRESS, BATH
PRINTED IN THE UNITED STATES OF AMERICA

Preface

During the last decade the methods of algebraic topology have invaded extensively the domain of pure algebra, and initiated a number of internal revolutions. The purpose of this book is to present a unified account of these developments and to lay the foundations of a full-fledged theory.

The invasion of algebra has occurred on three fronts through the construction of cohomology theories for groups, Lie algebras, and associative algebras. The three subjects have been given independent but parallel developments. We present herein a single cohomology (and also a homology) theory which embodies all three; each is obtained from it by a suitable specialization.

This unification possesses all the usual advantages. One proof replaces three. In addition an interplay takes place among the three specializations; each enriches the other two.

The unified theory also enjoys a broader sweep. It applies to situations not covered by the specializations. An important example is Hilbert's theorem concerning chains of syzygies in a polynomial ring of n variables. We obtain his result (and various analogous new theorems) as a theorem of homology theory.

The initial impetus which, in part, led us to these investigations was provided by a problem of topology. Nearly thirty years ago, Künneth studied the relations of the homology groups of a product space to those of the two factors. He obtained results in the form of numerical relations among the Betti numbers and torsion coefficients. The problem was to strengthen these results by stating them in a group-invariant form. The first step is to convert this problem into a purely algebraic one concerning the homology groups of the tensor product of two (algebraic) complexes. The solution we shall give involves not only the tensor product of the homology groups of the two complexes, but also a second product called their *torsion* product. The torsion product is a new operation derived from the tensor product. The point of departure was the discovery that the process of deriving the torsion product from the tensor product could be generalized so as to apply to a wide class of functors. In particular, the process could be iterated and thus a sequence of functors could be obtained from a single functor. It was then observed that the resulting sequence possessed the formal properties usually encountered in homology theory.

In greater detail, let Λ be a ring, A a Λ-module with operators on the right (i.e. a right Λ-module) and C a left Λ-module. A basic operation is the formation of the tensor product $A \otimes_\Lambda C$. This is the group generated by pairs $a \otimes c$ with the relations consisting of the two distributive laws and the condition $a\lambda \otimes c = a \otimes \lambda c$. It is important to consider the behavior of this construction in relation to the usual concepts of algebra: homomorphisms, submodules, quotient modules, etc.

To facilitate the discussion of this behavior we adopt diagrammatic methods. A sequence of Λ-modules and Λ-homomorphisms

$$A_m \to A_{m+1} \to \cdots \to A_n \qquad m + 1 < n$$

is said to be *exact* if, for each consecutive two homomorphisms, the image of the first is the kernel of the following one. In particular we shall consider exact sequences

$$(1) \qquad 0 \to A' \to A \to A'' \to 0.$$

In such an exact sequence A' may be regarded as a submodule of A with A'' as the quotient module.

If an exact sequence of right Λ-modules is tensored with a fixed left Λ-module C, the resulting sequence of groups and homomorphisms is, in general, no longer exact. However, some measure of exactness is preserved. In particular, if the sequence (1) is tensored with C, the following portion is always exact:

$$(2) \qquad A' \otimes_\Lambda C \to A \otimes_\Lambda C \to A'' \otimes_\Lambda C \to 0.$$

We describe this property by saying that the tensor product is a *right exact* functor.

The kernel K of the homomorphism on the left in the sequence (2) is in general not zero. In case A is a free module, it can be shown that (up to natural isomorphisms) K depends only on A'' and C. We define the *torsion product* $\mathrm{Tor}_1^\Lambda(A'',C)$ to be the kernel in this case. In the general case there is a natural homomorphism

$$\mathrm{Tor}_1^\Lambda(A'',C) \to A' \otimes_\Lambda C$$

with image K. Continuing in this way we obtain an infinite exact sequence

$$(3) \qquad \cdots \to \mathrm{Tor}_{n+1}^\Lambda(A'',C) \to \mathrm{Tor}_n^\Lambda(A',C)$$
$$\to \mathrm{Tor}_n^\Lambda(A,C) \to \mathrm{Tor}_n^\Lambda(A'',C) \to \cdots$$

which terminates on the right with the sequence (2) above, provided that we set

$$\text{Tor}_0^\Lambda(A,C) = A \otimes_\Lambda C. \tag{4}$$

The homomorphisms in (3) which pass from index $n+1$ to n are called *connecting homomorphisms.*

The condition that A be free in the definition of $\text{Tor}(A'',C)$ is unnecessarily restrictive. It suffices that A be *projective*, i.e. that every homomorphism of A into a quotient B/B' admit a factorization $A \to B \to B/B'$.

The inductive definition of the sequence (3) as described above is cumbersome, and does not exhibit clearly the connection with homology theory. This is remedied by a direct construction as follows. If A is a module, then an exact sequence

$$\cdots \to A_n \to A_{n-1} \to \cdots \to A_1 \to A_0 \to A \to 0$$

is called a *projective resolution* of A if each A_i, $i = 0,1,2,\ldots$ is projective. Tensoring with C gives a sequence

$$\cdots \to A_n \otimes_\Lambda C \to \cdots \to A_0 \otimes_\Lambda C \tag{5}$$

which may not be exact but which is a complex (the composition of two consecutive homomorphisms is zero). The n-th homology group of the complex (5) is precisely $\text{Tor}_n^\Lambda(A,C)$. Using the second definition of Tor, the sequence (3) is constructed in the usual manner as the homology sequence of an exact sequence of complexes

$$0 \to X' \otimes_\Lambda C \to X \otimes_\Lambda C \to X'' \otimes_\Lambda C \to 0$$

where X', X, X'' are appropriate projective resolutions of A', A, A''.

A basic property of Tor is

$$\text{Tor}_n^\Lambda(A,C) = 0 \text{ if } n > 0 \text{ and } A \text{ is projective.} \tag{6}$$

In fact, this property, the exactness of (3), property (4) and the usual formal properties of functors suffice as an axiomatic description of the functors Tor_n^Λ.

The description of $\text{Tor}_n^\Lambda(A,C)$ given above favored the variable A and treated C as a constant. If the reversed procedure is adopted, the same functors $\text{Tor}_n^\Lambda(A,C)$ are obtained. This "symmetry" of the two variables in $A \otimes_\Lambda C$ is emphasized by adopting a definition of Tor which uses simultaneously projective resolutions of both A and C. This symmetry should not be confused with the symmetry resulting from the natural isomorphism $A \otimes_\Lambda C \approx C \otimes_\Lambda A$ which is valid only when Λ is commutative.

Another functor of at least as great importance as the tensor product is given by the group $\operatorname{Hom}_\Lambda(A,C)$ of all Λ-homomorphisms of the left Λ-module A into the left Λ-module C. This functor is contravariant in the variable A, covariant in the variable C and is *left exact* in that when applied to an exact sequence (1), it yields an exact sequence

$$(2') \qquad 0 \to \operatorname{Hom}_\Lambda(A'',C) \to \operatorname{Hom}_\Lambda(A,C) \to \operatorname{Hom}_\Lambda(A',C).$$

A similar discussion to that above leads to an exact sequence

$$(3') \qquad \cdots \to \operatorname{Ext}^n_\Lambda(A'',C) \to \operatorname{Ext}^n_\Lambda(A,C) \to \operatorname{Ext}^n_\Lambda(A',C) \to \operatorname{Ext}^{n+1}_\Lambda(A'',C) \to \cdots$$

which is a continuation of (2'), provided that we set

$$(4') \qquad \operatorname{Ext}^0_\Lambda(A,C) = \operatorname{Hom}_\Lambda(A,C).$$

These properties together with the property

$$(6') \qquad \operatorname{Ext}^n(A,C) = 0 \text{ if } n > 0 \text{ and } A \text{ is projective}$$

and the usual formal properties of functors suffice as an axiomatic description of the functors $\operatorname{Ext}^n_\Lambda(A,C)$.

The description above favored A as a variable while keeping C constant. Again symmetry prevails, and identical results are obtained by treating A as a constant and varying C. In this case however, instead of projective modules and projective resolutions, we employ the dual notions of injective modules and injective resolutions. A module C is *injective* if every homomorphism $B' \to C$ admits an extension $B \to C$ for each module B containing B' as a submodule. An injective resolution of C is an exact sequence

$$0 \to C \to C^0 \to C^1 \to \cdots \to C^n \to C^{n+1} \to \cdots$$

with C^i injective for $i = 0,1,2, \ldots$.

With the functors Tor and Ext introduced we can now show how the cohomology theories of groups, Lie algebras and associative algebras fit into a uniform pattern.

Let Π be a multiplicative group and C an (additive) abelian group with Π as a group of left operators. The integral group ring $Z(\Pi)$ is defined and C may be regarded as a left $Z(\Pi)$-module. The group Z of rational integers also may be regarded as a $Z(\Pi)$-module with each element of Π acting as the identity on Z. The *cohomology groups of* Π *with coefficients in* C are then

$$H^q(\Pi,C) = \operatorname{Ext}^q_{Z(\Pi)}(Z,C).$$

These cohomology groups were first introduced by Eilenberg-MacLane (*Proc. Nat. Acad. Sci. U.S.A.* 29 (1943), 155–158) in connection with a topological application. Subsequently they found a number of topological and algebraic applications; some of these will be considered in Ch. XIV and XVI. Quite recently, the theory for finite groups has been greatly enriched by the efforts of Artin and Tate; Ch. XII deals with these new developments. This theory has had its most striking application in the subject of Galois theory and class field theory. As this is a large and quite separate topic we shall not attempt an exposition here, although we do prove nearly all the results of the cohomology theory of groups needed for this application.

Let $\mathfrak{g}$ be a Lie algebra over a commutative ring K and let C be a (left) representation space for $\mathfrak{g}$. The enveloping (associative) algebra $\mathfrak{g}^e$ is then defined and C is regarded as a left $\mathfrak{g}^e$-module. The ground ring K with the trivial representation of $\mathfrak{g}$ also is a left $\mathfrak{g}^e$-module. The *cohomology groups of* $\mathfrak{g}$ *with coefficients in* C are then

$$H^q(\mathfrak{g},C) = \operatorname{Ext}^q_{\mathfrak{g}^e}(K,C).$$

This theory, implicit in the work of Elie Cartan, was first explicitly formulated by Chevalley-Eilenberg (*Trans. Am. Math. Soc.* 63 (1948), 85–124). We shall give an account of this theory in Ch. XIII; however we do not enter into its main applications to semi-simple Lie algebras and compact Lie groups.

Let Λ be an associative algebra (with a unit element) over a commutative ring K, and let A be a two-sided Λ-module. We define the enveloping algebra $\Lambda^e = \Lambda \otimes_K \Lambda^*$ where Λ^* is the "opposite" algebra of Λ. A may now be regarded as a left Λ^e-module. The algebra Λ itself also is a two-sided Λ-module and thus a left Λ^e-module. The cohomology groups are

$$H^q(\Lambda,A) = \operatorname{Ext}^q_{\Lambda^e}(\Lambda,A).$$

This theory, closely patterned after the cohomology theory of groups, was initiated by Hochschild (*Ann. of Math.* 46 (1945), 58–67). A fairly complete account of existing results is given in Ch. IX.

In all three cases above, homology groups also are defined using the functors Tor.

So far we have mentioned only the functors $A \otimes_\Lambda C$ and $\operatorname{Hom}_\Lambda(A,C)$ and their derived functors Tor and Ext. It has been found useful to consider other functors besides these two; Ch. II–V develop such a theory for arbitrary *additive* functors. Both procedures that led to the definition of Tor are considered. The slow but elementary iterative procedure leads to the notion of *satellite functors* (Ch. III). The faster, homological method using resolutions leads to the *derived functors* (Ch. v). In most important cases (including the functors ⊗ and Hom) both procedures yield identical results.

Beginning with Ch. vi we abandon general functors and confine our attention to the special functors Tor and Ext and their composites. The main developments concerning homology theory are grouped in Ch. viii–xiii.

The last three chapters (xv–xvii) are devoted to the method of spectral sequences, which has been a major tool in recent developments in algebraic topology. In Ch. xv we give the general theory of spectral sequences, while the subsequent two chapters give applications to questions studied earlier in the book.

There is an appendix by David A. Buchsbaum outlining a more abstract method of treating the subject of satellites and derived functors.

Each chapter is preceded by a short introduction and is followed by a list of exercises of varied difficulty. There is no general bibliography; references are made in the text, whenever needed. Crossreferences are made as follows: Theorem 2.1 (or Proposition 2.1 or Lemma 2.1) of Chapter x is referred to as 2.1 if the reference is in Chapter x, and as x,2.1 if the reference is outside of that chapter. Similarly viii,3,(8) refers to formula (8) of § 3 of Chapter viii.

We owe expressions of gratitude to the John Simon Guggenheim Memorial Foundation who made this work possible by a fellowship grant to one of the authors. We received help from several colleagues: D. A. Buchsbaum and R. L. Taylor read the manuscript carefully and contributed many useful suggestions; G. P. Hochschild and J. Tate helped with Chapter xii; J. P. Serre and N. E. Steenrod offered valuable criticism and suggestions. Special thanks are due to Miss Alice Krikorian for her patience shown in typing the manuscript.

H. CARTAN
S. EILENBERG

University of Paris
Columbia University
September, 1953

Contents

HOMOLOGICAL ALGEBRA

CHAPTER I

Rings and Modules

Introduction. After some preliminaries concerning rings, modules, homomorphisms, direct sums, direct products, and exact sequences, the notions of projective and injective modules are introduced. These notions are fundamental for this book. The basic results here are that each module may be represented as a quotient of a projective module and also as a submodule of an injective one.

In § 4–7 we consider special classes of rings, namely: semi-simple rings, hereditary rings, semi-hereditary rings, and Noetherian rings. It will be seen later (Ch. VII) that for integral domains the hereditary (semi-hereditary) rings are precisely the Dedekind (Prüfer) rings.

1. PRELIMINARIES

Let Λ be a ring with a unit element $1 \neq 0$. We shall consider (left) *modules* over Λ, i.e. abelian groups A with an operation $\lambda a \in A$, for $\lambda \in \Lambda$, $a \in A$ such that

$$\lambda(a_1 + a_2) = \lambda a_1 + \lambda a_2, \qquad (\lambda_1 + \lambda_2)a = \lambda_1 a + \lambda_2 a,$$
$$(\lambda_1\lambda_2)(a) = \lambda_1(\lambda_2 a), \qquad 1a = a.$$

We shall denote by 0 the module containing the zero element alone.

In the special case $\Lambda = Z$ is the ring of rational integers, the modules over Z are simply abelian groups. If Λ is a (commutative) field, they are the vector spaces over Λ.

Given two modules A and B (over the same ring Λ), a *homomorphism* (or linear transformation, or mapping) of A into B is a function f defined on A with values in B, such that $f(x + y) = fx + fy$; $f(\lambda x) = \lambda(fx)$; $x,y \in A$, $\lambda \in \Lambda$. We then write $f\colon A \to B$, or $A \to B$ if there is no ambiguity as to the definition of f. The *kernel* of f is the submodule of A consisting of all $x \in A$ such that $fx = 0$; it will be denoted by Ker (f) or Ker $(A \to B)$. The *image* of f is the submodule of B consisting of all elements of the form fx, $x \in A$; it will be denoted by Im (f) or Im $(A \to B)$.

We also define the *coimage* and *cokernel* of f as follows:

$$\text{Coim}\,(f) = A/\text{Ker}\,(f),$$
$$\text{Coker}\,(f) = B/\text{Im}\,(f).$$

Of course, f induces an isomorphism Coim $(f) \approx$ Im (f) and because of this isomorphism the coimage is very seldom employed.

A homomorphism $f\colon A \to B$ as is called a *monomorphism* if $\operatorname{Ker} f = 0$; f is called an *epimorphism* if Coker $f = 0$ or equivalently if $\operatorname{Im} f = B$. If f is both an epimorphism and a monomorphism then f is an *isomorphism* (notation: $f\colon A \approx B$).

Let A be a module and $\{A_\alpha\}$ a (finite or infinite) family of modules (all over the same ring Λ) with homomorphisms

$$A_\alpha \xrightarrow{i_\alpha} A \xrightarrow{p_\alpha} A_\alpha$$

such that $p_\alpha i_\alpha =$ identity, $p_\beta i_\alpha = 0$ if $\beta \neq \alpha$. We shall say that $\{i_\alpha, p_\alpha\}$ is a *direct family of homomorphisms.*

If we assume that each $x \in A$ can be written as a finite sum $x = \sum i_\alpha x_\alpha$, $x_\alpha \in A_\alpha$, it follows readily that A is isomorphic with the direct sum $\sum A_\alpha$. We therefore say that the family $\{i_\alpha, p_\alpha\}$ yields a *representation of A as a direct sum* of the modules A_α. In this case the mappings $\{p_\alpha\}$ can be defined using the $\{i_\alpha\}$ alone.

If we assume that for each family $\{x_\alpha\}$, $x_\alpha \in A_\alpha$, there is a unique $x \in A$ with $p_\alpha x = x_\alpha$, it follows readily that A is isomorphic with the direct product $\prod A_\alpha$. We therefore say that the family $\{i_\alpha, p_\alpha\}$ yields a *representation of A as a direct product* of the modules A_α. In this case the homomorphisms $\{i_\alpha\}$ can be defined using the $\{p_\alpha\}$ alone.

If the family $\{A_\alpha\}$ is finite, the notions of direct sum and direct product coincide. A finite direct family yields a direct sum (or direct product) representation if and only if $\sum i_\alpha p_\alpha =$ identity.

A sequence of homomorphisms

$$A_m \to A_{m+1} \to \cdots \to A_n, \qquad m + 1 < n$$

is said to be *exact* if for each $m < q < n$ we have $\operatorname{Im}(A_{q-1} \to A_q) = \operatorname{Ker}(A_q \to A_{q+1})$. Thus $A \to B$ is a monomorphism if and only if $0 \to A \to B$ is exact and an epimorphism if and only if $A \to B \to 0$ is exact. We shall also allow sequences which extend to infinity to the left or to the right or in both directions.

In particular, we shall consider exact sequences

$$(*) \qquad 0 \to A' \to A \to A'' \to 0.$$

Since $A' \to A$ is a monomorphism we may regard A' as a submodule of A. Since $A \to A''$ is an epimorphism with A' as kernel, we may regard A' as the quotient module A/A'. Thus (*) may be replaced by

$$0 \to A' \to A \to A/A' \to 0.$$

We shall say that the exact sequence (*) *splits* if Im $(A' \to A)$ is a direct summand of A. In this case, there exist homomorphisms $A'' \to A \to A'$ which together with the homomorphisms $A' \to A \to A''$ yield a direct sum representation of A.

Let F be a module and X a subset of F. We shall say that F is *free* with X as *base* if every $x \in F$ can be written uniquely as a finite sum $\sum \lambda_i x_i$, $\lambda_i \in \Lambda$, $x_i \in X$. If X is any set we may define F_X as the set of all formal finite sums $\sum \lambda_i x_i$. If we identify $x \in X$ with $1x \in F_X$, then F_X is free with base X.

In particular, if A is a module we may consider F_A. The identity mapping of the base of F_A onto A extends then to a homomorphism $F_A \to A$. If R_A denotes the kernel of this homomorphism, we obtain an exact sequence

$$0 \to R_A \to F_A \to A \to 0.$$

A diagram

$$\begin{array}{ccc} A & \longrightarrow & B \\ \downarrow & & \downarrow \\ C & \longrightarrow & D \end{array}$$

of modules and homomorphisms, is said to be *commutative* if the compositions $A \to B \to D$ and $A \to C \to D$ coincide. Similarly the diagram

$$\begin{array}{ccc} A & \longrightarrow & B \\ & \searrow \quad \swarrow & \\ & C & \end{array}$$

is commutative, if $A \to B \to C$ coincides with $A \to C$.

We shall have occasion to consider larger diagrams involving several squares and triangles. We shall say that such a diagram is commutative, if each component square and triangle is commutative.

PROPOSITION 1.1. (*The* "5 *lemma*"). *Consider a commutative diagram*

$$\begin{array}{ccccccccc} A_2 & \xrightarrow{f_2} & A_1 & \xrightarrow{f_1} & A_0 & \xrightarrow{f_0} & A_{-1} & \xrightarrow{f_{-1}} & A_{-2} \\ \downarrow{\scriptstyle h_2} & & \downarrow{\scriptstyle h_1} & & \downarrow{\scriptstyle h_0} & & \downarrow{\scriptstyle h_{-1}} & & \downarrow{\scriptstyle h_{-2}} \\ B_2 & \xrightarrow[g_2]{} & B_1 & \xrightarrow[g_1]{} & B_0 & \xrightarrow[g_0]{} & B_{-1} & \xrightarrow[g_{-1}]{} & B_{-2} \end{array}$$

with exact rows. If

(1) Coker $h_2 = 0$, Ker $h_1 = 0$, Ker $h_{-1} = 0$,

then Ker $h_0 = 0$. *If*

(2) Coker $h_1 = 0$, Coker $h_{-1} = 0$, Ker $h_{-2} = 0$,

then Coker $h_0 = 0$.

Proof. Assume (1) and let $a \in \operatorname{Ker} h_0$. Then $g_0 h_0 a = 0$ so that $h_{-1} f_0 a = 0$. It follows that $f_0 a = 0$ and therefore $a = f_1 a'$ for some $a' \in A_1$. Then $g_1 h_1 a' = h_0 f_1 a' = h_0 a = 0$ so that $h_1 a' = g_2 b$ for some $b \in B_2$. Then $b = h_2 a''$ for some $a'' \in A_2$. We have

$$h_1 f_2 a'' = g_2 h_2 a'' = g_2 b = h_1 a'$$

which implies $a' = f_2 a''$. It follows that $a = f_1 a' = f_1 f_2 a'' = 0$.

The other half is proved similarly.

2. PROJECTIVE MODULES

A module P will be called *projective* if given any homomorphism $f\colon P \to A''$ and any epimorphism $g\colon A \to A''$ there is a homomorphism $h\colon P \to A$ with $gh = f$. In the language of diagrams this means that every diagram

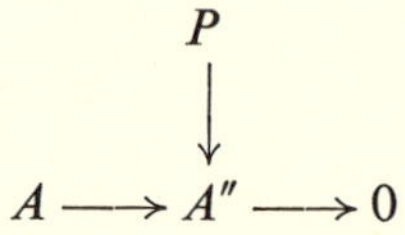

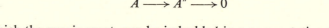

in which the row is exact, can be imbedded in a commutative diagram

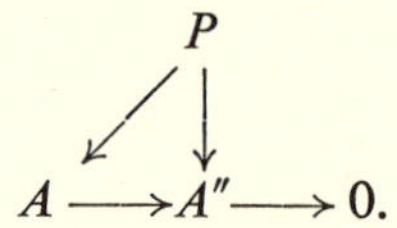

Proposition 2.1. *A direct sum of modules is projective if and only if each summand is projective.*

Proof. Let $\{i_\alpha, p_\alpha\}$ be a representation of P as a direct sum of modules $\{P_\alpha\}$. Let $g\colon A \to A''$ be an epimorphism. Assume P projective and let $f\colon P_\alpha \to A''$. Then $fp_\alpha\colon P \to A''$, so that there is an $h\colon P \to A$ with $gh = fp_\alpha$. It follows that $ghi_\alpha = fp_\alpha i_\alpha = f$, so that P_α is projective. Suppose now that all the P_α are projective, and let $f\colon P \to A''$. Then $fi_\alpha\colon P_\alpha \to A''$, so that there is an $h_\alpha\colon P_\alpha \to A$ with $gh_\alpha = fi_\alpha$. The homomorphisms h_α yield a single homomorphism $h\colon P \to A$ with $hi_\alpha = h_\alpha$ for each index α. Then $ghi_\alpha = fi_\alpha$, which implies $gh = f$. Thus P is projective.

Theorem 2.2. *In order that P be projective it is necessary and sufficient that P be a direct summand of a free module.*

Proof. Let $0 \to R_P \to F_P \to P \to 0$ be the exact sequence of § 1. If P is projective then there is a map $P \to F_P$ such that the composed map $P \to F_P \to P$ is the identity. Thus the sequence splits and P is a direct summand of the free module F_P. It remains to be proved that each direct summand of a free module is projective. By 2.1 it suffices to prove that every free module is projective. Let then F be free with base $\{x_\alpha\}$, let $f\colon F \to A''$ and let $g\colon A \to A''$ be an epimorphism. For each x_α select $y_\alpha \in A$ with $gy_\alpha = fx_\alpha$. Then the homomorphism $h\colon F \to A$ such that $hx_\alpha = y_\alpha$ satisfies $gh = f$. Thus F is projective.

THEOREM 2.3. *Each module A can be imbedded in an exact sequence $0 \to M \to P \to A \to 0$ with P projective (i.e. each module is a quotient of a projective module).*

Indeed $0 \to R_A \to F_A \to A \to 0$ is such an exact sequence.

PROPOSITION 2.4. *In order that P be projective it is necessary and sufficient that all exact sequences $0 \to A' \to A \to P \to 0$ split.*

PROOF. If P is projective, then, since $A \to P$ is an epimorphism, there is a homomorphism $P \to A$ such that $P \to A \to P$ is the identity. Thus the sequence splits. Conversely if each sequence splits, then in particular the sequence $0 \to R_P \to F_P \to P \to 0$ splits. Thus P is a direct summand of F_P and therefore, by 2.2, P is projective.

PROPOSITION 2.5. *Every exact sequence $0 \to A' \to A \to A'' \to 0$ can be imbedded in a commutative diagram*

$$\begin{array}{ccccccccc}
 & & 0 & & 0 & & 0 & & \\
 & & \downarrow & & \downarrow & & \downarrow & & \\
0 & \longrightarrow & M' & \longrightarrow & M & \longrightarrow & M'' & \longrightarrow & 0 \\
 & & \downarrow & & \downarrow & & \downarrow & & \\
0 & \longrightarrow & P' & \longrightarrow & P & \longrightarrow & P'' & \longrightarrow & 0 \\
 & & \downarrow & & \downarrow & & \downarrow & & \\
0 & \longrightarrow & A' & \longrightarrow & A & \longrightarrow & A'' & \longrightarrow & 0 \\
 & & \downarrow & & \downarrow & & \downarrow & & \\
 & & 0 & & 0 & & 0 & &
\end{array}$$

in which all rows and columns are exact, the middle row splits and consists of projective modules. In fact, the exact sequences

$$0 \to M' \to P' \to A' \to 0, \qquad 0 \to M'' \to P'' \to A'' \to 0,$$

with P' and P'' projective, may be given in advance.

PROOF. We define P as the direct sum $P' + P''$ and the maps $P' \to P$ and $P \to P''$ as

$$p' \to (p', 0), \qquad (p', p'') \to p''.$$

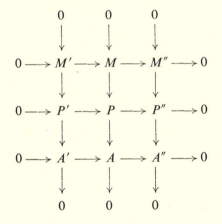

Since P'' is projective and $A \to A''$ is an epimorphism there is a map $h''\colon P'' \to A$ which when composed with $A \to A''$ yields $P'' \to A''$. Let h' be the composition $P' \to A' \to A$. We define $h\colon P \to A$ by setting

$$h(p',p'') = h'p' + h''p''.$$

Then h maps P onto A and the lower two squares are commutative. We define M as the kernel of h. Then the definition of the map in the upper row is forced by the commutativity conditions and the proof of exactness of the upper row is straightforward.

3. INJECTIVE MODULES

A module Q will be called *injective* if given any module A, a submodule A' and a homomorphism $A' \to Q$, there is an extension $A \to Q$. In the language of diagrams this means that every diagram

$$\begin{array}{ccccc} 0 & \longrightarrow & A' & \longrightarrow & A \\ & & \downarrow & & \\ & & Q & & \end{array}$$

in which the row is exact, can be imbedded in a commutative diagram

$$\begin{array}{ccccc} 0 & \longrightarrow & A' & \longrightarrow & A \\ & & \downarrow & \swarrow & \\ & & Q & & \end{array}$$

PROPOSITION 3.1. *A direct product of modules is injective if and only if each factor is injective.*

PROOF. Let $\{i_\alpha, p_\alpha\}$ be a representation of Q as a direct product of modules $\{Q_\alpha\}$. Let A be a module and A' a submodule of A. Assume Q injective and let $f\colon A' \to Q_\alpha$. Then $i_\alpha f\colon A' \to Q$ and therefore there is an extension $g\colon A \to Q$ of $i_\alpha f$. Then $p_\alpha g\colon A \to Q_\alpha$ is an extension of $f\colon A' \to Q_\alpha$. Thus each Q_α is injective. Assume now that all the Q_α are injective and let $f\colon A' \to Q$. Then each $p_\alpha f\colon A' \to Q_\alpha$ admits an extension $g_\alpha\colon A \to Q_\alpha$. The homomorphisms g_α yield $g\colon A \to Q$ with $p_\alpha g = g_\alpha$. Thus for each $x \in A'$ we have $p_\alpha g x = g_\alpha x = p_\alpha f x$ for all α, and therefore $fx = gx$. Consequently Q is injective.

THEOREM 3.2. *In order that a module Q be injective it is necessary and sufficient that for each left ideal I of Λ and each homomorphism $f\colon I \to Q$ (with I regarded as a left Λ-module) there exists an element $g \in Q$ such that $f\lambda = \lambda g$ for all $\lambda \in I$.*

PROOF. Suppose Q is injective, then the homomorphism f has an extension g: $\Lambda \to Q$ and $f\lambda = g\lambda = \lambda g(1)$ for each $\lambda \in I$. Thus the condition is necessary. To prove sufficiency, consider a module A, a submodule A', and a homomorphism f: $A' \to Q$. Consider the family $\mathscr{F}$ of all pairs (A_1,f_1) where A_1 is a submodule of A containing A' and f_1: $A_1 \to Q$ is an extension of f. We introduce a partial order in $\mathscr{F}$ by setting $(A_1,f_1) < (A_2,f_2)$ if $A_1 \subset A_2$ and f_2 is an extension of f_1. The family $\mathscr{F}$ is obviously inductive and therefore by Zorn's lemma there is an element (A_0,f_0) of $\mathscr{F}$ which is maximal. We shall prove that $A_0 = A$. Assume to the contrary that $x \in A$ and x not $\in A_0$. The set of all $\lambda \in \Lambda$ such that $\lambda x \in A_0$ forms a left ideal I of Λ and the map f_0': $I \to Q$ defined by $f_0'\lambda = f_0(\lambda x)$ is a homomorphism. There is therefore an element $g \in Q$ such that $f_0(\lambda x) = \lambda g$ for all $\lambda \in I$. Setting

$$f'(a + \lambda x) = f_0 a + \lambda g, \qquad a \in A_0,\ \lambda \in \Lambda,$$

yields then a map f'' of the submodule $A_0 + \Lambda x$ of A which is an extension of f'. Thus (A_0,f_0) is not maximal.

THEOREM 3.3. *Each module A can be imbedded in an exact sequence $0 \to A \to Q \to N \to 0$ where Q is injective (i.e. each module is a submodule of an injective module).*

PROOF. For each module A we shall define a module $D(A)$ containing A with the following property:

(*) If I is a left ideal of Λ and f: $I \to A$, then there is an element $g \in D(A)$ such that $f(\lambda) = \lambda g$ for all $\lambda \in I$.

Let Φ be the set of all pairs (I,f) formed by a left ideal I of Λ and a homomorphism f: $I \to A$. Let F_Φ be the free module generated by the elements of Φ. Let $D(A)$ be the quotient of the direct sum $A + F_\Phi$ by the submodule generated by the elements

$$(f\lambda, -\lambda(I,f)) \qquad (I,f) \in \Phi,\ \lambda \in I.$$

The mapping $a \to (a,0)$ yields a homomorphism φ: $A \to D(A)$. If $\varphi a = 0$ then

$$\begin{aligned}\varphi a = (a,0) &= \textstyle\sum \mu_i(f_i\lambda_i, -\lambda_i(I_i,f_i))\\ &= \textstyle\sum (f_i(\mu_i\lambda_i), -\mu_i\lambda_i(I_i,f_i)).\end{aligned}$$

Therefore $\sum \mu_i\lambda_i(I_i,f_i) = 0$ in F_Φ, which implies $\sum f_i(\mu_i\lambda_i) = 0$. This implies $a = 0$. Thus φ is a monomorphism and, by identifying a and φa we may regard A as a submodule of $D(A)$.

We now prove that $D(A)$ has the property (*). Let f: $I \to A$ where I is a left ideal in Λ. Then $(I,f) \in \Phi$. Let g be the image in $D(A)$ of the element $(0,(I,f))$ of $A + F_\Phi$. Then for each $\lambda \in I$

$$f\lambda = (f\lambda,0) = (0,\lambda(I,f)) = \lambda g$$

as required.

Let now Ω be the least infinite ordinal number whose cardinal is larger than that of the ring Λ. We define $Q_\alpha(A)$ for $\alpha \leqq \Omega$ by transfinite induction as follows: $Q_1(A) = D(A)$; if $\alpha = \beta + 1$ then $Q_\alpha(A) = D(Q_\beta(A))$; if α is a limiting ordinal then $Q_\alpha(A)$ is the union of Q_β with $\beta < \alpha$. We now prove that $Q_\Omega(A)$ is injective. Indeed let $f\colon I \to Q_\Omega(A)$ where I is a left ideal of Λ. Then because of the choice of Ω we have $f(I) \subset Q_\alpha(A)$ for some $\alpha < \Omega$. Then by (*) there is an element $g \in D(Q_\alpha(A)) = Q_{\alpha+1}(A) \subset Q_\Omega(A)$ with $f(\lambda) = \lambda g$ for all $\lambda \in I$. Thus by 3.2, $Q_\Omega(A)$ is injective.

PROPOSITION 3.4. *In order that Q be injective it is necessary and sufficient that every exact sequence $0 \to Q \to A \to A' \to 0$ split.*

PROOF. If Q is injective, then, since $Q \to A$ is a monomorphism, there is a homomorphism $A \to Q$ such that $Q \to A \to Q$ is the identity. Thus the sequence splits. Conversely, if each sequence splits, we choose A to be an injective module containing Q, and $A' = A/Q$. Thus Q is a direct factor of A and therefore, by 3.1, Q is injective.

PROPOSITION 3.5. *Every exact sequence $0 \to A' \to A \to A'' \to 0$ can be imbedded in a commutative diagram*

$$\begin{array}{ccccccccc}
 & & 0 & & 0 & & 0 & & \\
 & & \downarrow & & \downarrow & & \downarrow & & \\
0 & \longrightarrow & A' & \longrightarrow & A & \longrightarrow & A'' & \longrightarrow & 0 \\
 & & \downarrow & & \downarrow & & \downarrow & & \\
0 & \longrightarrow & Q' & \longrightarrow & Q & \longrightarrow & Q'' & \longrightarrow & 0 \\
 & & \downarrow & & \downarrow & & \downarrow & & \\
0 & \longrightarrow & N' & \longrightarrow & N & \longrightarrow & N'' & \longrightarrow & 0 \\
 & & \downarrow & & \downarrow & & \downarrow & & \\
 & & 0 & & 0 & & 0 & &
\end{array}$$

in which all rows and columns are exact, the middle row splits and consists of injective modules. In fact, the exact sequences $0 \to A' \to Q' \to N' \to 0$, $0 \to A'' \to Q'' \to N'' \to 0$, with Q' and Q'' injective, may be given in advance.

The proof is similar to that of 2.5.

Injective modules (under a different terminology) were considered by R. Baer (*Bull. Am. Math. Soc.* 46 (1940), 800–806) who with minor variants has proved 3.2 and 3.3.

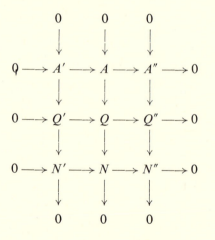

4. SEMI-SIMPLE RINGS

A module A is called *simple* if it is $\neq 0$ and it contains no submodules except A and 0. A module A is called *semi-simple* if it is a direct sum of simple modules.

PROPOSITION 4.1. *In order that a module A be semi-simple it is necessary and sufficient that each submodule of A be a direct summand.*

PROOF. Let $A = \sum S_i$, $i \in I$ be a direct sum of simple submodules S_i. For each $J \subset I$ let $S_J = \sum S_i$, $i \in J$. Let B be a submodule of A and let J be a maximal subset of I such that $S_J \cap B = 0$. For i not $\in J$ we then have $(S_J + S_i) \cap B \neq 0$ so that $(S_J + B) \cap S_i \neq 0$. Since S_i is simple it follows that $S_i \subset S_J + B$. This implies $A = S_J + B$ and since $S_J \cap B = 0$, it follows that B is a direct summand of A.

Suppose now that every submodule of A is a direct summand. It follows readily that every submodule of A also has the same property.

We first show that every non-zero submodule C of A contains a simple module. Indeed let $c \in C$, $c \neq 0$ and let D be a maximal submodule of C not containing c. Then C is the direct sum of D and a submodule E which we will prove is simple. Indeed let F be a proper submodule of E, $F \neq 0$. Then E is the direct sum of F and a submodule $G \neq 0$. Thus $C = D + F + G$ is a direct sum and either $D + F$ or $D + G$ does not contain c, contrary to the maximal character of D.

Now, let $\{S_\alpha\}$ be a maximal family of simple submodules of A such that $B = \sum S_\alpha$ is a direct sum of the modules S_α. Clearly such a family exists. Then A is the direct sum of B and a submodule C. If $C \neq 0$ then C contains a simple module, thus contradicting the maximal character of $\{S_\alpha\}$. Thus $A = B$ and A is semi-simple.

THEOREM 4.2. *For each ring* Λ (*with unit element* $1 \neq 0$), *the following properties are equivalent*:

(a) Λ *is semi-simple as a left* Λ*-module.*
(b) *Every left ideal of* Λ *is a direct summand of* Λ.
(c) *Every left ideal of* Λ *is injective.*
(d) *All left modules over* Λ *are semi-simple.*
(e) *All exact sequences* $0 \to A' \to A \to A'' \to 0$ *of left* Λ*-modules split.*
(f) *All left* Λ*-modules are projective.*
(g) *All left* Λ*-modules are injective.*

PROOF. The equivalence of (a) and (b) was proved in 4.1.

The equivalence of (d) and (e) follows from 4.1. The equivalence of (e) and (f) follows from 2.4, while the equivalence of (e) and (g) follows from 3.4. Thus (d) — (g) are equivalent.

The implication (g) $\Rightarrow$ (c) is obvious. If the ideal I of Λ is injective then by 3.4 the exact sequence $0 \to I \to \Lambda \to \Lambda/I \to 0$ splits, so that I

is a direct summand of Λ. Thus (c) $\Rightarrow$ (b). Finally if each ideal I of Λ is a direct summand then each homomorphism $f\colon I \to A$ into any module A admits an extension $\bar{f}\colon \Lambda \to A$ so that $f\lambda = \lambda\bar{f}(1)$ for all $\lambda \in I$. Thus by 3.2 the module A is injective. This proves (b) $\Rightarrow$ (g) and concludes the proof.

It is a classical result that a ring Λ is semi-simple (as a left Λ-module) if and only if Λ is the direct product of a finite number of rings each of which is a full matrix algebra over a (not necessarily commutative) field (see for instance B. L. van der Waerden, *Moderne Algebra*, vol. 2, 2nd edn., Berlin, 1940, p. 160). This implies that Λ is semi-simple as a left Λ-module if and only if it is semi-simple as a right Λ-module. Consequently conditions (a) — (g) could equally well be stated for right ideals and right modules.

5. HEREDITARY RINGS

PROPOSITION 5.1. *In order that a module P be projective, it is necessary and sufficient that every diagram*

$$\begin{array}{ccccc} & & P & & \\ & & \downarrow & & \\ Q & \longrightarrow & Q'' & \longrightarrow & 0 \end{array}$$

in which the row is exact and Q is injective, can be imbedded in a commutative diagram

$$\begin{array}{ccccc} & & P & & \\ & \swarrow & \downarrow & & \\ Q & \longrightarrow & Q'' & \longrightarrow & 0 \end{array}$$

PROOF. The necessity of the condition is obvious. To prove sufficiency, consider a module A, a submodule A' with $A'' = A/A'$ and a homomorphism $f\colon P \to A''$. We may regard A as a submodule of an injective module Q. Then A'' is a submodule of $Q'' = Q/A'$. By the condition above there is then a homomorphism $g\colon P \to Q$ which when combined with $Q \to Q''$ yields $P \to A'' \to Q''$. It follows that the values of g lie in A. This yields $g'\colon P \to A$ which when combined with $A \to A''$ yields $f\colon P \to A''$. Thus P is projective.

PROPOSITION 5.2. *In order that a module Q be injective, it is necessary and sufficient that every diagram*

$$\begin{array}{ccccc} 0 & \longrightarrow & P' & \longrightarrow & P \\ & & \downarrow & & \\ & & Q & & \end{array}$$

in which the row is exact and P is projective, can be imbedded in a commutative diagram

$$\begin{array}{ccccc} 0 & \longrightarrow & P' & \longrightarrow & P \\ & & \downarrow & \swarrow & \\ & & Q & & \end{array}$$

PROOF. The necessity of the condition is obvious. To prove sufficiency, consider a module A, a submodule A' and a homomorphism $f\colon A' \to Q$. Represent A as a quotient of a projective module P by a submodule M. If P' is the counter-image of A' in P then $A' = P'/M$. The composite homomorphism $P' \to A' \to Q$ can then be extended to a homomorphism $g\colon\ P \to Q$. But g maps M into zero and therefore yields a homomorphism $h\colon\ A \to Q$ which is an extension of $f\colon\ A' \to Q$. Thus Q is injective. The above proof is dual to that of 5.1.

A ring Λ will be called *left hereditary* if every (left) ideal of Λ is a projective module.

THEOREM 5.3. *If Λ is left hereditary then every submodule of a free module is the direct sum of modules each of which is isomorphic with a left ideal of Λ.* (I. Kaplansky, *Trans. Am. Math. Soc.* 72 (1952), 327–340).

PROOF. Let F be a free module with a well ordered base $\{x_\alpha\}$. We denote by F_α (or $\bar{F}_\alpha$) the submodule of F consisting of elements which can be expressed by means of generators x_β with $\beta < \alpha$ (or $\beta \leqq \alpha$). Let A be a submodule of F. Each element $a \in A \cap \bar{F}_\alpha$ is of the form $a = b + \lambda x_\alpha$ with $b \in F_\alpha$, $\lambda \in \Lambda$. The mapping $a \to \lambda$ maps $A \cap \bar{F}_\alpha$ onto a left ideal I_α of Λ and has $A \cap F_\alpha$ as kernel. Since I_α is projective, it follows that $A \cap \bar{F}_\alpha$ is the direct sum of $A \cap F_\alpha$ and a submodule C_α isomorphic with I_α. We shall show that A is a direct sum of the modules C_α.

Firstly, the relation $c_1 + \cdots + c_n = 0$ with $c_i \in C_{\alpha_i}$, $\alpha_1 < \cdots < \alpha_n$, implies that $c_i = 0$; indeed, the sum of $A \cap F_{\alpha_n}$ and C_{α_n} being direct, we have $c_1 + \cdots + c_{n-1} = 0$, $c_n = 0$; the assertion then follows by recursion on n. Secondly, A is the sum of the modules C_α; assume to the contrary $A \neq \sum_\alpha C_\alpha$. Then there is a least index β such that there is an element $a \in A \cap \bar{F}_\beta$ which is not in $\sum_\alpha C_\alpha$. Since $a = b + c$ with $b \in A \cap F_\beta$, $c \in C_\beta$ it follows that b is not in $\sum_\alpha C_\alpha$. However $b \in A \cap \bar{F}_\gamma$ for some $\gamma < \beta$, thus contradicting the minimality of β.

If Λ is a principal ideal ring, then each ideal I of Λ is isomorphic with Λ, thus I is free and Λ is hereditary. Since a direct sum of free modules is free, 5.3 implies the well known result that a submodule of a free module over a principal ideal ring is free.

THEOREM 5.4. *For each ring Λ, the following properties are equivalent:*

(a) *Λ is left hereditary.*

(b) *Each submodule of a projective left Λ-module is projective.*

(c) *Each quotiént of an injective left Λ-module is injective.*

PROOF. (a) $\Rightarrow$ (b). Let A be a submodule of a projective module P. By 2.2, P is a submodule of a free module. Thus, by 5.3, A is the direct sum of projective modules. Consequently, by 2.1, A is projective.

(b) $\Rightarrow$ (a). Since Λ is free, and therefore projective, each submodule of Λ, i.e. each left ideal of Λ, is projective.

In order to prove the equivalence of (b) and (c) consider a diagram

$$\begin{array}{ccccc} P & \longleftarrow & P' & \longleftarrow & 0 \\ & & \downarrow{\scriptstyle f} & & \\ Q & \longrightarrow & Q'' & \longrightarrow & 0 \end{array}$$

where the rows are exact, P is projective and Q is injective. Suppose now that (b) holds. Then P' is projective. There is then a map $P' \to Q$ such that f is the composition $P' \to Q \to Q''$. Since Q is injective there is a map $P \to Q$ such that $P' \to P \to Q$ yields $P' \to Q$. Thus $P' \to P \to Q \to Q''$ yields f. This implies by 5.2 that Q'' is injective. Thus (b) $\Rightarrow$ (c).

Now, assume (c). Then Q'' is injective, so that there is a map $P \to Q''$ such that $P' \to P \to Q''$ yields f. Since P is projective the map $P \to Q''$ may be factored into $P \to Q \to Q''$. Then the composition $P' \to P \to Q \to Q''$ yields f. This implies, by 5.1, that P' is projective. Thus (c) $\Rightarrow$ (b).

6. SEMI-HEREDITARY RINGS

A Λ-module A is said to be finitely generated if there exists a sequence $a_1, \dots, a_n \in A$ such that each element of A has the form $\lambda_1 a_1 + \cdots + \lambda_n a_n$, $\lambda_1, \dots, \lambda_n \in \Lambda$.

The ring Λ will be called *left semi-hereditary* if each finitely generated (left) ideal of Λ is a projective module.

PROPOSITION 6.1. *If Λ is left semi-hereditary then every finitely generated submodule of a free left Λ-module is the direct sum of a finite number of modules each of which is isomorphic with a finitely generated ideal of Λ.*

PROOF. Let $\{x_\alpha\}$ be a base for the free module F and let A be a finitely generated submodule of F. Then A must be contained in a submodule of F generated by a finite number of the elements x_α. Thus we may assume that F has a finite base $(x_1, \dots, x_n)$.

We proceed by induction with respect to n. Let B be the submodule

of those elements of A which can be expressed using $x_1, \ldots, x_{n-1}$. Then each $a \in A$ can be written uniquely as $a = \lambda x_n + b$, $\lambda \in \Lambda$, $b \in B$ (note that $B = 0$ if $n = 1$). The mapping $a \to \lambda$ maps A onto an ideal I of Λ, the kernel of the mapping being B. There results an exact sequence

$$0 \to B \to A \to I \to 0.$$

It follows that the ideal I is finitely generated and therefore is a projective module. Thus, by 2.4, the exact sequence splits, and A is isomorphic with the direct sum of I and B. This implies that B is finitely generated, and therefore by the inductive assumption, satisfies the conclusion of 6.1. It follows that A also satisfies the conclusion.

PROPOSITION 6.2. *For each ring Λ the following conditions are equivalent*:

(a) *Λ is left semi-hereditary.*

(b) *Each finitely generated submodule of a projective left Λ-module is projective.*

PROOF. The implication (a) $\Rightarrow$ (b) follows from 6.1 and the facts that each projective module is a submodule of a free module and that the direct sum of projective modules is projective. The implication (b) $\Rightarrow$ (a) is obvious since Λ itself is free and thus projective.

The definition of right hereditary and right semi-hereditary rings is entirely similar. It is an open question whether a left hereditary (or semi-hereditary) ring also is right hereditary (or semi-hereditary).

7. NOETHERIAN RINGS

A module A is called *Noetherian* if each submodule of A is finitely generated. A ring Λ is called left (right) Noetherian if it is Noetherian as a left (right) Λ-module.

PROPOSITION 7.1. *If Λ is left Noetherian then each finitely generated left Λ-module A is Noetherian.*

PROOF. We must show that each submodule B of A is finitely generated.

Let $x_1, \ldots, x_n$ be a system of generators for A. If $n = 1$, then $A \approx \Lambda/I$ for some left ideal I. Therefore $B \approx J/I$ for some left ideal J containing I. Since J is finitely generated, so is B. We now proceed by induction and assume that the proposition is proved for modules A generated by fewer than n elements. Assume $n > 1$ and let A' denote the submodule of A generated by x_1. There results an exact sequence $0 \to A' \to A \to A'' \to 0$ with both A' and A'' generated by fewer than n elements. This exact sequence induces an exact sequence $0 \to B' \to B \to B'' \to 0$ with $B' \subset A'$, $B'' \subset A''$. Thus B' and B'' are finitely generated, and therefore B is also finitely generated.

We shall now construct an example (due to J. Dieudonné) of a ring Λ which is left Noetherian without being right Noetherian. Let Λ be the ring generated by the elements 1, x, y with relations $yx = 0$, $yy = 0$. Let Γ be the subring of Λ generated by 1 and x. Every element of Λ may then be written uniquely as $\gamma_1 + \gamma_2 y$ where $\gamma_1, \gamma_2 \in \Gamma$.

The ring Γ is the ring of polynomials in the indeterminate x with integer coefficients, and is well known to be Noetherian. Since Λ regarded as a left Γ-module is finitely generated it follows from 7.1 that Λ is Noetherian as a left Γ-module and thus also as a left Λ-module.

Let I denote the subgroup of Λ generated by the elements $x^n y$ ($n \geqq 0$). Since $Ix = Iy = 0$, it follows that I is a right ideal and that any system of right Λ-generators for I is also a system of right Z-generators for I. Thus I is not finitely Λ-generated (as a right Λ-module). Therefore Λ is not right Noetherian.

EXERCISES

1. Let A_1, A_2 be submodules of a module A and let $A_{12} = A_1 \cap A_2$. Show that the diagram

$$\begin{array}{ccccccccc}
 & & 0 & & 0 & & 0 & & \\
 & & \downarrow & & \downarrow & & \downarrow & & \\
0 & \longrightarrow & A_{12} & \longrightarrow & A_1 & \longrightarrow & A_1/A_{12} & \longrightarrow & 0 \\
 & & \downarrow & & \downarrow & & \downarrow & & \\
0 & \longrightarrow & A_2 & \longrightarrow & A & \longrightarrow & A/A_2 & \longrightarrow & 0 \\
 & & \downarrow & & \downarrow & & \downarrow & & \\
0 & \longrightarrow & A_2/A_{12} & \longrightarrow & A/A_1 & \longrightarrow & A/(A_1 + A_2) & \longrightarrow & 0 \\
 & & \downarrow & & \downarrow & & \downarrow & & \\
 & & 0 & & 0 & & 0 & &
\end{array}$$

with the maps induced by inclusion, is commutative and has exact rows and columns.

2. Let $0 \to A' \to A \to A'' \to 0$ be an exact sequence of left Λ-modules. Show that if A' and A'' are finitely generated then so is A. If Λ is left Noetherian, then the converse also holds.

3. Let A be the direct sum of modules A_α. Show that A is finitely generated if and only if each A_α is finitely generated and $A_\alpha = 0$ for all but a finite number of indices α.

4. Let A_1 and A_2 be submodules of a module A. Show that if $A_1 + A_2$ and $A_1 \cap A_2$ are finitely generated, then so are A_1 and A_2.

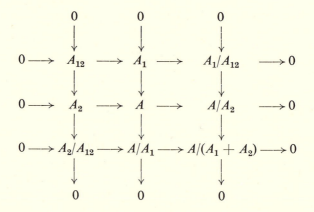

5. Consider the ring $Z_n = Z/nZ$, where Z denotes the ring of integers and n is an integer $(1 < n < \infty)$. For each divisor r of n consider the ideal rZ_n and define an exact sequence

$$0 \to r'Z_n \to Z_n \to rZ_n \to 0,$$

where $r' = n/r$. Show that this sequence splits if and only if $(r,r') = 1$. The Z_n-module rZ_n is projective if and only if $(r,n/r) = 1$. Give examples of projective modules which are not free.

6. For any integer n, prove the equivalence of the following conditions:

(a) the ring Z_n is semi-simple.

(b) the ring Z_n is hereditary.

(c) n is a product of distinct primes.

7. Show that for every ring Λ the following properties are equivalent:

(a) Every left ideal of Λ is a free Λ-module.

(b) Every submodule of a free left Λ-module is Λ-free.

8. Let Λ be a left Noetherian ring. Show that the direct limit of injective left Λ-modules is injective. [Hint: use 3.2.]

CHAPTER II

Additive Functors

Introduction. We consider functors (in the sense of Eilenberg-MacLane (*Trans. Am. Math. Soc.* 58 (1945), 231–294)) defined for Λ-modules and whose values are in the category of Γ-modules, where Λ and Γ are two given rings. We only consider functors which satisfy an additivity property reflecting the fact that homomorphisms of modules can be added. Functors of several variables, some covariant, some contravariant are also treated. The two basic examples of such functors are $A \otimes C$ (the tensor product) and $\operatorname{Hom}(A,C)$.

In § 4 we discuss the extent to which functors may preserve exactness. It turns out that $\operatorname{Hom}(A,C)$ is a left exact functor; this will give rise (in Ch. III, V, VI) to right satellites and right derived functors of $\operatorname{Hom}(A,C)$. Similarly the functor $A \otimes C$ is right exact; we shall later study its left satellites and its left derived functors.

The associativity relations of § 5 are quite elementary but of great importance in the sequel.

Given a Λ-module A it is frequently necessary to "restrict" the operators to a smaller ring or to "extend" the operators to a larger ring (by a suitable enlargement of A). In § 6 we set up the basic notions involved in such a *change of rings*. We shall return to these questions later (VI, 4; XVI, 5). There will be numerous applications to homology theory of groups and Lie algebras.

1. DEFINITIONS

Let Λ_1, Λ be any two rings. Suppose that for each Λ_1-module A a Λ-module $T(A)$ is given and that to each Λ_1-homomorphism $\varphi: A \to A'$ a Λ-homomorphism $T(\varphi): T(A) \to T(A')$ is given such that

(1) if $\varphi: A \to A$ is the identity, then $T(\varphi)$ is the identity,

(2) $T(\varphi'\varphi) = T(\varphi')T(\varphi)$ for $\varphi: A \to A'$, $\varphi': A' \to A''$.

We then say that the pair of functions $T(A)$, $T(\varphi)$ forms a *covariant functor* T on the category of Λ_1-modules with values in the category of Λ-modules. In the case of a *contravariant functor* we have $T(\varphi): T(A') \to T(A)$ and $T(\varphi'\varphi) = T(\varphi)T(\varphi')$.

In the sequel we shall have to consider functors in many variables, some covariant some contravariant. To simplify the notation we define explicitly a functor in only two variables, covariant in the first and contravariant in the second.

Let Λ_1, Λ_2, Λ be rings. We assume that for each Λ_1-module A and each Λ_2-module C a Λ-module $T(A,C)$ is given. Further, for each pair of homomorphisms $\varphi\colon A \to A'$, $\psi\colon C' \to C$, homomorphisms

$$T(\varphi,C)\colon T(A,C) \to T(A',C),\ T(A,\psi)\colon T(A,C) \to T(A,C')$$

are given subject to the following conditions:

(3) $T(\varphi,C)$ and $T(A,\psi)$ are identity maps if $\varphi\colon A \to A$, $\psi\colon C \to C$ are identity maps,

(4) $T(\varphi'\varphi,C) = T(\varphi',C)T(\varphi,C)$ and $T(A,\psi\psi') = T(A,\psi')T(A,\psi)$ for $\varphi'\colon A' \to A''$, $\psi'\colon C'' \to C'$,

(5) The following diagram is commutative

$$\begin{array}{ccc} T(A,C) & \xrightarrow{T(\varphi,C)} & T(A',C) \\ {\scriptstyle T(A,\psi)}\downarrow & & \downarrow{\scriptstyle T(A',\psi)} \\ T(A,C') & \xrightarrow[T(\varphi,C')]{} & T(A',C') \end{array}$$

The composite mapping $T(A,C) \to T(A',C')$ is denoted by $T(\varphi,\psi)$.

Clearly, by fixing C, T becomes a covariant functor of A and by fixing A, T becomes a contravariant functor of C.

We shall only be concerned with functors which are *additive*, i.e. satisfy

$$T(\varphi_1 + \varphi_2,C) = T(\varphi_1,C) + T(\varphi_2,C)$$

$$T(A,\psi_1 + \psi_2) = T(A,\psi_1) + T(A,\psi_2)$$

where $\varphi_1,\varphi_2\colon A \to A'$, $\psi_1,\psi_2\colon C' \to C$ and addition denotes addition of homomorphisms. In particular if φ and ψ are zero homomorphisms then $T(\varphi,C)$ and $T(A,\psi)$ also are zero homomorphisms. It follows that if one of the modules A or C is zero, then the identity map $T(A,C) \to T(A,C)$ is zero and therefore that $T(A,C)$ is the zero module.

PROPOSITION 1.1. *If the homomorphisms*

$$A_\alpha \xrightarrow{i_\alpha} A \xrightarrow{p_\alpha} A_\alpha,\quad C_\beta \xrightarrow{j_\beta} C \xrightarrow{q_\beta} C_\beta$$

($\alpha = 1, \ldots, m$; $\beta = 1, \ldots, n$) *yield direct sum decompositions of A and C, then the homomorphisms*

$$T(A_\alpha,C_\beta) \xrightarrow{T(i_\alpha,q_\beta)} T(A,C) \xrightarrow{T(p_\alpha,j_\beta)} T(A_\alpha,C_\beta)$$

yield a direct sum decomposition of $T(A,C)$.

PROOF. We have $T(p_{\alpha'}, j_{\beta'})T(i_\alpha, q_\beta) = T(p_{\alpha'} i_\alpha, q_\beta j_{\beta'})$. This yields the identity if $(\alpha', \beta') = (\alpha, \beta)$ and zero otherwise. Further

$$\begin{aligned}\textstyle\sum_{\alpha,\beta} T(i_\alpha, q_\beta)T(p_\alpha, j_\beta) &= \textstyle\sum_{\alpha,\beta} T(i_\alpha p_\alpha, j_\beta q_\beta)\\ &= T(\textstyle\sum_\alpha i_\alpha p_\alpha, \sum_\beta j_\beta q_\beta) = \text{identity}.\end{aligned}$$

Thus the conditions for a direct sum are satisfied.

COROLLARY 1.2. *If*

$$0 \to A' \to A \to A'' \to 0, \quad 0 \to C' \to C \to C'' \to 0$$

are split exact sequences, then

$$0 \to T(A',C) \to T(A,C) \to T(A'',C) \to 0$$

$$0 \to T(A,C'') \to T(A,C) \to T(A,C') \to 0$$

also are split exact sequences.

Let T_1 and T_2 be two functors, both covariant in A and contravariant in C. A *natural transformation* f: $T_1 \to T_2$ is a family of homomorphisms $f(A,C)$: $T_1(A,C) \to T_2(A,C)$ such that the diagram

$$\begin{array}{ccc} T_1(A,C) & \xrightarrow{f(A,C)} & T_2(A,C) \\ {\scriptstyle T_1(\phi,\psi)}\downarrow & & \downarrow{\scriptstyle T_2(\varphi,\psi)} \\ T_1(A',C') & \xrightarrow[f(A',C')]{} & T_2(A',C') \end{array}$$

is commutative for all φ: $A \to A'$, ψ: $C' \to C$. If each $f(A,C)$ maps $T_1(A,C)$ isomorphically onto $T_2(A,C)$ then f is called a *natural equivalence* or a *natural isomorphism*.

2. EXAMPLES

Our first example is the functor Hom (A,C). Let A and C be two (left) Λ-modules. We shall denote as usual by Hom (A,C) the group of all Λ-homomorphisms $A \to C$. Hom (A,C) is regarded as an abelian group (i.e. a Z-module where Z is the ring of integers). We usually write $\mathrm{Hom}_\Lambda (A,C)$ to indicate that we are considering Λ-homomorphisms.

Given Λ-homomorphisms

$$\varphi\colon\ A' \to A, \qquad \psi\colon\ C \to C'$$

we define

$$\mathrm{Hom}\,(\varphi,\psi)\colon\ \mathrm{Hom}\,(A,C) \to \mathrm{Hom}\,(A',C')$$

by setting for $\alpha \in$ Hom (A,C)

$$\mathrm{Hom}\,(\varphi,\psi)\alpha = \psi\alpha\varphi.$$

With this definition it is clear that Hom (A,C) is an (additive) functor contravariant in A and covariant in C.

Each element $c \in C$ determines a Λ-homomorphism φ_c: $\Lambda \to C$ (with Λ regarded as a left Λ-module) by setting $\varphi_c(\lambda) = \lambda c$. This establishes an isomorphism of C (regarded as an abelian group) with $\text{Hom}_\Lambda (\Lambda,C)$. Since Λ is also a right Λ-module it follows that $\text{Hom}_\Lambda (\Lambda,C)$ may be regarded as a left Λ-module (see next §); then $\text{Hom}_\Lambda (\Lambda,C) \approx C$ is a Λ-isomorphism. We shall frequently identify $\text{Hom}_\Lambda (\Lambda,C)$ with C under this isomorphism.

The functor $\text{Hom}_\Lambda (A,C)$ may also be defined when A and C are right Λ-modules.

Our next example is the tensor product $A \otimes_\Lambda C$ where A is a right Λ-module and C is a left Λ-module. We recall the definition. Let F be the free abelian group generated by the pairs (a,c) with $a \in A$, $c \in C$, and let R be the subgroup of F generated by elements of the form

$$(a + a',c) - (a,c) - (a',c), \qquad (a,c + c') - (a,c) - (a,c'),$$
$$(a\lambda,c) - (a,\lambda c), \qquad (\lambda \in \Lambda).$$

Then $A \otimes_\Lambda C$ is defined as the quotient group F/R, regarded as an abelian group (i.e. as a module over the ring Z of integers). The image in $A \otimes_\Lambda C$ of the element (a,c) of F is denoted by $a \otimes_\Lambda c$ or by $a \otimes c$. We then have the formal rules

$$(a + a') \otimes c = a \otimes c + a' \otimes c, \quad a \otimes (c + c') = a \otimes c + a \otimes c',$$
$$a\lambda \otimes c = a \otimes \lambda c.$$

If we regard A and C as abelian groups, we may form also the tensor product $A \otimes_Z C$, and it is clear that $A \otimes_\Lambda C$ is the quotient of $A \otimes_Z C$ by the subgroup generated by the elements $a\lambda \otimes_Z c - a \otimes_Z \lambda c$.

The function φ: $A \times C \to A \otimes_\Lambda C$ defined by $\varphi(a,c) = a \otimes c$ is bilinear and satisfies $\varphi(a\lambda,c) = \varphi(a,\lambda c)$. Furthermore, any function f: $A \times C \to D$ (where D is an abelian group) which is bilinear and satisfies $f(a\lambda,c) = f(a,\lambda c)$ admits a unique factorization $f = g\varphi$ where g: $A \otimes C \to D$ is a homomorphism. This last property could be used as an axiomatic definition of $A \otimes_\Lambda C$.

Given Λ-homomorphisms

$$\varphi\colon\ A \to A', \qquad \psi\colon\ C \to C'$$

there exists a unique homomorphism (of abelian groups)

$$\varphi \otimes \psi\colon\ A \otimes_\Lambda C \to A' \otimes_\Lambda C'$$

satisfying

$$(\varphi \otimes \psi)(a \otimes b) = (\varphi a) \otimes (\psi b).$$

With this definition it is clear that $A \otimes_\Lambda C$ is an additive functor covariant in both variables.

The mappings $\lambda \otimes c \to \lambda c$ and $a \otimes \lambda \to a\lambda$ yield natural isomorphisms $\Lambda \otimes_\Lambda C \approx C$ and $A \otimes_\Lambda \Lambda \approx A$. We shall frequently regard these as identifications.

3. OPERATORS

Very frequently the modules A and C, in addition to being Λ_1- and Λ_2-modules respectively, will have some other operators compatible with the module structure. It is usually possible to transfer these operators to $T(A,C)$.

For example, suppose that A in addition to being a (left) Λ_1-module, also is a (left) Γ-module where Γ is a ring, and that the operators of Λ_1 and Γ commute (i.e. that $\lambda(\gamma a) = \gamma(\lambda a)$ for $a \in A$, $\lambda \in \Lambda_1$, $\gamma \in \Gamma$). We then say that A is a Λ_1-Γ-bimodule. Each $\gamma \in \Gamma$ induces a Λ_1-endomorphism γ_A: $A \to A$ and thereby induces a Λ-endomorphism $T(\gamma_A,C)$ of $T(A,C)$. Thus $T(A,C)$ becomes a Λ-Γ-bimodule. If C is a Λ_2-Γ-bimodule with Γ operating on C on the left, then $T(A,C)$, because of the contravariance of C, becomes a Λ-Γ-bimodule with Γ operating on the right. Similarly a Λ_1-Γ-homomorphism φ: $A \to A'$ yields a Λ-Γ-homomorphism $T(\varphi,C)$, etc.

The group (i.e. Z-module) $\text{Hom}_\Lambda(A,C)$ is defined when A and C both are left Λ-modules. We indicate this situation by the symbol $({}_\Lambda A, {}_\Lambda C)$. If in addition either A or C is a Λ-Γ-bimodule then $\text{Hom}_\Lambda(A,C)$ becomes a Γ-module. The following four cases are possible:

$({}_{\Lambda\text{-}\Gamma}A, {}_\Lambda C)$,	$\text{Hom}_\Lambda(A,C)$ is a right Γ-module,
$({}_\Lambda A_\Gamma, {}_\Lambda C)$,	$\text{Hom}_\Lambda(A,C)$ is a left Γ-module,
$({}_\Lambda A, {}_{\Lambda\text{-}\Gamma}C)$,	$\text{Hom}_\Lambda(A,C)$ is a left Γ-module,
$({}_\Lambda A, {}_\Lambda C_\Gamma)$,	$\text{Hom}_\Lambda(A,C)$ is a right Γ-module.

If Γ is commutative, the difference between left and right Γ-modules disappears and the four cases reduce to two. If Γ is a subring of Λ contained in the center of Λ then A and C are automatically Λ-Γ-bimodules and, in this case, all four cases coincide since for $\alpha \in \text{Hom}_\Lambda(A,C)$ $a \in A$ and $\gamma \in \Gamma$ we have

$$(\gamma\alpha)(a) = \alpha(a\gamma) = \alpha(\gamma a) = \gamma(\alpha a).$$

Thus $\text{Hom}_\Lambda(A,C)$ may always be regarded as a module over the center of Λ. If Λ is commutative, then $\text{Hom}_\Lambda(A,C)$ is a Λ-module. A similar discussion applies starting with the situation described by the symbol (A_Λ, C_Λ).

The tensor product $A \otimes_\Lambda C$ is an abelian group (i.e. a Z-module) and is defined when A is a right Λ-module and C is a left Λ-module, a situation that we shall describe by the symbol $(A_\Lambda, {}_\Lambda C)$. If in addition either A or C is a Λ-Γ-bimodule then $A \otimes_\Lambda C$ becomes a Γ-module. The following four cases are possible:

$({}_\Gamma A_\Lambda, {}_\Lambda C)$, $A \otimes_\Lambda C$ is a left Γ-module,

$(A_{\Lambda\text{-}\Gamma}, {}_\Lambda C)$, $A \otimes_\Lambda C$ is a right Γ-module,

$(A_\Lambda, {}_{\Gamma\text{-}\Lambda} C)$, $A \otimes_\Lambda C$ is a left Γ-module,

$(A_\Lambda, {}_\Lambda C_\Gamma)$, $A \otimes_\Lambda C$ is a right Γ-module.

If Γ is commutative the difference between left and right Γ-modules disappears and the four cases reduce to two. If Γ is in the center of Λ then A and C are automatically Λ-Γ-bimodules and in this case, all four cases coincide since

$$(\gamma a) \otimes c = (a\gamma) \otimes c = a \otimes (\gamma c) = a \otimes (c\gamma).$$

Thus $A \otimes_\Lambda C$ may always be regarded as a module over the center of Λ. If Λ is commutative, then $A \otimes_\Lambda C$ is a Λ-module.

PROPOSITION 3.1. *If Λ is a commutative ring, then there exists a unique homomorphism f: $A \otimes_\Lambda C \to C \otimes_\Lambda A$ such that $f(a \otimes c) = c \otimes a$. This homomorphism is an isomorphism and establishes a natural equivalence of the functors $T(A,C) = A \otimes_\Lambda C$ and $T_1(A,C) = C \otimes_\Lambda A$.*

The proof is straightforward.

4. PRESERVATION OF EXACTNESS

A functor $T(A,C)$, covariant in A and contravariant in C, is called *exact* if whenever

$$A' \to A \to A'', \qquad C' \to C \to C''$$

are exact,

$$T(A',C) \to T(A,C) \to T(A'',C), \qquad T(A,C'') \to T(A,C) \to T(A,C')$$

also are exact.

PROPOSITION 4.1. *In order that T be exact it is necessary and sufficient that for all exact sequences*

$$0 \to A' \to A \to A'' \to 0, \qquad 0 \to C' \to C \to C'' \to 0,$$

the sequences

$$0 \to T(A',C) \to T(A,C) \to T(A'',C) \to 0$$

$$0 \to T(A,C'') \to T(A,C) \to T(A,C') \to 0$$

be exact.

PROOF. The necessity of the condition is obvious. Assume, then, that the condition holds, and suppose $A' \to A \to A''$ exact. Let $B' = \text{Ker}(A' \to A)$, $B = \text{Ker}(A \to A'')$, $B'' = \text{Im}(A \to A'')$. Then the sequences $0 \to B' \to A' \to B \to 0$, $0 \to B \to A \to B'' \to 0$, $0 \to B'' \to A'' \to A''/B'' \to 0$ are exact. Therefore the sequences

$$T(A',C) \to T(B,C) \to 0, \qquad 0 \to T(B'',C) \to T(A'',C)$$
$$T(B,C) \to T(A,C) \to T(B'',C)$$

are exact. This implies that $T(A',C) \to T(A,C) \to T(A'',C)$ is exact, as required. The proof with respect to the second variable is similar.

PROPOSITION 4.2. *If the rings Λ_1 and Λ_2 are semi-simple then any (additive) functor $T(A,C)$, defined for Λ_1-modules A and Λ_2-modules C, is exact.*

PROOF. Let $0 \to A' \to A \to A'' \to 0$ be an exact sequence. By I,4.2, this sequence splits. Therefore by 1.2 the sequence $0 \to T(A',C) \to T(A,C) \to T(A'',C) \to 0$ is exact. A similar reasoning applies to the second variable. It now follows from 4.1 that T is exact.

Functors that are exact are encountered very rarely. Most of the interesting functors that we shall consider preserve exactness only partially. To classify these various kinds of functors we consider arbitrary exact sequences $0 \to A' \to A \to A'' \to 0$ and $0 \to C' \to C \to C'' \to 0$. We say that T is *half exact* if

$$T(A',C) \to T(A,C) \to T(A'',C),$$
$$T(A,C'') \to T(A,C) \to T(A,C'),$$

are exact. We say that T is *right exact* if

$$T(A',C) \to T(A,C) \to T(A'',C) \to 0$$
$$T(A,C'') \to T(A,C) \to T(A,C') \to 0,$$

are exact. We say that T is *left exact* if

$$0 \to T(A',C) \to T(A,C) \to T(A'',C)$$
$$0 \to T(A,C'') \to T(A,C) \to T(A,C')$$

are exact.

PROPOSITION 4.3. *For each functor T the following conditions are equivalent:*

(a) *T is right exact,*

(b) *for any exact sequences $A' \to A \to A'' \to 0$, $0 \to C' \to C \to C''$ the sequences*

$$T(A',C) \to T(A,C) \to T(A'',C) \to 0$$
$$T(A,C'') \to T(A,C) \to T(A,C') \to 0$$

are exact,

(c) *for any exact sequences* $A' \to A \to A'' \to 0$, $0 \to C' \to C \to C''$ *the sequence*

$$T(A',C) + T(A,C'') \xrightarrow{\varphi} T(A,C) \longrightarrow T(A'',C') \longrightarrow 0$$

is exact, where the first term is a direct sum, and the homomorphism φ *has as its coordinates the maps* $T(A',C) \to T(A,C)$, $T(A,C'') \to T(A,C)$.

PROOF. (a) $\Rightarrow$ (b). Let $B = \operatorname{Ker}(A' \to A)$, $B' = \operatorname{Im}(A' \to A)$. Then $0 \to B \to A' \to B' \to 0$ and $0 \to B' \to A \to A'' \to 0$ are exact. Consequently $T(A',C) \to T(B',C) \to 0$ and $T(B',C) \to T(A,C) \to T(A'',C) \to 0$ are exact. This implies the exactness of $T(A',C) \to T(A,C) \to T(A'',C) \to 0$. The proof for the second variable is similar.

(b) $\Rightarrow$ (c). This proof is obtained by familiar "diagram chasing" in the commutative diagram

$$\begin{array}{ccccccc}
T(A',C'') & \longrightarrow & T(A,C'') & \longrightarrow & T(A'',C'') & \longrightarrow & 0 \\
\downarrow & & \downarrow & & \downarrow & & \\
T(A',C) & \longrightarrow & T(A,C) & \longrightarrow & T(A'',C) & \longrightarrow & 0 \\
\downarrow & & \downarrow & & \downarrow & & \\
T(A',C') & \longrightarrow & T(A,C') & \longrightarrow & T(A'',C') & \longrightarrow & 0 \\
\downarrow & & \downarrow & & \downarrow & & \\
0 & & 0 & & 0 & &
\end{array}$$

in which the rows and columns are exact.

(c) $\Rightarrow$ (b) is proved by applying (c) in the following two cases: $C'' = 0$, $C' = C$ and $A' = 0$, $A = A''$.

The implication (b) $\Rightarrow$ (a) is obvious.

PROPOSITION 4.3a. *For any functor T the following conditions are equivalent*:

(a) *T is left exact*,

(b) *for any exact sequences* $0 \to A' \to A \to A''$, $C' \to C \to C'' \to 0$ *the sequences*

$$0 \to T(A',C) \to T(A,C) \to T(A'',C)$$

$$0 \to T(A,C'') \to T(A,C) \to T(A,C')$$

are exact,

(c) *for any exact sequences* $0 \to A' \to A \to A''$, $C' \to C \to C'' \to 0$ *the sequence*

$$0 \longrightarrow T(A',C'') \longrightarrow T(A,C) \xrightarrow{\psi} T(A'',C) + T(A,C')$$

is exact, where the last term is a direct sum and the homomorphism ψ *has as coordinates the maps* $T(A,C) \to T(A'',C)$, $T(A,C) \to T(A,C')$.

The proof is analogous to that of 4.3. We also leave to the reader the statements and proofs of analogous propositions for other variances and for functors of a larger number of variables.

PROPOSITION 4.4. *The functor* Hom_Λ *is left exact.*

PROOF. Consider an exact sequence

$$(1) \qquad 0 \longrightarrow A' \xrightarrow{i} A \xrightarrow{p} A'' \longrightarrow 0.$$

We must show that the induced sequence

$$0 \longrightarrow \mathrm{Hom}\,(A'',C) \xrightarrow{p'} \mathrm{Hom}\,(A,C) \xrightarrow{i'} \mathrm{Hom}\,(A',C)$$

is exact. We already know that $i'p' = 0$, and therefore p' defines a homomorphism

$$u\colon \mathrm{Hom}\,(A'',C) \to \mathrm{Ker}\, i'.$$

It suffices to prove that u is an isomorphism. To this end we define a homomorphism

$$v\colon \mathrm{Ker}\, i' \to \mathrm{Hom}\,(A'',C)$$

as follows: given $f \in \mathrm{Hom}\,(A,C)$ such that $i'f = 0$ we have $fi = 0$; define $(vf)a''$ for $a'' \in A''$ to be fa where $a \in A$ is any element with $pa = a''$. It follows readily that uv and vu are identity maps, so that u is an isomorphism.

The left exactness with respect to the variable C is proved similarly.

PROPOSITION 4.5. *The functor* $\otimes_\Lambda$ *is right exact.*

PROOF. Consider an exact sequence (1) as above. We shall show that the induced sequence

$$A' \otimes C \xrightarrow{i'} A \otimes C \xrightarrow{p'} A'' \otimes C \longrightarrow 0$$

is exact. Since $p'i' = 0$ we have a homomorphism

$$u\colon \mathrm{Coker}\, i' \to A'' \otimes C,$$

and it suffices to show that u is an isomorphism. We define a homomorphism

$$v\colon A'' \otimes C \to \mathrm{Coker}\, i'$$

as follows. Given $a'' \in A$, $c \in C$, choose $a \in A$ with $pa = a''$ and denote by $\varphi(a'',c)$ the image of the element $a \otimes c$ in Coker i'. Clearly $\varphi(a'',c)$ is independent of the choice of a, is bilinear and satisfies $\varphi(a''\lambda,c) = \varphi(a'',\lambda c)$. Thus there is a unique homomorphism v such that $v(a'' \otimes c) = \varphi(a'',c)$. Since uv and vu are obviously identity maps, u is an isomorphism.

PROPOSITION 4.6. *A Λ-module A is projective if and only if the functor $T(C) = \mathrm{Hom}_\Lambda\,(A,C)$ is exact. A Λ-module C is injective if and only if the functor $U(A) = \mathrm{Hom}_\Lambda\,(A,C)$ is exact.*

PROOF. Since T is left exact it follows that T is exact if and only if for every epimorphism $C \to C''$ the mapping $\operatorname{Hom}(A,C) \to \operatorname{Hom}(A,C'')$ is an epimorphism. This however is immediately equivalent with A being projective. The second half of 4.6 is proved similarly.

5. COMPOSITE FUNCTORS

Functors may be composed exactly as functions. For instance, let $T(A,C)$ be a functor defined for Λ_1-modules A and Λ_2-modules C and with Λ-modules as values, let $U(D,E)$ be a functor defined for Λ_3-modules D and Λ_4-modules E and with Λ_1-modules as values. We define the composite functor V as follows

$$V(D,E,C) = T(U(D,E),C).$$

$$V(\delta,\varepsilon,\gamma) = T(U(\delta,\varepsilon),\gamma).$$

With respect to the variable C, V and T have the same variance; with respect to the variables D and E, V has the same (the opposite) variance as U if A is a covariant (contravariant) variable in T.

If both U and T are exact, then so is V. If one of the functors U or T is exact and the other is half exact, then V is half exact.

If A is a covariant variable of T and both T and U are right (left) exact, then V also is right (left) exact. If A is a contravariant variable of T, T is right (left) exact and U is left (right) exact, then V is right (left) exact. The proof of these facts uses the characterizations 4.3(b) and 4.3a(b) of right and left exact functors.

Using the functors $\otimes$ and Hom various functors of three variables may be obtained by composition. It turns out that various relations hold between these.

We begin with the situation described by the symbol $(A_\Lambda, {}_\Lambda B_\Gamma, {}_\Gamma C)$, i.e. A is a right Λ-module, C is a left Γ-module, and B is a Λ-Γ-bimodule with Λ operating on the left and Γ on the right. Then $A \otimes_\Lambda B$ is a right Γ-module and $B \otimes_\Gamma C$ is a left Λ-module, so that the groups

$$(A \otimes_\Lambda B) \otimes_\Gamma C, \qquad A \otimes_\Lambda (B \otimes_\Gamma C)$$

are defined.

PROPOSITION 5.1. *There exists a unique homomorphism*

$$r\colon (A \otimes_\Lambda B) \otimes_\Gamma C \to A \otimes_\Lambda (B \otimes_\Gamma C)$$

such that $r((a \otimes b) \otimes c) = a \otimes (b \otimes c)$. *The homomorphism* r *is an isomorphism and establishes a natural equivalence of functors. It expresses the associativity of the tensor product.*

Next consider the situation described by the symbol $({}_\Lambda A, {}_\Gamma B_\Lambda, {}_\Gamma C)$.

Then $B \otimes_\Lambda A$ and C are left Γ-modules, while A and $\mathrm{Hom}_\Gamma(B,C)$ are left Λ-modules. Hence the groups

$$\mathrm{Hom}_\Lambda(A, \mathrm{Hom}_\Gamma(B,C)), \qquad \mathrm{Hom}_\Gamma(B \otimes_\Lambda A, C)$$

are defined.

PROPOSITION 5.2. *There exists a unique homomorphism*

$$s\colon \mathrm{Hom}_\Lambda(A, \mathrm{Hom}_\Gamma(B,C)) \to \mathrm{Hom}_\Gamma(B \otimes_\Lambda A, C)$$

such that for each $\varphi\colon A \to \mathrm{Hom}_\Gamma(B,C)$ *we have* $(s\varphi)(b \otimes a) = (\varphi a)b$. *This homomorphism is an isomorphism and establishes a natural equivalence of functors.*

The next case $(A_\Lambda, {}_\Lambda B_\Gamma, C_\Gamma)$ differs from the above only in that all right operators have been changed to left ones and vice-versa.

PROPOSITION 5.2′. *There exists a unique homomorphism*

$$s'\colon \mathrm{Hom}_\Lambda(A, \mathrm{Hom}_\Gamma(B,C)) \to \mathrm{Hom}_\Gamma(A \otimes_\Lambda B, C)$$

such that for each $\varphi\colon A \to \mathrm{Hom}_\Gamma(B,C)$ *we have* $(s'\varphi)(a \otimes b) = (\varphi a)b$. *This homomorphism is an isomorphism and establishes a natural equivalence of functors.*

The proofs of 5.1-5.2′ are straightforward and are left to the reader. We shall frequently regard the isomorphisms r, s, and s' as identifications.

According to the rules given earlier the functor appearing in 5.1 is covariant in all three variables and is right exact. The functors appearing in 5.2 and 5.2′ are contravariant in A and B, covariant in C and are left exact.

PROPOSITION 5.3. *In the situation* $(A_\Lambda, {}_\Lambda B_\Gamma)$ *if* A *is* Λ*-projective and* B *is* Γ*-projective then* $A \otimes_\Lambda B$ *is* Γ*-projective.*

PROOF. Let C be any right Γ-module. Then by 4.6, $\mathrm{Hom}_\Gamma(B,C)$ is an exact functor of C. Therefore, again by 4.6, $\mathrm{Hom}_\Lambda(A, \mathrm{Hom}_\Gamma(B,C))$ is an exact functor of C. Thus applying 5.2′ we deduce that $\mathrm{Hom}_\Gamma(A \otimes_\Lambda B, C)$ is an exact functor of C. It thus follows from 4.6 that $A \otimes_\Lambda B$ is Γ-projective.

A similar proposition holds also in the situation $({}_\Gamma B_\Lambda, {}_\Lambda A)$.

A similar proposition in the case $({}_\Gamma A_\Lambda, {}_\Gamma C)$ will be established later (see VI, 1.4).

6. CHANGE OF RINGS

In all of this section we shall consider two rings Λ and Γ and a ring homomorphism

$$\varphi\colon \Lambda \to \Gamma \qquad (\varphi 1 = 1).$$

Every left Γ-module A may be treated as a left Λ-module, by setting

$$\lambda a = (\varphi\lambda)a, \qquad \lambda \in \Lambda,\ a \in A.$$

Similarly for right modules. In particular Γ itself may be regarded as a left or a right Λ-module.

Conversely suppose A is a right Λ-module. We place ourselves in the situation described by the symbol $(A_\Lambda, {}_\Lambda\Gamma_\Gamma)$ (i.e. we regard Γ as a left Λ-module and right Γ-module). We then form the right Γ-module

$$A_{(\varphi)} = A \otimes_\Lambda \Gamma$$

which we call the *covariant φ-extension* of A. If A is a left Λ-module we are in the case $({}_\Gamma\Gamma_\Lambda, {}_\Lambda A)$ and we define the left Γ-module ${}_{(\varphi)}A$ as $\Gamma \otimes_\Lambda A$. Thus $A_{(\varphi)}$(resp. ${}_{(\varphi)}A$) is a covariant right exact functor of A.

Again assuming A is a right Λ-module, we can place ourselves in the situation $({}_\Gamma\Gamma_\Lambda, A_\Lambda)$. Then

$$A^{(\varphi)} = \mathrm{Hom}_\Lambda(\Gamma, A)$$

is a right Γ-module called the *contravariant φ-extension* of A. If A is a left Λ-module, we are in the case $({}_\Lambda\Gamma_\Gamma, {}_\Lambda A)$ and ${}^{(\varphi)}A$ defined as above also is a left Γ-module. Thus $A^{(\varphi)}$ (resp. ${}^{(\varphi)}A$) is a covariant left exact functor of A.

Let A be a right Λ-module. We define the Λ-homomorphism

$$A \to A_{(\varphi)}$$

as $A \otimes \varphi$: $A \otimes_\Lambda \Lambda \to A \otimes_\Lambda \Gamma$. Similarly when A is a left Λ-module. We define also the Λ-homomorphism

$$A^{(\varphi)} \to A$$

as $\mathrm{Hom}(\varphi, A)$: $\mathrm{Hom}_\Lambda(\Gamma, A) \to \mathrm{Hom}_\Lambda(\Lambda, A)$.

We are now going to apply the identities of section 5 in the following four cases:

Case 1. $(A_\Lambda, {}_\Lambda\Gamma_\Gamma, {}_\Gamma C)$. Setting $B = \Gamma$ in 5.1, yields the identity

(1) $$A \otimes_\Lambda C = A_{(\varphi)} \otimes_\Gamma C.$$

Case 2. $(A_\Gamma, {}_\Gamma\Gamma_\Lambda, {}_\Lambda C)$. Again 5.1 yields the identity

(2) $$A \otimes_\Lambda C = A \otimes_\Gamma ({}_{(\varphi)}C).$$

Case 3. $({}_\Lambda A, {}_\Gamma\Gamma_\Lambda, {}_\Gamma C)$. Setting $B = \Gamma$ in 5.2 yields the identity

(3) $$\mathrm{Hom}_\Lambda(A, C) = \mathrm{Hom}_\Gamma({}_{(\varphi)}A, C).$$

Case 4. $({}_\Gamma A, {}_\Lambda\Gamma_\Gamma, {}_\Lambda C)$. Again 5.2 yields the identity

(4) $$\mathrm{Hom}_\Lambda(A, C) = \mathrm{Hom}_\Gamma(A, {}^{(\varphi)}C).$$

We also could consider two other cases 3′ and 4′ given by the symbols $(A_\Lambda, {}_\Lambda\Gamma_\Gamma, C_\Gamma)$ and $(A_\Gamma, {}_\Gamma\Gamma_\Lambda, C_\Lambda)$ and apply 5.2′. We obtain

$$\text{(3}'\text{)} \qquad \mathrm{Hom}_\Lambda (A,C) = \mathrm{Hom}_\Gamma (A_{(\varphi)},C),$$

$$\text{(4}'\text{)} \qquad \mathrm{Hom}_\Lambda (A,C) = \mathrm{Hom}_\Gamma (A,C^{(\varphi)}).$$

PROPOSITION 6.1. *If a right Λ-module A is Λ-projective then $A_{(\varphi)}$ is Γ-projective. Similarly for a left Λ-module.*

PROOF. Assume A is a Λ-projective right Λ-module. Then by 4.6, $\mathrm{Hom}_\Lambda (A,C)$ is an exact functor of the variable C. Thus, by (3′), $\mathrm{Hom}_\Gamma (A_{(\varphi)},C)$ is an exact functor of C and therefore $A_{(\varphi)}$ is Γ-projective, again by 4.6.

Similarly, using (4) or (4′) we prove:

PROPOSITION 6.1a. *If a right Λ-module C is Λ-injective then $C^{(\varphi)}$ is Γ-injective. Similarly for a left Λ-module.*

Assume from now on that A is a right Γ-module. We then define the Γ-homomorphism

$$g\colon\ A_{(\varphi)} \to A$$

by $g(a \otimes \gamma) = a\gamma$. The composition $A \longrightarrow A_{(\varphi)} \xrightarrow{g} A$ is the identity, which proves that g is an epimorphism and Ker g is a direct summand of $A_{(\varphi)}$ as a Λ-module. If A is Γ-projective then Ker g is a direct summand of $A_{(\varphi)}$ as a Γ-module.

DEFINITION. A Γ-module A is said to be *φ-projective* if Ker g is a direct summand as a Γ-module; i.e. if the exact sequence $0 \to \mathrm{Ker}\, g \to A_{(\varphi)} \xrightarrow{g} A \to 0$ of Γ-modules splits.

PROPOSITION 6.2. *If a Γ-module A is Λ-projective and φ-projective, then A is Γ-projective. If Γ is Λ-projective and A is Γ-projective, then A is Λ-projective.*

PROOF. If A is Λ-projective, then $A_{(\varphi)}$ is Γ-projective by 6.1. If further A is φ-projective, then A is isomorphic with a direct summand of $A_{(\varphi)}$ (as a Γ-module) and therefore A is Γ-projective.

Assume Γ is Λ-projective; then $C^{(\varphi)}$ is an exact functor of the (right) Λ-module C; if further A is Γ-projective, then $\mathrm{Hom}_\Gamma (A,C^{(\varphi)})$ is an exact functor of C; by (4′) this means that $\mathrm{Hom}_\Lambda (A,C)$ is an exact functor of C, thus A is Λ-projective.

Assume that C is a right Γ-module. We define the Γ-homomorphism

$$h\colon\ C \to C^{(\varphi)}$$

which to each $c \in C$ assigns the homomorphism $hc\colon\ \gamma \to c\gamma$. The composition $C \to C^{(\varphi)} \to C$ is the identity, which proves that h is a monomorphism and Im h is a direct summand of $C^{(\varphi)}$ as a Λ-module. If C is Γ-injective then Im h is a direct summand as a Γ-module.

DEFINITION. A Γ-module C is said to be *φ-injective* if Im h is a direct summand as a Γ-module, i.e. if the exact sequence $0 \to C \xrightarrow{h} C^{(\varphi)} \to \operatorname{Coker} h \to 0$ of Γ-modules splits.

PROPOSITION 6.2a. *If a Γ-module C is Λ-injective and φ-injective, then C is Γ-injective. If Γ is Λ-projective and C is Γ-injective, then C is Λ-injective.*

The proof is dual to that of 6.2.

PROPOSITION 6.3. *For any right Λ-module A, the module $A_{(\varphi)}$ is φ-projective and the module $A^{(\varphi)}$ is φ-injective. Similar results hold for left Λ-modules.*

PROOF. We shall only consider the module $A_{(\varphi)}$ where A is a right Λ-module. We define the homomorphisms

$$\Gamma \xrightarrow{\alpha} \Gamma \otimes_\Lambda \Gamma \xrightarrow{\beta} \Gamma.$$

by $\alpha\gamma = 1 \otimes \gamma$, $\beta(\gamma_1 \otimes \gamma_2) = \gamma_1\gamma_2$. These are left Λ- and right Γ-homomorphisms. Since $\beta\alpha =$ identity we obtain right Γ-homomorphisms

$$A \otimes_\Lambda \Gamma \xrightarrow{\alpha'} A \otimes_\Lambda (\Gamma \otimes_\Lambda \Gamma) \xrightarrow{\beta'} A \otimes_\Lambda \Gamma.$$

with $\beta'\alpha' =$ identity. However if we rewrite $A \otimes_\Lambda (\Gamma \otimes_\Lambda \Gamma)$ as $(A \otimes_\Lambda \Gamma) \otimes_\Lambda \Gamma = A_{(\varphi)} \otimes_\Lambda \Gamma$ we find that β' coincides with g: $(A_{(\varphi)})_{(\varphi)} \to A_{(\varphi)}$. Thus $A_{(\varphi)}$ is φ-projective.

As an application of 6.1a we give a new method for imbedding any left Γ-module A into an injective Γ-module (see Theorem I,3.3). We assume that the problem is already solved for the ring Z of rational integers (see remark at the end of VII,5) and we consider the natural homomorphism φ: $Z \to \Gamma$. Assume that we have a Z-monomorphism $A \to Q$ where Q is Z-injective. We then have the Γ-monomorphism $A \to {}^{(\varphi)}A$ and the Γ-homomorphism ${}^{(\varphi)}A \to {}^{(\varphi)}Q$ which also is a monomorphism since Hom is left exact. There results a Γ-monomorphism $A \to {}^{(\varphi)}Q$. However by 6.1a, ${}^{(\varphi)}Q$ is Γ-injective. This proof was communicated to us by B. Eckmann. A similar proof was also found by H. A. Forrester.

EXERCISES

1. Show that $A + B$ and $A \otimes A$ are not additive functors; however $A \otimes B + B \otimes A$ and $A + A$ are additive functors.

2. For a fixed family $\{A_\alpha\}$ of right Λ-modules define the functors

$$U(C) = (\prod_\alpha A_\alpha) \otimes_\Lambda C, \qquad V(C) = \prod_\alpha (A_\alpha \otimes_\Lambda C)$$

of the left Λ-module C. Show that V is right exact, and that V is exact if and only if for each α the functor $A_\alpha \otimes_\Lambda C$ is an exact functor of C.

Define a natural transformation $f\colon U \to V$ such that $\{a_\alpha\} \otimes c \to \{a_\alpha \otimes c\}$, and show that if C is finitely generated then $f\colon U(C) \to V(C)$ is an epimorphism.

Assume that Λ is left Noetherian and C is finitely generated. Prove that $U(C) \to V(C)$ is an isomorphism. [Hint: use an exact sequence $0 \to N \to F \to C \to 0$ where F is free on a finite base.]

3. Let $g\colon T \to U$ be a natural transformation of functors, and let $\tilde{T} = \operatorname{Ker} g$, $\tilde{U} = \operatorname{Coker} g$. Show that

$$\begin{aligned}
&(T \text{ half exact}) \text{ and } (U \text{ left exact}) &&\Rightarrow \tilde{T} \text{ half exact}\\
&(T \text{ left exact}) \text{ and } (U \text{ left exact}) &&\Rightarrow \tilde{T} \text{ left exact}\\
&(T \text{ right exact}) \text{ and } (U \text{ half exact}) &&\Rightarrow \tilde{U} \text{ half exact}\\
&(T \text{ right exact}) \text{ and } (U \text{ right exact}) &&\Rightarrow \tilde{U} \text{ right exact.}
\end{aligned}$$

4. Consider the situation described by the symbol $({}_\Lambda A, B_\Gamma, {}_\Lambda C_\Gamma)$. Define a natural transformation

$$t\colon \operatorname{Hom}_\Lambda(A, \operatorname{Hom}_\Gamma(B,C)) \to \operatorname{Hom}_\Gamma(B, \operatorname{Hom}_\Lambda(A,C))$$

and show that it is an isomorphism.

5. In the situation $({}_\Lambda A, {}_\Lambda C_\Gamma)$ show that if A is Λ-projective and C is Γ-injective, then $\operatorname{Hom}_\Lambda(A,C)$ is Γ-injective.

6. Let Λ be a commutative ring, and A and C finitely generated Λ-modules. Show that $A \otimes_\Lambda C$ is a finitely generated Λ-module. Assume that Λ is Noetherian and show that $\operatorname{Hom}_\Lambda(A,C)$ is a finitely generated Λ-module.

7. Let Λ be a ring such that there exists a ring homomorphism $\varphi\colon \Lambda \to K$ into a (not necessarily commutative) field K. Show that for a free left Λ-module F, any two bases have the same cardinal number. [Hint: consider the left K-module ${}_{(\varphi)}F$.] Show that for a commutative ring Λ a homomorphism φ, as above, always exists.

CHAPTER III

Satellites

Introduction. With each functor T of one variable (covariant or contravariant) we associate a right satellite functor S^1T and a left satellite functor $S^{-1}T = S_1T$. By iteration, we then obtain satellites S^nT for any integer n ($-\infty < n < \infty$) with $S^0T = T$. If the functor T is half exact, then each exact sequence

$$0 \to A' \to A \to A'' \to 0$$

gives rise to an unlimited exact sequence involving all the satellites of T. It is in this way that we are led to the important notion of a "connected sequence of functors" (§ 4) which yields an axiomatic description of satellites (§ 5).

It is in the nature of the definition of satellites, that it applies only to one variable at a time and that higher order satellites have to be obtained by iteration. This is in sharp contrast with the theory of derived functors (Ch. v) which uses homology methods and yields the derived functors of arbitrary degree all at once. The later developments in this book will be dominated by the theory of derived functors, and because of this a thorough knowledge of this chapter is not indispensable. However, the reader will find it well worth his trouble to familiarize himself with the technique of proofs based on diagrams, as well as with the notion of a "connected sequence of functors" that is useful throughout.

1. DEFINITION OF SATELLITES

Consider a diagram

$$\begin{array}{ccccccccc} 0 & \longrightarrow & M & \xrightarrow{\alpha} & P & \xrightarrow{\beta} & A & \longrightarrow & 0 \\ & & & & & & \downarrow{\scriptstyle g} & & \\ 0 & \longrightarrow & M_1 & \xrightarrow{\alpha_1} & P_1 & \xrightarrow{\beta_1} & A_1 & \longrightarrow & 0 \end{array} \tag{1}$$

where both rows are exact and P is projective. There is then a homomorphism $f\colon P \to P_1$ with $\beta_1 f = g\beta$. The homomorphism f defines uniquely a homomorphism $f'\colon M \to M_1$ with $\alpha_1 f' = f\alpha$.

Let now T be a covariant (additive) functor of one variable. We have then a commutative diagram

$$\begin{array}{ccc} T(M) & \xrightarrow{T(\alpha)} & T(P) \\ {\scriptstyle T(f')}\downarrow & & \downarrow{\scriptstyle T(f)} \\ T(M_1) & \xrightarrow[T(\alpha_1)]{} & T(P_1) \end{array} \tag{2}$$

It follows that $T(f')$ induces a homomorphism

$$\vartheta_1(g)\colon \operatorname{Ker} T(\alpha) \to \operatorname{Ker} T(\alpha_1).$$

If T is contravariant then all the arrows in (2) should be reversed. Then $T(f')$ induces

$$\vartheta^1(g)\colon \operatorname{Coker} T(\alpha_1) \to \operatorname{Coker} T(\alpha).$$

PROPOSITION 1.1. *The homomorphisms $\vartheta_1(g)$ and $\vartheta^1(g)$ are independent of the choice of f, satisfy the additivity conditions $\vartheta_1(g + \bar{g}) = \vartheta_1(g) + \vartheta_1(\bar{g})$, $\vartheta^1(g + \bar{g}) = \vartheta^1(g) + \vartheta^1(\bar{g})$, and the transitivity conditions $\vartheta_1(g_1g) = \vartheta_1(g_1)\vartheta_1(g)$, $\vartheta^1(g_1g) = \vartheta^1(g)\vartheta^1(g_1)$.*

The transitivity conditions refer to a diagram

$$\begin{array}{ccccccccc} 0 & \longrightarrow & M & \longrightarrow & P & \longrightarrow & A & \longrightarrow & 0 \\ & & & & & & \downarrow{\scriptstyle g} & & \\ 0 & \longrightarrow & M_1 & \longrightarrow & P_1 & \longrightarrow & A_1 & \longrightarrow & 0 \\ & & & & & & \downarrow{\scriptstyle g_1} & & \\ 0 & \longrightarrow & M_2 & \longrightarrow & P_2 & \longrightarrow & A_2 & \longrightarrow & 0 \end{array}$$

with exact rows and P and P_1 both projective.

PROOF. In view of the exactness of the bottom row of (1) the homomorphism f can only be replaced by a homomorphism $\bar{f} = f + \alpha_1 h$ where $h\colon P \to M_1$. Then f' gets replaced by $\bar{f}' = f' + h\alpha$. Thus in the covariant case $T(\bar{f}') = T(f') + T(h)T(\alpha)$. Hence $T(\bar{f}')$ and $T(f')$ have the same effect when applied to the kernel of $T(\alpha)$, thus showing the uniqueness of $\vartheta_1(g)$. In the contravariant case we have $T(\bar{f}') = T(f') + T(\alpha)T(h)$. Hence $T(\bar{f}')$ and $T(f')$ coincide modulo the image of $T(\alpha)$, thus showing the uniqueness of $\vartheta^1(g)$. In order to prove the additivity and transitivity it suffices to select any f, $\bar{f}$, f_1 for g, $\bar{g}$, g_1 respectively and then use $f + \bar{f}$ and $f_1 f$ for $g + \bar{g}$ and $g_1 g$.

Next we consider a diagram

$$\begin{array}{ccccccccc} 0 & \longrightarrow & A_1 & \xrightarrow{\beta_1} & Q_1 & \xrightarrow{\alpha_1} & N_1 & \longrightarrow & 0 \\ & & \downarrow{\scriptstyle g} & & & & & & \\ 0 & \longrightarrow & A & \xrightarrow{\beta} & Q & \xrightarrow{\alpha} & N & \longrightarrow & 0 \end{array} \tag{1a}$$

where both rows are exact and Q is *injective*. Then there is a homomorphism $f\colon Q_1 \to Q$ with $f\beta_1 = \beta g$. The homomorphism f defines uniquely a homomorphism $f'\colon N_1 \to N$ with $f'\alpha_1 = \alpha f$.

If T is covariant, there results a commutative diagram

$$\begin{array}{ccc} T(Q_1) & \xrightarrow{T(\alpha_1)} & T(N_1) \\ {\scriptstyle T(f)}\downarrow & & \downarrow{\scriptstyle T(f')} \\ T(Q) & \xrightarrow[T(\alpha)]{} & T(N) \end{array} \tag{2a}$$

which yields a homomorphism

$$\vartheta^1(g)\colon \text{ Coker } T(\alpha_1) \to \text{Coker } T(\alpha).$$

In the contravariant case the arrows in (2a) should be reversed. There results a homomorphism

$$\vartheta_1(g)\colon \text{ Ker } T(\alpha) \to \text{Ker } T(\alpha_1).$$

PROPOSITION 1.1a. *The homomorphisms $\vartheta^1(g)$ and $\vartheta_1(g)$ are independent of the choice of f, satisfy the additivity conditions $\vartheta^1(g + \bar{g}) = \vartheta^1(g) + \vartheta^1(\bar{g})$, $\vartheta_1(g + \bar{g}) = \vartheta_1(g) + \vartheta_1(\bar{g})$, and the transitivity conditions $\vartheta^1(g_1 g) = \vartheta^1(g_1)\vartheta^1(g)$, $\vartheta_1(g_1 g) = \vartheta_1(g)\vartheta_1(g_1)$.*

The proof is entirely analogous to that of 1.1 and will not be repeated.

We are now ready to proceed with the definition of the main object of this chapter.

Let A be a module and let

$$0 \to M \to P \to A \to 0 \tag{3}$$

$$0 \to A \to Q \to N \to 0 \tag{4}$$

be exact sequences with P projective and Q injective. Such exact sequences exist by I,2.3 and I,3.3.

Let T be a covariant functor. We define

$$S_1T(A) = \text{Ker}\,(T(M) \to T(P)) \tag{5}$$

$$S^1T(A) = \text{Coker}\,(T(Q) \to T(N)) \tag{6}$$

thereby obtaining exact sequences

$$0 \to S_1T(A) \to T(M) \to T(P) \tag{5'}$$

$$T(Q) \to T(N) \to S^1T(A) \to 0. \tag{6'}$$

A priori, these definitions depend upon the choice of the sequences (3) and (4). Let $\bar{S}_1T(A)$ and $\bar{S}^1T(A)$ denote the modules obtained using another pair of sequences $0 \to \bar{M} \to \bar{P} \to A \to 0$ and $0 \to A \to \bar{Q} \to \bar{N} \to 0$

with $\bar{P}$ projective and $\bar{Q}$ injective. Then the maps $\vartheta_1(g)$ and $\vartheta^1(g)$ taken for $A = A_1$ and $g =$ identity, yield maps

$$S_1T(A) \to \bar{S}_1T(A), \qquad \bar{S}_1T(A) \to S_1T(A),$$
$$S^1T(A) \to \bar{S}^1T(A), \qquad \bar{S}^1T(A) \to S^1T(A),$$

which in view of the transitivity conditions of 1.1 and 1.1a are inverses of each other. Thus the modules (5) and (6) are unique up to natural isomorphisms. In order to remove all logical difficulties from the definitions (5) and (6) it suffices to assign to each A particular sequences (3) and (4), for instance, those constructed in the proofs of I,2.3 and I,3.3.

If $g\colon A \to A_1$, then the maps $\vartheta_1(g)$ and $\vartheta^1(g)$ yield maps

(7) $$S_1T(g)\colon\ S_1T(A) \to S_1T(A_1)$$

(8) $$S^1T(g)\colon\ S^1T(A) \to S^1T(A_1).$$

The conclusions of 1.1 and 1.1a then show that (5)–(8) yield covariant (additive) functors S_1T and S^1T called the *left satellite* of T and the *right satellite* of T respectively. These new functors act on the same categories of modules as T.

If T is contravariant then the above formulae are replaced by

(5a) $$S_1T(A) = \operatorname{Ker}(T(N) \to T(Q))$$

(6a) $$S^1T(A) = \operatorname{Coker}(T(P) \to T(M))$$

(5′a) $$0 \to S_1T(A) \to T(N) \to T(Q)$$

(6′a) $$T(P) \to T(M) \to S^1T(A) \to 0$$

(7a) $$S_1T(g)\colon\ S_1T(A_1) \to S_1T(A)$$

(8a) $$S^1T(g)\colon\ S^1T(A_1) \to S^1T(A).$$

The left satellite S_1T and the right satellite S^1T are then contravariant.

The definition of the satellites may be iterated by setting

$$S_{n+1}T = S_1(S_nT), \qquad S_0T = T,$$
$$S^{n+1}T = S^1(S^nT), \qquad S^0T = T.$$

It will be convenient to arrange all the left and all the right satellites into a *single sequence* $\{S^nT\}$, $-\infty < n < \infty$ as follows:

$$S_nT = S^{-n}T.$$

Several properties of the satellites are clear from the definitions.

PROPOSITION 1.2. *If T isright exact then $S^nT = 0$ for all $n > 0$. If T is left exact then $S^nT = 0$ for all $n < 0$. If T is exact then $S^nT = 0$ for all $n \neq 0$.*

PROPOSITION 1.3. *If T is covariant (contrav.) and A is projective (injective) then $S^nT(A) = 0$ for all $n < 0$. If A is injective (projective) then $S^nT(A) = 0$ for $n > 0$.*

Indeed for A projective we can take $P = A$, $M = 0$ while for A injective we can take $Q = A$, $N = 0$.

PROPOSITION 1.4. *Let $0 \to M \to P \to A \to 0$, $0 \to A \to Q \to N \to 0$ be exact sequences with P projective and Q injective. If T is covariant then*

$$S_{n+1}T(A) = S_nT(M), \qquad S^{n+1}T(A) = S^nT(N), \qquad n > 0.$$

If T is contravariant then

$$S_{n+1}T(A) = S_nT(N), \qquad S^{n+1}T(A) = S^nT(M), \qquad n > 0.$$

This is an immediate consequence of 1.3 and the exact sequences (5′), (6′), (5′a) and (6′a).

PROPOSITION 1.5. *If the functor T is defined for modules over a hereditary ring Λ, then $S^nT = 0$ for $|n| > 1$.*

Indeed in this case M is projective and N is injective. Thus 1.4 and 1.3 yield the conclusion.

2. CONNECTING HOMOMORPHISMS

Throughout this section we shall consider exact sequences

(1) $$0 \to A' \to A \to A'' \to 0$$

and commutative diagrams

(2) $$\begin{array}{ccccccccc} 0 & \longrightarrow & A' & \longrightarrow & A & \longrightarrow & A'' & \longrightarrow & 0 \\ & & \downarrow & & \downarrow & & \downarrow & & \\ 0 & \longrightarrow & B' & \longrightarrow & B & \longrightarrow & B'' & \longrightarrow & 0 \end{array}$$

with exact rows.

Let $0 \to M \to P \to A'' \to 0$ be an exact sequence with P projective. Taking $g =$ identity we then obtain a diagram

$$\begin{array}{ccccccccc} 0 & \longrightarrow & M & \longrightarrow & P & \longrightarrow & A'' & \longrightarrow & 0 \\ & & & & & & \downarrow{\scriptstyle g} & & \\ 0 & \longrightarrow & A' & \longrightarrow & A & \longrightarrow & A'' & \longrightarrow & 0 \end{array}$$

as considered in § 1. If T is a covariant functor we obtain a map

$$\vartheta_1(g)\colon \operatorname{Ker}(T(M) \to T(P)) \to \operatorname{Ker}(T(A') \to T(A)).$$

This defines a map

(3) $$\Theta_1\colon S_1T(A'') \to T(A')$$

whose composition with $T(A') \to T(A)$ is zero. Similarly using an exact sequence $0 \to A' \to Q \to N \to 0$ with Q injective yields a homomorphism

(3a) $$\Theta^1: T(A'') \to S^1T(A')$$

whose composition with $T(A) \to T(A'')$ is zero.

For T contravariant we obtain similar homomorphisms

(3') $$\Theta_1: S_1T(A') \to T(A'')$$

(3'a) $$\Theta^1: T(A') \to S^1T(A'').$$

It follows readily from 1.1 and 1.1a that these homomorphisms are independent of the choice of the auxiliary sequences $0 \to M \to P \to A'' \to 0$ etc. We thus obtain an infinite sequence

(4)
$$\cdots \longrightarrow S^{n-1}T(A'') \xrightarrow{\Theta} S^nT(A') \longrightarrow S^nT(A) \longrightarrow S^nT(A'') \xrightarrow{\Theta} S^{n+1}T(A') \longrightarrow \cdots$$

defined for all integers n. For T contravariant A'' and A' should be interchanged.

PROPOSITION 2.1. *The diagram* (2) *induces a commutative diagram*

$$\begin{array}{ccccccccccc} \cdots \to & S^{n-1}T(A'') & \to & S^nT(A') & \to & S^nT(A) & \to & S^nT(A'') & \to & S^{n+1}T(A') & \to \cdots \\ & \downarrow & & \downarrow & & \downarrow & & \downarrow & & \downarrow & \\ \cdots \to & S^{n-1}T(B'') & \to & S^nT(B') & \to & S^nT(B) & \to & S^nT(B'') & \to & S^{n+1}T(B') & \to \cdots \end{array}$$

For T contravariant all arrows should be reversed and the indices lowered.

PROOF. Only the commutativity relations in the squares involving the maps Θ need to be established. We shall only carry out the proof in the case

$$\begin{array}{ccc} S_1T(A'') & \longrightarrow & T(A') \\ \downarrow & & \downarrow \\ S_1T(B'') & \longrightarrow & T(B') \end{array}$$

for T covariant. Let $0 \to M \to P \to A'' \to 0$, $0 \to \bar{M} \to \bar{P} \to B'' \to 0$ be exact sequences with P and $\bar{P}$ projective. We thus obtain a diagram

$$\begin{array}{ccccccccc} 0 & \longrightarrow & M & \longrightarrow & P & \longrightarrow & A'' & \longrightarrow & 0 \\ & & & & & & \downarrow & & \\ 0 & \longrightarrow & B' & \longrightarrow & B & \longrightarrow & B'' & \longrightarrow & 0 \end{array}$$

which as above yields a map $S_1T(A'') \to T(B')$. In view of 1.1 this map coincides with the compositions $S_1T(A'') \to T(A') \to T(B')$ and $S_1T(A'') \to S_1T(B'') \to T(B')$ as desired.

PROPOSITION 2.2. *The composition of any two consecutive homomorphisms in the sequence* (4) *is zero.*

PROOF. Since the composition $A' \to A \to A''$ is zero it follows that the composition $S^nT(A') \to S^nT(A) \to S^nT(A'')$ is zero. Next we consider the compositions

(5) $$S^nT(A) \to S^nT(A'') \to S^{n+1}T(A').$$

For $n = 0$ this has been observed at the time Θ^1 was defined. Thus, by iteration, the composition (5) is zero for $n \geqq 0$. Thus it suffices to consider $n < 0$, which reduces to the case

(6) $$S_1T(A) \to S_1T(A'') \to T(A').$$

This composite map is obtained from a diagram

$$\begin{array}{ccccccccc} 0 & \longrightarrow & M & \longrightarrow & P & \longrightarrow & A & \longrightarrow & 0 \\ & & & & & & \downarrow & & \\ 0 & \longrightarrow & M'' & \longrightarrow & P'' & \longrightarrow & A'' & \longrightarrow & 0 \\ & & & & & & \downarrow & & \\ 0 & \longrightarrow & A' & \longrightarrow & A & \longrightarrow & A'' & \longrightarrow & 0. \end{array}$$

It therefore suffices to show that the map ϑ induced by the diagram

$$\begin{array}{ccccccccc} 0 & \longrightarrow & M & \longrightarrow & P & \longrightarrow & A & \longrightarrow & 0 \\ & & & & & & \downarrow & & \\ 0 & \longrightarrow & A' & \longrightarrow & A & \longrightarrow & A'' & \longrightarrow & 0 \end{array}$$

is zero. To see this, choose the vertical map $P \to A$ to coincide with the horizontal one. Then the induced map $M \to A'$ is zero.

The proof that the compositions

$$S^nT(A'') \to S^{n+1}T(A') \to S^{n+1}T(A)$$

are zero is similar.

3. HALF EXACT FUNCTORS

The main objective of this section is to prove the following

THEOREM 3.1. *Let*

(1) $$0 \longrightarrow A' \xrightarrow{\varphi'} A \xrightarrow{\varphi} A'' \longrightarrow 0$$

be an exact sequence. If T is a covariant half exact functor then the sequence

(2)
$$\cdots \to S^{n-1}T(A'') \to S^nT(A') \to S^nT(A) \to S^nT(A'') \to S^{n+1}T(A') \to \cdots$$

is exact. For T contravariant A' and A'' should be interchanged.

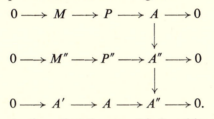

The proof will be preceded by two lemmas concerning homomorphisms derived from certain diagrams. These will also be useful in later sections.

LEMMA 3.2. *Let*

$$\begin{array}{ccccc} A' & \longrightarrow & A & \xrightarrow{\varphi} & A'' \\ \downarrow{\scriptstyle f'} & & \downarrow{\scriptstyle f} & & \downarrow{\scriptstyle f''} \\ C' & \xrightarrow[\psi']{} & C & \longrightarrow & C'' \end{array}$$

be a commutative diagram with exact rows. There result homomorphisms

$$\text{Ker } f' \rightarrow \text{Ker } f \rightarrow \text{Ker } f'' \tag{3}$$

$$\text{Coker } f' \rightarrow \text{Coker } f \rightarrow \text{Coker } f''. \tag{4}$$

If $\text{Ker } \psi' = 0$ *then* (3) *is exact. If* $\text{Coker } \varphi = 0$ *then* (4) *is exact.*

The proof is left to the reader.

Next we consider the commutative diagram

$$\begin{array}{ccccccccc} & & A' & \longrightarrow & A & \xrightarrow{\varphi} & A'' & \longrightarrow & 0 \\ & & \downarrow{\scriptstyle f'} & & \downarrow{\scriptstyle f} & & \downarrow{\scriptstyle f''} & & \\ 0 & \longrightarrow & C' & \xrightarrow[\psi']{} & C & \longrightarrow & C'' & & \end{array} \tag{5}$$

with exact rows. Given any element $x \in \text{Ker } f''$ we can find elements $a \in A$ and $c' \in C'$ with $\varphi a = x$ and $\psi' c' = fa$. The element $y \in \text{Coker} f'$ determined by c' can easily be seen to be a function of x only. We thus obtain a homomorphism

$$\text{Ker } f'' \rightarrow \text{Coker } f'.$$

LEMMA 3.3. *The sequence*

$$\text{Ker } f' \rightarrow \text{Ker } f \rightarrow \text{Ker } f'' \rightarrow \text{Coker } f' \rightarrow \text{Coker } f \rightarrow \text{Coker } f'' \tag{6}$$

is exact.

The verification is left to the reader.

It should be noted that the homomorphisms considered above are natural in the following sense. If $(\bar{5})$ is another diagram like (5) and we have a map of the diagram (5) into $(\bar{5})$ then there results a map of the exact sequence (6) into the exact sequence $(\bar{6})$.

PROOF of 3.1. We apply I,2.5 to the exact sequence (1). We obtain a commutative diagram

$$\begin{array}{ccccccc} & & T(M') & \longrightarrow & T(M) & \longrightarrow & T(M'') \\ & & \downarrow{\scriptstyle f'} & & \downarrow{\scriptstyle f} & & \downarrow{\scriptstyle f''} \\ 0 & \longrightarrow & T(P') & \longrightarrow & T(P) & \longrightarrow & T(P'') \end{array}$$

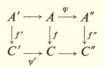

with exact rows. By 3.2 there results an exact sequence

$$\operatorname{Ker} f' \to \operatorname{Ker} f \to \operatorname{Ker} f''$$

which implies that

$$S_1T(A') \to S_1T(A) \to S_1T(A'')$$

is exact.

Next we consider an exact sequence $0 \to M \to P \to A \to 0$ with P projective. Let R be the kernel of the composed map $P \to A \to A''$. There result exact sequences $0 \to R \to P \to A'' \to 0$ and $0 \to M \to R \to A' \to 0$. We obtain a commutative diagram

$$\begin{array}{ccccccc} & & T(M) & \longrightarrow & T(R) & \longrightarrow & T(A') \\ & & \downarrow{\scriptstyle f'} & & \downarrow{\scriptstyle f} & & \downarrow \\ 0 & \longrightarrow & T(P) & \longrightarrow & T(P) & \longrightarrow & 0 \end{array}$$

with exact rows. Thus, by 3.2, the sequence

$$\operatorname{Ker} f' \to \operatorname{Ker} f \to T(A')$$

is exact. Since $\operatorname{Ker} f \to T(A')$ is easily seen to coincide with $S_1T(A'') \to T(A')$ it follows that

$$S_1T(A) \to S_1T(A'') \to T(A')$$

is exact.

Finally we consider an exact sequence

$$0 \longrightarrow M \xrightarrow{\psi'} P \xrightarrow{\psi} A'' \longrightarrow 0 \tag{7}$$

with P projective. We denote by R the submodule of the direct sum $A + P$ consisting of all pairs (a,p) with $\varphi(a) = \psi(p)$. We define the homomorphisms $R \to A$ and $R \to P$ by $(a,p) \to a$, $(a, p) \to p$ and the homomorphisms $A' \to R$ and $M \to R$ by $a' \to (\varphi'a',0)$, $m \to (0,\psi'm)$. There results a commutative diagram

$$\begin{array}{ccccccccc} & & & & 0 & & 0 & & \\ & & & & \downarrow & & \downarrow & & \\ & & 0 & \longrightarrow & M & \longrightarrow & M & \longrightarrow & 0 \\ & & \downarrow & & \downarrow & & \downarrow & & \\ 0 & \longrightarrow & A' & \longrightarrow & R & \longrightarrow & P & \longrightarrow & 0 \\ & & \downarrow & & \downarrow & & \downarrow & & \\ 0 & \longrightarrow & A' & \longrightarrow & A & \longrightarrow & A'' & \longrightarrow & 0 \\ & & \downarrow & & \downarrow & & \downarrow & & \\ & & 0 & & 0 & & 0 & & \end{array}$$

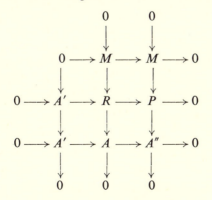

with exact rows and columns. Further, since P is projective, the middle row splits. We thus obtain a commutative diagram

$$\begin{array}{ccccccc} & & 0 & \longrightarrow & T(M) & \longrightarrow & T(M) & \longrightarrow 0 \\ & & \downarrow & & \downarrow & & \downarrow & \\ 0 & \longrightarrow & T(A') & \longrightarrow & T(R) & \longrightarrow & T(P) & \end{array}$$

with exact rows. An application of 3.3 yields an exact sequence

$$\operatorname{Ker}(T(M) \longrightarrow T(P)) \xrightarrow{u} T(A') \xrightarrow{v} \operatorname{Coker}(T(M) \longrightarrow T(R)).$$

Since the sequence $T(M) \to T(R) \to T(A)$ is exact, it follows that $\operatorname{Ker} v = \operatorname{Ker}(T(A') \to T(A))$, so that

$$\operatorname{Ker}(T(M) \longrightarrow T(P)) \xrightarrow{u} T(A') \longrightarrow T(A)$$

is exact. We must verify that u coincides with the connecting homomorphism Θ_1: $S_1T(A'') \to T(A')$. We first consider the case when the sequence (1) coincides with (7). In this case, it can be easily seen that the homomorphism

$$\bar{u}: \operatorname{Ker}(T(M) \to T(P)) \to T(M)$$

is defined by inclusion. In the general case we consider maps

$$\begin{array}{ccccccccc} 0 & \longrightarrow & M & \longrightarrow & P & \longrightarrow & A'' & \longrightarrow & 0 \\ & & \downarrow{\scriptstyle \gamma'} & & \downarrow{\scriptstyle \gamma} & & \downarrow{\scriptstyle \gamma''} & & \\ 0 & \longrightarrow & A' & \longrightarrow & A & \longrightarrow & A'' & \longrightarrow & 0 \end{array}$$

where γ'' is the identity map. The naturality property of u then yields a commutative diagram

$$\begin{array}{ccc} \operatorname{Ker}(T(M) \longrightarrow T(P)) & \xrightarrow{\bar{u}} & T(M) \\ \downarrow & & \downarrow{\scriptstyle T(\gamma')} \\ \operatorname{Ker}(T(M) \longrightarrow T(P)) & \xrightarrow[u]{} & T(A') \end{array}$$

where the first vertical map is the identity. It follows that $u = T(\gamma')\bar{u} = \Theta_1$. We have thus proved the exactness of

$$S_1T(A'') \to T(A') \to T(A).$$

Summarizing, we have established the exactness of the sequence

$$S_1T(A') \to S_1T(A) \to S_1T(A'') \to T(A') \to T(A) \to T(A'').$$

By a dual argument we show the exactness of

$$T(A') \to T(A) \to T(A'') \to S^1T(A') \to S^1T(A) \to S^1T(A'').$$

In particular, S_1T and S^1T are also shown to be half exact. The exactness of the sequence (2) now follows by iteration.

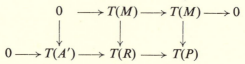

4. CONNECTED SEQUENCES OF FUNCTORS

A *connected* sequence of covariant functors is a family $T = \{T^n\}$ of covariant functors, n running through *all* integers, together with connecting homomorphisms $T^n(A'') \to T^{n+1}(A')$ defined for each exact sequence $0 \to A' \to A \to A'' \to 0$. The following two conditions are imposed:

(c.1) *The composition of any two consecutive homomorphisms in the sequence* $\cdots \to T^{n-1}(A'') \to T^n(A') \to T^n(A) \to T^n(A'') \to T^{n+1}(A') \cdots$ *is zero.*

(c.2) *If*

$$\begin{array}{ccccccccc} 0 & \longrightarrow & A' & \longrightarrow & A & \longrightarrow & A'' & \longrightarrow & 0 \\ & & \downarrow & & \downarrow & & \downarrow & & \\ 0 & \longrightarrow & B' & \longrightarrow & B & \longrightarrow & B'' & \longrightarrow & 0 \end{array}$$

is a commutative diagram with exact rows then the following diagram is commutative

$$\begin{array}{ccc} T^n(A'') & \longrightarrow & T^{n+1}(A') \\ \downarrow & & \downarrow \\ T^n(B'') & \longrightarrow & T^{n+1}(B') \end{array}$$

Actually, condition (c.2) follows from (c.1). A similar definition for contravariant functors is obtained by postulating connecting homomorphisms $T^n(A') \to T^{n+1}(A'')$. Thus in the sequence of (c.1) the roles of A' and A'' get interchanged.

The satellites S^nT of any (additive covariant or contravariant) functor together with the connecting homomorphisms defined in § 2 form a connected sequence of functors that will be denoted by ST.

Let $T = \{T^n\}$ be a connected sequence of covariant functors, let

$$0 \to A^0 \to \cdots \to A^p \to 0 \tag{S}$$

be an exact sequence of modules and let Z^i denote the kernel of $A^i \to A^{i+1}$. This yields exact sequences

$$0 \to Z^i \to A^i \to Z^{i+1} \to 0$$

which lead to homomorphisms

$$T^{n-i-1}(Z^{i+1}) \to T^{n-i}(Z^i) \qquad 0 < i < p.$$

Since $Z^1 \approx A^0$ and $Z^p = A^p$, we obtain by composition a homomorphism

$$T^{n-p}(A^p) \to T^{n-1}(A^0)$$

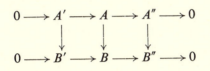

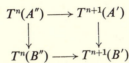

called the *iterated connecting homomorphism.* This homomorphism obviously commutes with the homomorphisms induced by a mapping of the exact sequence (S) into another such exact sequence. For contravariant functors the iterated homomorphism is $T^{n-p}(A^0) \to T^{n-1}(A^p)$.

Using the iterated connecting homomorphism we shall establish a curious anticommutativity relation resulting from a commutative diagram

$$\begin{array}{ccccccccc} & & 0 & & 0 & & 0 & & \\ & & \downarrow & & \downarrow & & \downarrow & & \\ 0 & \longrightarrow & A' & \longrightarrow & A & \longrightarrow & A'' & \longrightarrow & 0 \\ & & \downarrow & & \downarrow & & \downarrow & & \\ 0 & \longrightarrow & B' & \longrightarrow & B & \longrightarrow & B'' & \longrightarrow & 0 \\ & & \downarrow & & \downarrow & & \downarrow & & \\ 0 & \longrightarrow & C' & \longrightarrow & C & \longrightarrow & C'' & \longrightarrow & 0 \\ & & \downarrow & & \downarrow & & \downarrow & & \\ & & 0 & & 0 & & 0 & & \end{array}$$

with exact rows and columns.

PROPOSITION 4.1. *If $T = \{T^n\}$ is a connected sequence of covariant additive functors, then the diagram*

$$\begin{array}{ccc} T^{n-1}(C'') & \longrightarrow & T^n(C') \\ \downarrow & & \downarrow \\ T^n(A'') & \longrightarrow & T^{n+1}(A') \end{array}$$

is anticommutative, i.e. the two composite homomorphisms $T^{n-1}(C'') \to T^{n+1}(A')$ differ in sign.

For contravariant functors interchange A' and C''.

PROOF. We shall denote by Φ and Ψ the composite homomorphisms

$$T^{n-1}(C'') \to T^n(C') \to T^{n+1}(A')$$

$$T^{n-1}(C'') \to T^n(A'') \to T^{n+1}(A').$$

These are obviously the homomorphisms induced by the exact sequences

$$0 \to A' \to B' \to C \to C'' \to 0$$

$$0 \to A' \to A \to B'' \to C'' \to 0.$$

Now using the commutative diagram

$$\begin{array}{ccc} A' & \xrightarrow{\alpha} & A \\ {\scriptstyle\beta}\downarrow & & \downarrow{\scriptstyle\gamma} \\ B' & \xrightarrow[\delta]{} & B \end{array}$$

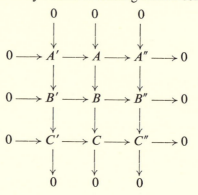

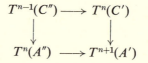

we define maps

$$A' \xrightarrow{\pi} A + B' \xrightarrow{\tau} B$$

by setting

$$\pi a' = (\alpha a', \beta a'), \qquad \tau(a,b') = \gamma a - \delta b'.$$

Then it is easy to verify that the sequence

$$0 \to A' \to A + B' \to B \to C'' \to 0$$

is exact. Further, using the projections $(a,b') \to a$, $(a,b') \to -b'$ and the map ε: $A' \to A'$, $\varepsilon(a') = -a'$, we obtain a commutative diagram

$$\begin{array}{ccccccccccc}
0 & \longrightarrow & A' & \longrightarrow & A & \longrightarrow & B'' & \longrightarrow & C'' & \longrightarrow & 0 \\
 & & \| & & \uparrow & & \uparrow & & \| & & \\
0 & \longrightarrow & A' & \longrightarrow & A + B' & \longrightarrow & B & \longrightarrow & C'' & \longrightarrow & 0 \\
 & & \downarrow{\scriptstyle \varepsilon} & & \downarrow & & \downarrow & & \| & & \\
0 & \longrightarrow & A' & \longrightarrow & B' & \longrightarrow & C & \longrightarrow & C'' & \longrightarrow & 0.
\end{array}$$

This implies $T^{n+1}(\varepsilon)\Psi = \Phi$. Since T^{n+1} is additive and $\varepsilon(a') = -a'$ it follows that $-\Psi = \Phi$.

COROLLARY 4.2. *For any additive covariant functor T and any integer n, the diagram*

$$\begin{array}{ccc}
S^{n-1}T(C'') & \longrightarrow & S^nT(C') \\
\downarrow & & \downarrow \\
S^nT(A'') & \longrightarrow & S^{n+1}T(A')
\end{array}$$

is anticommutative. For contravariant functors interchange C'' and A'.

5. AXIOMATIC DESCRIPTION OF SATELLITES

We shall give here an axiomatic description of the connected sequence ST of the satellites of a functor T.

Let $T = \{T^n\}$, $U = \{U^n\}$ be connected sequences of covariant functors. A map Φ: $T \to U$ is a sequence of natural transformations φ^n: $T^n \to U^n$ which commute with the connecting homomorphisms; i.e. the diagram

$$\begin{array}{ccc}
T^n(A'') & \longrightarrow & T^{n+1}(A') \\
\downarrow{\scriptstyle \varphi^n} & & \downarrow{\scriptstyle \varphi^{n+1}} \\
U^n(A'') & \longrightarrow & U^{n+1}(A')
\end{array}$$

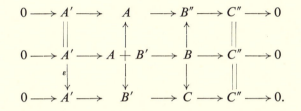

is commutative for each exact sequence $0 \to A' \to A \to A'' \to 0$. For T and U contravariant A' and A'' should be interchanged. If, for each n, φ^n is an equivalence, then Φ is called an isomorphism. We shall also consider maps $\Phi\colon T \to U$ defined only for $n \geqq 0$ or for $n \leqq 0$.

THEOREM 5.1. *Any connected sequence $T = \{T^n\}$ of covariant functors, satisfying the following two conditions:*

(c.3) *if* $0 \to M \to P \to A \to 0$ *is exact with P projective, then* $0 \to T^n(A) \to T^{n+1}(M) \to T^{n+1}(P)$ *is exact for $n < 0$.*

(c.4) if $0 \to A \to Q \to N \to 0$ *is exact with Q injective, then* $T^{n-1}(Q) \to T^{n-1}(N) \to T^n(A) \to 0$ *is exact for $n > 0$,*

is isomorphic with the connected sequence ST^0 of the satellites of the functor T^0.

The theorem is a consequence of the following more detailed proposition.

PROPOSITION 5.2. *Let $T = \{T^n\}$, $U = \{U^n\}$ be connected sequences of covariant functors and let $\varphi^0\colon T^0 \to U^0$ be a natural transformation. If U satisfies axiom* (c.3) *then φ^0 admits a unique extension to a map $\Phi\colon T \to U$ defined for all $n \leqq 0$. If T satisfies axiom* (c.4) *then φ^0 admits a unique extension to a map $\Phi\colon T \to U$ defined for all $n \geqq 0$.*

PROOF. Assume that U satisfies axiom (c.3). Suppose that $\varphi^q\colon T^q \to U^q$ are already defined for $n < q \leqq 0$ and properly commute with the connecting homomorphisms. For a given module A select arbitrarily an exact sequence $0 \to M \to P \to A \to 0$ with P projective. This yields a commutative diagram

$$\begin{array}{ccccccc}
 & & T^n(A) & \xrightarrow{\partial'} & T^{n+1}(M) & \longrightarrow & T^{n+1}(P) \\
 & & & & \big\downarrow{\scriptstyle \varphi^{n+1}(M)} & & \big\downarrow{\scriptstyle \varphi^{n+1}(P)} \\
0 & \longrightarrow & U^n(A) & \xrightarrow{\partial} & U^{n+1}(M) & \longrightarrow & U^{n+1}(P).
\end{array}$$

The bottom row is exact (axiom (c.3) for U) while in the top row the composition is zero (axiom (c.1) for T). There results a *unique* homomorphism $\varphi^n(A)\colon T^n(A) \to U^n(A)$ which, inserted into the diagram, leaves it commutative.

Consider now $f\colon A_1 \to A$ and let $0 \to M_1 \to P_1 \to A_1 \to 0$ be the sequence used to define $\varphi^n(A_1)$. Then since P_1 is projective we may find homomorphisms $g\colon P_1 \to P$ and $h\colon M_1 \to M$ such that the diagram:

$$\begin{array}{ccccccccc}
0 & \longrightarrow & M_1 & \longrightarrow & P_1 & \longrightarrow & A_1 & \longrightarrow & 0 \\
 & & \big\downarrow{\scriptstyle h} & & \big\downarrow{\scriptstyle g} & & \big\downarrow{\scriptstyle f} & & \\
0 & \longrightarrow & M & \longrightarrow & P & \longrightarrow & A & \longrightarrow & 0
\end{array}$$

is commutative. It follows that

$$\begin{aligned}\partial U^n(f)\varphi^n(A_1) &= U^{n+1}(h)\partial_1'\varphi^n(A_1) = U^{n+1}(h)\varphi^{n+1}(M_1)\partial_1' \\ &= \varphi^{n+1}(M)T^{n+1}(h)\partial_1' = \varphi^{n+1}(M)\partial' T^n(f) \\ &= \partial\varphi^n(A)T^n(f).\end{aligned}$$

Since $\partial\colon\ U^n(A) \to U^{n+1}(M)$ has zero kernel, we obtain

$$U^n(f)\varphi^n(A_1) = \varphi^n(A)T^n(f).$$

This proves that φ^n is natural, and incidentally implies that it is independent of the choice of the auxiliary sequence $0 \to M \to P \to A \to 0$.

To verify that φ^n commutes with the connecting homomorphisms, consider an exact sequence $0 \to A' \to A \to A'' \to 0$ and let $0 \to M'' \to P'' \to A'' \to 0$ be exact with P'' projective. Then there exist maps $f\colon\ P'' \to A$ and $g\colon\ M'' \to A'$ such that the diagram

$$\begin{array}{ccccccccc} 0 & \longrightarrow & M'' & \longrightarrow & P'' & \longrightarrow & A'' & \longrightarrow & 0 \\ & & \big\downarrow{\scriptstyle g} & & \big\downarrow{\scriptstyle f} & & \big\updownarrow & & \\ 0 & \longrightarrow & A' & \longrightarrow & A & \longrightarrow & A'' & \longrightarrow & 0 \end{array}$$

is commutative. This yields a commutative diagram:

$$\begin{array}{ccccc} T^n(A'') & \longrightarrow & T^{n+1}(M'') & \longrightarrow & T^{n+1}(A') \\ \big\downarrow & & \big\downarrow & & \big\downarrow \\ U^n(A'') & \longrightarrow & U^{n+1}(M'') & \longrightarrow & U^{n+1}(A') \end{array}$$

which implies the requisite commutativity relation. This proves the first part of 5.2. The proof of the second part is dual and will be omitted.

Passing to *contravariant functors*, axioms (c.3) and (c.4) should be replaced by:

(c.3′) *if* $0 \to A \to Q \to N \to 0$ *is exact with* Q *injective, then* $0 \to T^n(A) \to T^{n+1}(N) \to T^{n+1}(Q)$ *is exact for* $n < 0$;

(c.4′) *if* $0 \to M \to P \to A \to 0$ *is exact with* P *projective, then* $T^{n-1}(P) \to T^{n-1}(M) \to T^n(A) \to 0$ *is exact for* $n > 0$.

Otherwise 5.1 and 5.2. remain unchanged.

COROLLARY 5.3. *Given a natural transformation of functors*

$$\varphi\colon\ T \to U,$$

there exists a unique map $\Phi\colon\ ST \to SU$ *extending* φ. *Thus the corresponding natural transformations*

$$\varphi^n\colon\ S^nT \to S^nU \qquad (-\infty < n < +\infty)$$

commute with the connecting homomorphisms.

As a rule the functors considered are defined on the category $\mathcal{M}_\Lambda$ of *all* Λ-modules. However there may arise situations where it is convenient to consider functors defined only on suitable subcategories $\mathcal{M}$ of $\mathcal{M}_\Lambda$. For instance, let T be a half-exact covariant functor, and let $\mathcal{M}$ denote the subcategory of $\mathcal{M}_\Lambda$ consisting of all modules A such that $S_nT(A) = 0$ for $n > 0$ and of all maps of one such module into another. Clearly all projective modules are in $\mathcal{M}$ and if $0 \to A' \to A \to A'' \to 0$ is exact and $A, A'' \in \mathcal{M}$, then $A' \in \mathcal{M}$. On this category $\mathcal{M}$, the functor T is left exact. It can be easily seen that all that was said about left satellites of covariant functors and right satellites of contravariant functors remains valid for functors considered only on the category $\mathcal{M}$ above.

6. COMPOSITE FUNCTORS

Let $V = TU$ be a functor obtained by composition of two functors each of one variable. We shall consider the sequence of functors TSU defined by $(TSU)^n = TS^{\varepsilon n}U$, where $\varepsilon = +1$ or -1 depending upon whether T is covariant or contravariant. It is easily seen that TSU is a connected sequence of functors which for $n = 0$ coincides with SV. Thus 5.2 implies maps

$$\lambda\colon\ TSU \to SV \qquad \text{defined for } n \leqq 0$$

$$\rho\colon\ SV \to TSU \qquad \text{defined for } n \geqq 0.$$

Specifically, we obtain natural transformations for $n \geqq 0$

$$\lambda_n\colon\ T(S_nU) \to S_nV, \qquad \rho^n\colon\ S^nV \to T(S^nU), \qquad T \text{ covariant}$$

$$\lambda_n\colon\ T(S^nU) \to S_nV, \qquad \rho^n\colon\ S^nV \to T(S_nU), \qquad T \text{ contravariant.}$$

These homomorphisms commute with the connecting homomorphisms and yield the identity for $n = 0$.

PROPOSITION 6.1. *If T is left exact then λ_n are isomorphisms. If T is right exact then ρ^n are isomorphisms.*

PROOF. Assume T covariant and left exact. Let $0 \to M \to P \to A \to 0$ be exact with P projective. Then $0 \to S_nU(A) \to S_{n-1}U(M) \to S_{n-1}U(P)$ is exact for $n > 0$. Since T is left exact it follows that

$$0 \to TS_nU(A) \to TS_{n-1}U(M) \to TS_{n-1}U(P)$$

is exact. Thus TSU satisfies axiom (c.3). It therefore follows from 5.2 that λ_n are isomorphisms. The other cases are proved similarly.

If the functor U is exact then TSU collapses to the single term $TU = V$ and λ_n, ρ^n are zero for $n \neq 0$. In this case we obtain another connected sequence $(ST)U$ composed of the composite functors $\{(S^nT)U\}$. As above we obtain maps

$$\sigma_n\colon (S_nT)U \to S_nV, \qquad \tau^n\colon S^nV \to (S^nT)U$$

defined for $n \geqq 0$. The maps σ_n (resp τ^n) becomes isomorphisms whenever $(ST)U$ satisfies axiom (c.3) (resp. axiom (c.4)).

As an application consider a ring homomorphism: $\varphi\colon \Lambda \to \Gamma$. Any Γ-module A may be regarded as a Λ-module by setting $\lambda a = (\varphi\lambda)a$. This yields a covariant and exact functor U defined for Γ-modules whose values are Λ-modules. If T is an additive functor on the category of Λ-modules then $T' = TU$ is a functor on the category of Γ-modules. We thus have the natural homomorphisms

$$\sigma_n\colon (S_nT)' \to S_n(T'), \qquad \tau^n\colon S^n(T') \to (S^nT)', \qquad n \geqq 0.$$

PROPOSITION 6.2. *If Γ regarded as a Λ-module is projective then σ_n are isomorphisms for T covariant and τ^n are isomorphisms for T contravariant.*

PROOF. Assume T covariant (resp. contravariant). It suffices then to show that $(S_nT)'$ satisfies axiom (c.3) (resp. (c.4)). This is an immediate consequence of II,6.2.

7. SEVERAL VARIABLES

So far we considered only satellites of functors of one variable. Let $T(A,C)$ be a functor of two variables. Then for a fixed value of C we obtain a functor $T_C(A) = T(A,C)$ of the variable A alone. The resulting satellites $S^nT_C(A)$ will be denoted by

(1) $$S^n_{(1)}T(A,C),$$

the subscript 1 indicating that we consider the satellites with respect to the first variable. A map $\psi\colon C \to C'$ induces a natural transformation $T_C(A) \to T_{C'}(A)$ (if the variable C is contravariant the arrow is reversed) which induces a natural transformation of satellites. It follows that (1) may be regarded as a functor of the two variables A and C. Similarly we introduce the satellites $S^n_{(2)}\,T(A,C)$ with respect to the variable C.

We shall consider exact sequences

(2) $$0 \to M \to P \to A \to 0$$

(3) $$0 \to A \to Q \to N \to 0$$

(4) $$0 \to M' \to P' \to C \to 0$$

(5) $$0 \to C \to Q' \to N' \to 0$$

with P, P' projective and Q, Q' injective.

Assume now that T is covariant in both A and C. We obtain then a commutative diagram

$$
\begin{array}{ccccccc}
 & & 0 & & & & \\
 & & \downarrow & & & & \\
 & & S^{-1}_{(2)}S^{-1}_{(1)}T(A,C) & & & & \\
 & & \downarrow & & & & \\
0 & \longrightarrow & S^{-1}_{(1)}T(A,M') & \longrightarrow & T(M,M') & \longrightarrow & T(P,M') \\
 & & \downarrow & & \downarrow & & \\
0 & \longrightarrow & S^{-1}_{(1)}T(A,P') & \longrightarrow & T(M,P') & &
\end{array}
$$

with exact rows. This yields the exact sequence

(6) $$0 \to S^{-1}_{(2)}S^{-1}_{(1)}T(A,C) \to T(M,M') \to T(P,M') + T(M,P').$$

Interchanging the roles of the variables we obtain a similar sequence

(7) $$0 \to S^{-1}_{(1)}S^{-1}_{(2)}T(A,C) \to T(M,M') \to T(P,M') + T(M,P').$$

The sequences (6) and (7) yield a natural equivalence

(8) $$S^{-1}_{(2)}S^{-1}_{(1)} \approx S^{-1}_{(1)}S^{-1}_{(2)}.$$

For the right satellites we obtain similar exact sequences

(6a) $$T(Q,N') + T(Q',N) \to T(N,N') \to S^{1}_{(2)}S^{1}_{(1)}T(A,C) \to 0$$

(7a) $$T(Q,N') + T(Q',N) \to T(N,N') \to S^{1}_{(1)}S^{1}_{(2)}T(A,C) \to 0$$

which imply

(8a) $$S^{1}_{(2)}S^{1}_{(1)} \approx S^{1}_{(1)}S^{1}_{(2)}.$$

The above was for T covariant in both variables. If for instance C is a contravariant variable of T then in the above exact sequences P' and M' should be interchanged with Q' and N'. The isomorphisms (8) and (8a) remain valid.

Formulae (8) and (8a) yield by iteration:

THEOREM 7.1. *If T is any (additive) functor of two variables then the following natural equivalences hold*

(9) $$S^{m}_{(2)}S^{n}_{(1)} \approx S^{n}_{(1)}S^{m}_{(2)}$$

for m, n both $\geqq 0$ or both $\leqq 0$.

The conclusion is false for m,n of opposite signs.

We leave to the reader the discussion of the behavior of (9) with respect to the connecting homomorphisms (on either variable).

EXERCISES

1. Show that for any functor T of one variable, the connected sequence of functors ST is characterized (up to an isomorphism) by the following properties:

(i) $S^0T = T$.

(ii) For every connected sequence of functors U, every map φ^0: $U^0 \to T$ admits a unique extension φ: $U \to ST$ defined for $n \leqq 0$.

(iii) For every connected sequence of functors U, every map ψ^0: $T \to U^0$ admits a unique extension ψ: $ST \to U$ defined for $n \geqq 0$.

2. Consider a natural transformation of covariant functors

$$g\colon\ T \to U,$$

where T is right exact and U is left exact. Consider the new functors

$$\tilde{T} = \operatorname{Ker} g, \qquad \tilde{U} = \operatorname{Coker} g.$$

For each sequence of modules

$$0 \to A' \to A \to A'' \to 0,$$

consider the commutative diagram (cf. 5.3)

$$\begin{array}{ccccccccccccccccc}
\cdots \to & S_1T(A') & \to & S_1T(A) & \to & S_1T(A'') & \to & T(A') & \to & T(A) & \to & T(A'') & \to & 0 & \to & 0 & \to \cdots \\
& \downarrow & & \downarrow & & \downarrow & & \downarrow & & \downarrow & & \downarrow & & \downarrow & & \downarrow & \\
\cdots \to & 0 & \to & 0 & \to & 0 & \to & U(A') & \to & U(A) & \to & U(A'') & \to & S^1U(A') & \to & S^1U(A) & \to \cdots
\end{array}$$

Applying 3.2 and 3.3 to suitable portions of this diagram, define a sequence

$$\cdots \to S_1T(A') \to S_1T(A) \to S_1T(A'') \to \tilde{T}(A') \to \tilde{T}(A) \to \tilde{T}(A'') \to$$
$$\to \tilde{U}(A') \to \tilde{U}(A) \to \tilde{U}(A'') \to S^1U(A') \to S^1U(A) \to S^1U(A'') \to \cdots$$

and prove that this sequence is exact.

Examine the case when T and U are contravariant functors.

3. Consider an exact sequence of covariant (resp. contravariant) functors and natural transformations

$$T \to U \to V \to 0$$

(i.e. for each module A the sequence

$$T(A) \to U(A) \to V(A) \to 0$$

is exact). We assume that $T(A) \to U(A)$ is a *monomorphism* whenever A is projective (resp. injective). Then define a natural transformation

φ: $S_1V \to T$ in the following way: assuming for example that all functors are covariant, consider an exact sequence of modules $0 \to M \to P \to A \to 0$ with P projective. Using the diagram

$$\begin{array}{ccccccc} & & T(M) & \longrightarrow & U(M) & \longrightarrow & V(M) \longrightarrow 0 \\ & & \downarrow{\scriptstyle f'} & & \downarrow{\scriptstyle f} & & \downarrow{\scriptstyle f''} \\ 0 & \longrightarrow & T(P) & \longrightarrow & U(P) & \longrightarrow & V(P) \end{array}$$

define an exact sequence

(1) $\quad S_1T(A) \to S_1U(A) \to S_1V(A) \to \text{Coker } f' \to \text{Coker } f \to \text{Coker } f''.$

Using the natural mapping $\text{Coker } f' \to T(A)$, define $S_1V(A) \to T(A)$, which yields the desired transformation φ.

Now define φ_n: $S_{n+1}V \to S_nT$ for any $n \geqq 0$, and prove that, in the sequence

(Σ) $\quad \cdots \to S_{n+1}T \to S_{n+1}U \to S_{n+1}V \to S_nT \to \cdots \to S_1V \to T \to U \to V \to 0,$

the composition of any two consecutive homomorphisms is zero.

4. In the situation of Exer. 3, assume now that T is right exact and U is half exact. Then prove that (Σ) is an exact sequence.

[Hint: first, using the sequence (1) of Exer. 3, prove that

$$S_1T \to S_1U \to S_1V \to T \to U \to V \to 0$$

is an exact sequence. Then, by induction on $n \geqq 1$, prove that

$$S_nT \to S_nU \to S_nV \to S_{n-1}T \to \cdots \to S_1V \to T \to U \to V \to 0$$

is an exact sequence].

5. Translate Exer. 3 and 4 for the dual case of an exact sequence

$$0 \to T \to U \to V$$

such that $U(A) \to V(A)$ is an epimorphism whenever A is injective (in the case of covariant functors), resp. projective (in the case of contravariant functors). The sequence (Σ'), dual of (Σ), will be exact if V is left exact and U is half exact.

CHAPTER IV

Homology

Introduction. In this chapter we present all the algebraic tools of homology theory that will be needed later, with the exception of spectral sequences that will be treated in Ch. XV. The treatment here differs from the standard one in that great care is taken to maintain all symmetries and thus keep the system self-dual at all times. For example, the homology module $H(A)$ is usually defined as a quotient module of the module of "cycles" $Z(A)$, which is the kernel of the differentiation operator $d\colon A \rightarrow A$. We introduce the "dual" $Z'(A)$ which is the cokernel of d and show that $H(A)$ is equally well defined as a submodule of $Z'(A)$. The reader will have ample opportunities to convince himself that the preservation of this kind of a duality is indispensable.

§ 3–5 are concerned with graded and multiply graded modules and complexes. In § 6 we introduce a sign convention which causes a large number of signs usually present in algebraic topology to disappear from the symbolism.

The known homomorphisms $\alpha\colon\ H(A) \otimes H(C) \rightarrow H(A \otimes C)$ and $\alpha'\colon\ H(\mathrm{Hom}\,(A,C)) \rightarrow \mathrm{Hom}\,(H(A),H(C))$ are studied in § 6 and § 7 and are generalized to other functors. As an application we give in § 8 an elementary version (not involving spectral sequences) of the Künneth exact sequences. For the functors $\otimes$ and Hom, these results will be made more explicit in VI,3. A more advanced treatment must wait until Ch. XVII.

1. MODULES WITH DIFFERENTIATION

A Λ-module A with differentiation is a Λ-module A together with a Λ-endomorphism $d\colon\ A \rightarrow A$ such that $dd = 0$. We introduce the following notations

$$Z(A) = \mathrm{Ker}\, d, \qquad Z'(A) = \mathrm{Coker}\, d,$$
$$B(A) = \mathrm{Im}\, d, \qquad B'(A) = \mathrm{Coim}\, d.$$

Note that the differentiation d induces an isomorphism $\delta\colon\ B'(A) \approx B(A)$ but nevertheless there will be situations in which it is not convenient to identify B with B'. The operator d admits the following factorization

$$A \longrightarrow Z'(A) \longrightarrow B'(A) \xrightarrow{\delta} B(A) \longrightarrow Z(A) \longrightarrow A$$

The map $B(A) \to Z(A)$ is a monomorphism, valid because $dd = 0$. For the same reason the map $A \to B'(A)$ induces an epimorphism $Z'(A) \to B'(A)$.

This factorization yields a map

$$\tilde{d}\colon Z'(A) \to Z(A)$$

and a sequence

$$0 \to B(A) \to Z(A) \to Z'(A) \to B'(A) \to 0$$

which can easily be seen to be exact. Further we have the following equalities

$$\operatorname{Coker} \tilde{d} = Z(A)/B(A) = \operatorname{Ker}(Z'(A) \to B'(A)) = \operatorname{Ker} \tilde{d}.$$

This module is denoted by $H(A)$ and is called the homology module of A. We thus obtain an exact sequence

(1) $$0 \to H(A) \to Z'(A) \to Z(A) \to H(A) \to 0$$

and a commutative diagram

(2) $$\begin{array}{ccccccccc} & & & & 0 & & 0 & & \\ & & & & \downarrow & & \downarrow & & \\ 0 & \longrightarrow & B & \longrightarrow & Z & \longrightarrow & H & \longrightarrow & 0 \\ & & \updownarrow & & \downarrow & & \downarrow & & \\ 0 & \longrightarrow & B & \longrightarrow & A & \longrightarrow & Z' & \longrightarrow & 0 \\ & & & & \downarrow & & \downarrow & & \\ & & & & B' & \longleftrightarrow & B' & & \\ & & & & \downarrow & & \downarrow & & \\ & & & & 0 & & 0 & & \end{array}$$

with exact rows and columns.

A *mapping* or *map* $f\colon A \to A'$ of modules with differentiation is a Λ-homomorphism $f\colon A \to A'$ such that $df = fd$, where d is used to denote the differentiations in A and A'. It induces mappings $f\colon Z(A) \to Z(A')$, $\dots$, $f\colon H(A) \to H(A')$. If $f,g\colon (A,d) \to (A',d)$ are two such maps, a *homotopy* $s\colon f \simeq g$ is a Λ-homomorphism $s\colon A \to A'$ such that $ds + sd = g - f$. Homotopic maps f and g induce the same homomorphism $H(A) \to H(A')$.

Given an exact sequence

(3) $$0 \to A' \to A \to A'' \to 0$$

of modules with differentiation, we obtain a commutative diagram

$$\begin{array}{ccccccc} & & Z'(A') & \longrightarrow & Z'(A) & \longrightarrow & Z'(A'') & \longrightarrow & 0 \\ & & \downarrow & & \downarrow & & \downarrow & & \\ 0 & \longrightarrow & Z(A') & \longrightarrow & Z(A) & \longrightarrow & Z(A'') & & \end{array}$$

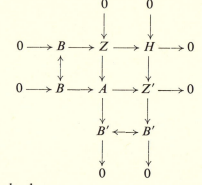

with exact rows, where the vertical maps are $\tilde{d}$. Applying III,3.3 we obtain an exact sequence

$$(4) \qquad H(A') \longrightarrow H(A) \longrightarrow H(A'') \xrightarrow{\Delta} H(A') \longrightarrow H(A) \longrightarrow H(A'')$$

with the connecting homomorphism defined in III,3. Explicitly Δ may be described as follows: given $h \in H(A'')$ choose $x \in Z'(A)$ which is mapped onto h; then $\tilde{d}x \in Z(A)$ is the image of an element $z \in Z(A')$ and Δh is determined by the element z. Composing Δ with the maps $H(A') \to Z'(A')$ and $Z(A'') \to H(A'')$ we obtain homomorphisms

$$H(A'') \to Z'(A'), \qquad Z(A'') \to H(A');$$

it is then easy to verify:

THEOREM 1.1. *For each exact sequence* (3) *the sequences*

$$\cdots \to H(A'') \to H(A') \to H(A) \to H(A'') \to Z'(A') \to Z'(A) \to Z'(A'') \to 0$$

$$0 \to Z(A') \to Z(A) \to Z(A'') \to H(A') \to H(A) \to H(A'') \to H(A') \to \cdots$$

are exact.

If

$$\begin{array}{ccccccccc} 0 & \longrightarrow & A' & \longrightarrow & A & \longrightarrow & A'' & \longrightarrow & 0 \\ & & \downarrow & & \downarrow & & \downarrow & & \\ 0 & \longrightarrow & C' & \longrightarrow & C & \longrightarrow & C'' & \longrightarrow & 0 \end{array}$$

is a commutative diagram (of modules with differentiation), with exact rows, then the vertical maps induce homomorphisms of the exact sequences associated with the top row into the corresponding exact sequences associated with the bottom row.

REMARK. The two exact sequences displayed in the theorem, coincide in their main part (4). One frequently employs the "exact" triangle

$$\begin{array}{ccc} H(A') & \longrightarrow & H(A) \\ & \nwarrow \quad \swarrow & \\ & H(A'') & \end{array}$$

to indicate this main part.

It should be noted that any Λ-module A may be regarded as a module with differentiation, by taking $d = 0$. In this case $Z(A) = Z'(A) = H(A) = A$, $B'(A) = B(A) = 0$. For any module A with differentiation the modules $Z(A)$, $Z'(A)$, $H(A)$, $B'(A)$, $B(A)$ will be regarded as modules with zero differentiation.

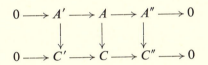

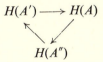

2. THE RING OF DUAL NUMBERS

The ring of dual numbers $\Gamma = (\Lambda,d)$ over a ring Λ is defined as the free Λ-module with generators 1 and d (1 being the unit element of Λ) and with multiplication defined by

$$(\lambda + \lambda' d)(\mu + \mu' d) = \lambda\mu + (\lambda\mu' + \lambda'\mu)d, \qquad \lambda,\lambda',\mu,\mu' \in \Lambda.$$

In particular $dd = 0$, $\lambda d = d\lambda$.

It is immediately clear that a Λ-module A with differentiation as defined in the preceding section is precisely a Γ-module. A map of (Λ,d)-modules is a Γ-homomorphism. It further follows that $Z(A)$, $Z'(A)$, $B(A)$, $B'(A)$ and $H(A)$ yield covariant functors defined on the category of left Γ-modules with values in the category of left Λ-modules. Each Λ-module may be regarded as a Γ-module (by setting $da = 0$ for all $a \in A$). In particular, Λ may be regarded as a left or right Γ-module. We observe the following identities

$$Z'(A) = \Lambda \otimes_\Gamma A, \qquad Z(A) = \mathrm{Hom}_\Gamma(\Lambda,A)$$

which are consequences of the identifications $A = \Lambda \otimes_\Lambda A$ and $A = \mathrm{Hom}_\Lambda(\Lambda,A)$. This again justifies the fact (contained in 1.1) that Z' is right exact and Z is left exact.

PROPOSITION 2.1. *If*

$$\begin{array}{ccccccccc}
 & & 0 & & 0 & & 0 & & \\
 & & \downarrow & & \downarrow & & \downarrow & & \\
0 & \longrightarrow & A' & \longrightarrow & A & \longrightarrow & A'' & \longrightarrow & 0 \\
 & & \downarrow & & \downarrow & & \downarrow & & \\
0 & \longrightarrow & B' & \longrightarrow & B & \longrightarrow & B'' & \longrightarrow & 0 \\
 & & \downarrow & & \downarrow & & \downarrow & & \\
0 & \longrightarrow & C' & \longrightarrow & C & \longrightarrow & C'' & \longrightarrow & 0 \\
 & & \downarrow & & \downarrow & & \downarrow & & \\
 & & 0 & & 0 & & 0 & &
\end{array}$$

is a commutative diagram of modules with differentiation with exact rows and columns, then the diagram

$$\begin{array}{ccc}
H(C'') & \longrightarrow & H(C') \\
\downarrow & & \downarrow \\
H(A'') & \longrightarrow & H(A')
\end{array}$$

is anticommutative. The same holds with $H(C'')$ *replaced by* $Z(A'')$ *or* $H(A')$ *replaced by* $Z'(A')$*, or both.*

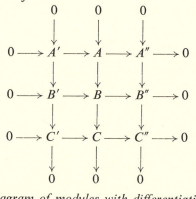

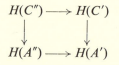

PROOF. Setting $T^n = H$ we obtain a connected sequence of functors defined on the category of Γ-modules. The anticommutativity in question is then a direct consequence of the general proposition III,4.1. The other cases are proved similarly using other connected sequences as indicated in the preceding section.

THEOREM 2.2. *The satellites of the functors Z', Z, H are as follows*

$$S^nZ' = H \text{ for } n < 0, \qquad S^nZ' = 0 \text{ for } n > 0,$$

$$S^nZ = 0 \text{ for } n < 0, \qquad S^nZ = H \text{ for } n > 0,$$

$$S^nH = H \text{ for all } n.$$

PROOF. We use the axiomatic description of satellites. Theorem 1.1, and the commutativity relation following 1.1 imply that it suffices to prove that $H(A) = 0$ if A is Γ-projective or Γ-injective. This is a consequence of 2.4 below.

We denote by η the inclusion mapping $\Lambda \to \Gamma$.

PROPOSITION 2.3. *For any left Γ-module A, the following conditions are equivalent:*

(a) *A is η-projective.*

(a′) *A is η-injective.*

(b) *There is a Λ-endomorphism s: $A \to A$ such that $ds + sd =$ identity.*

(c) *There is a Λ-module B with $A \approx {}_{(\eta)}B$.*

(c′) *There is a Λ-module B with $A \approx {}^{(\eta)}B$.*

PROOF. Given any Λ-module B we denote by B^x the Γ-module $B + B$ with $d(b_1,b_2) = (0,b_1)$. It is easy to see that

$${}_{(\eta)}B = \Gamma \otimes_\Lambda B \approx B^x \approx \operatorname{Hom}_\Lambda(\Gamma,B) = {}^{(\eta)}B.$$

Assume that an endomorphism s as required in (b) is given. Let $B = B(A)$ and define φ: $A \to B^x$ by setting $\varphi a = (da,dsa)$. Then $\varphi da = (dda,dsda) = (0,da) = d\varphi a$. If $\varphi a = 0$ then $a = dsa + sda = 0$. If $b \in B$ then $dsb = b$ and $dssb = 0$, therefore $\varphi b = (0,b)$ and $\varphi(sb) = (b,0)$. Thus φ is a Γ-isomorphism. This proves the relations (b) $\Rightarrow$ (c), (b) $\Rightarrow$ (c′). The implications (c) $\Rightarrow$ (a) and (c′) $\Rightarrow$ (a′) follow from II,6.3. There remains to be shown that (a) $\Rightarrow$ (b) and (a′) $\Rightarrow$ (b).

If we identify ${}_{(\eta)}A$ with A^x we find that the natural mapping ${}_{(\eta)}A \to A$ becomes the mapping f: $A^x \to A$ given by $f(a_1,a_2) = a_1 + da_2$. If A is η-projective then there exists a Γ-mapping g: $A \to A^x$ with $fg =$ identity. Let $ga = (ta,sa)$. The condition $dg = gd$ yields $ta = sda$ while $fga = a$ yields $ta + dsa = a$. Thus $dsa + sda = a$ as required. The proof that (a′) $\Rightarrow$ (b) is similar.

COROLLARY 2.4. *If the* Γ*-module* A *is* η*-projective* (η*-injective*), *then* $H(A) = 0$.

Indeed, by (b) above we have $a = dsa$ if $da = 0$.

The proof of the following proposition is left to the reader.

PROPOSITION 2.5. *A* Γ*-module* A *is* Γ*-projective if and only if* $A \approx_{(\eta)} B$ *where* B *is a* Λ*-projective module. Similarly* A *is* Γ*-injective if and only if* $A \approx^{(\eta)} B$ *where* B *is* Λ*-injective.*

3. GRADED MODULES, COMPLEXES

A *grading* in a module A is defined by a family of submodules A^n (n running through all integers) such that A is the direct sum $\sum_n A^n$. Each $a \in A$ has then a unique representation $a = \sum a^n$, $a^n \in A^n$ where only a finite number of a^n's is different from zero; we call a^n the homogeneous component of degree n of a. Each element of A^n is called homogeneous of degree n. The element 0 is homogeneous of degree n for all n.

A graded module A is called *positive* if $A^n = 0$ for $n < 0$, it is called *negative* if $A^n = 0$ for $n > 0$. We systematically use the notation $A_n = A^{-n}$; this is particularly convenient if A is negative.

A submodule B of a graded module A is called homogeneous if $B = \sum B^n$ where $B^n = B \cap A^n$. The quotient A/B may then be regarded as a graded module by setting

$$(A/B)^n = (A^n + B)/B \approx A^n/B^n.$$

It will be convenient to identify A/B with $\sum A^n/B^n$.

Let A and C be two graded Λ-modules. A Λ-homomorphism $f\colon A \to C$ will be said to have degree p if $f(A^n) \subset C^{n+p}$ for all n. The induced map $f^n\colon A^n \to C^{n+p}$ is called the n-th component of f. The modules Ker (f) and Im (f) are homogeneous submodules of A and C respectively; Coim (f) and Coker (f) are graded using the convention for quotient modules. The mapping Coim (f) $\to$ Im (f) induced by f has degree p; thus, despite the fact that this mapping is a Λ-isomorphism, Coim (f) and Im (f) should not be identified.

A Λ-*complex* is a graded Λ-module A together with an endomorphism $d\colon A \to A$ of degree 1 such that $dd = 0$. Thus a complex is completely determined by a sequence

$$\cdots \longrightarrow A^{n-1} \xrightarrow{d^{n-1}} A^n \xrightarrow{d^n} A^{n+1} \longrightarrow \cdots$$

such that $d^n d^{n-1} = 0$. Note that we are using the word "complex" to denote what is usually called a "cochain complex." A "chain complex" may be obtained by lowering indices. Since a complex is a special instance of a module with a differentiation operator, the various definitions of § 1 apply. The various modules, $Z(A), \ldots, H(A)$ are all graded. The main diagrams (1) and (2) of § 1 take then the following form

$$0 \to H^n \to Z'^n \to Z^{n+1} \to H^{n+1} \to 0 \tag{1'}$$

$$\begin{array}{ccccccccc}
 & & & & 0 & & 0 & & \\
 & & & & \downarrow & & \downarrow & & \\
0 & \longrightarrow & B^n & \longrightarrow & Z^n & \longrightarrow & H^n & \longrightarrow & 0 \\
 & & \updownarrow & & \downarrow & & \downarrow & & \\
0 & \longrightarrow & B^n & \longrightarrow & A^n & \longrightarrow & Z'^n & \longrightarrow & 0 \\
 & & & & \downarrow & & \downarrow & & \\
 & & & & B'^n & \longleftrightarrow & B'^n & \longrightarrow & 0 \\
 & & & & \downarrow & & \downarrow & & \\
 & & & & 0 & & 0 & &
\end{array} \tag{2'}$$

Let A and C be complexes. A *map* f: $A \to C$ is a homomorphism of degree 0 of the graded modules such that $fd = df$, i.e. $f^{n+1}d^n = d^nf^n$ where the same letter d has been used for the differentiation operators in A and C. A map f induces homomorphisms of the diagrams (1′) and (2′) of the complex A into the corresponding diagrams of the complex C.

Let f,g: $A \to C$ be maps of complexes. A *homotopy* s: $f \simeq g$ is a homomorphism s: $A \to C$ of degree -1 such that $ds + sd = g - f$, i.e. $d^{n-1}s^n + s^{n+1}d^n = g^n - f^n$. If f and g are homotopic, they induce the same homomorphisms $H^n(A) \to H^n(C)$.

An exact sequence $0 \to A' \to A \to A'' \to 0$ of complexes and maps, yields as before homomorphisms

$$H^n(A'') \to H^{n+1}(A')$$

which induce the connecting homomorphisms

$$H^n(A'') \to Z'^{n+1}(A')$$
$$Z^n(A'') \to H^{n+1}(A').$$

The sequences

$$\cdots \to H^{n-2}(A'') \to H^{n-1}(A') \to H^{n-1}(A) \to H^{n-1}(A'') \to Z'^n(A') \to Z'^n(A) \to Z'^n(A'') \to 0$$

$$0 \to Z^n(A') \to Z^n(A) \to Z^n(A'') \to H^{n+1}(A') \to H^{n+1}(A) \to H^{n+1}(A'') \to H^{n+2}(A') \to \cdots$$

are exact. In particular, we obtain the exact sequence

$$\cdots \to H^{n-1}(A'') \to H^n(A') \to H^n(A) \to H^n(A'') \to H^{n+1}(A') \to \cdots$$

unlimited in both directions. This last sequence is usually referred to as the homology (or rather cohomology) sequence.

4. DOUBLE GRADINGS AND COMPLEXES

A *double grading* (or bi-grading) in a module A consists of a family of submodules $A^{n,m}$ ((n,m) running through all pairs of integers) such that A is the direct sum $\sum_{n,m} A^{n,m}$. The elements of $A^{n,m}$ are called bihomogeneous of bidegree (n,m). We define the *associated graded* module (also denoted by A) by setting

$$A^p = \sum_{n+m=p} A^{n,m}$$

An element of bidegree (n,m) has thus degree $n + m$.

The bigraded module is said to be *positive* if $A^{n,m} = 0$ for $n < 0$ or $m < 0$. It is said to be negative if $A^{n,m} = 0$ for $n > 0$ or $m > 0$. We write $A_{n,m} = A^{-n,-m}$; this notation is particularly useful if A is negative.

A submodule B of A is bihomogeneous if $B = \sum B^{n,m}$, where $B^{n,m} = B \cap A^{n,m}$. As before we identify A/B with $\sum A^{n,m}/B^{n,m}$.

Let A, C be bigraded modules. A homomorphism $f\colon A \to C$ has bidegree (p,q) if for all n,m

$$f(A^{n,m}) \subset C^{n+p,m+q}.$$

The induced map $f^{n,m}\colon A^{n,m} \to C^{n+p,m+q}$ is called the (n,m)-component of f. The remarks made in § 3 about Ker (f), Im (f) etc. apply equally in this case. A map $f\colon A \to C$ of bidegree (p,q) induces a map of degree $p + q$ of the associated graded modules.

A *double complex* A is a doubly graded module together with two differentiation operators d_1 and d_2 of degree (1,0) and (0,1) respectively, which anticommute. Thus we have

$$(1)\begin{cases} d_1^{n,m}\colon A^{n,m} \to A^{n+1,m}, \qquad d_2^{n,m}\colon A^{n,m} \to A^{n,m+1} \\ d_1^{n,m} d_1^{n-1,m} = 0, \qquad d_2^{n,m} d_2^{n,m-1} = 0 \\ d_2^{n+1,m} d_1^{n,m} + d_1^{n,m+1} d_2^{n,m} = 0. \end{cases}$$

These conditions can be expressed by means of the anticommutative diagram

$$
\begin{array}{ccccccc}
 & & \vdots & & \vdots & & \\
 & & \uparrow & & \uparrow & & \\
\cdots \longrightarrow & & A^{n+1,m} & \xrightarrow{d_2^{n+1,m}} & A^{n+1,m+1} & & \longrightarrow \cdots \\
 & & {\scriptstyle d_1^{n,m}}\uparrow & & \uparrow{\scriptstyle d_1^{n,m+1}} & & \\
\cdots \longrightarrow & & A^{n,m} & \xrightarrow[d_2^{n,m}]{} & A^{n,m+1} & & \longrightarrow \cdots \\
 & & \uparrow & & \uparrow & & \\
 & & \vdots & & \vdots & &
\end{array}
$$

The (singly) graded module associated with A is now converted into a (single) complex by defining a *total* differentiation operator d: $A^p \to A^{p+1}$ which on $A^{n,m}$ is equal to

(3) $$d_1^{n,m} + d_2^{n,m}.$$

Thus

(4) $$d(A^{n,m}) \subset A^{n+1,m} + A^{n,m+1}.$$

Conditions (1) then imply $dd = 0$. Conversely any differentiation d in the associated graded module satisfying (4) defines uniquely the operators d_1 and d_2 which yield a double complex for which d is the total differentiation.

The modules $Z^n(A), \ldots, H^n(A)$ where A is a double complex are always understood as those defined for the associated (single) complex using the total differentiation operator.

Let A and C be double complexes. A *map* f: $A \to C$ is a map of bidegree (0,0) of the doubly graded modules, which commutes with the first and second differentiation operators in A and C. Clearly f induces a map of the associated (single) complexes.

Let f,g: $A \to C$ be two maps of double complexes. A homotopy (s_1,s_2): $f \simeq g$ consists of a pair of homorphisms s_1,s_2: $A \to C$ of bidegree $(-1,0)$ and $(0,-1)$ respectively such that

$$d_1s_1 + s_1d_1 + d_2s_2 + s_2d_2 = g - f$$

$$s_1d_2 + d_2s_1 = 0, \qquad s_2d_1 + d_1s_2 = 0$$

where d_1 and d_2 are the first and second differentiations in A and C. Passing to the associated single complexes we define s^p: $A^p \to C^{p-1}$ as

$$s_1^{n,m} + s_2^{n,m}$$

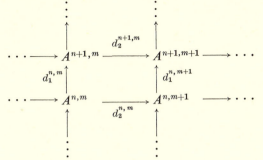

on $A^{n,m}$. We thus obtain a homotopy $s\colon f \simeq g$ satisfying

$$s(A^{n,m}) \subset C^{n-1,m} + C^{n,m-1}.$$

Conversely every such homotopy uniquely determines a pair (s_1,s_2) as above.

All the concepts introduced above admit an immediate extension to n-graded modules and n-tuple complexes. For instance a quadruple complex consists of a 4-graded module $A = \sum A^{n,m,p,q}$ and four differentiation operators

$$d_1^{n,m,p,q}\colon\ A^{n,m,p,q} \to A^{n+1,m,p,q}$$

$$d_2^{n,m,p,q}\colon\ A^{n,m,p,q} \to A^{n,m+1,p,q}$$

$$d_3^{n,m,p,q}\colon\ A^{n,m,p,q} \to A^{n,m,p+1,q}$$

$$d_4^{n,m,p,q}\colon\ A^{n,m,p,q} \to A^{n,m,p,q+1}$$

each having square zero and anticommuting with one another. The total differentiation operator d on the associated (singly) graded module is $d = d_1 + d_2 + d_3 + d_4$.

Instead of passing directly from the n-tuple complex to the associated single complex, we can pass to m-complexes for $m < n$ by a suitable grouping of the indices. For instance in the case of the quadruple complex described above, we can obtain a double complex by grouping the first index with the third and the second with the fourth:

$$A^{r,s} = \sum A^{n,m,p,q} \qquad n+p=r,\ m+q=s,$$

$$\delta_1^{r,s}\colon\ A^{r,s} \to A^{r+1,s}, \qquad \delta_2^{r,s}\colon\ A^{r,s} \to A^{r,s+1}$$

with δ_1 and δ_2 defined as $d_1 + d_3$ and $d_2 + d_4$. The original quadruple complex and the double complex just constructed have the *same* associated single complex.

5. FUNCTORS OF COMPLEXES

Let $T(A_1, \ldots, A_r)$ be a functor of r variables, some covariant, some contravariant, where A_i is a Λ_i-module and $T(A_1, \ldots, A_r)$ is a Λ-module. Suppose now that each A_i is a graded Λ_i-module. We define an r-graded module $T(A_1, \ldots, A_r)$ by setting

$$T^{n_1,\ldots,n_r}(A_1, \ldots, A_r) = T(A_1^{\varepsilon_1 n_1}, \ldots, A_r^{\varepsilon_r n_r})$$

where $\varepsilon_i = +1$ or -1 depending whether the variable A_i is covariant or contravariant. From this r-graded module we may pass to a singly graded module by defining $T^n(A_1, \ldots, A_r)$ as the direct sum of the modules $T^{n_1,\ldots,n_r}(A_1, \ldots, A_r)$ for all $(n_1, \ldots, n_r)$ such that $n_1 + \cdots + n_r = n$.

One should be cautious not to confuse the r-graded module $T(A_1, \ldots, A_r)$ with the module $T(|A_1|, \ldots, |A_r|)$ where $|A_i|$ is the non-graded module underlying A_i. It is clear, for instance, that Hom (A,C) and Hom $(|A|,|C|)$ differ not only in the fact that the first one is 2-graded; they actually differ as modules. The situation is somewhat similar to that encountered with topological groups. Let A and C be topological abelian groups, $|A|$, $|C|$ the underlying discrete groups and let Hom (A,C) be the group of all continuous homomorphisms $A \to C$. Then Hom (A,C) and Hom $(|A|,|C|)$ are distinct.

Let $A'_1, \ldots, A'_r$ be another sequence of graded modules and consider maps f_i: $A_i \to A'_i$ (resp. f_i: $A'_i \to A_i$) if the i-th variable of T is covariant (resp. contravariant). Let f_i have degree p_i. We define the map

$$T(f_1, \ldots, f_r): T(A_1, \ldots, A_r) \to T(A'_1, \ldots, A'_r)$$

of r-degree $(p_1, \ldots, p_r)$, by defining the map $T^{n_1, \ldots, n_r}(f_1, \ldots, f_r)$ on $T^{n_1, \ldots, n_r}(A_1, \ldots, A_r)$ as

$$(-1)^{\varepsilon} T(f_1^{l_1}, \ldots, f_r^{l_r}): T(A_1^{\varepsilon_1 n_1}, \ldots, A_r^{\varepsilon_r n_r}) \to T(A_1'^{\varepsilon_1(n_1+p_1)}, \ldots, A_r'^{\varepsilon_r(n_r+p_r)})$$

where $\varepsilon = \sum n_i p_j$ for $i < j$, $l_i = n_i$ if T is covariant in A_i and $l_i = -(n_i + p_i)$ if T is contravariant in A_i.

If g_i: $A'_i \to A''_i$ (resp. g_i: $A''_i \to A'_i$) are maps of degrees q_i, and $h_i = g_i f_i$: $A_i \to A''_i$ (resp. $h_i = f_i g_i$: $A''_i \to A_i$), then it is easy to verify the rule

$$T(h_1, \ldots, h_r) = (-1)^{\eta} T(g_1, \ldots, g_r) T(f_1, \ldots, f_r)$$

where $\eta = \sum p_i q_j$ for $i < j$.

Suppose now that each A_i is a complex with differentiation d_i. Then setting

$$\delta_i = T(A_1, \ldots, d_i, \ldots, A_r)$$

we find that $\delta_1, \ldots, \delta_r$ anticommute and define $T(A_1, \ldots, A_r)$ as an r-tuple complex. If f_i are maps of complexes (each f_i has then degree zero) then $T(f_1, \ldots, f_r)$ (this time not involving any signs) is a map of r-tuple complexes. If s_i: $f_i \simeq f'_i$ are homotopies for $i = 1, \ldots, r$, then setting $t_i = T(A_1, \ldots, s_i, \ldots, A_r)$ we obtain a homotopy

$$(t_1, \ldots, t_r): T(f_1, \ldots, f_r) \simeq T(f'_1, \ldots, f'_r).$$

To illustrate the above definitions we consider the tensor product $A \otimes C$ of graded modules. This is by definition the doubly graded module $\Sigma_{n,m} A^n \otimes C^m$. If f: $A \to A'$, g: $C \to C'$ are maps of degrees p and q then

$$(f \otimes g): A \otimes C \to A' \otimes C'$$

is the map of degree $p + q$ defined for $a \in A^n$, $c \in C^m$ as

$$(f \otimes g)(a \otimes c) = (-1)^{nq} fa \otimes gc.$$

The sign is due to the interchange of the symbols g and a. If A and C are complexes with differentiations d_1 and d_2, then $A \otimes C$ is a double complex with differentiations $d_1 \otimes C$ and $A \otimes d_2$. The total differentiation in $A \otimes C$ is $d = d_1 \otimes C + A \otimes d_2$ and we have

$$d(a \otimes c) = (d_1 a) \otimes c + (-1)^n a \otimes (d_2 c).$$

6. THE HOMOMORPHISM α

In the remaining sections of this chapter we study certain relations between $H(T(A,C))$ and $T(H(A),H(C))$ that play a fundamental role in later chapters.

We consider as a typical case, a functor T of two variables, covariant in the first variable and contravariant in the second. If A and C are complexes, then $T(A,C)$ is a double complex, and thus $T(A,C)$ may be regarded also as a complex.

We consider the commutative diagrams

$$\begin{array}{ccc} Z(A) & \longrightarrow & H(A) \\ \downarrow & & \downarrow \\ A & \longrightarrow & Z'(A) \end{array} \qquad \begin{array}{ccc} Z(C) & \longrightarrow & H(C) \\ \downarrow & & \downarrow \\ C & \longrightarrow & Z'(C) \end{array}$$

They induce a commutative diagram

$$(1) \qquad \begin{array}{ccc} T(Z(A),Z'(C)) & \xrightarrow{\xi} & T(H(A),H(C)) \\ \eta \downarrow & & \downarrow \tau \\ H(T(A,C)) & \xrightarrow[\zeta]{} & T(Z'(A),Z(C)) \end{array}$$

Actually all four modules in the diagram should have the operator H in front; however in three of the modules the differentiation is zero so that H may be omitted.

PROPOSITION 6.1. *If T is right exact, there exists a unique homomorphism of degree zero*

$$\alpha: \ T(H(A),H(C)) \longrightarrow H(T(A,C))$$

which when inserted in (1), *leaves the diagram commutative. The homomorphism α is natural relative to maps $A \to A'$ and $C' \to C$, and if A and C have zero differentiations then α is the identity. The last two properties characterize α uniquely.*

PROOF. Since T is right exact, ξ is an epimorphism. Thus there exists at most one α with $\alpha\xi = \eta$. For such an α we have $\zeta\alpha\xi = \zeta\eta = \tau\xi$ so that $\zeta\alpha = \tau$. To show that such an α exists, it suffices to prove that $\operatorname{Ker}\xi \subset \operatorname{Ker}\eta$. Since T is right exact it follows from II,4.3 that $\operatorname{Ker}\xi$ is the sum of the images

$$T(B(A),Z'(C)) \to T(Z(A),Z'(C)) \leftarrow T(Z(A),B'(C)).$$

Thus to prove $\operatorname{Ker}\xi \subset \operatorname{Ker}\eta$ amounts to showing that the homomorphisms

$$T(B(A),Z'(C)) \to H(T(A,C)) \leftarrow T(Z(A),B'(C))$$

are zero. These homomorphisms admit factorizations

$$T(B(A),Z'(C)) \xrightarrow{\beta} H(T(A,Z'(C))) \longrightarrow H(T(A,C))$$

$$T(Z(A),B'(C)) \xrightarrow{\gamma} H(T(Z(A),C)) \longrightarrow H(T(A,C))$$

and it suffices to show that β and γ are zero. To show this we factor the differentiation operators in $T(A,Z'(C))$ and $T(Z(A),C)$ as follows

$$T(A,Z'(C)) \to T(B'(A),Z'(C)) \xrightarrow{T(\delta_A,\, Z'(C))} T(B(A),Z'(C)) \xrightarrow{\beta'} T(A,Z'(C))$$

$$T(Z(A),C) \to T(Z(A),B(C)) \xrightarrow{T(Z(A),\, \delta_C)} T(Z(A),B'(C)) \xrightarrow{\gamma'} T(Z(A),C)$$

where δ is the map $B' \to B$ induced by d. Since T is right exact, the left hand homomorphisms are epimorphisms. Since the middle homomorphisms are isomorphisms it follows that the image of β' is in $B(T(A,Z'(C)))$ and the image of γ' is in $B(T(Z(A),C))$. Thus β and γ are zero.

The naturality of α and the fact that α is the identity if A and C have derivation zero are obvious. To prove the last assertion assume that another family $\bar{\alpha}$ of homomorphisms is given satisfying these two conditions. The maps $Z(A) \to A$ and $C \to Z'(C)$ then induce a commutative diagram

$$\begin{array}{ccc} T(Z(A),Z'(C)) & \xrightarrow{\xi} & T(H(A),H(C)) \\ \downarrow & & \downarrow \bar{\alpha} \\ T(Z(A),Z'(C)) & \xrightarrow[\eta]{} & H(T(A,C)) \end{array}$$

Therefore $\bar{\alpha}\xi = \eta = \alpha\xi$. Since ξ is an epimorphism, it follows that $\bar{\alpha} = \alpha$.

PROPOSITION 6.1a. *If T is left exact, there exists a unique homomorphism of degree zero*

$$\alpha'\colon\ H(T(A,C)) \to T(H(A),H(C))$$

which when inserted in (1), *leaves the diagram commutative. The homomorphism* α' *is natural relative to maps* $A \to A', C' \to C$ *of degree zero, and if* A *and* C *have zero differentiation then* α' *is the identity. These last two properties characterize* α' *uniquely.*

The proof is dual to the preceding one.

PROPOSITION 6.2. *If* T *is right exact and the sequences*

$$0 \to H(A) \to Z'(A) \to B'(A) \to 0$$

$$0 \to B(C) \to Z(C) \to H(C) \to 0$$

split, then α *has kernel zero and its image is a direct summand of* $H(T(A,C))$.

PROOF. Composing the splitting homomorphisms $Z'(A) \to H(A)$ and $H(C) \to Z(C)$ with the natural maps $A \to Z'(A)$ and $Z(C) \to C$ yields maps $\beta\colon\ A \to H(A)$ and $\bar{\gamma}\colon\ H(C) \to C$ such that the induced maps $\beta_*\colon H(A) \to H(A)$, $\bar{\gamma}_*\colon\ H(C) \to H(C)$ are identities. There results a commutative diagram

$$\begin{array}{ccc} T(H(A),H(C)) & \xrightarrow{\ \alpha\ } & H(T(A,C)) \\ \big\downarrow & & \big\downarrow{\scriptstyle \delta} \\ T(H(A),H(C)) & \longrightarrow & T(H(A),H(C)) \end{array}$$

where the vertical maps are induced by $T(\beta,\bar{\gamma})$. Thus $\delta\alpha =$ identity, which implies the conclusion of 6.2.

PROPOSITION 6.2a. *If* T *is left exact and the sequences*

$$0 \to B(A) \to Z(A) \to H(A) \to 0$$

$$0 \to H(C) \to Z'(C) \to B'(C) \to 0$$

split, then α' *is an epimorphism and its kernel is a direct summand of* $H(T(A,C))$.

The proof is dual to that of 6.2.

7. THE HOMOMORPHISM α (CONTINUATION)

We shall establish here some less elementary properties of the homomorphisms α and α', that will be needed later. We begin by establishing a commutativity relation with the connecting homomorphisms for homology.

We consider a functor $T(A,C)$ where C is a covariant variable and A denotes all the remaining variables, some of which may precede the variable C and some of which may follow it. We shall assume that each of the variables is a complex.

We now assume that we have an exact sequence of complexes

(1) $$0 \longrightarrow C' \xrightarrow{\psi} C \xrightarrow{\varphi} C'' \longrightarrow 0$$

(φ and ψ having degree zero) such that the sequence

(2) $$0 \to T(A,C') \to T(A,C) \to T(A,C'') \to 0$$

is exact.

We then have connecting homomorphisms

(3) $$\delta\colon\ H(C'') \to H(C')$$

(4) $$\Delta\colon\ H(T(A,C'')) \to H(T(A,C')).$$

PROPOSITION 7.1. *If T is right exact, the following diagram is commutative*

(5) $$\begin{array}{ccc} T(H(A),H(C'')) & \xrightarrow{T(H(A),\delta)} & T(H(A),H(C')) \\ \downarrow{\alpha_2} & & \downarrow{\alpha_1} \\ H(T(A,C'')) & \xrightarrow[\Delta]{} & H(T(A,C')) \end{array}$$

If T is left exact, the same holds with the vertical arrows reversed and replaced by α'_2, α'_1.

For T contravariant in C, we must interchange C' and C'' in (2), in (4) and in (5).

REMARK. The homomorphism $T(H(A),\delta)$ in diagram (5) involves a sign (see § 5).

PROOF. For the sake of brevity we shall use a notation as if all the variables of A were covariant. Thus for instance, if we write $T(Z(A),Z(C))$ we actually replace each covariant variable A_i by $Z(A_i)$ and each contravariant variable by $Z'(A_i)$.

The proof is based on an alternative description of the homomorphism δ. We denote by X the kernel of the composed homomorphism

$$C \xrightarrow{d_C} C \xrightarrow{\varphi} C''.$$

We then obtain the homomorphisms

$$H(C'') \xleftarrow{\mu''} Z(C'') \xleftarrow{\tau''} X \xrightarrow{\tau'} Z(C') \xrightarrow{\mu'} H(C')$$

where μ' and μ'' are natural factorization homomorphisms, τ' is defined by d_C since $d_C(X) \subset \operatorname{Im} \psi$, and τ'' is defined by φ since $\varphi(X) = Z(C'')$. Thus τ'' and μ'' are epimorphisms and it is easy to see that

(6) $$\mu'\tau' = \delta\mu''\tau''.$$

Quite similarly we define Y as the kernel of the composed homomorphism

$$(7) \qquad T(A,C) \xrightarrow{d} T(A,C) \longrightarrow T(A,C'')$$

and obtain

$$HT(A,C'') \xleftarrow{\rho''} ZT(A,C'') \xleftarrow{\sigma''} Y \xrightarrow{\sigma'} ZT(A,C') \xrightarrow{\rho'} HT(A,C')$$

with

$$(8) \qquad \rho'\sigma' = \Delta\rho''\sigma''.$$

If we compose the map $T(Z(A),X) \to T(A,C)$ with (7) we obtain zero, thus we have a map

$$\Theta\colon\ T(Z(A),X) \to Y.$$

We consider the diagram

$$\begin{array}{ccccccccc}
T(H(A),H(C'')) & \longleftarrow & T(Z(A),Z(C'')) & \longleftarrow & T(Z(A),X) & \longrightarrow & T(Z(A),Z(C')) & \longrightarrow & T(H(A),H(C')) \\
\downarrow{\scriptstyle \alpha_2} & & \downarrow & & \downarrow{\scriptstyle \Theta} & & \downarrow & & \downarrow{\scriptstyle \alpha_1} \\
H(T(A,C'')) & \underset{\rho''}{\longleftarrow} & Z(T(A,C'')) & \underset{\sigma''}{\longleftarrow} & Y & \underset{\sigma'}{\longrightarrow} & Z(T(A,C')) & \underset{\rho'}{\longrightarrow} & H(T(A,C'))
\end{array}$$

where the maps in the upper row are

$$T(\mu,\mu''), \qquad T(Z(A),\tau''), \qquad T(Z(A),\tau'), \qquad T(\mu,\mu').$$

The commutativity of the extreme two squares follows from the definition of α_1 and α_2. The commutativity in the remaining two squares is an easy consequence of the definition of Θ. We now compute using (6) and (8)

$$\begin{aligned}
\Delta\alpha_2 T(\mu,\mu''\tau'') = \Delta\rho''\sigma''\Theta = \rho'\sigma'\Theta &= \alpha_1 T(\mu,\mu'\tau') \\
&= \alpha_1 T(\mu,\delta\mu''\tau'') = \alpha_1 T(H(A),\delta)T(\mu,\mu''\tau'').
\end{aligned}$$

Since μ,μ'',τ'' are epimorphisms and T is right exact, it follows that $T(\mu,\mu'',\tau'')$ is an epimorphism. This proves that (5) is commutative.

For T left exact we consider instead of X the cokernel X' of $C' \xrightarrow{\psi} C \xrightarrow{d_C} C$.

THEOREM 7.2. *If the functor T is exact, α and α' are isomorphisms and are inverses of each other.*

PROOF. We observe that (using the notation of diagram (1) of the preceding section)

$$\tau\xi = \zeta\eta = \tau\alpha'\alpha\xi$$

and since $\operatorname{Ker}\tau = 0 = \operatorname{Coker}\xi$, it follows that $\alpha'\alpha =$ identity. There remains to be shown that α is an isomorphism. This is clear if all the variables have zero differentiation. The proof is carried out by induction with respect to the number of variables that have a non-zero differentiation. We denote one of these variables by C and write $T(A,C)$ where all the remaining variables have been lumped into a single symbol A. We assume T covariant in C and consider the exact sequence

$$0 \to Z(C) \to C \to B(C) \to 0$$

which yields the homology sequence

$$\cdots \longrightarrow B(C) \xrightarrow{\delta} Z(C) \longrightarrow H(C) \longrightarrow B(C) \xrightarrow{\delta} Z(C) \longrightarrow \cdots.$$

Since T is exact we obtain an exact sequence

$$T(H(A),B(C)) \xrightarrow{\delta'} T(H(A),Z(C)) \longrightarrow T(H(A),H(C)) \longrightarrow T(H(A),B(C)) \xrightarrow{\delta'} T(H(A),Z(C))$$

where $\delta' = T(H(A),\delta)$.

Now, applying 7.1 and the naturality of α we obtain a commutative diagram

$$\begin{array}{ccccccccc} T(H(A),B(C)) & \xrightarrow{\delta'} & T(H(A),Z(C)) & \xrightarrow{i} & T(H(A),H(C)) & \xrightarrow{j} & T(H(A),B(C)) & \xrightarrow{\delta'} & T(H(A),Z(C)) \\ \downarrow{\alpha_2} & & \downarrow{\alpha_1} & & \downarrow{\alpha} & & \downarrow{\alpha_2} & & \downarrow{\alpha_1} \\ H(T(A,B(C))) & \longrightarrow & H(T(A,Z(C))) & \longrightarrow & H(T(A,C)) & \longrightarrow & H(T(A,B(C))) & \longrightarrow & H(T(A,Z(C))) \end{array}$$

The lower row is the homology sequence of

$$0 \to T(A,Z(C)) \to T(A,C) \to T(A,B(C)) \to 0.$$

Since the rows of the diagram are exact, and since, by the inductive assumption, α_1 and α_2 are isomorphisms, it follows from I,1.1 (the "5 lemma") that α also is an isomorphism.

PROPOSITION 7.3. *Let $T(A,C)$ be right exact and covariant in C. If*

$$T(A,Z(C)) \to T(A,C)$$

is a monomorphism, and

$$\alpha\colon\ T(H(A),B(C)) \to HT(A,B(C))$$

is an epimorphism, then the sequence

$$(9)\cdots \longrightarrow H(T(A,C)) \xrightarrow{k^*} H(T(A,B(C))) \xrightarrow{i^*} H(T(A,Z(C))) \xrightarrow{j^*} H(T(A,C)) \longrightarrow \cdots$$

induced by the natural maps

$$(10)\qquad C \xrightarrow{k} B(C) \xrightarrow{i} Z(C) \xrightarrow{j} C$$

is exact. For T contravariant in C, replace $B(C)$ and $Z(C)$ by $B'(C)$ and $Z'(C)$ and reverse the arrows in (10).

PROOF. Since the sequence

$$0 \to T(A,Z(C)) \to T(A,C) \to T(A,B(C)) \to 0$$

is exact, we obtain a homology sequence like (9) but with i^* replaced by the connecting homomorphism Δ. It therefore suffices to show that $\Delta = i^*$.

By 7.1, we have the commutative diagram

$$\begin{array}{ccc} T(H(A),B(C)) & \xrightarrow{T(H(A),\delta)} & T(H(A),Z(C)) \\ \downarrow{\scriptstyle \alpha} & & \downarrow{\scriptstyle \alpha_1} \\ HT(A,B(C)) & \xrightarrow[\Delta]{} & HT(A,Z(C)) \end{array}$$

where $\delta\colon B(C) \to Z(C)$ is the connecting homomorphism induced by the exact sequence $0 \to Z(C) \to C \to B(C) \to 0$. It is clear that $\delta = i$. Since by the naturality of α we have $i^*\alpha = \alpha_1 T(H(A),\delta)$ it follows that $\Delta\alpha = i^*\alpha$. Thus $\Delta = i^*$ since α is an epimorphism.

PROPOSITION 7.3a. *Let T be left exact and covariant in C. If*

$$T(A,C) \to T(A,Z'(C))$$

is an epimorphism, and

$$\alpha\colon\ HT(A,B'(C)) \to T(H(A),B'(C))$$

is a monomorphism, then the sequence

(9a) $\cdots \to H(T(A,C)) \to H(T(A,Z'(C))) \to H(T(A,B'(C))) \to H(T(A,C)) \to \cdots$

induced by the natural maps

(10a) $$C \to Z'(C) \to B'(C) \to C$$

is exact. For T contravariant in C, replace $B'(C)$ and $Z'(C)$ by $B(C)$ and $Z(C)$ and reverse the arrows in (10a).

The preceding results may be sharpened in the following way. For each complex A, let $\mathcal{M}(A)$ denote the category consisting of the complexes A, $B(A)$, $B'(A)$, $Z(A)$, $Z'(A)$, $H(A)$, the identity maps, the maps occurring in diagram (1) of § 1, the maps $B'(A) \to B(A)$ and of their compositions. The conditions "T is right exact," "T is left exact" and "T is exact" that occurred before may be replaced by "T is right exact on the categories $\mathcal{M}(A)$, $\mathcal{M}(C)$", etc.

We shall say that a complex *splits* if the sequences in diagram (1) of § 1 split.

PROPOSITION 7.4. *If the complexes A and C split, then α and α' are defined, are isomorphisms and are inverses of each other.*

This follows directly from 7.2 since the functor $T(A,C)$ is exact on the categories $\mathcal{M}(A)$ and $\mathcal{M}(C)$.

8. KÜNNETH RELATIONS

We shall consider a functor T of any number of variables. We shall denote one of the variables by C and use the symbol A for all the remaining variables. We shall use the symbols S_1T and S^1T to denote the satellites of T with respect to the variable C. We shall assume that all the variables in T are complexes.

THEOREM 8.1. *Let T be right exact and covariant in C. If the homomorphisms*

$$\alpha_1: \ T(H(A),B(C)) \to HT(A,B(C))$$

$$\alpha_2: \ T(H(A),Z(C)) \to HT(A,Z(C))$$

are isomorphisms, and if

$$S_1T(A,B(C)) = 0 = S_1T(H(A),Z(C)), \tag{1}$$

then we have an exact sequence

$$0 \longrightarrow T(H(A),H(C)) \xrightarrow{\alpha} H(T(A,C)) \xrightarrow{\beta} S_1T(H(A),H(C)) \longrightarrow 0 \tag{2}$$

with β of degree 1. If T is contravariant in C, we replace $B(C)$ and $Z(C)$ by $B'(C)$ and $Z'(C)$.

PROOF. We consider the commutative diagram

$$\begin{array}{ccccccccc}
T(H(A),B(C)) & \to & T(H(A),Z(C)) & \to & T(H(A),H(C)) & \to & 0 & & \\
\downarrow{\scriptstyle \alpha_1} & & \downarrow{\scriptstyle \alpha_2} & & \downarrow{\scriptstyle \alpha} & & & & \\
H(T(A,B(C))) & \to & H(T(A,Z(C))) & \to & H(T(A,C)) & \to & H(T(A,B(C))) & \to & H(T(A,Z(C))) \\
 & & & & & & \downarrow{\scriptstyle \alpha_1^{-1}} & & \downarrow{\scriptstyle \alpha_2^{-1}} \\
 & & & & 0 \to S_1T(H(A),H(C)) & \to & T(H(A),B(C)) & \to & T(H(A),Z(C))
\end{array}$$

Since $S_1T(A,B(C)) = 0$, $T(A,Z(C)) \to T(A,C)$ is a monomorphism. Thus 7.3 implies that the middle row in the diagram above is exact. Since $S_1T(H(A),Z(C)) = 0$, the other two rows also are exact. It follows easily that there is a unique homomorphism

$$\beta: \ H(T(A,C)) \to S_1T(H(A),H(C))$$

which when inserted into the diagram, leads to a commutative diagram. The exactness of (2) then follows readily from the diagram above.

The exact sequence (2) is natural in the following sense. Let A', C' be another pair satisfying the conditions of 8.1, and f: $A \to A'$, g: $C \to C'$ be maps of complexes (actually f is a family consisting of one mapping

f_i: $A_i \to A_i'$ for each covariant variable in A and f_i: $A_i' \to A_i$ for each contravariant variable in A). Then the diagram

$$\begin{array}{ccccccccc} 0 & \longrightarrow & T(H(A),H(C)) & \longrightarrow & H(T(A,C)) & \longrightarrow & S_1T(H(A),H(C)) & \longrightarrow & 0 \\ & & \downarrow & & \downarrow & & \downarrow & & \\ 0 & \longrightarrow & T(H(A'),H(C')) & \longrightarrow & H(T(A',C')) & \longrightarrow & S_1T(H(A'),H(C')) & \longrightarrow & 0 \end{array}$$

is commutative.

It should further be remarked that if T is a functor of one variable C, then the conditions concerning α_1 and α_2 are automatically satisfied and conditions (1) become

$$(1') \qquad S_1T(B(C)) = 0 = S_1T(Z(C)).$$

Since α has degree 0 and β has degree 1 the exact sequence (2) may be rewritten as

$$0 \longrightarrow \sum_{p+q=n} T(H^p(A),H^q(C)) \xrightarrow{\alpha} H^n(T(A,C)) \xrightarrow{\beta} \sum_{p+q=n+1} S_1T(H^p(A),H^q(C)) \longrightarrow 0$$

REMARK. The only property of S_1T that was used above is that for each exact sequence $0 \to C' \to C \to C'' \to 0$ the sequence

$$S_1T(C) \to S_1T(C'') \to T(C') \to T(C)$$

is exact (assuming T covariant in C).

THEOREM 8.1a. *Let T be left exact and covariant in C. If the homomorphisms*

$$\alpha_1': \ H(T(A,B'(C))) \to T(H(A),B'(C))$$

$$\alpha_2': \ H(T(A,Z'(C))) \to T(H(A),Z'(C))$$

are isomorphisms, and if

$$(1a) \qquad S^1T(A,B'(C)) = 0 = S^1T(H(A),Z'(C)),$$

then we have an exact sequence

$$(2a) \quad 0 \longrightarrow S^1T(H(A),H(C)) \xrightarrow{\beta'} H(T(A,C)) \xrightarrow{\alpha'} T(H(A),H(C)) \longrightarrow 0$$

with β' of degree 1. If T is contravariant in C, we replace $B'(C)$ and $Z'(C)$ by $B(C)$ and $Z(C)$.

EXERCISES

1. Let A and A' be Λ-complexes. A *map* f: $A \to A'$ of degree u is defined as a homomorphism of degree u such that $df = (-1)^u fd$. If g: $A \to A'$ is another map of degree u, then a *homotopy* s: $f \simeq g$ is a homomorphism s: $A \to A'$ of degree $u - 1$ such that $ds + (-1)^u sd = g - f$.

Show that if $f\colon A \to A'$ and $f'\colon A' \to A''$ are maps of degree u and v respectively, then $f'f\colon A \to A''$ is a map of degree $u + v$. If further $s\colon f \simeq g$, $s'\colon f' \simeq g'$ then

$$s'f + (-1)^v g's\colon f'f \simeq g'g.$$

2. Extend the above definitions to double and n-tuple complexes. In particular show that if $f\colon A \to A'$, $g\colon C \to C'$ are maps of degrees u and v, then

$$f \otimes g\colon A \otimes C \to A' \otimes C', \quad \operatorname{Hom}(f,g)\colon \operatorname{Hom}(A',C) \to \operatorname{Hom}(A,C')$$

are maps of bidegrees (u,v). Draw appropriate conclusions if $s\colon f \simeq f'$, $t\colon g \simeq g'$.

3. Let A' and A'' be complexes and let $f\colon A'' \to A'$ be a map of degree 0. In the direct sum $A = A' + A''$ introduce the grading $A^n = A'^n + A''^{n+1}$ and the differentiation operator d_f given by

$$d_f(a',a'') = (da' + fa'', -da'').$$

Show that with this differentiation the homomorphisms

$$(1) \qquad 0 \longrightarrow A' \xrightarrow{\psi} A \xrightarrow{\varphi} A'' \longrightarrow 0$$

given by $\psi a' = (a',0)$, $\varphi(a',a'') = a''$ are maps (ψ is of degree 0, and φ of degree 1). Prove that any differentiation operator in A with this property has the form d_f (for some map f of degree zero).

The exact sequence (1) gives rise to an exact homology sequence

$$\cdots \longrightarrow H^n(A') \xrightarrow{\psi^*} H^n(A) \xrightarrow{\varphi^*} H^{n+1}(A'') \xrightarrow{\delta} H^{n+1}(A') \longrightarrow \cdots.$$

Show that δ coincides with the map induced by f.

4. Denote the complex A of Exer. 3 by (A',A'',f). Let (C',C'',g) be another such complex and let $h'\colon A' \to C'$, $h''\colon A'' \to C''$ be maps (of degree zero). Show that a map $h\colon (A',A'',f) \to (C',C'',g)$ of degree 0 compatible with h' and h'', exists if and only if $gh'' \simeq h'f$. Show that each homotopy $s\colon gh'' \simeq h'f$ uniquely determines such a map h and vice-versa.

In particular, a map $(A',A'',f) \to C$ is given by a map $h\colon A' \to C$ and a homotopy $s\colon 0 \simeq hf$. A map $A \to (C',C'',g)$ is given by a map $h\colon A \to C''$ and a homotopy $s\colon gh \simeq 0$.

5. Let A and C be graded Λ-modules. Denote by $M^u(A,C)$ the group of all Λ-homomorphisms $A \to C$ of degree u. Assume that A and C are complexes and consider the subgroup $\operatorname{Map}^u(A,C)$ of all maps of degree u. Further let $\operatorname{Map}^u_0(A,C)$ denote the subgroup of maps homotopic to zero.

Convert the graded group $M(A,C) = \sum_u M^u(A,C)$ into a complex by setting

$$(dg)a = g(da) + (-1)^{u+1}d(ga), \qquad a \in A, \; g \in M^u(A,C).$$

Prove the equalities

$$Z^u(M(A,C)) = \mathrm{Map}^u(A,C), \qquad B^u(M(A,C)) = \mathrm{Map}^u_0(A,C).$$

6. Consider an exact sequence of Λ-complexes

$$0 \longrightarrow A' \xrightarrow{\psi} A \xrightarrow{\varphi} A'' \longrightarrow 0$$

with $\psi \in \mathrm{Map}^0(A',A)$, $\varphi \in \mathrm{Map}^1(A,A'')$, and assume that each of the exact sequences $0 \to A'^n \to A^n \to A''^{n+1} \to 0$ splits. Establish the exact sequence of complexes

$$0 \to M(A'',C) \to M(A,C) \to M(A',C) \to 0,$$

and using 1.1 obtain an exact sequence

$$\mathrm{Map}^u(A,C) \longrightarrow \mathrm{Map}^u(A',C) \xrightarrow{\delta} \mathrm{Map}^u(A'',C)/\mathrm{Map}^u_0(A'',C).$$

Assuming that A is given in the form (A',A'',f) of Exer. 3, show that the map δ is the one induced by f. Compare with the last part of Exer. 4.

Carry out a similar discussion with an exact sequence

$$0 \to C' \to C \to C'' \to 0.$$

7. Let A and C be double complexes such that $A^{p,q} = 0 = C^{p,q}$ if $p < 0$, and let A' and C' denote the double complex obtained from A and C by setting the second differentiation equal to zero. Show that if a map f: $A \to C$ induces an isomorphism $H(A') \approx H(C')$, then f also induces an isomorphism $H(A) \approx H(C)$.

[Hint: observe that $A' = \sum_r F^r(A)/F^{r+1}(A)$, where $F^r(A) = \sum_p \sum_{q \geqq r} A^{p,q}$; similarly for the complex C. Then, for a given n, prove, by a descending induction on r, that $H^n(F^r(A)) \to H^n(F^r(C))$ is an isomorphism; this being true for $r > n$, use the 5-lemma (I,1.1) for each step of the induction. Prove finally that $H^n(A) \to H^n(C)$ is an isomorphism.]

8. Show that a right exact functor T is exact, if and only if the map α: $T(H(A),H(C)) \to H(T(A,C))$ is an isomorphism for any complexes A and C. Establish a similar proposition for left exact functors.

CHAPTER V

Derived Functors

Introduction. This chapter is central and should be studied carefully. First we define for each module A certain complexes which are called projective (or injective) resolutions of A. Then given a functor $T(A,C)$ we replace A and C by projective or injective resolutions X and Y (depending upon the variances of the variables). We then obtain a double complex $T(X,Y)$. The homology groups of this double complex are independent of the choice of X and Y and are the left derived functors $L_nT(A,C)$ (or the right derived functors $R^nT(A,C)$ depending upon the case) of the functor T. There are connecting homomorphisms which link these functors for different values of n, and which lead to various exact sequences. This is done in § 1–4.

The fundamental properties of these derived functors are studied in § 5–9. The last § 10 is a digression intended primarily to prepare the ground for Ch. XII on finite groups.

1. COMPLEXES OVER MODULES; RESOLUTIONS

In what follows it will be convenient to regard a Λ-module A as a complex with $A^0 = A$, $A^n = 0$ for $n \neq 0$ and differentiation zero. Thus A coincides with $Z(A)$, $Z'(A)$ and $H(A)$, while $B(A)$ and $B'(A)$ are zero.

A *left complex* X over A is a negative complex X (i.e. $X^n = 0$ for $n > 0$) and a map ε: $X \to A$ called *the augmentation.* Since $A^n = 0$ for $n \neq 0$, the map ε actually reduces to a single map $X^0 \to A$ subject to the condition that the composition $X^{-1} \to X^0 \to A$ be zero. The left complex X is called *projective* if all X^n are projective, it is called *acyclic* if ε induces an isomorphism $H(X) \approx A$. This last condition is equivalent to the requirement that the sequence

$$\cdots \longrightarrow X_n \xrightarrow{d_n} X_{n-1} \longrightarrow \cdots \xrightarrow{d_1} X_0 \xrightarrow{\varepsilon} A \longrightarrow 0$$

be exact. Note that we have lowered the indices to avoid writing negative numbers. This will be done systematically with left complexes.

A left complex X over A which is both projective and acyclic will be called a *projective resolution* of A.

Let $f\colon\ A \to A'$ be a homomorphism of modules, and let X, X' be left complexes over A, A' with augmentations $\varepsilon, \varepsilon'$. A map $F\colon\ X \to X'$ such that the diagram

$$\begin{array}{ccc} X & \xrightarrow{F} & X' \\ {\scriptstyle\varepsilon}\downarrow & & \downarrow{\scriptstyle\varepsilon'} \\ A & \xrightarrow[f]{} & A' \end{array}$$

is commutative, is called a *map over f.*

PROPOSITION 1.1. *Let X be a projective left complex over A, X' an acyclic left complex over A' and let $f\colon\ A \to A'$. There is then a map $F\colon\ X \to X'$ over f, and any two such maps are homotopic* (see IV,3).

PROOF. In this proof as well as in various proofs in the following section it will be convenient to use the following property of projective modules which is immediately derivable from the definition. Consider a diagram

$$\begin{array}{ccccc} & & P & & \\ & & \downarrow{\scriptstyle\tau} & & \\ A'' & \xrightarrow[\psi]{} & A & \xrightarrow[\varphi]{} & A' \end{array}$$

in which the row is exact and P is projective. If $\varphi\tau = 0$ then τ admits a factorization $\psi\sigma$ where $\sigma\colon\ P \to A''$.

We now begin with the construction of the map $F\colon\ X \to X'$. Consider the diagram

$$\begin{array}{ccccc} & & X_0 & & \\ & & \downarrow{\scriptstyle f\varepsilon} & & \\ X_0' & \xrightarrow[\varepsilon']{} & A' & \longrightarrow & 0 \end{array}$$

Since X_0 is projective there is an $F_0\colon\ X_0 \to X_0'$ with $\varepsilon' F_0 = f\varepsilon$. Next consider the diagram

$$\begin{array}{ccccc} & & X_1 & & \\ & & \downarrow{\scriptstyle F_0 d_1} & & \\ X_1' & \xrightarrow[d_1']{} & X_0' & \xrightarrow[\varepsilon']{} & A' \end{array}$$

Since $\varepsilon' F_0 d_1 = f\varepsilon d_1 = 0$ there is a map $F_1\colon\ X_1 \to X_1'$ with $d_1' F_1 = F_0 d_1$. Assume by induction that $F_n\colon\ X_n \to X_n'$ are already defined for $n < p$ $(p > 1)$ and satisfy $d_n' F_n = F_{n-1} d_n$ for $n > 0$. Consider the diagram

$$\begin{array}{ccccc} & & X_p & & \\ & & \downarrow{\scriptstyle F_{p-1} d_p} & & \\ X_p' & \xrightarrow[d_p']{} & X_{p-1}' & \xrightarrow[d_{p-1}']{} & X_{p-2}' \end{array}$$

Since $d'_{p-1}F_{p-1}d_p = F_{p-2}d_{p-1}d_p = 0$ there is a map $F_p\colon X_p \to X'_p$ with $d'_pF_p = F_{p-1}d_p$.

Now suppose that F, F' are two maps of X into X' over f. Consider the diagram

$$\begin{array}{ccccc} & & X_0 & & \\ & & \downarrow{\scriptstyle\tau} & & \\ X'_1 & \xrightarrow[d'_1]{} & X'_0 & \xrightarrow[\varepsilon']{} & A \end{array}$$

where $\tau = F'_0 - F_0$. Since $\varepsilon'\tau = \varepsilon'F'_0 - \varepsilon'F_0 = f\varepsilon - f\varepsilon = 0$ there is a map $s_0\colon X_0 \to X'_1$ with $d'_1s_0 = F'_0 - F_0$. Assume that $s_n\colon X_n \to X'_{n+1}$ are already defined for $n < p$ $(p > 0)$ and satisfy $d'_{n+1}s_n + s_{n-1}d_n = F'_n - F_n$ for $n > 0$. Consider the diagram

$$\begin{array}{ccccc} & & X_p & & \\ & & \downarrow{\scriptstyle\tau} & & \\ X'_{p+1} & \xrightarrow[d'_{p+1}]{} & X'_p & \xrightarrow[d'_p]{} & X'_{p-1} \end{array}$$

with $\tau = F'_p - F_p - s_{p-1}d_p$. Since $d'_p\tau$, upon calculation gives 0, there is a map $s_p\colon X_p \to X'_{p+1}$ with $\tau = d'_{p+1}s_p$, i.e. with $d'_{p+1}s_p + s_{p-1}d_p = F'_p - F_p$.

PROPOSITION 1.2. *For each module there exists a projective resolution. If X and X' are projective resolutions of A and A', and $f\colon A \to A'$ is a homomorphism, then there exists a map $F\colon X \to X'$ over f. Any two maps $F, F'\colon X \to X'$ over the same homomorphism $A \to A'$ are homotopic.*

PROOF. The existence proof consists in a successive application of I,2.3; given A, choose exact sequences

$$\begin{array}{l} 0 \longrightarrow Z_0 \longrightarrow X_0 \xrightarrow{\varepsilon} A \longrightarrow 0 \\ 0 \longrightarrow Z_1 \longrightarrow X_1 \longrightarrow Z_0 \longrightarrow 0 \\ \cdots\cdots\cdots\cdots\cdots\cdots \\ 0 \longrightarrow Z_n \longrightarrow X_n \longrightarrow Z_{n-1} \longrightarrow 0 \\ \cdots\cdots\cdots\cdots\cdots\cdots \end{array}$$

with X_n projective. Then define d_n as the composition $X_n \to Z_{n-1} \to X_{n-1}$. This yields a projective resolution of A. The second and third part of 1.2 are consequences of 1.1.

Note that actually the above proof yields a projective resolution X of A with the modules X_n not only projective but free.

It follows from 1.2 that any two projective resolutions X and X' of the same module A have the same *homotopy type*, i.e. there exist maps $F\colon X \to X'$ and $F'\colon X' \to X$ over the identity map of A, such that the compositions $F'F$ and FF' are homotopic to identity maps.

We now briefly and without proofs carry out a similar discussion for right complexes.

A *right complex* X over A is a positive complex X and an augmentation map $\varepsilon : A \to X$. The right complex is called *injective* if all X^n are injective; it is called acyclic if ε induces an isomorphism $A \approx H(X)$, or equivalently if the sequence

$$0 \longrightarrow A \xrightarrow{\varepsilon} X^0 \xrightarrow{d^0} X^1 \longrightarrow \cdots \longrightarrow X^n \xrightarrow{d^n} X^{n+1} \longrightarrow \cdots$$

is exact. If X is both injective and acyclic it is called an *injective resolution* of A.

Let f: $A \to A'$ be a map of modules, and let X, X' be right complexes over A, A'. A map F: $X \to X'$ such that the diagram

$$\begin{array}{ccc} A & \xrightarrow{f} & A' \\ {\scriptstyle\varepsilon}\downarrow & & \downarrow{\scriptstyle\varepsilon'} \\ X & \xrightarrow[F]{} & X' \end{array}$$

is commutative, is called a map over f.

PROPOSITION 1.1a. *Let X be an acyclic right complex over A, X' an injective right complex over A', and let f: $A \to A'$. There is then a map F: $X \to X'$ over f, and any two such maps are homotopic.*

PROPOSITION 1.2a. *For each module there exists an injective resolution. If X and X' are injective resolutions of A and A', and f: $A \to A'$ is a homomorphism, then there exists a map F: $X \to X'$ over f. Any two F, F': $X \to X'$ over the same map $A \to A'$ are homotopic.*

PROPOSITION 1.3. *If Λ is left Noetherian and A is a finitely generated left Λ-module, then A has a Λ-projective resolution X such that each X_n is free on a finite base.*

PROOF. We use the notation of the proof of 1.2. Since A is finitely generated we may choose X_0 to be free on a finite base. Then since Λ is left Noetherian, Z_0 is finitely generated. Thus X_1 may be chosen free on a finite base, etc.

2. RESOLUTIONS OF SEQUENCES

Let

(1) $$0 \longrightarrow A' \xrightarrow{\psi} A \xrightarrow{\varphi} A'' \longrightarrow 0$$

be an exact sequence, and let

(2) $$0 \longrightarrow X' \xrightarrow{\Psi} X \xrightarrow{\Phi} X'' \longrightarrow 0$$

be an exact sequence, where X', X, X'' are left complexes over A', A, A'' respectively, Ψ is a map over ψ and Φ is a map over φ. If X', X, X'' are projective resolutions of A', A, A'' respectively, then we say that (2) is a *projective resolution* of (1).

PROPOSITION 2.1. *If X' and X'' are projective, then so is X. If X' and X'' are acyclic, then so is X. If X' and X'' are projective resolutions of A' and A'', then* (2) *is a projective resolution of* (1).

PROOF. For each index n, the sequence $0 \to X'_n \to X_n \to X''_n \to 0$ is exact. If X''_n is projective, then the sequence splits and X_n is isomorphic to the direct sum $X'_n + X''_n$. If X'_n also is projective then X_n is projective.

Suppose now that X' and X'' are acyclic. This means that the sequences

$$\cdots \to X'_n \to X'_{n-1} \to \cdots \to X'_0 \to A' \to 0 \to \cdots$$

$$\cdots \to X''_n \to X''_{n-1} \to \cdots \to X''_0 \to A'' \to 0 \to \cdots$$

are exact. Let $\bar{X}'$ and $\bar{X}''$ denote the complexes defined by these sequences and let $\bar{X}$ denote the similar complex defined using X. Since $H(\bar{X}') \to H(\bar{X}) \to H(\bar{X}'')$ is exact, it follows that $H(\bar{X}) = 0$, i.e. X is an acyclic left complex over A.

We shall say that the exact sequence (2) is *normal* if, for each index n, the exact sequence $0 \to X'_n \to X_n \to X''_n \to 0$ splits. This for instance is always the case if X'' is projective. If the sequence (2) is normal, we may replace X_n by the direct sum $X'_n + X''_n$ and assume that

$$\Psi x'_n = (x'_n, 0), \qquad \Phi(x'_n, x''_n) = x''_n.$$

With this representation we have

$$d_n(x'_n, x''_n) = (d'_n x'_n + \Theta_n x''_n,\ d''_n x''_n)$$

$$\varepsilon(x'_0, x''_0) = \psi\varepsilon' x'_0 + \sigma x''_0$$

where

$$\sigma\colon\ X''_0 \to A$$

$$\Theta_n\colon\ X''_n \to X'_{n-1}, \qquad n > 0$$

The homomorphisms σ and Θ_n satisfy the conditions

$$(3) \qquad \begin{cases} \varepsilon'' = \varphi\sigma & \\ \psi\varepsilon'\Theta_1 + \sigma d''_1 = 0 & \\ d'_{n-1}\Theta_n + \Theta_{n-1}d''_n = 0 & n > 1 \end{cases}$$

which are the translations of the conditions $\varepsilon''\Phi_0 = \varphi\varepsilon$, $\varepsilon d_1 = 0$ and $d_{n-1}d_n = 0$ respectively. This description of a normal sequence (2) will be called the *normal form* of (2).

PROPOSITION 2.2. *Given an exact sequence* (1), *an acyclic left complex* X' *over* A' *and a projective left complex* X'' *over* A'', *there exists a left complex* X *over* A *and maps* Ψ, Φ *over the maps* ψ, φ, *such that the sequence* (2) *is exact. If* X' *and* X'' *are projective resolutions of* A' *and* A'', *then* X *is a projective resolution of* A.

PROOF. The second part follows from 2.1. To prove the first part, it suffices to find homomorphisms $\sigma\colon X_0'' \to A$ and $\Theta_n\colon X_n'' \to X_{n-1}'$ satisfying (3). Consider the diagram

$$\begin{array}{ccccc} & & X_0'' & & \\ & & \downarrow{\scriptstyle \varepsilon''} & & \\ A & \xrightarrow{\varphi} & A'' & \longrightarrow & 0 \end{array}$$

Since X_0'' is projective there is a $\sigma\colon X_0'' \to A$ with $\varphi\sigma = \varepsilon''$. Next consider the diagram

$$\begin{array}{ccccc} & & X_1'' & & \\ & & \downarrow{\scriptstyle -\sigma d_1''} & & \\ X_0' & \xrightarrow{\psi\varepsilon'} & A & \xrightarrow{\varphi} & A'' \end{array}$$

Since the row is exact, X_1'' is projective, and $\varphi\sigma\, d_1'' = \varepsilon''\, d_1'' = 0$, there is a $\Theta_1\colon X_1'' \to X_0'$ with $\psi\,\varepsilon'\,\Theta_1 = -\sigma d_1''$. Next we consider the diagram

$$\begin{array}{ccccc} & & X_2'' & & \\ & & \downarrow{\scriptstyle -\Theta_1 d_2''} & & \\ X_1' & \xrightarrow{d_1'} & X_0' & \xrightarrow{\varepsilon'} & A' \end{array}$$

Since $-\psi\,\varepsilon'\,\Theta_1 d_2'' = \sigma d_1'' d_2'' = 0$ and since $\operatorname{Ker}\psi = 0$ it follows that $-\,\varepsilon'\,\Theta_1 d_2'' = 0$ so that there exists a homomorphism $\Theta_2\colon X_2'' \to X_1'$ with $d_1'\Theta_2 = -\Theta_1 d_2''$. For $p > 2$ we define Θ_p inductively using the diagram

$$\begin{array}{ccccc} & & X_p'' & & \\ & & \downarrow{\scriptstyle -\Theta_{p-1} d_p''} & & \\ X_{p-1}' & \longrightarrow & X_{p-2}' & \longrightarrow & X_{p-3}' \end{array}$$

PROPOSITION 2.3. *Let*

$$\begin{array}{ccccccccc} 0 & \longrightarrow & A' & \xrightarrow{\psi} & A & \xrightarrow{\varphi} & A'' & \longrightarrow & 0 \\ & & \downarrow{\scriptstyle f'} & & \downarrow{\scriptstyle f} & & \downarrow{\scriptstyle f''} & & \\ 0 & \longrightarrow & B' & \xrightarrow{\psi^*} & B & \xrightarrow{\varphi^*} & B'' & \longrightarrow & 0 \end{array}$$

be a commutative diagram with exact rows and let

$$0 \longrightarrow X' \xrightarrow{\Psi} X \xrightarrow{\Phi} X'' \longrightarrow 0, \qquad 0 \longrightarrow Y' \xrightarrow{\Psi^*} Y \xrightarrow{\Phi^*} Y'' \longrightarrow 0$$

be normal exact sequences of left complexes over the rows such that X'' is projective and Y' is acyclic.

Given maps F': $X' \to Y'$ and F'': $X'' \to Y''$ over f' and f'', there is a map F: $X \to Y$ over f such that the diagram

$$\begin{array}{ccccccccc} 0 & \longrightarrow & X' & \longrightarrow & X & \longrightarrow & X'' & \longrightarrow & 0 \\ & & \downarrow{\scriptstyle F'} & & \downarrow{\scriptstyle F} & & \downarrow{\scriptstyle F''} & & \\ 0 & \longrightarrow & Y' & \longrightarrow & Y & \longrightarrow & Y'' & \longrightarrow & 0 \end{array}$$

is commutative.

If G', G, G'' is another triple of maps over f', f, f'' with the same property and if s': $F' \simeq G'$ and s'': $F'' \simeq G''$ are homotopies, then there exists a homotopy s: $F \simeq G$ such that the diagram

$$\begin{array}{ccccccccc} 0 & \longrightarrow & X'_n & \longrightarrow & X_n & \longrightarrow & X''_n & \longrightarrow & 0 \\ & & \downarrow{\scriptstyle s'_n} & & \downarrow{\scriptstyle s_n} & & \downarrow{\scriptstyle s''_n} & & \\ 0 & \longrightarrow & Y'_{n+1} & \longrightarrow & Y_{n+1} & \longrightarrow & Y''_{n+1} & \longrightarrow & 0 \end{array}$$

is commutative for all $n \geqq 0$.

PROOF. We assume that X and Y are given in normal form with the maps σ^X, Θ^X_n, σ^Y, Θ^Y_n. The required map F: $X \to Y$ must then have the form

$$F_n(x'_n, x''_n) = (F'_n x'_n + \gamma_n x''_n,\ F''_n x''_n)$$

where γ_n: $X''_n \to Y'_n$ satisfies the conditions

$$(4) \qquad \begin{cases} \psi^* \varepsilon' \gamma_0 + \sigma^Y F_0 = f\sigma^X & \\ d'_n \gamma_n - \gamma_{n-1} d''_n = F'_{n-1}\Theta^X_n - \Theta^Y_n F''_n, & n > 0 \end{cases}$$

which are translations of the conditions $\varepsilon^Y F_0 = f\,\varepsilon^X$ and $d^Y_n F_n = F_{n-1} d^X_n$ for $n > 0$. Equations (4) allow us to define γ_n inductively by the same method as before.

We now turn to the part concerning the homotopy. The required homotopy s: $F \simeq G$ must have the form

$$s_n(x'_n, x''_n) = (s'_n x'_n + t_n x''_n,\ s''_n x''_n)$$

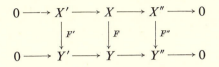

where t_n: $X''_n \to Y'_{n+1}$. The condition $d_{n+1}s_n + s_{n-1}d_n = G_n - F_n$ then becomes

$$d'_1 t_0 + \Theta^Y_1 s''_0 = \gamma^G_0 - \gamma^F_0$$

$$d'_{n+1}t_n + t_{n-1}d''_n + \Theta^Y_{n+1}s''_n + s'_{n-1}\Theta^X_n = \gamma^G_n - \gamma^F_n, \quad n > 0.$$

Again these equations are solved inductively for $t_0, t_1, \ldots$.

The analogs of the results of this section for right complexes are straightforward and will not be restated.

3. DEFINITION OF DERIVED FUNCTORS

As in Chs. III and IV we shall be concerned with additive functors T in any number of covariant and contravariant variables. We shall treat explicitly the case of a functor $T(A,C)$ covariant in the variable A and contravariant in the variable C. However it is understood that A may be replaced by any number of covariant variables, and C by any number of contravariant variables. In most definitions and results the number of variables is of no importance. In all other cases specific statements will be made.

Consider the (additive) functor $T(A,C)$ covariant in A, contravariant in C, where A is a Λ_1-module, C is a Λ_2-module and $T(A,C)$ is a Λ-module. Let X be a right complex over A and Y a left complex over C. Then $T(X,Y)$ is a double complex. With this double complex there is associated a single complex, also written $T(X,Y)$ which is a right complex over $T(A,C)$. If F: $X \to X'$ and G: $Y' \to Y$ are maps over f: $A \to A'$, g: $C' \to C$ then

$$T(F,G)\colon\ T(X,Y) \to T(X',Y')$$

is a map over

$$T(f,g)\colon\ T(A,C) \to T(A',C').$$

Homotopies $F \simeq F'$ and $G \simeq G'$ imply (see IV, 5) a homotopy $T(F,G) \simeq T(F',G')$.

Suppose now that in the above discussion X, X' are Λ_1-injective resolutions of A, A' and Y, Y' are Λ_2-projective resolutions of C, C'. Given the maps f: $A \to A'$, g: $C' \to C$, the existence of maps F: $X \to X'$ and G: $Y' \to Y$ over f and g is assured by 1.2 and 1.2a. If F': $X \to X'$ and G': $Y' \to Y$ is another pair of such maps, then by 1.2 and 1.2a there exist homotopies $F \simeq F'$ and $G \simeq G'$. Thus $T(F,G)$ and $T(F',G')$ are homotopic and therefore yield the same homomorphism of the respective homology modules. Thus the homomorphism

$$H(T(X,Y)) \to H(T(X',Y')) \tag{1}$$

depends only on the maps f and g and not on F and G. We denote the homomorphism (1) by $(RT)(f,g)$. It follows now readily that if $A = A'$, $C = C'$ and f and g are identity maps, then (1) is an isomorphism. Thus up to natural isomorphisms $HT(X,Y)$ is independent of the resolutions X and Y and may be written as $(RT)(A,C)$. These modules together with the maps $(RT)(f,g)$ yield a new (additive) functor RT, covariant in A and contravariant in C. The values of RT are graded Λ-modules. The components of degree n yield a functor R^nT called the *right n-th derived functor* of T. Since the complex $T(X,Y)$ was positive, we have $R^nT = 0$ for $n < 0$.

Let X be an acyclic right complex over A and Y an acyclic left complex over C. We shall define a natural homomorphism

$$H(T(X,Y)) \to RT(A,C). \tag{2}$$

Indeed, let X' be an injective resolution of A and Y' a projective resolution of C as used in defining $RT(A,C)$. By 1.1 and 1.1a, there exist maps F: $X \to X'$ and G: $Y' \to Y$ over the identity maps of A and C. Then $T(F,G)$ induces a homomorphism (2). Since, by 1.1 and 1.1a, F and G are unique up to a homotopy, it follows that (2) is independent of the choice of F and G.

Quite similarly if X is an injective right complex over A, and Y is a projective left complex over C, then we obtain a homomorphism

$$RT(A,C) \to H(T(X,Y)). \tag{3}$$

PROPOSITION 3.1. *If A is injective and C is projective, then $RT(A,C)$ coincides with $T(A,C)$, i.e. $R^nT(A,C) = 0$ for $n > 0$ and $R^0T(A,C) = T(A,C)$. If the functor T is exact, then the same holds for all modules A and C.*

PROOF. If A is injective then A (regarded as a complex) is its own injective resolution. Similarly C, if projective, is its own projective resolution. Thus $RT(A,C) = H(T(A,C)) = T(A,C)$.

Assume now that T is exact and let X be an injective resolution of A and Y a projective resolution of C. We consider the augmentations ε: $A \to X$ and μ: $Y \to C$ as maps of complexes. Applying the homomorphisms α' of IV,6 we obtain the commutative diagram

$$\begin{array}{ccc} T(A,C) & \longrightarrow & H(T(X,Y)) \\ \downarrow & & \downarrow \\ T(A,C) & \longrightarrow & T(H(X),H(Y)) \end{array}$$

The vertical maps α' are isomorphisms by IV,7.2. Since $A \to H(X)$ and $H(Y) \to C$ are also isomorphisms it follows that $T(A,C) \to H(T(X,Y)) = RT(A,C)$ is an isomorphism.

PROPOSITION 3.2. *If T is defined for modules over hereditary rings, then $R^nT = 0$ if n exceeds the number of variables in T.*

PROOF. Consider the case of 2 variables as above. Since A is a module over a hereditary ring the injective resolution X of A may be chosen with $X^n = 0$ for $n > 1$. Similarly the projective resolution Y of C may be chosen with $Y_n = 0$ for $n > 1$. Thus in the complex $T(X, Y)$ we have $T^n(X, Y) = 0$ for $n > 2$. Thus $R^nT = 0$ for $n > 2$.

In defining RT we took injective resolutions for all covariant variables and projective resolutions for all contravariant variables. If instead we take projective resolutions for all the covariant variables and injective resolutions for all contravariant variables we obtain a functor $LT = \sum L_nT$, where L_nT is the *n-th left derived functor* of T. We have $L_nT = 0$ for $n < 0$; the indices have been lowered to avoid negative numbers.

As before if X is an acyclic left complex over A and Y is an acyclic right complex over C, we have the homomorphism

(2a) $$LT(A,C) \to H(T(X, Y)).$$

If X is a projective left complex over A and Y is an injective right complex over C, then

(3a) $$H(T(X, Y)) \to LT(A,C).$$

PROPOSITION 3.1a. *If A is projective and C is injective then $LT(A,C)$ coincides with $T(A,C)$, i.e. $L_nT(A,C) = 0$ for $n > 0$ and $L_0T(A,C) = T(A,C)$. If the functor T is exact then the same holds for all modules A and C.*

PROPOSITION 3.2a. *If T is defined for modules over hereditary rings then $L_nT = 0$ if n exceeds the number of variables in T.*

4. CONNECTING HOMOMORPHISMS

Consider a functor $T(A,C)$ as in 3, and let

(1) $$0 \to A' \to A \to A'' \to 0$$

be an exact sequence. By 2.2 there exists a sequence

(2) $$0 \to X' \to X \to X'' \to 0$$

which is an injective resolution of (1). Let further Y be a projective resolution of C. Since for each degree the sequence (2) splits, it follows that the sequence of complexes

$$0 \to T(X', Y) \to T(X, Y) \to T(X'', Y) \to 0$$

is exact. There result connecting homomorphisms

$$H^n(T(X'', Y)) \to H^{n+1}(T(X', Y))$$

which yield homomorphisms

(3) $$R^nT(A'',C) \to R^{n+1}T(A',C).$$

The independence of (3) from the choice of (2) follows readily from 2.3.

We similarly define the connecting homomorphisms

(4) $$R^nT(A,C') \to R^{n+1}T(A,C'')$$

for each exact sequence

(5) $$0 \to C' \to C \to C'' \to 0.$$

PROPOSITION 4.1. *Let*

$$\begin{array}{ccccccccc} 0 & \to & A' & \to & A & \to & A'' & \to & 0 \\ & & \downarrow & & \downarrow & & \downarrow & & \\ 0 & \to & A'_1 & \to & A_1 & \to & A''_1 & \to & 0 \end{array} \qquad \begin{array}{ccccccccc} 0 & \to & C'_1 & \to & C_1 & \to & C''_1 & \to & 0 \\ & & \downarrow & & \downarrow & & \downarrow & & \\ 0 & \to & C' & \to & C & \to & C'' & \to & 0 \end{array}$$

be commutative diagrams with exact rows. Then the following diagrams are commutative.

$$\begin{array}{ccc} R^nT(A'',C) & \to & R^{n+1}T(A',C) \\ \downarrow & & \downarrow \\ R^nT(A''_1,C) & \to & R^{n+1}T(A'_1,C) \end{array} \qquad \begin{array}{ccc} R^nT(A,C') & \to & R^{n+1}T(A,C'') \\ \downarrow & & \downarrow \\ R^nT(A,C'_1) & \to & R^{n+1}T(A,C''_1) \end{array}$$

$$\begin{array}{ccc} R^nT(A'',C) & \to & R^{n+1}T(A',C) \\ \downarrow & & \downarrow \\ R^nT(A'',C_1) & \to & R^{n+1}T(A',C_1) \end{array} \qquad \begin{array}{ccc} R^nT(A,C') & \to & R^{n+1}T(A,C'') \\ \downarrow & & \downarrow \\ R^nT(A_1,C') & \to & R^{n+1}T(A_1,C'') \end{array}$$

The diagram

$$\begin{array}{ccc} R^nT(A'',C') & \to & R^{n+1}T(A',C') \\ \downarrow & & \downarrow \\ R^{n+1}T(A'',C'') & \to & R^{n+2}T(A',C'') \end{array}$$

is anticommutative. The sequences

(6) $$\cdots \to R^nT(A',C) \to R^nT(A,C) \to R^nT(A'',C) \to R^{n+1}T(A',C) \to \cdots$$

(7) $$\cdots \to R^nT(A,C'') \to R^nT(A,C) \to R^nT(A,C') \to R^{n+1}T(A,C'') \to \cdots$$

are exact.

PROOF. The first four commutativity relations are trivial consequences of the definitions and of 2.3. The exactness of the sequences follows from the fact that these are homology sequences of suitable exact sequences of complexes. It remains to verify the anticommutativity relation.

Let then $0 \to X' \to X \to X'' \to 0$ be an injective resolution of (1) and $0 \to Y' \to Y \to Y'' \to 0$ a projective resolution of (5). There results a commutative diagram

$$\begin{array}{ccccccccc} & & 0 & & 0 & & 0 & & \\ & & \downarrow & & \downarrow & & \downarrow & & \\ 0 & \to & T(X',Y'') & \to & T(X,Y'') & \to & T(X'',Y'') & \to & 0 \\ & & \downarrow & & \downarrow & & \downarrow & & \\ 0 & \to & T(X',Y) & \to & T(X,Y) & \to & T(X'',Y) & \to & 0 \\ & & \downarrow & & \downarrow & & \downarrow & & \\ 0 & \to & T(X',Y') & \to & T(X,Y') & \to & T(X'',Y') & \to & 0 \\ & & \downarrow & & \downarrow & & \downarrow & & \\ & & 0 & & 0 & & 0 & & \end{array}$$

Thus IV,2.1 yields the anti-commutative diagram

$$\begin{array}{ccc} H^nT(X'',Y') & \to & H^{n+1}T(X',Y') \\ \downarrow & & \downarrow \\ H^{n+1}T(X'',Y'') & \to & H^{n+2}T(X',Y'') \end{array}$$

as desired.

An immediate consequence of 4.1 is

COROLLARY 4.2. *R^0T is left exact. If $R^{n+1}T = 0$ then R^nT is right exact.*

If we consider left derived functors, then (3) and (4) above are replaced by

(3a) $$L_nT(A'',C) \to L_{n-1}T(A',C)$$

(4a) $$L_nT(A,C') \to L_{n-1}T(A,C'')$$

Propositions 4.1 and 4.2 remain valid with the obvious formal changes.

PROPOSITION 4.3. *Let X be an injective right complex over A and let $0 \to Y' \to Y \to Y'' \to 0$ be an exact sequence of projective left complexes over an exact sequence $0 \to C' \to C \to C'' \to 0$. Then the sequence $0 \to T(X,Y'') \to T(X,Y) \to T(X,Y') \to 0$ is exact and the diagram*

$$\begin{array}{ccc} R^nT(A,C') & \to & R^{n+1}T(A,C'') \\ \downarrow & & \downarrow \\ H^n(T(X,Y')) & \to & H^{n+1}(T(X,Y'')) \end{array}$$

is commutative.

PROOF. Since $0 \to Y'_n \to Y_n \to Y''_n \to 0$ splits for each n, it follows that $0 \to T(X,Y'') \to T(X,Y) \to T(X,Y') \to 0$ is exact. Let $0 \to Z' \to Z \to Z'' \to 0$ be a projective resolution of $0 \to C' \to C \to C'' \to 0$. Then by 2.3 there exist maps F', F, F'' over the respective identity maps, which yield a commutative diagram

$$\begin{array}{ccccccccc} 0 & \longrightarrow & Y' & \longrightarrow & Y & \longrightarrow & Y'' & \longrightarrow & 0 \\ & & \downarrow{\scriptstyle F'} & & \downarrow{\scriptstyle F} & & \downarrow{\scriptstyle F''} & & \\ 0 & \longrightarrow & Z' & \longrightarrow & Z & \longrightarrow & Z'' & \longrightarrow & 0 \end{array}$$

This implies the commutativity relation above.

Proposition 4.3 is only an example of similar propositions with other variances and with left derived functors. We leave it as an exercise to the reader to state and prove some of these.

The functors $RT = \{R^nT\}$ and $LT = \{L_nT\}$ are examples of what we shall call *multiply connected sequences* of functors. We consider a sequence of functors $\{T^n\}$, all of the same variables and the same variance. We suppose that with respect to each variable we have given connecting homomorphisms such that, (1°) with respect to each variable separately $\{T^n\}$ is a connected sequence of functors, (2°) with the notation of 4.1 the diagrams

$$\begin{array}{ccc} T^n(A'',C) & \rightarrow & T^{n+1}(A',C) \\ \downarrow & & \downarrow \\ T^n(A'',C_1) & \rightarrow & T^{n+1}(A',C_1) \end{array} \qquad \begin{array}{ccc} T^n(A,C') & \rightarrow & T^{n+1}(A,C'') \\ \downarrow & & \downarrow \\ T^n(A_1,C') & \rightarrow & T^{n+1}(A_1,C'') \end{array}$$

are commutative. We then say that $\{T^n\}$ with the given connecting homomorphisms constitute a multiply connected sequence of functors. We do not postulate any anticommutativity relation between the connecting homomorphisms for the different variables. If, with respect to each variable, the connecting homomorphisms yield exact sequences like (6) and (7), then we say that the multiply connected sequence $\{T^n\}$ is *exact*.

Let $\{T^n\}$, $\{U^n\}$ be two multiply connected sequences of functors. A *homomorphism* $\Phi: \{T^n\} \rightarrow \{U^n\}$ is a sequence of natural transformations $\Phi^n: T^n \rightarrow U^n$ which properly commute with the connecting homomorphisms.

An example of such a homomorphism can be obtained by considering a natural transformation $\varphi: T \rightarrow U$ of functors. If X and Y are appropriate resolutions of the variables A and C of T and U, then φ induces homomorphisms $T(X, Y) \rightarrow U(X, Y)$ which in turn induce homomorphisms of the homology modules $R^nT(A,C) \rightarrow R^nU(A,C)$. These clearly commute with the connecting homomorphisms. The same applies to the left derived functors.

PROPOSITION 4.4. (*Isomorphism criterion.*) *Let* $\Phi: \{T^n\} \rightarrow \{U^n\}$ *be a homomorphism of multiply connected exact sequences of functors. We assume that* $\Phi^0: T^0 \rightarrow U^0$ *is a natural equivalence.*

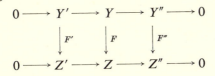

If the homomorphism

$$T^n(A_1, \ldots, A_p) \to U^n(A_1, \ldots, A_p) \tag{8}$$

is an isomorphism whenever $n > 0$ and all the covariant variables are injective and all the contravariant variables are projective, then (8) *is an isomorphism for $n > 0$ and for any variables.*

Similarly if (8) *is an isomorphism for $n < 0$ whenever all the covariant variables are projective and all the contravariant variables are injective, then* (8) *is an isomorphism for $n < 0$ and for any variables.*

PROOF. We shall only consider the case $n > 0$. Let p be the number of variables in T^n and U^n. We first consider the case $p = 1$. Assume that the functors are contravariant and that we already have shown that Φ^i: $T^i \to U^i$ yields isomorphisms for $0 \leqq i < n$. Consider an exact sequence $0 \to M \to P \to A \to 0$ where P is projective. We obtain a commutative diagram

$$\begin{array}{ccccccccc} T^{n-1}(P) & \longrightarrow & T^{n-1}(M) & \longrightarrow & T^n(A) & \longrightarrow & T^n(P) & \longrightarrow & T^n(M) \\ \downarrow{\scriptstyle \varphi_1} & & \downarrow{\scriptstyle \varphi_2} & & \downarrow{\scriptstyle \varphi_3} & & \downarrow{\scriptstyle \varphi_4} & & \downarrow{\scriptstyle \varphi_5} \\ U^{n-1}(P) & \longrightarrow & U^{n-1}(M) & \longrightarrow & U^n(A) & \longrightarrow & U^n(P) & \longrightarrow & U^n(M) \end{array}$$

with exact rows. We know that φ_1, φ_2, φ_4 are isomorphisms. This implies by I,1.1 (the "5 lemma") that φ_3 has kernel zero. Since this holds for all A it follows that φ_5 also has kernel zero. Thus by another application of the "5 lemma" φ_3 is an isomorphism.

The case when the variable A is covariant is treated similarly using an exact sequence $0 \to A \to Q \to N \to 0$ with Q injective.

Suppose now that the proposition is already established if the number of variables is $p - 1$. Suppose now that T^n and U^n are functors of p variables and that the last variable is contravariant. We replace the last variable A_p by a fixed projective module and treat T^n and U^n as functors of the $p - 1$ remaining variables. It follows from the inductive hypothesis that (8) is an isomorphism in this case. We now fix the variables $A_1, \ldots, A_{p-1}$ and regard T^n and U^n as functors of A_p alone. Since (8) is an isomorphism whenever A_p is projective, the result follows from the case $p = 1$ already treated.

PROPOSITION 4.5. *Let $\{T^n\}$ be an exact multiply connected sequence of functors. If the exact sequence $0 \to A' \to A \to A'' \to 0$ splits, then the connecting homomorphisms relative to this sequence are zero.*

PROOF. Assume that A is a covariant variable of T, then (omitting all other variables) we know that $T^n(A) \to T^n(A'')$ is an epimorphism and $T^{n+1}(A') \to T^{n+1}(A)$ is a monomorphism. Consequently $T^n(A'') \to T^{n+1}(A')$ is zero.

5. THE FUNCTORS R^0T AND L_0T

Let X be an injective resolution of A, and Y a projective resolution of C. The augmentation maps $A \to X$, $Y \to C$ induce a map

$$T(A,C) \to T(X,Y)$$

where A, C and $T(A,C)$ are regarded as complexes consisting of elements of degree 0 only. There results a natural transformation

$$\tau^0\colon\ T \to R^0T.$$

PROPOSITION 5.1. *The map τ^0 is a natural equivalence if and only if T is left exact.*

PROOF. By 4.2, R^0T is left exact, thus if τ^0 is an equivalence, then T also is left exact. Suppose now that T is left exact. By II,4.3a the sequence

$$0 \to T(A,C) \to T(X^0,Y_0) \to T(X^1,Y_0) + T(X^0,Y_1)$$

is exact. However the kernel of the last homomorphism is precisely $H^0T(X,Y) = R^0T(A,C)$. Thus τ^0 is an isomorphism.

PROPOSITION 5.2. *Let T be a left exact functor. Then T is exact if and only if $R^1T = 0$.*

PROOF. If T is exact, then $R^1T = 0$ by 3.1. If $R^1T = 0$ then, by 4.2, R^0T is exact so that T is exact by 5.1.

THEOREM 5.3. *The mapping $R^n\tau^0\colon\ R^nT \to R^nR^0T$ induced by $\tau^0\colon\ T \to R^0T$ is an equivalence for all $n \geqq 0$.*

PROOF. We first consider the case $n = 0$. We observe that $R^0\tau^0$ coincides with the mapping $\tau^0\colon\ R^0T \to R^0(R^0T)$. Since R^0T is left exact, it follows from 5.1 that this map is an equivalence.

Next we observe (using 3.1) that both R^nR^0T and R^nT yield zero if $n > 0$ and all covariant variables are replaced by injective modules and all contravariant variables by projective modules. Since $R^n\tau^0$ is a map of multiply connected sequences of functors, the conclusion follows from the isomorphism criterion 4.4.

If X is a projective resolution of A and Y is an injective resolution of C then the augmentations $X \to A$ and $C \to Y$ yield a map $T(X,Y) \to T(A,C)$ thus defining a natural map

$$\sigma_0\colon\ L_0T \to T.$$

PROPOSITION 5.1a. *The map σ_0 is an equivalence if and only if T is right exact.*

PROPOSITION 5.2a. *Let T be a right exact functor. Then T is exact if and only if $L_1T = 0$.*

THEOREM 5.3a. *The mapping $L_n\sigma_0\colon\ L_nL_0T \to L_nT$ induced by $\sigma_0\colon\ L_0T \to T$ is an equivalence.*

Theorem 5.3 shows that the right derived functors are of real interest only if T is left exact. Indeed R^nT may always be replaced by R^nT' with $T' = R^0T$ which is left exact. Similarly the left derived functors are mainly interesting for functors which are right exact.

REMARK. If we regard T as a connected sequence of functors with the functor T in degree zero and the zero functor in all other degrees, then τ^0 and σ_0 may be regarded as homomorphisms of connected sequences of functors

$$LT \xrightarrow{\sigma_0} T \xrightarrow{\tau^0} RT.$$

6. COMPARISON WITH SATELLITES

In this section we limit ourselves to functors of *one variable*.

THEOREM 6.1. *The natural maps*

$$\sigma_0\colon L_0T \to T, \qquad \tau^0\colon T \to R^0T$$

admit unique extensions to maps

$$\sigma_n\colon L_nT \to S_nT, \qquad \tau^n\colon S^nT \to R^nT$$

of connected sequences of functors. If T is right exact, then σ_n is an isomorphism; if T is left exact, then τ^n is an isomorphism.

PROOF. The existence and uniqueness of σ_n and τ^n follow from III,5.2. If T is right exact, then σ_0 is an isomorphism by 5.1a, and 4.4 implies the same for σ_n. Similarly if T is left exact.

PROPOSITION 6.2. *If the ring Λ is hereditary, then σ_n and τ^n are isomorphisms for $n \geqq 1$.*

PROOF. We first prove that $L_nT = 0 = S_nT$ for $n \geqq 2$, and similarly $S^nT = 0 = R^nT$ for $n \geqq 2$. Assume for example that T is covariant; then each module A has a projective resolution of the form

$$0 \to X_1 \to X_0 \to A \to 0;$$

it follows that $L_nT(A) = 0$ for $n \geqq 2$. Moreover, we have

$$S_nT(A) = \operatorname{Ker}(S_{n-1}T(X_1) \to S_{n-1}T(X_0)),$$

and $S_{n-1}T(X_1) = 0$ for $n \geqq 2$, since X_1 is projective. The proof is similar for the other cases. It remains now to be proved that

$$\sigma_1\colon L_1T \to S_1T \quad \text{and} \quad \tau^1\colon S^1T \to R^1T$$

are isomorphisms. We shall give the proof for σ_1, assuming T covariant

Consider the commutative diagram

$$\begin{array}{ccc} L_1L_0T & \xrightarrow{\alpha} & S_1L_0T \\ \downarrow{\scriptstyle\beta} & & \downarrow{\scriptstyle\gamma} \\ L_1T & \xrightarrow{\sigma_1} & S_1T \end{array}$$

where β and γ are induced by σ_0: $L_0T \to T$, and α is the homomorphism σ_1 applied to the functor L_0T. Since, by 4.2, L_0T is right exact, it follows from 6.1 that α is an isomorphism. By 5.3a, β is an isomorphism. In order to prove that σ_1 is an isomorphism, it suffices to show that γ is an isomorphism. Consider the commutative diagram

$$\begin{array}{ccccccc} 0 \longrightarrow & S_1L_0T(A) & \longrightarrow & L_0T(X_1) & \longrightarrow & L_0T(X_0) \\ & \downarrow{\scriptstyle\gamma} & & \downarrow{\scriptstyle u} & & \downarrow{\scriptstyle v} \\ 0 \longrightarrow & S_1T(A) & \longrightarrow & T(X_1) & \longrightarrow & T(X_0) \end{array}$$

Since X_1 and X_0 are projective, u and v are isomorphisms by 3.1. The conclusion follows.

PROPOSITION 6.3. *For any functor T and any n,*

$$R^nS^1T = 0, \qquad L_nS_1T = 0.$$

PROOF. Assume T covariant, and let X be an injective resolution of A. Then $R^nS^1T(A) = H^n(S^1T(X))$. Since X^n is injective, we have $S^1T(X^n) = 0$ for any n; thus $R^nS^1T = 0$. The other cases are proved similarly.

7. COMPUTATIONAL DEVICES

We shall give here a number of propositions which will be useful in computing the derived functors or the connecting homomorphisms. All the propositions here being auxiliary in nature, we shall limit ourselves to stating only the cases needed in the sequel. Restatements for other cases are left to the reader.

Let X be a projective resolution of A. Given $i > 0$ we denote $A^{(i)} = \operatorname{Im}(X_i \to X_{i-1})$. We then obtain exact sequences

(1) $$0 \to A^{(i)} \to X_{i-1} \to \cdots \to X_0 \to A \to 0,$$

(2) $$\cdots \to X_n \to \cdots \to X_i \to A^{(i)} \to 0.$$

We may regard the sequence (2) as a projective resolution $X^{(i)}$ of $A^{(i)}$ with augmentation $\varepsilon^{(i)}$: $X_i \to A^{(i)}$ induced by the map $X_i \to X_{i-1}$. Of course, the sequence (2) needs to be renumbered, before it may be regarded as a projective resolution of $A^{(i)}$. There is a natural mapping

(3) $$X \to X^{(i)}$$

which maps the module X_n of X identically into the module X_n of $X^{(i)}$ for $n \geqq i$, and maps X_n into zero for $n < i$. The map (3) thus lowers the degree by i and commutes with the differentiation.

PROPOSITION 7.1. *Let T be a contravariant functor in one variable. With the notations above the sequence* (1) *induces an iterated connecting homomorphism*

$$\delta\colon\ R^nT(A^{(i)}) \to R^{n+i}T(A)$$

while the map (3) *induces a homomorphism*

$$\gamma\colon\ R^nT(A^{(i)}) \to R^{n+i}T(A).$$

These maps are related by the rule

$$\delta = (-1)^t\gamma, \qquad t = ni + \frac{i(i+1)}{2}.$$

PROOF. We first consider the case $i = 1$. We construct a projective resolution

$$0 \to X^{(1)} \to Y \to X \to 0$$

of the exact sequence

$$0 \to A^{(1)} \to X_0 \to A \to 0$$

given by the following diagram:

$$\begin{array}{ccccccccc}
 & & \vdots & & \vdots & & \vdots & & \\
 & & \downarrow & & \downarrow & & \downarrow & & \\
0 & \longrightarrow & X_{n+2} & \longrightarrow & X_{n+2}+X_{n+1} & \longrightarrow & X_{n+1} & \longrightarrow & 0 \\
 & & \downarrow{\scriptstyle d} & & \downarrow & & \downarrow{\scriptstyle d} & & \\
0 & \longrightarrow & X_{n+1} & \longrightarrow & X_{n+1}+X_n & \longrightarrow & X_n & \longrightarrow & 0 \\
 & & \downarrow & & \downarrow & & \downarrow & & \\
 & & \vdots & & \vdots & & \vdots & & \\
0 & \longrightarrow & X_1 & \longrightarrow & X_1+X_0 & \longrightarrow & X_0 & \longrightarrow & 0 \\
 & & \downarrow{\scriptstyle \varepsilon^{(1)}} & & \downarrow{\scriptstyle \eta} & & \downarrow{\scriptstyle \varepsilon} & & \\
0 & \longrightarrow & A^{(1)} & \longrightarrow & X_0 & \longrightarrow & A & \longrightarrow & 0
\end{array}$$

The horizontal maps are

$$x_{n+1} \to (x_{n+1},0), \qquad (x_{n+1},x_n) \to x_n, \quad n \geqq 0, \quad x_{n+1} \in X_{n+1}, \quad x_n \in X_n.$$

The vertical maps in the middle column (i.e. in the complex Y) are

$$(x_{n+2},x_{n+1}) \to (dx_{n+2} + (-1)^{n+1}x_{n+1},dx_{n+1}) \qquad n \geqq 0$$

$$\eta(x_1,x_0) = dx_1 + x_0.$$

If we apply the functor T to this diagram (reversing all arrows) and compute the connecting homomorphism $H^n(T(X^{(1)})) \to H^{n+1}(T(X))$ we find that the result differs from the homomorphism induced by $X \to X^{(1)}$ by the sign $(-1)^{n+1}$. This is the desired result for $i = 1$.

The general case now follows easily by induction. The connecting homomorphism δ and the map γ both admit factorizations

$$R^nT(A^{(i)}) \underset{\gamma'}{\overset{\delta'}{\longrightarrow}} R^{n+i-1}T(A^{(1)}) \underset{\gamma''}{\overset{\delta''}{\longrightarrow}} R^{n+i}T(A), \qquad i > 1,$$

and we already know that

$$\delta' = (-1)^{t'}\gamma', \qquad \delta'' = (-1)^{n+i}\gamma'', \qquad t' = n(i-1) + \frac{(i-1)i}{2}.$$

This implies the final result.

PROPOSITION 7.2. *Let*

$$0 \to X_n \to \cdots \to X_0 \to A \to 0 \qquad n > 0 \tag{4}$$

be an exact sequence with $X_{n-1}, \ldots, X_0$ *projective. If T is a covariant functor of one variable, then the iterated connecting homomorphism yields isomorphisms*

$$L_{p+n}T(A) \approx L_pT(X_n) \qquad \textit{for } p > 0$$

and the exact sequence

$$0 \to L_nT(A) \to L_0T(X_n) \to L_0T(X_{n-1}) \qquad \textit{for } p = 0.$$

For T contravariant, we have the isomorphisms

$$R^pT(X_n) \approx R^{p+n}T(A) \qquad \textit{for } p > 0$$

and the exact sequence

$$R^0T(X_{n-1}) \to R^0T(X_n) \to R^nT(A) \to 0 \qquad \textit{for } p = 0.$$

PROOF. For $n = 1$, the conclusions follow directly from the exact sequences for the derived functors. For $n > 1$ we break the sequence (4) into exact sequences $0 \to X_n \to X_{n-1} \to X'_{n-1} \to 0$ and $0 \to X'_{n-1} \to \cdots \to X_0 \to A \to 0$. Applying 7.2 to each of these sequences separately yields the desired result.

PROPOSITION 7.3. *Let*

$$0 \to X_n \to \cdots \to X_0 \to A \to 0, \qquad 0 \to C \to Y^0 \to \cdots \to Y^n \to 0$$

be exact sequences with $X_{n-1}, \ldots, X_0$ *projective and* $Y^{n-1}, \ldots, Y^0$ *injective. We denote by X and Y the acyclic complexes over A and C given by these exact sequences. If $T(A,C)$ is a left exact functor contravariant in A and covariant in C, then the natural homomorphism* (v,3,(2))

$$H^k(T(X,Y)) \to R^kT(A,C)$$

is an isomorphism for $k \leq n$.

PROOF. Let $\bar{X}$ be a projective resolution of A such that $\bar{X}_k = X_k$ for $k < n$. Similarly let $\bar{Y}$ be an injective resolution of C such that $\bar{Y}^k = Y^k$ for $k < n$. There result exact sequences

$$0 \to X' \to \bar{X} \to X \to 0, \qquad 0 \to Y \to \bar{Y} \to Y' \to 0$$

where X' and Y' are acyclic complexes and $X'_k = 0 = Y'^k$ for $k < n$.

Since T is left exact, II,4.3a yields an exact sequence

$$0 \to T(X,Y) \to T(\bar{X},\bar{Y}) \to T(X',\bar{Y}) + T(\bar{X},Y').$$

If we denote by N the cokernel of $T(X,Y) \to T(\bar{X},\bar{Y})$ we obtain exact sequences

$$0 \to T(X,Y) \to T(\bar{X},\bar{Y}) \to N \to 0 \tag{5}$$

$$0 \to N \to T(X',\bar{Y}) + T(\bar{X},Y'). \tag{6}$$

The desired isomorphisms $H^kT(X,Y) \approx H^kT(\bar{X},\bar{Y})$ for $k \leqq n$ will follow from the homology sequence of (5), if we prove that $H^k(N) = 0$ for $k \leqq n$. For $k < n$ this follows from the fact that 0 is the only homogeneous element of N of degree $< n$. To show that $H^n(N) = 0$ it suffices to show that the differentiation $N^n \to N^{n+1}$ is a monomorphism. In view of (6) it suffices to prove the same fact about the complex $Z = T(X',\bar{Y}) + T(\bar{X},Y')$. The component of degree n in Z is $T(X'_n,\bar{Y}^0) + T(\bar{X}^0,Y'^n)$. Since $X'_{n+1} \to X'_n \to 0$ and $0 \to Y'^n \to Y'^{n+1}$ are exact and T is left exact it follows that

$$0 \to T(X'_n,\bar{Y}^0) \to T(X'_{n+1},\bar{Y}^0), \qquad 0 \to T(\bar{X}_0,Y'^n) \to T(\bar{X}_0,Y'^{n+1})$$

are exact. This proves that $Z^n \to Z^{n+1}$ is a monomorphism.

8. PARTIAL DERIVED FUNCTORS

Let T be a functor of p variables, some of which covariant, some contravariant. Let s be a subset of $\{1, \dots, p\}$; the variables whose indices are in s will be called *active*, the others will be called *passive*. If we fix all the passive variables, we obtain a functor T_s for which we may consider the derived functors R^nT_s; more explicitly these derived functors are the homology modules of the complex obtained from T by taking injective resolutions of the covariant active variables, projective resolutions of the contravariant active variables, and leaving the passive variables unresolved. We shall denote these *partial derived* functors by R^n_sT, and will regard them as functors of all the variables, both active and passive.

The homomorphism (2) of § 3 yields natural transformations

$$R^n_sT \to R^nT. \tag{1}$$

These homomorphisms commute with the connecting homomorphisms with respect to each of the active variables.

THEOREM 8.1. *Given any functor T, the following conditions are equivalent:*

(a) *The mapping* (1) *is an isomorphism for any* $n \geq 0$.

(b) *If all the covariant active variables are replaced by injective modules and all the contravariant active variables are replaced by projective modules, then T becomes an exact functor of the passive variables.*

PROOF. (a) $\Rightarrow$ (b). If we replace the active variables as stated in (b), then, by 3.1, $R^n_s T = 0$ for $n > 0$ and $T \approx R^0_s T$; thus (a) implies that $R^n T = 0$ for $n > 0$ and $T \approx R^0 T$, hence T becomes an exact functor of the passive variables.

(b) $\Rightarrow$ (a). We shall prove first that $R^0_s T \rightarrow R^0 T$ is an isomorphism. Denote simply by one letter A all active variables, by one letter C all passive variables, by one letter X a set of resolutions of all active variables, by one letter Y a set of resolutions of all passive variables (injective resolutions for covariant variables, projective resolutions for contravariant variables). Then $T(X,Y)$ may be regarded as a double complex over $T(A,C)$, with two differentiation operators d_1, d_2 corresponding respectively to the set X and to the set Y. Consider the commutative diagram

$$\begin{array}{ccccccc}
0 \longrightarrow & T^0(X,C) & \xrightarrow{\varepsilon_2^0} & T^{0,0}(X,Y) & \xrightarrow{d_2^{0,0}} & T^{0,1}(X,Y) \\
& \downarrow d_1^0 & & \downarrow d_1^{0,0} & & \\
0 \longrightarrow & T^1(X,C) & \xrightarrow{\varepsilon_2^1} & T^{1,0}(X,Y) & &
\end{array}$$

By (b), the rows are exact. This implies that ε_2^0 induces an isomorphism

$$R^0_s T(A,C) = \operatorname{Ker} d_1^0 \approx \operatorname{Ker} d_2^{0,0} \cap \operatorname{Ker} d_1^{0,0} = R^0 T(A,C).$$

It remains now to be proved that $R^n_s T \rightarrow R^n T$ is an isomorphism for $n > 0$. We regard $\{R^n T\}$ and $\{R^n_s T\}$ as multiply connected sequences of functors in the active variables. If $n > 0$ and all covariant (contravariant) active variables are injective (projective), then both $R^n T$ and $R^n_s T$ yield zero. It therefore follows from the isomorphism criterion 4.4 that (1) is an isomorphism.

For another proof of the part (b) $\Rightarrow$ (a), see Exer. 6.

Now let t be another subset of $\{1, \ldots, p\}$, containing s. The homomorphism (1) admits a factorization

$$R^n_s T \rightarrow R^n_t T \rightarrow R^n T \tag{2}$$

If (1) is an isomorphism, then it follows from 8.1 that the same is true for $R^n_s T \rightarrow R^n_t T$ and $R^n_t T \rightarrow R^n T$.

The connecting homomorphisms of $R_s^n T$ with respect to a passive variable are in general not defined. However, if the conditions of 8.1 are satisfied, then (1) is an isomorphism, and we can define these connecting homomorphisms using those of $R^n T$. We shall now show how under these conditions these connecting homomorphisms can be expressed directly. In view of the factorization (2), it suffices to consider the case when there is only one passive variable. Suppose this variable A is covariant, and let A' denote all the remaining (active) covariant variables, and C denote all the (active) contravariant variables. Let $0 \to A_1 \to A \to A_2 \to 0$ be an exact sequence; let further X' be an injective resolution of A', and Y a projective resolution of C. It follows from the condition of 8.1 that the sequence

$$0 \to T(A_1,X',Y) \to T(A,X',Y) \to T(A_2,X',Y) \to 0$$

is exact, and therefore yields a connecting homomorphism

$$R_s^n T(A_2,A',C) \to R_s^{n+1} T(A_1,A',C).$$

It remains to show that this connecting homomorphism is the one obtained from the connecting homomorphism of $R^n T$ using the isomorphism (1). To this end we choose an injective resolution $0 \to X_1 \to X \to X_2 \to 0$ of the sequence $0 \to A_1 \to A \to A_2 \to 0$ (see 2.2). There results a commutative diagram with exact rows

$$\begin{array}{ccccccccc} 0 \to & T(A_1,X',Y) & \to & T(A,X',Y) & \to & T(A_2,X',Y) & \to 0 \\ & \downarrow & & \downarrow & & \downarrow & \\ 0 \to & T(X_1,X',Y) & \to & T(X,X',Y) & \to & T(X_2,X',Y) & \to 0 \end{array}$$

Passing to homology, we obtain the commutative diagram

$$\begin{array}{ccc} R_s^n T(A_2,A',C) & \to & R_s^{n+1} T(A_1,A',C) \\ \downarrow & & \downarrow \\ R^n T(A_2,A',C) & \to & R^{n+1} T(A_1,A',C) \end{array}$$

which proves our assertion.

In view of 8.1 we introduce the following definition. A functor T will be called *right balanced* if (1°) when any one of the covariant variables of T is replaced by an injective module, T becomes an exact functor in the remaining variables; (2°) when any one of the contravariant variables in T is replaced by a projective module, T becomes an exact functor of the remaining variables.

It follows from 8.1 that for a right balanced functor, the derived functors $R^n T$ may be identified with the partial derived functors $R_s^n T$ taken with respect to any non empty set s of active variables.

A similar discussion applies to left derived functors. The mapping (1) is replaced by

(1a) $$L_nT(A,C) \to L_n^sT(A,C)$$

in 8.1 and in the definition of a *left balanced* functor, we interchange the words "projective" and "injective" throughout.

It will be shown in the next chapter that $A \otimes C$ is left balanced and Hom (A,C) is right balanced. We know no balanced functors that are not obtained in a trivial way from these two. In particular, we have no example of a balanced functor of three variables.

9. SUMS, PRODUCTS, LIMITS

Let

(1) $$A_\alpha \xrightarrow{i_\alpha} A \xrightarrow{p_\alpha} A_\alpha$$

(2) $$C_\beta \xrightarrow{j_\beta} C \xrightarrow{q_\beta} C_\beta$$

be direct families as defined in I,1. Then as we have already seen in the proof of II,1.1, we obtain a direct family

(3) $$T(A_\alpha,C_\beta) \xrightarrow{T(i_\alpha,q_\beta)} T(A,C) \xrightarrow{T(p_\alpha,j_\beta)} T(A_\alpha,C_\beta).$$

We recall that as usual T is assumed covariant in A and contravariant in C.

We introduce the following four types of functors:

Type $L\Sigma$—if (1) is a direct sum and (2) is a direct product, then (3) is a direct sum.

Type $R\Pi$—if (1) is a direct product and (2) is a direct sum, then (3) is a direct product.

Similar definitions can be made for functors with any number of variables.

PROPOSITION 9.1. *The functor* Hom_Λ (A,C) *is of type* $R\Pi$.

PROOF. We assume that (1) is a direct sum and (2) is a direct product. We must show that the direct family

(4) $$\mathrm{Hom}\,(A_\alpha,C_\beta) \xrightarrow{\mathrm{Hom}\,(p_\alpha,j_\beta)} \mathrm{Hom}\,(A,C) \xrightarrow{\mathrm{Hom}\,(i_\alpha,q_\beta)} \mathrm{Hom}\,(A_\alpha,C_\beta)$$

is a direct product. Let $\varphi_{\alpha\beta} \in \mathrm{Hom}\,(A_\alpha,C_\beta)$ be a family of homomorphisms.

Let $a \in A$. Since (1) is a direct sum we have $a = \sum i_\alpha a_\alpha$, $a_\alpha \in A_\alpha$ and only a finite number of a_α's is $\neq 0$. Since (2) is a direct product, there is for each a_α a single $c_\alpha \in C$ with $q_\beta c_\alpha = \varphi_{\alpha\beta} a_\alpha$. Setting $\varphi a = \sum c_\alpha$ yields a homomorphism $\varphi\colon\ A \to C$ with $q_\beta \varphi i_\alpha = \varphi_{\alpha\beta}$. The uniqueness of φ is clear from the construction. This proves that (4) is a direct product representation.

PROPOSITION 9.2. *The functor $A \otimes_\Lambda C$ is of type $L\Sigma$.*

PROOF. Let (1) and (2) be direct sum representations. Since each element of $A \otimes C$ is a sum of a finite number of elements of the form $a \otimes c$ and since a and c are finite sums $a = \sum i_\alpha a_\alpha$, $c = \sum j_\beta c_\beta$, it follows that each element of $A \otimes C$ is a finite sum of elements of the form $(i_\alpha \otimes j_\beta)(a_\alpha \otimes c_\beta)$. This proves that

$$A_\alpha \otimes C_\beta \to A \otimes C \to A_\alpha \otimes C_\beta$$

is a direct sum representation of $A \otimes C$.

Next we consider the functors $Z(A)$, $Z'(A)$ and $H(A)$, where A is a module with differentiation or a complex. Let

$$A_\alpha \xrightarrow{i_\alpha} A \xrightarrow{p_\alpha} A_\alpha \tag{5}$$

be a direct family. Then

$$Z(A_\alpha) \xrightarrow{Z(i_\alpha)} Z(A) \xrightarrow{Z(p_\alpha)} Z(A_\alpha) \tag{6}$$

$$Z'(A_\alpha) \xrightarrow{Z'(i_\alpha)} Z'(A) \xrightarrow{Z'(p_\alpha)} Z'(A_\alpha) \tag{7}$$

$$H(A_\alpha) \xrightarrow{H(i_\alpha)} H(A) \xrightarrow{H(p_\alpha)} H(A_\alpha) \tag{8}$$

also are direct families. It is trivial to verify that if (5) is a direct sum or direct product representation, then the same is true for (6)–(8). We thus obtain

PROPOSITION 9.3. *The functors Z, Z' and H are of type $L\Sigma$ and $R\Pi$, in other words the functors Z, Z' and H commute with direct sums and direct products.*

THEOREM 9.4. *If the functor T is of type $L\Sigma$ or $L\Pi$ then the same is true for the left derived functors L_nT. If T is of type $R\Sigma$ or $R\Pi$ then the same is true for the right derived functors R^nT.*

PROOF. Assume that (1) is a direct sum representation. Let X_α be a projective resolution of A_α and let X be the direct sum of the complexes X_α. Clearly we may regard X as a left complex over A. By I,2.1, X is projective, and, by 9.3, X is acyclic. Thus X is a projective resolution of A. Similarly if (2) is a direct product representation and Y_β are injective resolutions of C_β, then the direct product Y of the complexes Y_β is an injective resolution of C.

Suppose now that T is of type $L\Sigma$ (or type $L\Pi$). Then

$$T(X_\alpha, Y_\beta) \to T(X, Y) \to T(X_\alpha, Y_\beta)$$

is a direct sum (or direct product) representation. By 9.3, the same applies to

$$H(T(X_\alpha, Y_\beta)) \to H(T(X, Y)) \to H(T(X_\alpha, Y_\beta))$$

i.e: to

$$LT(A_\alpha, C_\beta) \to LT(A, C) \to LT(A_\alpha, C_\beta).$$

The second half of 9.4 is proved similarly.

REMARK. The analogue of 9.4 holds also for satellites of functors of one variable. If T is of type $L\Sigma$ or $L\Pi$ then the same holds for S_nT, $n > 0$. If T is of type $R\Sigma$ or $R\Pi$ then the same holds for S^nT, $n > 0$.

Part of the results established above carries over with direct sums replaced by direct limits and direct products replaced by inverse limits. Let $A = \underrightarrow{\operatorname{Lim}}\, A_\alpha$ be a direct limit of modules A_α and let $C = \underleftarrow{\operatorname{Lim}}\, C_\beta$ be an inverse limit of modules C_β. Then $T(A_\alpha, C_\beta)$ forms a direct system of modules and we have a natural homomorphism

$$\underrightarrow{\operatorname{Lim}}\, T(A_\alpha, C_\beta) \to T(A, C).$$

If this homomorphism always is an isomorphism we say that the functor T is of type $L\Sigma^*$. Similarly if $A = \underleftarrow{\operatorname{Lim}}\, A_\alpha$, $C = \underrightarrow{\operatorname{Lim}}\, C_\beta$ then $T(A_\alpha, C_\beta)$ forms an inverse system of modules and we have

$$T(A, C) \to \underleftarrow{\operatorname{Lim}}\, T(A_\alpha, C_\beta).$$

If this map always is an isomorphism, we say that T is of type $R\Pi^*$.

PROPOSITION 9.1*. *The functor* $\operatorname{Hom}_\Lambda (A, C)$ *is of type* $R\Pi^*$.

Since this proposition is not used in the sequel, the proof is left as an exercise to the reader.

PROPOSITION 9.2*. *The functor* $A \otimes_\Lambda C$ *is of type* $L\Sigma^*$.

PROOF. Let A be the direct limit of the modules A_α with the maps $\varphi_{\alpha'\alpha}\colon A_\alpha \to A_{\alpha'}$ for $\alpha < \alpha'$. Similarly let $C = \underrightarrow{\operatorname{Lim}}\, C_\beta$ with $\psi_{\beta'\beta}\colon C_\beta \to C_{\beta'}$ for $\beta < \beta'$. The modules $A_\alpha \otimes C_\beta$ with the maps $\varphi_{\alpha'\alpha} \otimes \psi_{\beta'\beta}$ form a direct system of modules indexed by pairs (α, β) with direct limit D. The maps $\varphi_\alpha \otimes \psi_\beta\colon A_\alpha \otimes C_\beta \to A \otimes C$ induce a map $\mu\colon D \to A \otimes C$. We must show that μ is an isomorphism (onto). To this end we shall define a map $\xi\colon A \otimes C \to D$ and show that $\mu\xi$ and $\xi\mu$ are identity maps. Let

$x \in A, y \in C$. There exist then indices α and β such that $x = \varphi_\alpha(x_\alpha), y = \psi_\beta(c_\beta)$ for some $x_\alpha \in A_\alpha$, $y_\beta \in C_\beta$. Let $\chi_{\alpha,\beta}$: $A_\alpha \otimes C_\beta \to D$ be the natural projection. Then the element $u(x,y) = \chi_{\alpha,\beta}(x_\alpha \otimes y_\beta) \in D$ is independent of choice of $\alpha,\beta,x_\alpha,y_\beta$. Further $u(x,y)$ is bilinear and satisfies $u(x\lambda,y) = u(x,\lambda y)$ for $\lambda \in \Lambda$. Thus there exists a unique homomorphism ξ: $A \otimes C \to D$ with $\xi(x \otimes y) = u(x,y)$. The verification that $\mu\xi$ and $\xi\mu$ are identity maps is trivial.

PROPOSITION 9.3*. *The functors Z,Z' and H are of type $L\Sigma^*$, i.e. they commute with direct limits.*

PROOF. Let A be the direct limit of modules with differentiation A_α with maps $\varphi_{\alpha'\alpha}$: $A_\alpha \to A_{\alpha'}$. Then the modules $H(A_\alpha)$ with maps $H(\varphi_{\alpha'\alpha})$ form a direct system of modules with limit D. The homomorphisms

$$H(\varphi_\alpha)\colon\ H(A_\alpha) \to H(A)$$

yield a map μ: $D \to H(A)$ which we shall show is an isomorphism.

Let $a \in H(A)$ and let $x \in Z(A)$ be an element of the coset a. There is then an index α such that $x = \varphi_\alpha x_\alpha$ for some $x_\alpha \in A_\alpha$. Since $0 = da = d\varphi_\alpha x_\alpha = \varphi_\alpha dx_\alpha$ there is an index $\alpha' > \alpha$ such that $\varphi_{\alpha'\alpha} dx_\alpha = 0$. Setting $x_{\alpha'} = \varphi_{\alpha'\alpha} x_\alpha$ we have $dx_{\alpha'} = 0$ so that $x_{\alpha'}$ determines an element of $H(A_{\alpha'})$ which in turn determines an element $\xi(a)$ of D. It is easy to see that $\xi(a)$ is independent of the choice of x,α,α' etc. and yields a map ξ: $H(A) \to D$ which is the inverse of μ. The proofs for the functors Z and Z' are similar but simpler.

REMARK. The functor Z is also of type $R\Pi^*$, however the functors Z' and H are not of type $R\Pi^*$, i.e. do not commute with inverse limits.

THEOREM 9.4*. *If T is a covariant functor (in any number of variables) of type $L\Sigma^*$, then the same is true of the left derived functors L_nT.*

The proof is an immediate consequence of

LEMMA 9.5*. *If $A = \underrightarrow{\text{Lim}}\, A_\alpha$ then there exist projective resolutions X_α of A_α forming a direct system such that $X = \underrightarrow{\text{Lim}}\, X_\alpha$ is a projective resolution of A.*

PROOF. Let $X_{0,\alpha}$ be the free module F_{A_α} generated by the elements of A_α, and let $X_0 = F_A$. The maps $\varphi_{\alpha'\alpha}$: $A_\alpha \to A_{\alpha'}$ induce maps $X_{0,\alpha} \to X_{0,\alpha'}$ and X_0 may be identified with the limit $\underrightarrow{\text{Lim}}\, X_{0,\alpha}$. Let R_α be the kernel of the natural map $X_{0,\alpha} \to A_\alpha$. Then R_α forms a direct system of modules with $R = \text{Ker}\,(X_0 \to A)$ as limit. We now repeat the argument with A_α replaced by R_α. The complexes X_α are thus constructed by iteration.

REMARK. The reason why we restricted ourselves to covariant functors in 9.4* is that we have no analogue of 9.5* for injective resolutions and inverse limits.

10. THE SEQUENCE OF A MAP

Let $f\colon T \to U$ be a natural transformation of functors. As usual, we shall treat the case when T and U are functors of two variables covariant in the first and contravariant in the second. We denote by $\tilde{f}$ the map

$$\tilde{f}\colon L_0T \to R^0U$$

obtained from the commutative diagram

$$\begin{array}{ccccc} L_0T & \longrightarrow & T & \longrightarrow & R^0T \\ \downarrow & & \downarrow & & \downarrow \\ L_0U & \longrightarrow & U & \longrightarrow & R^0U \end{array}$$

and introduce the functors

$$L_0f = \operatorname{Ker} \tilde{f}, \qquad R^0f = \operatorname{Coker} \tilde{f}.$$

The sequence of functors

(1) $$\dots, L_nT, \dots, L_1T, L_0f, R^0f, R^1U, \dots, R^nU, \dots$$

will be called the *derived sequence of the map f.*

Before we define the connecting homomorphisms in the sequence (1) we establish the following

LEMMA 10.1. *Consider a commutative diagram*

$$\begin{array}{ccccccccccccccc} \cdots \longrightarrow A_1' \longrightarrow A_1 \longrightarrow & A_1'' & \longrightarrow & A_0' & \longrightarrow & A_0 & \longrightarrow & A_0'' & \longrightarrow & 0 & & & & \\ & \downarrow & & \downarrow{\scriptstyle\varphi'} & & \downarrow{\scriptstyle\varphi} & & \downarrow{\scriptstyle\varphi''} & & \downarrow & & & & \\ & 0 & \longrightarrow & B_0' & \longrightarrow & B_0 & \longrightarrow & B_0'' & \longrightarrow & B_1' & \longrightarrow & B_1 & \longrightarrow & B_1'' \longrightarrow \cdots \end{array}$$

with exact rows. Denote

$$\tilde{A}_0 = \operatorname{Ker} \varphi, \qquad \tilde{B}_0 = \operatorname{Coker} \varphi.$$

and similarly with ′ and ″. Then with the connecting homomorphism $\tilde{A}_0'' \to \tilde{B}_0'$ *defined in* III,3 *the sequence*

$$\cdots \to A_1' \to A_1 \to A_1'' \to \tilde{A}_0' \to \tilde{A}_0 \to \tilde{A}_0'' \to \tilde{B}_0' \to \tilde{B}_0 \to \tilde{B}_0'' \to B_1' \to B_1 \to B_1'' \to \cdots$$

is exact.

Indeed, the exactness of $\tilde{A}_0' \to \tilde{A}_0 \to \tilde{A}_0'' \to \tilde{B}_0' \to \tilde{B}_0 \to \tilde{B}_0''$ is asserted by III,3.3. The remaining parts of the proof are trivial.

Now consider an exact sequence $0 \to A' \to A \to A'' \to 0$. We obtain a commutative diagram

$$\begin{array}{cccccc} \cdots \to L_1T(A'',C) \to & L_0T(A',C) \to & L_0T(A,C) \to & L_0T(A'',C) \to & 0 & \\ \downarrow & \downarrow & \downarrow & \downarrow & \downarrow & \\ 0 \quad \to & R^0U(A',C) \to & R^0U(A,C) \to & R^0U(A'',C) \to & R^1U(A',C) \to \cdots & \end{array}$$

with exact rows. Applying 10.1 to the above diagram we obtain connecting homomorphisms for the sequence (1) (with respect to the first variable). Moreover these connecting homomorphisms yield an exact sequence for each exact sequence $0 \to A' \to A \to A'' \to 0$. The same applies to the second variable. We thus obtain

PROPOSITION 10.2. *The derived sequence* (1) *of a map* $f\colon T \to U$ *is a multiply connected exact sequence of functors.*

Naming (1) a connected sequence of functors is not quite precise since the functors are not properly indexed. However, it is clear that they can be so re-indexed. Any such re-indexing would destroy the notational symmetry between left and right derived functors, and because of this we prefer to leave (1) with its indices as they are.

Let

$$\begin{array}{ccc} T & \xrightarrow{f} & U \\ {\scriptstyle\varphi}\downarrow & & \downarrow{\scriptstyle\psi} \\ T' & \xrightarrow[g]{} & U' \end{array}$$

be a commutative diagram of natural transformations of functors. Then the pair (φ,ψ) defines a map of the derived sequence of f into that of g.

PROPOSITION 10.3. *Suppose that* $\varphi\colon T(A,C) \to T'(A,C)$ *is an isomorphism whenever* A *is projective and* C *is injective, and that* $\psi\colon U(A,C) \to U'(A,C)$ *is an isomorphism whenever* A *is injective and* C *is projective. Then the pair* (φ,ψ) *induces an isomorphism of the derived sequence of* f *onto that of* g.

PROOF. By 4.4, the hypotheses imply that φ and ψ induce isomorphisms

$$L_nT \approx L_nT', \qquad R^nU \approx R^nU'.$$

This implies the result.

In the special case of the identity map $f\colon T \to T$ we introduce the notation

$$\tilde{L}_0T = \operatorname{Ker}(L_0T \to R^0T), \qquad \tilde{R}^0T = \operatorname{Coker}(L_0T \to R^0T).$$

The derived sequence is then

$$(2) \qquad \ldots, L_nT, \ldots, L_1T, \tilde{L}_0T, \tilde{R}^0T, R^1T, \ldots, R^nT, \ldots$$

and is called the *derived sequence* of the functor T.

We now turn to the problem of computing the derived sequence of a map $f\colon T \to U$ using resolutions of the variables.

Let X be a left complex over a module A, Y a right complex over a module C and let $\varphi\colon A \to C$ be a map. We denote by (X,φ,Y) the complex

$$\cdots \to X_n \to \cdots \to X_0 \to Y^0 \to \cdots \to Y^n \to \cdots$$

where $X_0 \longrightarrow A \xrightarrow{\varphi} C \longrightarrow Y^0$ is obtained by composing φ with the augmentation maps. Strictly speaking, (X,φ,Y) is a complex only after the modules are renumbered.

PROPOSITION 10.4. *Let $0 \to X' \to X \to X'' \to 0$ be an exact sequence of left complexes over the exact sequence $0 \to A' \to A \to A'' \to 0$, let $0 \to Y' \to Y \to Y'' \to 0$ be an exact sequence of right complexes over the exact sequence $0 \to C' \to C \to C'' \to 0$ and let*

$$\begin{array}{ccccccccc} 0 & \longrightarrow & A' & \longrightarrow & A & \longrightarrow & A'' & \longrightarrow & 0 \\ & & \downarrow{\scriptstyle \varphi'} & & \downarrow{\scriptstyle \varphi} & & \downarrow{\scriptstyle \varphi''} & & \\ 0 & \longrightarrow & C' & \longrightarrow & C & \longrightarrow & C'' & \longrightarrow & 0 \end{array}$$

be a commutative diagram. Then the homology sequence of the exact sequence

$$0 \to (X',\varphi',Y') \to (X,\varphi,Y) \to (X'',\varphi'',Y'') \to 0$$

coincides with the exact sequence obtained from the diagram

$$\begin{array}{cccccccccc} \cdots \to H_1(X) \to & H_1(X'') \to & H_0(X') \to & H_0(X) \to & H_0(X'') \to & 0 \\ & \downarrow & \downarrow & \downarrow & \downarrow & \downarrow \\ & 0 & \to H^0(Y') \to & H^0(Y) \to & H^0(Y'') \to & H^1(Y') \\ & & & & & \to H^1(Y) \to \cdots \end{array}$$

using 10.1.

PROOF. Clearly $H_n(X) = H_n(X,\varphi,Y)$ and $H^n(Y) = H^n(X,\varphi,Y)$ for $n > 0$. Furthermore

$$\begin{aligned} H_0(X,\varphi,Y) &= \operatorname{Ker}(X_0 \to Y^0)/\operatorname{Im}(X_1 \to X_0) \\ &= \operatorname{Ker}(X_0/\operatorname{Im}(X_1 \to X_0) \to Y^0) \\ &= \operatorname{Ker}(\operatorname{Coker}(X_1 \to X_0) \to \operatorname{Ker}(Y^0 \to Y^1)) \\ &= \operatorname{Ker}(H_0(X) \to H^0(Y)), \end{aligned}$$

and similarly $H^0(X,\varphi,Y)$ coincides with $\operatorname{Coker}(H_0(X) \to H^0(Y))$. It remains to be verified that the connecting homomorphisms agree. The only one for which this fact is not evident is the connecting homomorphism

$$H_0(X'',\varphi'',Y'') \to H^0(X',\varphi',Y'). \tag{3}$$

By definition (see IV,1) this homomorphism is defined from the diagram

$$\begin{array}{ccccccc} & Z'_0(X',\varphi',Y') \to & Z'_0(X,\varphi,Y) \to & Z'_0(X'',\varphi'',Y'') \to 0 \\ & \downarrow & \downarrow & \downarrow \\ 0 \to & Z^0(X',\varphi',Y') \to & Z^0(X,\varphi,Y) \to & Z^0(X'',\varphi'',Y'') \end{array}$$

This diagram is identical with

$$\begin{array}{ccccccc} & H_0(X') & \to & H_0(X) & \to & H_0(X'') & \to 0 \\ & \downarrow & & \downarrow & & \downarrow & \\ 0 \to & H^0(Y') & \to & H^0(Y) & \to & H^0(Y'') & \end{array}$$

which proves that (3) coincides with the connecting homomorphism obtained by 10.1.

We now return to the map $f\colon T \to U$. Let X, $\bar{X}$ be respectively projective and injective resolutions of a module A, and Y, $\bar{Y}$ similar resolutions of a module C. It follows from 10.4 that the complex

$$(T(X,\bar{Y}),f,U(\bar{X},Y)) \tag{4}$$

has as homology groups the values of the derived sequence of f for the pair (A,C). It further follows from 10.4 that the connecting homomorphisms also may be computed by this method. If the functor T is left balanced then in $T(X,\bar{Y})$ in (4) we may replace X by A or $\bar{Y}$ by C. Similarly if U is right balanced, then in $U(\bar{X},Y)$ in (4) we may replace $\bar{X}$ by A or Y by C.

EXERCISES

1. Let

$$\cdots \to X_n \to X_{n-1} \to \cdots \to X_1 \to X_0 \to A \to 0$$

be a projective resolution of a module A, and let Z_n denote the image of $X_{n+1} \to X_n$, as in the proof of 1.2. If T is a covariant functor of one variable, prove that

$$S_nT(A) = \operatorname{Ker}(T(Z_{n-1}) \to T(X_{n-1})), \qquad n \geqq 1$$

$$L_nT(A) = \operatorname{Ker}(T(X_n) \to T(X_{n-1}))/\operatorname{Im}(T(X_{n+1}) \to T(X_n)).$$

Then $T(X_n) \to T(Z_{n-1})$ induces a map

$$L_nT(A) \to S_nT(A), \qquad n \geqq 1.$$

Prove that this map is σ_n, as defined in 6.1. Prove again that σ_n is an isomorphism, whenever T is right exact, or whenever Λ is hereditary.

Examine the other similar cases.

2. Let T be a covariant, half exact functor of one variable (II,4). Given an exact sequence $0 \to A \to Q \to N \to 0$ with Q injective, establish an isomorphism

$$S^1S_1T(A) \approx \operatorname{Ker}(T(A) \to T(Q)).$$

Then prove that

$$\operatorname{Ker}(T(A) \to T(Q)) = \operatorname{Ker}(T(A) \to R^0T(A)),$$

and deduce an exact sequence of natural transformations of functors

$$0 \to S^1S_1T \to T \to R^0T.$$

If T is contravariant and half exact, the same sequence is obtained.

3. Apply Exer. III,5 to the exact sequence of Exer. 2. There results an exact sequence

$$0 \to S^1S_1T \to T \to R^0T \to S^2S_1T \to \cdots$$
$$\cdots \to S^{n+1}S_1T \to S^nT \to R^nT \to S^{n+2}S_1T \to \cdots$$

valid for any half exact functor T of one variable.

Establish the dual exact sequence.

4. If T is a right exact functor of one variable, then

$$R^nT \approx S^{n+2}L_1T \qquad \text{for } n > 0,$$

and there is an exact sequence

$$0 \to S^1L_1T \to T \to R^0T \to S^2L_1T \to 0$$

[Hint: observe that $L_1T \approx S_1T$, and $S^nT = 0$ for $n > 0$.]

Give the dual statements.

5. If T is a half exact functor of one variable, then for $n > 0$

$$S^nS_1S^1T \approx S^nT.$$

[Hint: replace T by S^1T in the exact sequence of Exer. 3.]

6. Give an alternative proof of the part (b) $\Rightarrow$ (a) of theorem 8.1, by applying Exer. IV,7 to the map $T(X,C) \to T(X,Y)$.

7. Let $T(A,C)$ be a right balanced functor of two variables, contravariant in A, covariant in C. Replacing A by a projective resolution X, and C by any acyclic right complex Y over C, prove that

$$R^nT(A,C) \approx H^n(T(X,Y)).$$

(Use Exer. IV,7 as in Exer. 6.) Examine the case when X is any acyclic left complex over A, and Y as an injective resolution of C. Examine the case of a left balanced functor; example: $A \otimes C$.

8. Consider an exact sequence

$$(1) \qquad 0 \to X_n \to \cdots \to X_0 \to A \to 0$$

which may be regarded as an acyclic complex X over A. For each contravariant functor T of one variable, there results a homomorphism

$$\begin{array}{ccc} T(X_n) & \longrightarrow & H^nT(X) \\ \downarrow{\scriptstyle \tau} & & \downarrow \\ R^0T(X_n) & \xrightarrow[\delta]{} & R^nT(A) \end{array}$$

where δ is the iterated connecting homomorphism corresponding to the sequence (1). Show that this diagram is commutative or anticommutative depending on whether $n(n + 1)/2$ is even or odd. [Hint: use 7.1.]

9. Let T be a half exact covariant functor of type $L\Sigma^*$ (i.e. commuting with direct limits). If $T(\Lambda/I) = 0$ for every (left) ideal I of Λ, then $T = 0$.

CHAPTER VI

Derived Functors of $\otimes$ *and Hom*

Introduction. The methods of Ch. v are applied to the functors $A \otimes C$ and Hom (A,C). The left derived functors of $A \otimes C$ are denoted by $\text{Tor}_n (A,C)$; the right derived functors of Hom (A,C) are written as $\text{Ext}^n (A,C)$. These are also the satellite functors with respect to each of the variables A or C. The particular notation chosen will be justified in VII,4 and XIV,1.

The notion of the *projective dimension* of a Λ-module A is introduced in § 2 and will be of considerable use later. It is analogous with the topological dimension of a space defined by homological methods. There is also the notion of *injective dimension* for a module and the notion of a *global dimension* for a ring. The semi-simple rings are precisely those of global dimension zero; the hereditary rings are precisely those of global dimension $\leqq 1$.

In § 3 we return to a more detailed study of the Künneth relations of IV,8. In § 4 we return to the questions concerning the "change of rings" initiated in II,6. These results will be applied to homology theory of groups (X,7) and of Lie algebras (XIII,4).

1. THE FUNCTORS Tor AND Ext

In this chapter we shall be concerned exclusively with the functors $A \otimes_\Lambda C$, $\text{Hom}_\Lambda (A,C)$ and their derived functors. The symbol Λ will be omitted whenever there is no danger of confusion.

We have already seen (II,4.4) that Hom (A,C) is left exact.

PROPOSITION 1.1 *The functor* Hom (A,C) *is right balanced.*

This is an immediate consequence of II,4.6.

We have already seen (II,4.5) that $A \otimes C$ is right exact.

PROPOSITION 1.1a. *The functor* $A \otimes C$ *is left balanced.*

PROOF. Let F be a free module. Then F is the direct sum of modules F_α each of which is isomorphic with the ring Λ. Consequently the functor $T(C) = F \otimes C$ is the direct sum of the functors $T_\alpha(C) = F_\alpha \otimes C$. Since each of the functors T_α is exact it follows that T is exact. Suppose now that A is a direct summand of F. Then the functor $T(C) = F \otimes C$ is the direct sum of the functor $T'(C) = A \otimes C$ and some other functor T''. Since T is exact, it follows that T' is exact. A similar argument applies to the other variable.

We now apply the results of Ch. v to the balanced functors $\mathrm{Hom}_\Lambda(A,C)$ and $A \otimes_\Lambda C$. The right derived functors of $\mathrm{Hom}_\Lambda(A,C)$ are denoted by $\mathrm{Ext}^n{}_\Lambda(A,C)$ (or $\mathrm{Ext}^n(A,C)$ with Λ omitted); in particular $\mathrm{Ext}^0_\Lambda(A,C) = \mathrm{Hom}_\Lambda(A,C)$. If X is a projective resolution of A and Y is an injective resolution of C then $\mathrm{Ext}_\Lambda(A,C) = \sum_{n \geqq 0} \mathrm{Ext}^n{}_\Lambda(A,C)$ can be computed as the homology module of any one of the complexes $\mathrm{Hom}_\Lambda(X, Y)$, $\mathrm{Hom}_\Lambda(X,C)$ or $\mathrm{Hom}_\Lambda(A, Y)$. In view of v,6.1, we can also compute Ext_Λ as the n-th satellite $S^n\,\mathrm{Hom}_\Lambda$ with respect to either of the two variables.

The left derived functors of $A \otimes_\Lambda C$ are denoted by $\mathrm{Tor}^\Lambda_n(A,C)$ (or $\mathrm{Tor}_n(A,C)$ with Λ omitted); in particular, $\mathrm{Tor}^\Lambda_0(A,C) = A \otimes_\Lambda C$. If X and Y are projective resolutions of A and C then $\mathrm{Tor}^\Lambda(A,C) = \sum_{n\geqq 0} \mathrm{Tor}^\Lambda_n(A,C)$ can be computed as the homology module of either of the complexes $X \otimes_\Lambda Y$, $X \otimes_\Lambda C$ or $A \otimes_\Lambda Y$. We can also compute Tor^Λ_n as the n-th satellite S_n of the functor $\otimes_\Lambda$ with respect to any variable.

We shall not study the left derived functors of Hom or the right derived functors of $\otimes$ (cf. Exer. vii, 2–6).

PROPOSITION 1.2. *The functors* Ext^n_Λ *are of type* $R\,\Pi$.

This is an immediate consequence of v,9.1 and v,9.4.

PROPOSITION 1.2a. *The functors* Tor^Λ_n *are of type* $L\Sigma$.

This follows from v,9.2 and v,9.4.

PROPOSITION 1.3. *The functors* Tor^Λ_n *are of type* $L\Sigma^*$ (*i.e. they commute with direct limits*).

This follows from v,9.2* and v,9.4*.

As an application of the fact that $\otimes_\Lambda$ is left balanced we prove:

PROPOSITION 1.4. *In the situation* $({}_\Gamma A_\Lambda, {}_\Gamma C)$ *if* A *is* Λ*-projective and* C *is* Γ*-injective then* $\mathrm{Hom}_\Gamma(A,C)$ *is* Λ*-injective.*

PROOF. Let B be a left Λ-module. Then $A \otimes_\Lambda B$ is an exact functor of B and therefore $\mathrm{Hom}_\Gamma(A \otimes_\Lambda B, C)$ is an exact functor of B. It follows from ii,5.2 that $\mathrm{Hom}_\Lambda(B, \mathrm{Hom}_\Gamma(A,C))$ is an exact functor of B. Thus by ii,4.6, $\mathrm{Hom}_\Gamma(A,C)$ is Λ-injective.

A similar proposition holds for $({}_\Lambda A_\Gamma, C_\Gamma)$.

The next two theorems often allow us to compute Tor_n and Ext^n in concretely given situations.

THEOREM 1.5. *Let* $0 \to M \xrightarrow{\alpha} P \to A \to 0$ *and* $0 \to C \to Q \xrightarrow{\beta} N \to 0$ *be exact sequences with* P *projective and* Q *injective. We then have the following natural isomorphisms:*

(1) $\mathrm{Ext}^n_\Lambda(A,C) \approx \mathrm{Ext}^{n-2}_\Lambda(M,N)$ for $n > 2$

(2) $\mathrm{Ext}^2_\Lambda(A,C) \approx \mathrm{Coker}\,(\mathrm{Hom}_\Lambda(\alpha,\beta))$

(3) $\mathrm{Ext}^1_\Lambda(A,C) \approx \mathrm{Ker}(\mathrm{Hom}_\Lambda(\alpha,\beta))/[\mathrm{Ker}(\mathrm{Hom}_\Lambda(\alpha,Q)) + \mathrm{Ker}(\mathrm{Hom}_\Lambda(P,\beta))]$.

PROOF. First consider the case $n > 2$. We then have the anticommutative diagram

$$\begin{array}{ccc} \operatorname{Ext}^{n-2}(M,N) & \to & \operatorname{Ext}^{n-1}(A,N) \\ \downarrow & & \downarrow \\ \operatorname{Ext}^{n-1}(M,C) & \to & \operatorname{Ext}^{n}(A,C) \end{array}$$

in which all four maps are isomorphisms. This yields two isomorphisms (1) differing in sign.

For $n = 2$ we consider the diagram

$$\begin{array}{ccccccc} \operatorname{Hom}(P,Q) & \to & \operatorname{Hom}(M,Q) & \to & 0 & & \\ \downarrow & & \downarrow & & \downarrow & & \\ \operatorname{Hom}(P,N) & \to & \operatorname{Hom}(M,N) & \to & \operatorname{Ext}^1(A,N) & \to & 0 \\ \downarrow & & \downarrow & & \downarrow & & \\ 0 & \to & \operatorname{Ext}^1(M,C) & \to & \operatorname{Ext}^2(A,C) & \to & 0 \\ & & \downarrow & & \downarrow & & \\ & & 0 & & 0 & & \end{array}$$

with exact rows and columns and which is commutative except for the lower right square which is anticommutative. This yields:

$$\operatorname{Ext}^2(A,C) \approx \operatorname{Ext}^1(A,N) \approx \operatorname{Coker}(\operatorname{Hom}(\alpha,N)) = \operatorname{Coker}(\operatorname{Hom}(\alpha,\beta)).$$

Replacing $\operatorname{Ext}^1(A,N)$ by $\operatorname{Ext}^1(M,C)$ will yield the opposite isomorphism.

Finally we consider the case $n = 1$. We use the exact sequence $0 \longrightarrow M \xrightarrow{\alpha} P \longrightarrow A \longrightarrow 0$ to define a left complex X over A with $X_0 = P$, $X_1 = M$ and $X_i = 0$ for $i > 1$. Similarly we use $0 \longrightarrow C \longrightarrow Q \xrightarrow{\beta} N \longrightarrow 0$ to define a right complex Y over C with $Y^0 = Q$, $Y^1 = N$ and $Y^i = 0$ for $i > 1$. An application of v,7.3 yields then the isomorphism $\operatorname{Ext}^1_\Lambda(A,C) \approx H^1(\operatorname{Hom}_\Lambda(X,Y))$. The complex $\operatorname{Hom}_\Lambda(X,Y)$ may be written explicitly as

$$\operatorname{Hom}(P,Q) \xrightarrow{d^0} \operatorname{Hom}(M,Q) + \operatorname{Hom}(P,N) \xrightarrow{d^1} \operatorname{Hom}(M,N)$$

where

$$d^0 f = (f\alpha, \beta f), \qquad d^1(g,h) = -\beta g + h\alpha$$

for $f\colon P \to Q$, $g\colon M \to Q$, $h\colon P \to N$.

Since Q is injective and P is projective it follows that $\operatorname{Hom}(\alpha,Q)$ and $\operatorname{Hom}(P,\beta)$ are epimorphisms. Therefore the element (g,h) of degree 1 may be written as $(f_1\alpha, \beta f_2)$ for some $f_1, f_2\colon P \to Q$. The element f_1 is determined uniquely modulo $\operatorname{Ker}(\operatorname{Hom}(\alpha,Q))$ while f_2 is determined uniquely modulo $\operatorname{Ker}(\operatorname{Hom}(P,\beta))$. Thus the congruence class

$$\varphi(g,h) = [f_2 - f_1]$$

modulo $\operatorname{Ker}(\operatorname{Hom}(\alpha,Q)) + \operatorname{Ker}(\operatorname{Hom}(P,\beta))$ is uniquely determined.

Since

$$d^1(g,h) = -\beta g + h\alpha = \beta(f_2 - f_1)\alpha$$

we have $d^1(g,h) = 0$ if and only if $f_2 - f_1 \in \operatorname{Ker}(\operatorname{Hom}(\alpha,\beta))$. If $(g,h) = d^0 f = (f\alpha,\beta f)$ then $\varphi(g,h) = [f - f] = 0$. If $f \in \operatorname{Ker}(\operatorname{Hom}(\alpha,\beta))$ then $\varphi(f\alpha,0) = (f\alpha,\beta 0) = [f]$. This shows that φ induces the isomorphism (3). Taking $[f_1 - f_2]$ instead of $[f_2 - f_1]$ replaces φ by $-\varphi$.

Quite analogously we prove

THEOREM 1.5a. *Let* $0 \to M \xrightarrow{\alpha} P \to A \to 0$ *and* $0 \to M' \xrightarrow{\beta} P' \to C \to 0$ *be exact sequences with P and P′ projective. We then have the following natural isomorphisms:*

(1a) $$\operatorname{Tor}_n^\Lambda(A,C) \approx \operatorname{Tor}_{n-2}^\Lambda(M,M') \qquad \text{for } n > 2$$

(2a) $$\operatorname{Tor}_2^\Lambda(A,C) \approx \operatorname{Ker}(\alpha \otimes_\Lambda \beta)$$

(3a) $$\operatorname{Tor}_1^\Lambda(A,C) \approx [\operatorname{Im}(\alpha \otimes_\Lambda P') \cap \operatorname{Im}(P \otimes_\Lambda \alpha')]/\operatorname{Im}(\alpha \otimes_\Lambda \beta).$$

We now take up the question of the commutativity of the functors Tor_n. For each ring Λ, the *opposite* ring Λ^* has elements λ^* in 1–1 correspondence with the elements $\lambda \in \Lambda$ and the multiplication is given by $\lambda_1^*\lambda_2^* = (\lambda_2\lambda_1)^*$. The ring $(\Lambda^*)^*$ clearly may be identified with Λ. If Λ is commutative then Λ^* and Λ may be identified. Any left (right) Λ-module A may be regarded as a right (left) Λ^*-module by setting $a\lambda^* = \lambda a$. Thus the situation $(A_\Lambda, {}_\Lambda C)$ leads to the situation $(C_{\Lambda^*}, {}_{\Lambda^*}A)$ and the mapping $a \otimes_\Lambda c \to c \otimes_{\Lambda^*} a$ yields an isomorphism $A \otimes_\Lambda C \approx C \otimes_{\Lambda^*} A$. If X and Y are Λ-projective resolutions of A and C then we may regard X and Y also as projective Λ^*-resolutions of A and C. We define the map $\varphi\colon X \otimes_\Lambda Y \to Y \otimes_{\Lambda^*} X$ by setting

$$\varphi(x \otimes y) = (-1)^{pq} y \otimes x, \qquad x \in X_p,\ y \in Y_q.$$

Then φ is an isomorphism of complexes. Passing to homology we obtain the isomorphism

(4) $$\operatorname{Tor}_n^\Lambda(A,C) \approx \operatorname{Tor}_n^{\Lambda^*}(C,A).$$

2. DIMENSION OF MODULES AND RINGS

We shall say that the left Λ-module A has *projective dimension* $\leqq n$ if A has a projective resolution X satisfying $X_k = 0$ for $k > n$. The least such integer n is called the projective dimension of A and denoted by $\operatorname{l.dim}_\Lambda A$. If no such integer exists the dimension is defined to be ∞. We shall also write $\dim_\Lambda A$ or $\dim A$ whenever no confusion can arise. The zero module has dimension -1. Projective modules are precisely those of projective dimension $\leqq 0$. The integer (or ∞) $\operatorname{r.dim}_\Lambda A$ for a right Λ-module is defined similarly.

PROPOSITION 2.1. *For each* Λ*-module* A *and each* $n \geqq 0$, *the following conditions are equivalent:*

(a) *A has projective dimension* $\leqq n$.

(b) $\operatorname{Ext}_\Lambda^{n+1}(A,C) = 0$ *for all (left or right)* Λ*-modules* C.

(c) $\operatorname{Ext}_\Lambda^n(A,C)$ *is a right exact functor of the variable* C.

(d) *Given an exact sequence* $0 \to X_n \to X_{n-1} \to \cdots \to X_0 \to A \to 0$ *with* X_k $(0 \leqq k < n)$ *projective, the module* X_n *is projective.*

PROOF. (a) $\Rightarrow$ (b). Let X be a projective resolution of A with $X_k = 0$ for $k > n$. Then $\operatorname{Ext}_\Lambda^{n+1}(A,C) = H^{n+1}(\operatorname{Hom}(X,C)) = 0$.

(b) $\Rightarrow$ (c) is immediate.

(c) $\Rightarrow$ (d) is trivial if $n = 0$. Assume $n > 0$ and let $C \to C''$ be an epimorphism. Applying the iterated connecting homomorphism we obtain the commutative diagram

$$\begin{array}{ccccccc}
\operatorname{Hom}_\Lambda(X_{n-1},C) & \to & \operatorname{Hom}_\Lambda(X_n,C) & \to & \operatorname{Ext}_\Lambda^n(A,C) & \to & 0 \\
\downarrow & & \downarrow & & \downarrow & & \\
\operatorname{Hom}_\Lambda(X_{n-1},C'') & \to & \operatorname{Hom}_\Lambda(X_n,C'') & \to & \operatorname{Ext}_\Lambda^n(A,C'') & \to & 0
\end{array}$$

in which, by v,7.2, the rows are exact. Since X_{n-1} is projective, the vertical map on the left is an epimorphism. The vertical map on the right is an epimorphism since $\operatorname{Ext}_\Lambda^n(A,\ )$ is supposed right exact. It follows easily that the middle vertical map also is an epimorphism. Thus $\operatorname{Hom}_\Lambda(X_n,C)$ is an exact functor of C, and therefore X_n is Λ-projective (II,4.6).

(d) $\Rightarrow$ (a). By an iterated use of I,2.3 we construct a sequence as described in (d). Since, by (d), X_n is projective this sequence yields a projective resolution and thus $\dim A \leqq n$.

COROLLARY 2.2. *In order that* $\operatorname{Ext}_\Lambda^1(A,Y) = 0$ *for all* Y *it is necessary and sufficient that* A *be* Λ*-projective.*

PROPOSITION 2.3. *If* $0 \to A' \to A \to A'' \to 0$ *is exact with* A *projective and* A'' *not projective then* $\dim A'' = 1 + \dim A'$.

This is an immediate consequence of the relation $\operatorname{Ext}_\Lambda^{n+1}(A',C) \approx \operatorname{Ext}_\Lambda^{n+2}(A'',C)$ for $n \geqq 0$.

PROPOSITION 2.4. *If* $0 \to A' \to A \to A'' \to 0$ *is exact,* $\dim A' \leqq n$ *and* $\dim A'' \leqq n$, *then* $\dim A \leqq n$.

This follows from the exactness of $\operatorname{Ext}_\Lambda^{n+1}(A'',C) \to \operatorname{Ext}_\Lambda^{n+1}(A,C) \to \operatorname{Ext}_\Lambda^{n+1}(A',C)$.

PROPOSITION 2.5. *Assume that* Λ *is left Noetherian and* A *is a finitely generated left* Λ*-module. Then* $\dim A \leqq n$ *if and only if* $\operatorname{Ext}_\Lambda^{n+1}(A,C) = 0$ *for all finitely generated left* Λ*-modules* C.

PROOF. The necessity of the condition follows from 2.1(b). To prove sufficiency, consider an exact sequence $0 \to M \to P \to A \to 0$ with P projective and finitely generated. Since Λ is left Noetherian, M is finitely generated. Assume first that $n = 0$. Then $\operatorname{Ext}_\Lambda^1(A,M) = 0$ so that

$\text{Hom}_\Lambda(A,P) \to \text{Hom}_\Lambda(A,A)$ is an epimorphism. It follows that the exact sequence $0 \to M \to P \to A \to 0$ splits and A is projective. Thus $\dim A = 0$. Assume now that $n > 0$ and the proposition is valid for $n-1$. Since $\text{Ext}^n_\Lambda(M,C) \approx \text{Ext}^{n+1}_\Lambda(A,C) = 0$ it follows that $\dim M \leqq n-1$, which, by 2.3, implies $\dim A \leqq n$.

The *injective dimension* of a module C is defined as the least integer n for which there is an injective resolution Y with $Y^k = 0$ for $k > n$. We do not introduce a symbol for the injective dimension.

PROPOSITION 2.1a. *For each Λ-module C and each $n \geqq 0$, the following conditions are equivalent:*

(a) *C has injective dimension $\leqq n$.*

(b) $\text{Ext}^{n+1}_\Lambda(A,C) = 0$ *for all Λ-modules A.*

(c) $\text{Ext}^n_\Lambda(A,C)$ *is a right exact functor of the variable A.*

(d) *Given an exact sequence $0 \to C \to Y^0 \to \cdots \to Y^{n-1} \to Y^n \to 0$ with Y^k $(0 \leqq k < n)$ injective, the module Y^n is injective.*

COROLLARY 2.2a. *In order that* $\text{Ext}^1_\Lambda(X,C) = 0$ *for all X it is necessary and sufficient that C be Λ-injective.*

Analogues of 2.3 and 2.4 also hold; however there is no analogue of 2.5.

THEOREM 2.6. *For each ring Λ and each $n \geqq 0$, the following conditions are equivalent:*

(a) *Each left Λ-module has projective dimension $\leqq n$.*

(b) *Each left Λ-module has injective dimension $\leqq n$.*

(c) $\text{Ext}^k_\Lambda = 0$ *for $k > n$.*

(d) $\text{Ext}^{n+1}_\Lambda = 0$.

(e) Ext^n_Λ *is right exact.*

Here Ext^k_Λ *is understood as a functor of left Λ-modules.*

PROOF. The implications (a) $\Rightarrow$ (c) $\Rightarrow$ (d) $\Leftrightarrow$ (e) are obvious. Moreover, it follows from 2.1 that (a) and (d) are equivalent, and from 2.1a that (b) and (d) are equivalent.

The least integer $n \geqq 0$ for which (a)–(e) hold will be called the *left global dimension* of Λ (notation: $\text{l.gl.dim}\,\Lambda$). The *right global dimension* of Λ may be defined similarly. We know no connection between the left and right global dimensions, except for the following:

COROLLARY 2.7. *For each ring Λ the following conditions are equivalent:*

(a) *Λ is semi-simple.*

(b) $\text{l.gl.dim}\,\Lambda = 0$.

(c) $\text{r.gl.dim}\,\Lambda = 0$.

This follows from the fact that left and right semi-simplicity coincide and are equivalent to all Λ-modules being projective.

It may be noted that l.gl. dim $\Lambda =$ r.gl. dim Λ^* and therefore for rings for which $\Lambda = \Lambda^*$ the left and right global dimensions coincide.

PROPOSITION 2.8. *A ring* Λ *is left hereditary if and only if* l.gl. dim $\Lambda \leqq 1$.

The proof follows immediately from 2.1.

PROPOSITION 2.9. *If* Λ *is (left or right) semi-hereditary then* $\mathrm{Tor}_n^\Lambda = 0$ *for* $n > 1$.

PROOF. Assume Λ left semi-hereditary, and let $0 \to M \to P \to C \to 0$ be an exact sequence of left Λ-modules with P projective. Then $\mathrm{Tor}_n^\Lambda(A,C) \approx \mathrm{Tor}_{n-1}^\Lambda(A,M)$. Since $\mathrm{Tor}_{n-1}^\Lambda$ commutes with direct limits it suffices to prove that $\mathrm{Tor}_{n-1}^\Lambda(A,M') = 0$ for any finitely generated submodule M' of M. However, by I,6.2 each such M' is projective which implies $\mathrm{Tor}_{n-1}^\Lambda(A,M') = 0$ for $n > 1$.

3. KÜNNETH RELATIONS

Let A be a right Λ-complex and C a left Λ-complex. We wish to establish connections between the graded groups $H(A \otimes C)$, $H(A) \otimes H(C)$ and $\mathrm{Tor}_1(H(A),H(C))$ where $\otimes = \otimes_\Lambda$. We make the following assumptions:

$$(1) \qquad \begin{cases} \mathrm{Tor}_1(B(A),B(C)) = 0 = \mathrm{Tor}_1(H(A),B(C)) \\ \\ \mathrm{Tor}_1(B(A),Z(C)) = 0 = \mathrm{Tor}_1(H(A),Z(C)). \end{cases}$$

We first consider the functor $T(D) = D \otimes B(C)$. Since $S_1T(B(A)) = S_1T(H(A)) = 0$ it follows from the exact sequence $0 \to B(A) \to Z(A) \to H(A) \to 0$ that $S_1T(Z(A)) = 0$. Thus applying IV,8.1 to the functor T we obtain that $\alpha_1\colon T(H(A)) \to H(T(A))$ is an isomorphism. Thus we have proved that

$$\alpha_1\colon H(A) \otimes B(C) \to H(A \otimes B(C))$$

is an isomorphism. Similarly

$$\alpha_2\colon H(A) \otimes Z(C) \to H(A \otimes Z(C))$$

is an isomorphism.

Now we consider the functor of two variables $A \otimes C$. Since $\mathrm{Tor}_1(B(A),B(C)) = \mathrm{Tor}_1(H(A),B(C)) = 0$ it follows from the exact sequence $0 \to B(A) \to Z(A) \to H(A) \to 0$ that $\mathrm{Tor}_1(Z(A),B(C)) = 0$. Then from the exact sequence $0 \to Z(A) \to A \to B(A) \to 0$ we deduce that $\mathrm{Tor}_1(A,B(C)) = 0$. Thus all the conditions IV,8.1 are satisfied and we obtain

THEOREM 3.1. *Under the conditions* (1) *we have an exact sequence*

(2) $$0 \longrightarrow H(A) \otimes H(C) \xrightarrow{\alpha} H(A \otimes C) \xrightarrow{\beta} \operatorname{Tor}_1(H(A),H(C)) \longrightarrow 0$$

where α *is of degree zero and* β *is of degree* 1. *Explicitly*

(2′)
$$0 \to \sum_{p+q=n} H_p(A) \otimes H_q(C) \to H_n(A \otimes C) \to \sum_{p+q=n-1} \operatorname{Tor}_1(H_p(A),H_q(C)) \to 0$$

If the ring Λ is (right or left) semi-hereditary conditions (1) are equivalent with

(1′) $$\operatorname{Tor}_1(B(A),C) = 0 = \operatorname{Tor}_1(H(A),C).$$

Indeed, the implication (1) → (1′) follows from the exact sequence $0 \to Z(C) \to C \to B(C) \to 0$. The implication (1′) → (1) follows from the fact that Tor_1 is left exact if Λ is semi-hereditary. In particular, (1′) is satisfied if C is projective.

THEOREM 3.2. *If* Λ *is left and right hereditary and the* Λ*-complexes* A *and* C *are projective, then the exact sequence* (2) *is valid and splits.*

PROOF. Since condition (1′) is satisfied the exact sequence (2) is valid. Since A and C are projective and Λ is hereditary it follows that $B'(A)$ and $B'(C)$ are projective. Therefore, the exact sequences

$$0 \to H(A) \to Z'(A) \to B'(A) \to 0$$

$$0 \to H(C) \to Z'(C) \to B'(C) \to 0$$

split. It then follows from IV,6.2 that the image of α is a direct summand of $H(A \otimes C)$ and thus (2) splits.

REMARK. Under the conditions of 3.2 the hypotheses (1) are satisfied also with the roles of A and C interchanged. This yields another exact sequence (2). The maps α of these two exact sequences are the same; however we do not know whether the maps β of these two exact sequences coincide.

If A, instead of being a Λ-complex is a Λ-module, then $A \otimes C$ may be regarded as a functor of the variable C alone. Since $B(A) = 0$ and $H(A) = A$ we obtain the following result.

THEOREM 3.3 (*Universal coefficient theorem for homology*). *If* A *is a right* Λ*-module and* C *is a left* Λ*-complex such that*

(3) $$\operatorname{Tor}_1(A,B(C)) = 0 = \operatorname{Tor}_1(A,Z(C))$$

then we have the exact sequence

(4) $$0 \longrightarrow A \otimes H_n(C) \xrightarrow{\alpha} H_n(A \otimes C) \xrightarrow{\beta} \operatorname{Tor}_1(A,H_{n-1}(C)) \longrightarrow 0.$$

If Λ is left or right semi-hereditary, condition (3) *is equivalent to*

$$(3') \qquad \mathrm{Tor}_1(A,C) = 0.$$

If Λ is left hereditary and C is Λ-projective, the exact sequence (4) *is valid and splits.*

We now rapidly state the analogous results for the functor Hom ($=\mathrm{Hom}_\Lambda$) where A and C are assumed to be left Λ-complexes. We assume

$$(1a) \qquad \begin{cases} \mathrm{Ext}^1(B(A),B'(C)) = 0 = \mathrm{Ext}^1(B(A),H(C)) \\ \mathrm{Ext}^1(Z(A),B'(C)) = 0 = \mathrm{Ext}^1(Z(A),H(C)). \end{cases}$$

THEOREM 3.1a. *Under the conditions* (1a) *we have an exact sequence*

$$(2a) \quad 0 \longrightarrow \mathrm{Ext}^1(H(A),H(C)) \xrightarrow{\beta'} H(\mathrm{Hom}(A,C)) \xrightarrow{\alpha'} \mathrm{Hom}(H(A),H(C)) \longrightarrow 0$$

with β' of degree 1 *and α' of degree* 0.

There is an analogous theorem under hypotheses dual to (1a).

If Λ is left hereditary, conditions (1a) are equivalent to

$$(1'a) \qquad \mathrm{Ext}^1(A,B'(C)) = 0 = \mathrm{Ext}^1(A,H(C)) = 0.$$

In particular, (1′a) always holds if A is projective.

THEOREM 3.2a. *If Λ is left hereditary, the complex A is projective and the complex C is injective, then the exact sequence* (2a) *is valid and splits.*

THEOREM 3.3a (*Universal coefficient theorem for cohomology*). *If A is a left Λ-complex and C is a left Λ-module such that*

$$(3a) \qquad \mathrm{Ext}^1(B(A),C) = 0 = \mathrm{Ext}^1(Z(A),C)$$

then we have the exact sequence

$$(4a) \quad 0 \longrightarrow \mathrm{Ext}^1(H_{n-1}(A),C) \xrightarrow{\beta'} H^n(\mathrm{Hom}(A,C)) \xrightarrow{\alpha'} \mathrm{Hom}(H_n(A),C) \longrightarrow 0.$$

If Λ is left hereditary, condition (3a) *is equivalent to*

$$(3'a) \qquad \mathrm{Ext}^1(A,C) = 0.$$

If Λ is left hereditary and A is projective, then the exact sequence (4a) *is valid and splits.*

We shall use the above result to derive certain associativity relations for Tor and Ext.

PROPOSITION 3.4. *In the situation* $(A_\Lambda, {}_\Lambda B_\Gamma, {}_\Gamma C)$ *assume that* B *is a module,* A *is a* Λ*-projective complex,* C *is a* Γ*-projective complex and* Λ *and* Γ *are left or right semi-hereditary. Then we have the exact sequences*

$$(5)\qquad 0 \longrightarrow H(A) \otimes_\Lambda H(B \otimes_\Gamma C) \xrightarrow{\alpha} H(A \otimes_\Lambda (B \otimes_\Gamma C)) \xrightarrow{\beta} \mathrm{Tor}_1^\Lambda (H(A), H(B \otimes_\Gamma C)) \longrightarrow 0$$

$$(6)\qquad 0 \longrightarrow H(A \otimes_\Lambda B) \otimes_\Gamma H(C) \xrightarrow{\alpha} H((A \otimes_\Lambda B) \otimes_\Gamma C) \xrightarrow{\beta} \mathrm{Tor}_1^\Gamma (H(A \otimes_\Lambda B), H(C)) \longrightarrow 0,$$

where α *is of degree* 0 *and* β *of degree* $+1$. *If further* $\Lambda = \Gamma$ *is commutative and* B *is a* Λ*-module (instead of a* Λ-Γ*-bimodule) then the sequences* (5) *and* (6) *split.*

PROOF. The exact sequence (6) follows directly from 3.1 since conditions (1′) (with A replaced by $A \otimes B$) are satisfied. The sequence (5) is established similarly. To prove the second half we denote by Y a projective resolution of B, and consider the augmentation map $Y \to B$. We then obtain a commutative diagram

$$\begin{array}{ccccccccc} 0 \to & H(Y) \otimes H(C) & \to & H(Y \otimes C) & \to & \mathrm{Tor}_1 (H(Y), H(C)) & \to 0 \\ & \downarrow & & \downarrow & & \downarrow & \\ 0 \to & B \otimes H(C) & \to & H(B \otimes C) & \to & \mathrm{Tor}_1 (B, H(C)) & \to 0 \end{array}$$

with exact rows. Since $H(Y) \to H(B) = B$ is an isomorphism it follows from I,1.1 (the "5 lemma") that $H(Y \otimes C) \to H(B \otimes C)$ is an isomorphism. Next we consider the commutative diagram

$$\begin{array}{ccccccccc} 0 \to & H(A) \otimes H(Y \otimes C) & \to & H(A \otimes (Y \otimes C)) & \to & \mathrm{Tor}_1 (H(A), H(Y \otimes C)) & \to 0 \\ & \downarrow & & \downarrow & & \downarrow & \\ 0 \to & H(A) \otimes H(B \otimes C) & \to & H(A \otimes (B \otimes C)) & \to & \mathrm{Tor}_1 (H(A), H(B \otimes C)) & \to 0 \end{array}$$

with exact rows. Since the two extreme vertical maps are isomorphisms it follows again from I,1.1 that the middle vertical map is an isomorphism. Since $Y \otimes C$ is projective by II,5.3, it follows from 3.2 that the top row in the diagram splits. Therefore the lower row, i.e. the sequence (5), also splits. The proof that (6) splits is similar.

PROPOSITION 3.5. *In the situation* $(A_\Lambda, {}_\Lambda B_\Gamma, {}_\Gamma C)$ *assume that* Λ *and* Γ *are left or right semi-hereditary. Then we have the natural isomorphism*

$$(7)\qquad \mathrm{Tor}_1^\Lambda (A, \mathrm{Tor}_1^\Gamma (B,C)) \approx \mathrm{Tor}_1^\Gamma (\mathrm{Tor}_1^\Lambda (A,B),C).$$

PROOF. Let X be a Λ-projective resolution of A and Y a Γ-projective resolution of C. Applying 3.4 to the triple (X,B,Y) we obtain

$$H_2(X \otimes_\Lambda (B \otimes_\Gamma Y)) \approx \mathrm{Tor}_1^\Lambda (A, \mathrm{Tor}_1^\Gamma (B,C))$$

$$H_2((X \otimes_\Lambda B) \otimes_\Gamma Y) \approx \mathrm{Tor}_1^\Gamma (\mathrm{Tor}_1^\Lambda (A,B),C)$$

which yields the desired result.

PROPOSITION 3.6. *If Λ is a commutative and hereditary ring, then for any Λ-modules A, B, C we have the (non-natural) isomorphism*

(8) $A \otimes \mathrm{Tor}_1(B,C) + \mathrm{Tor}_1(A,B \otimes C) \approx \mathrm{Tor}_1(A,B) \otimes C + \mathrm{Tor}_1(A \otimes B,C).$

PROOF. With X and Y as above, we again apply 3.4 to the triple (X,B,Y). We obtain exact sequences

$$0 \to A \otimes \mathrm{Tor}_1(B,C) \to H_1(X \otimes (B \otimes Y)) \to \mathrm{Tor}_1(A,B \otimes C) \to 0$$

$$0 \to \mathrm{Tor}_1(A,B) \otimes C \to H_1((X \otimes B) \otimes Y) \to \mathrm{Tor}_1(A \otimes B,C) \to 0.$$

Since, by 3.4, these exact sequences split, the result follows.

We now state (without proof) similar results involving Hom and Ext.

PROPOSITION 3.4a. *In the situation $(A_\Lambda, {}_\Lambda B_\Gamma, C_\Gamma)$ assume that A is a Λ-projective complex, C is a Γ-injective complex, and Λ and Γ are right hereditary. Then we have the exact sequences*

(5a) $0 \to \mathrm{Ext}^1_\Lambda(H(A),H(D)) \to H(\mathrm{Hom}_\Lambda(A,D)) \to \mathrm{Hom}_\Lambda(H(A),H(D)) \to 0$

(6a) $0 \to \mathrm{Ext}^1_\Gamma(H(E),H(C)) \to H(\mathrm{Hom}_\Gamma(E,C)) \to \mathrm{Hom}_\Gamma(H(E),H(C)) \to 0$

where $D = \mathrm{Hom}_\Gamma(B,C)$, $E = A \otimes_\Lambda B$.

If further $\Lambda = \Gamma$ is commutative and B is a Λ-module then the sequences (5a) *and* (6a) *split.*

PROPOSITION 3.5a. *In the situation $(A_\Lambda, {}_\Lambda B_\Gamma, C_\Gamma)$ assume that Λ and Γ are right hereditary. Then we have the natural isomorphism*

(7a) $$\mathrm{Ext}^1_\Lambda(A, \mathrm{Ext}^1_\Gamma(B,C)) \approx \mathrm{Ext}^1_\Gamma(\mathrm{Tor}^\Lambda_1(A,B),C).$$

PROPOSITION 3.6a. *If Λ is a commutative and hereditary ring then for any Λ-modules A, B, C we have the (non-natural) isomorphism*

(8a) $$\mathrm{Ext}^1(A, \mathrm{Hom}(B,C)) + \mathrm{Hom}(A, \mathrm{Ext}^1(B,C)) \approx \mathrm{Ext}^1(A \otimes B,C) + \mathrm{Hom}(\mathrm{Tor}_1(A,B),C).$$

In Ch. XVI we shall return to these questions and, using the method of spectral sequences, we shall obtain much more complete results.

4. CHANGE OF RINGS

We return to the discussion of the change of rings given by a ring homomorphism $\varphi\colon \Lambda \to \Gamma$, as initiated in II,6. The discussion breaks up into four cases, the situation in each case being indicated by an appropriate symbol.

Case 1. $(A_\Lambda, {}_\Lambda\Gamma_\Gamma, {}_\Gamma C)$. We have the relation

(1) $$A \otimes_\Lambda C = A_{(\varphi)} \otimes_\Gamma C.$$

Let X be a Λ-projective resolution of A. Then by II,6.1, $X_{(\varphi)}$ is a Γ-projective left complex over $A_{(\varphi)}$. Thus the homomorphism (3a) of V,3 yields a homomorphism

(i) $$H_n(X_{(\varphi)} \otimes_\Gamma C) \to \operatorname{Tor}_n^\Gamma (A_{(\varphi)}, C).$$

However, by (1)

$$H_n(X_{(\varphi)} \otimes_\Gamma C) = H_n(X \otimes_\Lambda C) = \operatorname{Tor}_n^\Lambda (A,C).$$

Thus we obtain a homomorphism

$$f_{1,n}\colon \operatorname{Tor}_n^\Lambda (A,C) \to \operatorname{Tor}_n^\Gamma (A_{(\varphi)},C),$$

which for $n = 0$ yields the identity (1). An alternative way of obtaining the same homomorphism consists in considering the functor

$$T(A,C) = A \otimes_\Lambda C = A_{(\varphi)} \otimes_\Gamma C.$$

Then

$$L_nT(A,C) = L_n^{(1)}T(A,C) = \operatorname{Tor}_n^\Lambda (A,C)$$

$$L_n^{(2)}T(A,C) = \operatorname{Tor}_n^\Gamma (A_\varphi \otimes_\Gamma C)$$

where $L_n^{(1)}T$ and $L_n^{(2)}T$ denote the partial left derived functors with respect to the first and second variable respectively. Then $f_{1.n}$ coincides with the homomorphism $L_nT \to L_n^{(2)}T$ of V,8.

PROPOSITION 4.1.1. *If* $\operatorname{Tor}_p^\Lambda (A,\Gamma) = 0$ *for all* $p > 0$ *then* $f_{1,n}$ *is an isomorphism.*

PROOF. Since

$$\operatorname{Tor}_p^\Lambda (A,\Gamma) = H_p(X \otimes_\Lambda \Gamma) = H_p(X_{(\varphi)})$$

the hypothesis implies that $X_{(\varphi)}$ is acyclic and thus is a projective resolution of $A_{(\varphi)}$. Thus (i) is an isomorphism and so is $f_{1,n}$.

COROLLARY 4.2.1. *In the situation* $(A_\Gamma, {}_\Gamma C)$ *assume that* Γ *and* C *are* Λ*-projective and* A *is* φ*-projective. Then* $\operatorname{Tor}_n^\Gamma (A,C) = 0$ *for* $n > 0$.

Indeed, since A is isomorphic to a direct summand of $A_{(\varphi)}$ it suffices to prove $\operatorname{Tor}_n^\Gamma (A_{(\varphi)},C) = 0$ for $n > 0$. However by 4.1.1, $\operatorname{Tor}_n^\Gamma (A_{(\varphi)},C) \approx \operatorname{Tor}_n^\Lambda (A,C)$ which is zero for $n > 0$ because C is Λ-projective.

Case 2. $(A_\Gamma, {}_\Gamma\Gamma_\Lambda, {}_\Lambda C)$. We have

(2) $$A \otimes_\Lambda C = A \otimes_\Gamma ({}_{(\varphi)}C).$$

Taking a Λ-projective resolution of C we obtain the homomorphism

$$f_{2,n}\colon \operatorname{Tor}_n^\Lambda (A,C) \to \operatorname{Tor}_n^\Gamma (A, {}_{(\varphi)}C)$$

which for $n = 0$ reduces to (2).

PROPOSITION 4.1.2. *If* $\operatorname{Tor}_p^\Lambda (\Gamma,C) = 0$ *for all* $p > 0$ *then* $f_{2,n}$ *is an isomorphism.*

COROLLARY 4.2.2. *In the situation* $(A_\Gamma, {}_\Gamma C)$ *assume that* Γ *and* A *are* Λ*-projective and* C *is* φ*-projective. Then* $\mathrm{Tor}_n^\Gamma(A,C) = 0$ *for* $n > 0$.

Case 3. $({}_\Lambda A, {}_\Gamma\Gamma_\Lambda, {}_\Gamma C)$. We have the identity

(3) $$\mathrm{Hom}_\Gamma({}_{(\varphi)}A, C) = \mathrm{Hom}_\Lambda(A,C).$$

Taking a Λ-projective resolution of A we obtain the homomorphism

$$f_{3,n}\colon \mathrm{Ext}_\Gamma^n({}_{(\varphi)}A, C) \to \mathrm{Ext}_\Lambda^n(A,C)$$

which for $n = 0$ reduces to (3).

PROPOSITION 4.1.3. *If* $\mathrm{Tor}_p^\Lambda(\Gamma, A) = 0$ *for all* $p > 0$ *then* $f_{3,n}$ *is an isomorphism.*

COROLLARY 4.2.3. *In the situation* $({}_\Gamma A, {}_\Gamma C)$ *assume that* Γ *is* Λ*-projective,* C *is* Λ*-injective and* A *is* φ*-projective. Then* $\mathrm{Ext}_\Gamma^n(A,C) = 0$ *for* $n > 0$.

Case 4. $({}_\Gamma A, {}_\Lambda\Gamma_\Gamma, {}_\Lambda C)$. We have the identity

(4) $$\mathrm{Hom}_\Gamma(A, {}^{(\varphi)}C) = \mathrm{Hom}_\Lambda(A,C).$$

Taking a Λ-injective resolution of C we obtain the homomorphism

$$f_{4,n}\colon \mathrm{Ext}_\Gamma^n(A, {}^{(\varphi)}C) \to \mathrm{Ext}_\Lambda^n(A,C)$$

which for $n = 0$ reduces to (4).

PROPOSITION 4.1.4. *If* $\mathrm{Ext}_\Lambda^p(\Gamma, C) = 0$ *for all* $p > 0$ *then* $f_{4,n}$ *is an isomorphism.*

COROLLARY 4.2.4. *In the situation* $({}_\Gamma A, {}_\Gamma C)$ *assume that* Γ *and* A *are* Λ*-projective and* C *is* φ*-injective. Then* $\mathrm{Ext}_\Gamma^n(A,C) = 0$ *for* $n > 0$.

We also could consider two other cases 3′ and 4′ given by the symbols $(A_\Lambda, {}_\Lambda\Gamma_\Gamma, C_\Gamma)$ and $(A_\Gamma, {}_\Gamma\Gamma_\Lambda, C_\Lambda)$ and apply the identification of II,5.2′. However these cases differ from cases 3 and 4 only by a complete interchange of right and left operators and give the same results except that in case 3′ we must replace $\mathrm{Tor}_p^\Lambda(\Gamma, A)$ by $\mathrm{Tor}_p^\Lambda(A, \Gamma)$.

We now place ourselves in the situation $(A_\Gamma, {}_\Gamma C)$, and define a homomorphism

$$\varphi_n\colon \mathrm{Tor}_n^\Lambda(A,C) \to \mathrm{Tor}_n^\Gamma(A,C)$$

as follows. We first define $\varphi_0\colon A \otimes_\Lambda C \to A \otimes_\Gamma C$ by $a \otimes_\Lambda c \to a \otimes_\Gamma c$. Then we consider Γ-projective resolutions X and Y of A and C. Regarded as Λ-modules X and Y are acyclic left complexes over A and C. Thus by V,3 we have the homomorphisms

(5) $$\mathrm{Tor}_n^\Lambda(A,C) \to H_n(X \otimes_\Lambda Y).$$

Further $\varphi_0\colon X \otimes_\Lambda Y \to X \otimes_\Gamma Y$ yields

(6) $$H_n(X \otimes_\Lambda Y) \to H_n(X \otimes_\Gamma Y) = \mathrm{Tor}_n^\Gamma(A,C).$$

We define φ_n as the composition of (5) and (6).

PROPOSITION 4.4. *If A is a right Γ-module and C is a left Γ-module then the diagram*

$$\begin{array}{ccc} \operatorname{Tor}_n^\Lambda(A,C) & \xrightarrow{f_{1,n}} & \operatorname{Tor}_n^\Gamma(A_{(\varphi)},C) \\ \downarrow{\scriptstyle f_{2,n}} & \searrow^{\varphi_n} & \downarrow{\scriptstyle g_n} \\ \operatorname{Tor}_n^\Gamma(A,{}_{(\varphi)}C) & \xrightarrow[g'_n]{} & \operatorname{Tor}_n^\Gamma(A,C) \end{array}$$

is commutative, where g_n and g'_n are induced by the maps $g\colon A_{(\varphi)} \to A$, $g'\colon {}_{(\varphi)}C \to C$.

The proof is left as an exercise to the reader.

Similarly in the situation $({}_\Gamma A, {}_\Gamma C)$ we define

$$\varphi^n\colon \operatorname{Ext}_\Gamma^n(A,C) \to \operatorname{Ext}_\Lambda^n(A,C)$$

using a Γ-projective resolution of A and Γ-injective resolution of C.

PROPOSITION 4.4a. *If A and C are left Γ-modules then the diagram*

$$\begin{array}{ccc} \operatorname{Ext}_\Gamma^n(A,C) & \xrightarrow{g_n} & \operatorname{Ext}_\Gamma^n({}_{(\varphi)}A,C) \\ \downarrow{\scriptstyle h_n} & \searrow^{\varphi^n} & \downarrow{\scriptstyle f_{3,n}} \\ \operatorname{Ext}_\Gamma^n(A,{}^{(\varphi)}C) & \xrightarrow[f_{4,n}]{} & \operatorname{Ext}_\Lambda^n(A,C) \end{array}$$

is commutative, where g_n, h_n are induced by the maps $g\colon {}_{(\varphi)}A \to A$, $h\colon C \to {}^{(\varphi)}C$.

REMARK. It is clear from the definition of the homomorphisms φ_n and φ^n that they commute with the connecting homomorphisms relative to either variable.

5. DUALITY HOMOMORPHISMS

We consider the situation described by the symbol $(A_\Lambda, {}_\Lambda B_\Gamma, C_\Gamma)$. In II,5.2′ we have established a natural isomorphism

$$\text{(1)} \qquad \operatorname{Hom}_\Lambda(A, \operatorname{Hom}_\Gamma(B,C)) \approx \operatorname{Hom}_\Gamma(A \otimes_\Lambda B, C).$$

Now, let X be a projective resolution of A; we have

$$\operatorname{Ext}_\Lambda(A, \operatorname{Hom}_\Gamma(B,C)) = H(\operatorname{Hom}_\Lambda(X, \operatorname{Hom}_\Gamma(B,C)) \approx H(\operatorname{Hom}_\Gamma(X \otimes_\Lambda B, C)).$$

An application of the homomorphism α' of IV,6.1a yields

$$H(\operatorname{Hom}_\Gamma(X \otimes_\Lambda B,C)) \xrightarrow{\alpha'} \operatorname{Hom}_\Gamma(H(X \otimes_\Lambda B),C) = \operatorname{Hom}_\Gamma(\operatorname{Tor}^\Lambda(A,B),C).$$

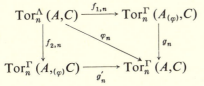

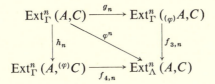

Thus we obtain a homomorphism

$$(2) \qquad \rho\colon \operatorname{Ext}_\Lambda (A, \operatorname{Hom}_\Gamma (B,C)) \to \operatorname{Hom}_\Gamma (\operatorname{Tor}^\Lambda (A,B),C)$$

which reduces to (1) in degree zero.

PROPOSITION 5.1. *If C is Γ-injective then ρ is an isomorphism.*

Indeed, the functor $T(D) = \operatorname{Hom}_\Gamma (D,C)$ is exact. Thus by IV,7.2 the homomorphism $\alpha'\colon H(T(X \otimes B)) \to T(H(X \otimes B))$ is an isomorphism.

The preceding result will be obtained, together with many others of a similar nature, in Ch. XVI by an application of spectral sequences. The homomorphism ρ and proposition 5.1 can also be found using the results of III,6 about satellites of composite functors.

Next we consider the situation described by the symbol $({}_\Lambda A, {}_\Lambda B_\Gamma, C_\Gamma)$ and define the homomorphism

$$(3) \qquad \sigma\colon \operatorname{Hom}_\Gamma (B,C) \otimes_\Lambda A \to \operatorname{Hom}_\Gamma (\operatorname{Hom}_\Lambda (A,B),C)$$

by setting

$$[\sigma(f \otimes a)]g = f(ga) \qquad f \in \operatorname{Hom}_\Gamma (B,C),\ g \in \operatorname{Hom}_\Lambda (A,B).$$

PROPOSITION 5.2. *If A is Λ-projective and finitely generated then σ is an isomorphism.*

PROOF. First consider the case $A = \Lambda$. Then σ is easily seen to be an isomorphism. Therefore, since the functors involved are additive, it follows that σ is an isomorphism if A is a free Λ-module F on a finite base. Consequently, again by a direct sum argument, σ is an isomorphism if A is a direct summand of F.

Now let X be a projective resolution of A. The homomorphism (3) combined with the homomorphism α' of IV,6 yield

$$H(\operatorname{Hom}_\Gamma (B,C) \otimes_\Lambda X) \to H(\operatorname{Hom}_\Gamma (\operatorname{Hom}_\Lambda (X,B),C)) \xrightarrow{\alpha'} \operatorname{Hom}_\Gamma (H(\operatorname{Hom}_\Lambda (X,B)),C).$$

We thus obtain a homomorphism

$$(4) \qquad \sigma\colon \operatorname{Tor}^\Lambda (\operatorname{Hom}_\Gamma (B,C),A) \to \operatorname{Hom}_\Gamma (\operatorname{Ext}_\Lambda (A,B),C)$$

which reduces to (3) in degree zero.

PROPOSITION 5.3. *If Λ is left Noetherian, A is finitely Λ-generated and C is Γ-injective then σ is an isomorphism.*

PROOF. Since C is Γ-injective, the functor $\operatorname{Hom}_\Gamma (D,C)$ is exact and therefore, by IV,7.2, the map α' above is an isomorphism. If Λ is left Noetherian and A is finitely Λ-generated, then, by V,1.3, the resolution X may be chosen to be composed of finitely generated projective modules. The result now follows from 5.2.

REMARK. Instead of assuming that Λ is left Noetherian and A is finitely generated it suffices to assume that A has a projective resolution composed of finitely generated modules.

PROPOSITION 5.4. *If Λ is left hereditary and left Noetherian and A is finitely generated then*

$$\sigma_1\colon \operatorname{Tor}_1^\Lambda(\operatorname{Hom}_\Gamma(B,C),A) \to \operatorname{Hom}_\Gamma(\operatorname{Ext}_\Lambda^1(A,B),C)$$

is an isomorphism.

PROOF. Consider an exact sequence

$$0 \to X_1 \to X_0 \to A \to 0$$

with X_0 projective and finitely generated. Since Λ is Noetherian and hereditary it follows that X_1 is projective and finitely generated so that we obtain a projective resolution X of A. By 5.2, the homomorphism

$$H_1(\operatorname{Hom}_\Gamma(B,C) \otimes_\Lambda X) \to H_1(\operatorname{Hom}_\Gamma(\operatorname{Hom}_\Lambda(X,B),C))$$

is an isomorphism. It therefore suffices to show that

$$\alpha'\colon H_1(\operatorname{Hom}_\Gamma(Y,C)) \to \operatorname{Hom}_\Gamma(H^1(Y),C)$$

is an isomorphism, where $Y = \operatorname{Hom}_\Lambda(X,B)$. We have the exact sequence

$$Y^0 \to Y^1 \to H^1(Y) \to 0$$

and since $\operatorname{Hom}_\Gamma$ is left exact there results an exact sequence

$$0 \longrightarrow \operatorname{Hom}_\Gamma(H^1(Y),C) \longrightarrow \operatorname{Hom}_\Gamma(Y^1,C) \xrightarrow{\gamma} \operatorname{Hom}_\Gamma(Y^0,C).$$

There results an isomorphism of $\operatorname{Hom}_\Gamma(H^1(Y),C)$ with $\operatorname{Ker}\gamma = H_1(\operatorname{Hom}_\Gamma(Y,C))$ which can easily be verified to coincide with α'.

Next we consider the case when $\Lambda = \Gamma = B$. In this case $\operatorname{Hom}_\Gamma(B,C)$ is identified with C so that (4) becomes

$$(4') \qquad \sigma\colon \operatorname{Tor}^\Lambda(C,A) \to \operatorname{Hom}_\Lambda(\operatorname{Ext}_\Lambda(A,\Lambda),C) \qquad ({}_\Lambda A, C_\Lambda).$$

PROPOSITION 5.5. *If Λ is left hereditary and left Noetherian then*

$$\sigma_1\colon \operatorname{Tor}_1^\Lambda(C,A) \to \operatorname{Hom}_\Lambda(\operatorname{Ext}_\Lambda^1(A,\Lambda),C)$$

is a monomorphism.

PROOF. Let A_α be a finitely generated submodule of A. We obtain the commutative diagram

$$\begin{array}{ccc} \operatorname{Tor}_1(C,A_\alpha) & \xrightarrow{\sigma_{1,\alpha}} & \operatorname{Hom}(\operatorname{Ext}^1(A_\alpha,\Lambda),C) \\ \downarrow{\scriptstyle i_\alpha} & & \downarrow{\scriptstyle j_\alpha} \\ \operatorname{Tor}_1(C,A) & \xrightarrow[\sigma_1]{} & \operatorname{Hom}(\operatorname{Ext}^1(A,\Lambda),C). \end{array}$$

Since Λ is hereditary, Ext^1_Λ is right exact by 2.8 and 2.6. Thus $\text{Hom}_\Gamma(\text{Ext}^1_\Lambda(A,\Lambda),C)$ is left exact and thus j_α is a monomorphism. Since, by 5.4, $\sigma_{1,\alpha}$ is a monomorphism it follows that $\sigma_1 i_\alpha$ is a monomorphism. Since Tor_1 commutes with direct limits, $\text{Tor}_1(C,A)$ is the union of the images of i_α for A_α running through all finitely generated submodules of A. Thus σ_1 is a monomorphism.

EXERCISES

1. Consider the ring $Z_n = Z/nZ$. For each divisor r of n, use Exer. I,5 to define an infinite exact sequence

$$\cdots \to Z_n \to Z_n \to Z_n \to rZ_n \to 0.$$

Show that the projective dimension of rZ_n, as a Z_n-module, is 0 or ∞ according as $(r,n/r) = 1$ or $\neq 1$. Conclude that the ring Z_n is either semi-simple or $\text{gl.dim}\, Z_n = \infty$.

2. Let Λ be a commutative Noetherian ring. If A and C are finitely generated Λ-modules, then $\text{Tor}^\Lambda_n(A,B)$ and $\text{Ext}^n_\Lambda(A,B)$ are finitely generated Λ-modules [cf. Exer. II,6.]

3. We define the *weak dimension* of a left Λ-module A (notation: $\text{w.dim}_\Lambda A$) as the highest integer n such that $\text{Tor}^\Lambda_n(C,A) \neq 0$ for some right Λ-module C.

If $\text{w.dim}_\Lambda A = 0$ (i.e. if $\text{Tor}^\Lambda_n(C,A) = 0$ for all C and all $n > 0$), we say that A is Λ-flat. Similar definitions are made for right Λ-modules.

(a) Show that $\text{w.dim}_\Lambda A \leqq \dim_\Lambda A$.

(b) Show that if Λ is left Noetherian and A is finitely generated, then

$$\text{w.dim}_\Lambda A = \dim_\Lambda A$$

[apply 5.3]. Similarly for right Λ-modules.

4. Show that if Λ is left Noetherian and $\{A_\alpha\}$ is a family of right Λ-modules, then ΠA_α is Λ-flat if and only if each A_α is [use II, Exer. 2].

5. For each right Λ-module A and left ideal I of Λ establish the equivalence of the following conditions:

(a) For each relation $\sum_i a_i\mu_i = 0$ ($a_i \in A$, $\mu_i \in I$) there exist elements $b_j \in A$, $\lambda_{ij} \in \Lambda$, finite in number, such that

$$a_i = \sum_j b_j\lambda_{ij}, \qquad \sum_i \lambda_{ij}\mu_i = 0.$$

(b) The map $A \otimes_\Lambda I \to A \otimes_\Lambda \Lambda = A$ is a monomorphism.

(c) $\text{Tor}^\Lambda_1(A,\Lambda/I) = 0$.

(d) For each exact sequence $0 \to N \to P \to A \to 0$ with P a Λ-projective module, we have $N \cap (PI) = NI$.

(e) There exists an exact sequence $0 \to N \to P \to A \to 0$ with P projective such that $N \cap (PI) = NI$.

6. For each right Λ-module A establish the equivalence of the following conditions:

(a) A is Λ-flat.

(b) $\text{Tor}_1^\Lambda(A,\Lambda/I) = 0$ for each left ideal I of Λ.

(c) For each relation $\sum_i a_i\mu_i = 0$ ($a_i \in A$, $\mu_i \in \Lambda$), there exist elements $b_j \in A$, $\lambda_{ij} \in \Lambda$, finite in number, such that

$$a_i = \sum_j b_j\lambda_{ij}, \qquad \sum_i \lambda_{ij}\mu_i = 0.$$

Condition (c) expresses the fact that each linear relation in A is a consequence of linear relations in Λ.

7. If a Λ-module A is a direct sum of Λ-modules A_α then

$$\dim A = \sup_\alpha \dim A_\alpha, \qquad \text{w.dim}\, A = \sup_\alpha \text{w.dim}\, A_\alpha.$$

State a similar result for the injective dimension of a direct product.

8. Let $\Lambda = \Lambda_1 + \cdots + \Lambda_n$ be a direct product of rings. Show that $\text{l.gl.dim}\, \Lambda = \sup_i \text{l.gl.dim}\, \Lambda_i$.

9. Let $\text{l.dim}_\Lambda A = n$, $0 \leqq n < \infty$. Show that $\text{Ext}_\Lambda^n(A,F) \neq 0$ for some free Λ-module F. If further Λ is left Noetherian and A is finitely generated, then $\text{Ext}_\Lambda^n(A,\Lambda) \neq 0$. [Hint: choose C with $\text{Ext}_\Lambda^n(A,C) \neq 0$, then consider an exact sequence $0 \to B \to F \to C \to 0$ with F free.]

10. Let $\varphi: \Lambda \to \Gamma$ be a ring homomorphism. Show that for each left Γ-module A

$$\begin{aligned} \text{w.dim}_\Lambda A &\leqq \text{w.dim}_\Gamma A && \text{if } \Gamma \text{ is left } \Lambda\text{-flat},\\ \dim_\Lambda A &\leqq \dim_\Gamma A && \text{if } \Gamma \text{ is left } \Lambda\text{-projective},\\ \text{inj.dim}_\Lambda A &\leqq \text{inj.dim}_\Gamma A && \text{if } \Gamma \text{ is right } \Lambda\text{-flat}. \end{aligned}$$

Show that for each left Λ-module A

$$\begin{aligned} \text{w.dim}_\Gamma({}_{(\varphi)}A) &\leqq \text{w.dim}_\Lambda A && \text{if } \Gamma \text{ is right } \Lambda\text{-flat},\\ \dim_\Gamma({}_{(\varphi)}A) &\leqq \dim_\Lambda A && \text{if } \Gamma \text{ is right } \Lambda\text{-flat},\\ \text{inj.dim}_\Gamma({}^{(\varphi)}A) &\leqq \text{inj.dim}_\Lambda A && \text{if } \Gamma \text{ is left } \Lambda\text{-projective}. \end{aligned}$$

[Hint: use 4,1.1–4,1.4.]

11. Let $\varphi: \Lambda \to \Gamma$ be a homomorphism of commutative rings such that Γ is Λ-flat. For Λ-modules A and C, establish a natural isomorphism

$$(A \otimes_\Lambda C)_{(\varphi)} \approx A_{(\varphi)} \otimes_\Gamma C_{(\varphi)}$$

of Γ-modules. Derive a natural isomorphism

$$(\text{Tor}_n^\Lambda(A,C))_{(\varphi)} \approx \text{Tor}_n^\Gamma(A,_{(\varphi)},C_{(\varphi)}).$$

Establish a natural homomorphism

$$(\operatorname{Hom}_\Lambda(A,C))_{(\varphi)} \to \operatorname{Hom}_\Gamma(A_{(\varphi)},C_{(\varphi)})$$

of Γ-modules, which becomes an isomorphism if A is Λ-projective and finitely generated. Derive a homomorphism

$$(\operatorname{Ext}^n_\Lambda(A,C))_{(\varphi)} \to \operatorname{Ext}^n_\Gamma(A_{(\varphi)},C_{(\varphi)})$$

which becomes an isomorphism if Λ is Noetherian and A finitely generated.

12. Consider a commutative diagram

$$\begin{array}{ccccccccc} 0 \to & A & \to & B & \to & C & \to 0 \\ & \downarrow f & & \downarrow g & & \downarrow h & \\ 0 \to & A' & \to & B' & \to & C' & \to 0 \end{array}$$

with exact rows. Show that if f and h are monomorphisms then so is g. Assume that the exact sequences split and show that if f and h are monomorphisms onto a direct summand (of A' or C'), then the same holds for g.

13. Let Λ be a (right and left) hereditary ring. Let A and A' be projective right Λ-complexes, and C and C' projective left Λ-complexes. Consider maps $\varphi\colon A \to A'$ and $\psi\colon C \to C'$ such that $\varphi_*\colon H(A) \to H(A')$ is a monomorphism of $H(A)$ onto a direct summand of $H(A')$, and $\psi_*\colon H(C) \to H(C')$ is a monomorphism of $H(C)$ onto a direct summand of $H(C')$. Show that $(\varphi \otimes \psi)_*\colon H(A \otimes_\Lambda C) \to H(A' \otimes_\Lambda C')$ is a monomorphism of $H(A \otimes_\Lambda C)$ onto a direct summand of $H(A' \otimes_\Lambda C')$. [Hint: use Exer. 12.]

14. In the situation $({}_\Lambda A, B_\Gamma, {}_\Lambda C_\Gamma)$, assume that C is a module, A is a Λ-projective complex, B is a Γ-projective complex, Λ is left hereditary and Γ is right hereditary. Then state and prove propositions analogous to 3.4, 3.5, 3.4a and 3.5a [use II, Exer. 4].

15. Let Λ be a commutative, hereditary ring; A, B, C, D being any Λ-modules, show that the following modules M and N are (not naturally) isomorphic:

$$M = \operatorname{Tor}(A,B) \otimes \operatorname{Tor}(C,D) + \operatorname{Tor}(\operatorname{Tor}(A,B), C \otimes D) + \operatorname{Tor}(A \otimes B, \operatorname{Tor}(C,D))$$

$$N = \operatorname{Tor}(\operatorname{Tor}(A,B),C) \otimes D + \operatorname{Tor}(\operatorname{Tor}(A,B) \otimes C, D) + \operatorname{Tor}(\operatorname{Tor}(A \otimes B, C), D).$$

Moreover, any permutation of A,B,C,D transforms M (resp. N) into a module isomorphic to M (resp. N).

[Hint: use resolutions of A, B, C, D.]

16. Let X and Y be two complexes over the ring Z of integers. Assume that X and Y have finite Z-bases for each degree. Show that

$$H(X \otimes Z_n) \otimes H(Y \otimes Z_n) \approx H(X \otimes Y \otimes Z_n),$$

Z_n denoting Z/nZ; but, if n is not a prime, there is no natural isomorphism.

[Hint: apply the Künneth formula to $H(X \otimes Z_n)$, $H(Y \otimes Z_n)$ and $H((X \otimes Z_n) \otimes Y)$. Then prove that

$$A \otimes \operatorname{Tor}_1(Z_n,C) \approx \operatorname{Tor}_1(A \otimes Z_n,C)$$

$$\operatorname{Tor}_1(A,Z_n) \otimes \operatorname{Tor}_1(Z_n,C) \approx \operatorname{Tor}_1(\operatorname{Tor}_1(A,Z_n),C)$$

whenever A and C are finitely generated abelian groups. Reduce the problem to the case of cyclic groups.]

17. Let $\Lambda = \underrightarrow{\operatorname{Lim}}\, \Lambda_\alpha$ be a direct limit of rings. Let A be a right Λ-module, such that $A = \underrightarrow{\operatorname{Lim}}\, A_\alpha$, A_α being a right Λ_α-module. In the same way, let C be a left Λ-module, $C = \underrightarrow{\operatorname{Lim}}\, C_\alpha$, C_α being a left Λ_α-module. Prove that

$$A \otimes_\Lambda C = \underrightarrow{\operatorname{Lim}}\, A_\alpha \otimes_{\Lambda_\alpha} C_\alpha.$$

Then, using v,9.5*, prove that

$$\operatorname{Tor}_n^\Lambda(A,C) = \underrightarrow{\operatorname{Lim}}\, \operatorname{Tor}_n^{\Lambda_\alpha}(A_\alpha,C_\alpha).$$

18. Consider a commutative diagram of Λ-modules and Λ-homomorphisms

$$\begin{array}{ccccccccc} 0 & \longrightarrow & M & \longrightarrow & P & \longrightarrow & A & \longrightarrow & 0 \\ & & \downarrow{\scriptstyle\gamma} & & \downarrow & & \downarrow{\scriptstyle\alpha} & & \\ 0 & \longrightarrow & C & \longrightarrow & Q & \longrightarrow & N & \longrightarrow & 0 \end{array}$$

with exact rows, P projective and Q injective. Let

$$\Theta^{(1)}\colon \operatorname{Hom}(M,C) \to \operatorname{Ext}^1(A,C)$$

$$\Theta^{(2)}\colon \operatorname{Hom}(A,N) \to \operatorname{Ext}^1(A,C)$$

be the connecting homomorphisms induced by the two rows. Prove that

$$\Theta^{(1)}(\gamma) + \Theta^{(2)}(\alpha) = 0.$$

[Hint: regard the top row as an acyclic left complex over A, and the bottom row as an acyclic right complex over C; then use the commutative diagram

$$\begin{array}{ccccc} H^1(\operatorname{Hom}(A,Y)) & \to & H^1(\operatorname{Hom}(X,Y)) & \to & H^1(\operatorname{Hom}(X,C)) \\ & \searrow & \downarrow & \swarrow & \\ & & \operatorname{Ext}^1(A,C) & & \end{array}$$

and apply Exer. v,8.]

19. For a right ideal I and left ideal J of the ring Λ, prove (using 1.5a) that

$$\operatorname{Tor}_1^\Lambda(\Lambda/I, \Lambda/J) \approx (I \cap J)/IJ$$

$$\operatorname{Tor}_2^\Lambda(\Lambda/I, \Lambda/J) \approx \operatorname{Ker}(I \otimes_\Lambda J \to IJ)$$

$$\operatorname{Tor}_n^\Lambda(\Lambda/I, \Lambda/J) \approx \operatorname{Tor}_{n-2}^\Lambda(I,J), \qquad n > 2$$

where IJ denotes the image of $I \otimes_\Lambda J \to \Lambda \otimes_\Lambda \Lambda \approx \Lambda$.

CHAPTER VII

Integral Domains

Introduction. Our main objective is to show how the notions introduced earlier apply to the case of modules over an integral domain Λ. The special case of abelian groups, i.e. of modules over the ring Z of rational integers, is treated in the last two sections.

The functor Tor has remarkable properties if the ring Λ is a Prüfer ring (i.e. an integral domain in which all finitely generated ideals are inversible in the field of fractions). For such a ring Λ we have $\text{Tor}_n = 0$ for $n \geqq 2$; $\text{Tor}_1(A,C)$ depends only on the torsion sub-modules of A and C; in order that $\text{Tor}_1(A, Y) = 0$ for all Y it is necessary and sufficient that A be torsion-free. These properties (studied in § 4) are the origin of the notation Tor and explain some of the peculiarities of the elementary Künneth relations (for the tensor product of two Z-complexes).

In the last two sections (§ 6–7) we study the relations of the functors Tor_1^Z and Ext_Z^1 with the Pontrjagin duality theory for compact abelian groups.

1. GENERALITIES

We shall assume throughout this chapter that Λ is an integral domain, i.e. a commutative ring with a unit element $\neq 0$ such that $\alpha, \beta \in \Lambda$, $\alpha \neq 0$ and $\beta \neq 0$ imply $\alpha\beta \neq 0$.

An element a of a module A is called a *torsion element* if $\lambda a = 0$ for some $\lambda \in \Lambda$, $\lambda \neq 0$. The torsion elements form a submodule tA of A. We say that A is a torsion module if $tA = A$. Clearly tA is a torsion module. We say that A is a *torsion-free* module if $tA = 0$. For instance A/tA is torsion-free. Any submodule of a torsion-free module is itself torsion-free.

If A is a direct sum $\Sigma_\alpha A_\alpha$, then $tA = \Sigma_\alpha(tA_\alpha)$.

PROPOSITION 1.1. *A projective module is torsion-free.*

PROOF. Λ being an integral domain, is torsion free. Since each free module is a direct sum of modules isomorphic with Λ, it follows that free modules are torsion-free. Since each projective module is a submodule of a free module, the conclusion follows.

An element a of a module A is called *divisible* if for each $\lambda \in \Lambda$, $\lambda \neq 0$, there is an element $b \in A$ with $a = \lambda b$. The divisible elements of A form a submodule δA of A. A module A is called *divisible* if $\delta A = A$. A quotient module of a divisible module by a submodule always is divisible. It can be easily proved that $\delta(A/\delta A) = 0$.

If A is a direct product $\Pi_\alpha A_\alpha$, then $\delta A = \Pi_\alpha(\delta A_\alpha)$.

PROPOSITION 1.2. *An injective module is divisible.*

PROOF. Let A be an injective module and let $a \in A, \lambda \in \Lambda, \lambda \neq 0$. Consider the ideal $I = \lambda\Lambda$. Since $\alpha\lambda = \beta\lambda$ implies $\alpha = \beta$, the formula $f(\alpha\lambda) = \alpha a$ defines a homomorphism $f\colon I \to A$. Since A is injective there exists, by I,3.2, a $b \in A$ with $f\beta = \beta b$ for all $\beta \in I$. Thus $a = f\lambda = \lambda b$ so that a is divisible.

PROPOSITION 1.3. *A torsion-free module is injective if and only if it is divisible.*

PROOF. The necessity of the condition follows from 1.2. To prove sufficiency assume A torsion-free and divisible. Consider a homomorphism $f\colon I \to A$ where I is a non zero ideal of Λ. Then for each $\lambda \in I, \lambda \neq 0$ there is a unique $a_\lambda \in A$ such that $f\lambda = \lambda a_\lambda$. If $\mu \in I, \mu \neq 0$, then

$$\mu\lambda a_\lambda = \mu f\lambda = f(\lambda\mu) = \lambda(f\mu) = \lambda\mu a_\mu$$

so that $a_\lambda = a_\mu = a$. Thus $f\lambda = \lambda a$ for all $\lambda \in I$ and A is injective by I,3.2.

Since Λ is commutative, it follows from II,3 that for any functor T defined for Λ-modules A whose values $T(A)$ are groups, we may regard $T(A)$ as a Λ-module. It also follows from II,3 that $A \otimes_\Lambda C$ may be regarded as a Λ-module with $\lambda(a \otimes c) = (\lambda a) \otimes c = a \otimes \lambda c$ and similarly for $\mathrm{Hom}_\Lambda(A,C)$. Thus $\mathrm{Tor}_n^\Lambda(A,C)$ and $\mathrm{Ext}_\Lambda^n(A,C)$ are Λ-modules. We shall write $\otimes$, Hom, Tor_n, Ext^n omitting the symbol Λ.

PROPOSITION 1.4. *If the functor T is covariant and right exact and A is divisible then $T(A)$ is divisible. If T is contravariant and left exact and A is divisible then $T(A)$ is torsion-free. If T is covariant and left exact and A is torsion-free then $T(A)$ is torsion-free. If T is contravariant and right exact and A is torsion-free then $T(A)$ is divisible. If A is divisible and torsion-free then $T(A)$ also is divisible and torsion-free for any functor T.*

PROOF. For each $\lambda \in \Lambda$ consider the Λ-endomorphism $\lambda\colon A \to A$ given by $a \longrightarrow \lambda a$. Then A is divisible if and only if $A \xrightarrow{\lambda} A \longrightarrow 0$ is exact for all $\lambda \neq 0$, and A is torsion-free if and only if $0 \longrightarrow A \xrightarrow{\lambda} A$ is exact. This implies the conclusions above, since the map $T(\lambda)\colon T(A) \to T(A)$ by definition coincides with $\lambda\colon T(A) \to T(A)$.

COROLLARY 1.5. *If either A or C is divisible then $A \otimes C$ is divisible. If A is divisible or C is torsion-free then* Hom (A,C) *is torsion-free.*

PROPOSITION 1.6. *If A is a finitely generated torsion module then for* *any functor T the module T(A) is a torsion module. If T is covariant and of type $L\Sigma^*$ (i.e. T commutes with direct limits) then for any torsion module A, T(A) is a torsion module.*

Proof. If A is a finitely generated torsion module then there is a $\lambda \in \Lambda$, $\lambda \neq 0$ such that the homomorphism λ: $A \to A$ is zero. Then λ: $T(A) \to T(A)$ also is zero and $T(A)$ is a torsion module. The second part of the proposition is immediate.

Corollary 1.7. *If A or C is a torsion module, then* $\mathrm{Tor}_n\,(A,C)$ *is a torsion module.*

This follows from VI,1.3 and from 1.6.

Proposition 1.8. *If A is a torsion module and C is divisible then* $A \otimes C = 0$. *If A is a torsion module and the module C is torsion-free, then* $\mathrm{Hom}\,(A,C) = 0$.

Proof. Consider $a \otimes c \in A \otimes C$. If A is a torsion module there is an element $\lambda \in \Lambda$, $\lambda \neq 0$ with $\lambda a = 0$. If C is divisible then there is a $c' \in C$ with $\lambda c' = c$. Thus $a \otimes c = a \otimes \lambda c' = a\lambda \otimes c' = 0$. The other half of the proposition is obvious.

Let A be a Λ-module. For each $\lambda \in \Lambda$ the mapping $a \to \lambda a$ is a Λ-homomorphism λ: $A \to A$. We denote

$$
\begin{aligned}
{}_{\lambda}A &= \mathrm{Ker}\,(\lambda\colon\ A \to A),\\
\lambda A &= \mathrm{Im}\,(\lambda\colon\ A \to A),\\
A_{\lambda} &= A/\lambda A = \mathrm{Coker}\ \lambda.
\end{aligned}
$$

Since Λ is an integral domain it follows that λ: $\Lambda \to \Lambda$ has zero kernel for $\lambda \neq 0$. Thus the sequence

$$0 \longrightarrow \Lambda \xrightarrow{\lambda} \Lambda \longrightarrow \Lambda_{\lambda} \longrightarrow 0$$

is exact. Since Λ is projective this sequence yields a projective resolution of Λ_{λ}. Thus we find

$$
\begin{aligned}
&\mathrm{Tor}_1\,(\Lambda_{\lambda},C) = {}_{\lambda}C, \qquad &&\mathrm{Tor}_n\,(\Lambda_{\lambda},C) = 0 \qquad &&\text{for } n > 1\\
&\mathrm{Ext}^1\,(\Lambda_{\lambda},C) = C_{\lambda}, \qquad &&\mathrm{Ext}^n\,(\Lambda_{\lambda},C) = 0 \qquad &&\text{for } n > 1.
\end{aligned}
$$

In some cases these formulae and direct sum properties facilitate the computation of Tor_n and Ext^n.

2. THE FIELD OF QUOTIENTS

We shall denote by Q the field of quotients of Λ and will write $K = Q/\Lambda$. Thus

$$0 \to \Lambda \to Q \to K \to 0$$

is an exact sequence. Since Q is torsion-free and divisible, it follows from 1.3 that Q is injective. K is divisible but in general is not injective.

Since Q is torsion free and divisible it follows from the last part of 1.4 that all the modules $Q \otimes A$, $\operatorname{Hom}(A,Q)$, $\operatorname{Ext}^n(Q,A)$ are torsion free and divisible and therefore injective.

Since Q is the union of its submodules $\frac{1}{\alpha}\Lambda$ for $\alpha \in \Lambda$, $\alpha \neq 0$, it follows that Q is the direct limit of the projective modules $\frac{1}{\alpha}\Lambda$. Since Tor_n and direct limits commute, we obtain

$$\operatorname{Tor}_n(Q,A) = 0 \qquad n > 0. \tag{1}$$

It follows that for any module A we have the exact sequence

$$0 \longrightarrow \operatorname{Tor}_1(K,A) \xrightarrow{\psi} A \xrightarrow{\varphi} Q \otimes A$$

where A has been identified with $\Lambda \otimes A$.

PROPOSITION 2.1. *The kernel of the homomorphism*

$$\varphi\colon A \to Q \otimes A$$

given by $a \to 1 \otimes a$ is the torsion submodule tA.

In view of the exact sequence above, we obtain the equivalent formulation

PROPOSITION 2.2. *The homomorphism*

$$\psi\colon \operatorname{Tor}_1(K,A) \to A$$

maps $\operatorname{Tor}_1(K,A)$ *isomorphically onto tA.*

PROOF. Since Q is the union of its submodules $\frac{1}{\alpha}\Lambda$ for $a \in \Lambda$, $\alpha \neq 0$, it follows from the fact that $\otimes$ and direct limits commute, that the kernel of φ is the union of the kernels of

$$\varphi_\alpha\colon A \to \left(\frac{1}{\alpha}\Lambda\right) \otimes A$$

where $\varphi_\alpha a = 1 \otimes a$. Consider the mapping

$$f_\alpha\colon \left(\frac{1}{\alpha}\Lambda\right) \otimes A \to \Lambda \otimes A$$

given by $f_\alpha\left(\frac{1}{\alpha}\lambda\right) \otimes a = \lambda a$. Since f_α is an isomorphism, the kernel of φ_α coincides with the kernel of $f_\alpha\varphi_\alpha\colon A \to A$. This latter map is $a \to \alpha a$. Thus the union of the kernels of φ_α is tA.

PROPOSITION 2.3. *If C is torsion-free then the sequence*

$$0 \to C \to Q \otimes C \to K \otimes C \to 0 \tag{2}$$

is exact. If A is a torsion module, then this exact sequence yields an isomorphism

$$\text{(3)} \qquad \text{Hom}\,(A,K \otimes C) \approx \text{Ext}^1\,(A,C).$$

In particular, taking $C = \Lambda$, we obtain the isomorphism

$$\text{Hom}\,(A,K) \approx \text{Ext}^1\,(A,\Lambda),$$

for any torsion module A.

PROOF. The exactness of (2) follows from 2.1 since C is torsion-free. If A is a torsion module then $\text{Hom}\,(A,Q \otimes C) = 0$, since $Q \otimes C$ is torsion-free. Since further $Q \otimes C$ is injective we have $\text{Ext}^1(A,Q \otimes C) = 0$. This implies (3).

PROPOSITION 2.4. *Every finitely generated torsion-free module admits a monomorphism into a free module with a finite base.*

PROOF. Since the module A is torsion-free we may regard A as a submodule of $Q \otimes A$. Let $(a_1, \ldots, a_n)$ be generators for A. Then $Q \otimes A$ regarded as a vector space over Q is finite dimensional and has a base $(e_1, \ldots, e_m)$. Then $a_i = \Sigma_j q_{ij} e_j$ where $q_{ij} \in Q$. Let $\lambda \in \Lambda$, $\lambda \neq 0$ be such that all $\lambda q_{ij} \in \Lambda$. Then

$$a_i = \Sigma_j (\lambda q_{ij})\,(\lambda^{-1} e_j)$$

so that A is contained in the Λ-submodule F of $Q \otimes A$ generated by the elements $\lambda^{-1} e_j$, $j = 1, \ldots, m$. Since F is free with $(\lambda^{-1}e_1, \ldots, \lambda^{-1}e_m)$ as base, the proposition follows.

PROPOSITION 2.5. *If P is a projective module, P′ a projective submodule of P and A is torsion-free, then the homomorphism*

$$P' \otimes A \to P \otimes A$$

induced by $P' \to P$ is a monomorphism. Equivalently

$$\text{Tor}_1\,(P/P',A) = 0.$$

PROOF. Since tensor products and direct limits commute we may limit our attention to the case when A is finitely generated. By 2.4 we may then regard A as a submodule of a free module F. There results a commutative diagram

$$\begin{array}{ccc} P' \otimes A & \longrightarrow & P \otimes A \\ \downarrow & & \downarrow \\ P' \otimes F & \longrightarrow & P \otimes F \end{array}$$

Since P', P and F are projective, both vertical and the lower horizontal homomorphisms are monomorphisms. Thus $P' \otimes A \to P \otimes A$ also is a monomorphism.

PROPOSITION 2.6. *Any module A admits a monomorphism into a divisible module.*

PROOF. Choose an exact sequence $0 \to M \to P \to A \to 0$ with P torsion-free (e.g. projective or free). Then, by 2.1, $P \to P \otimes Q$ is a monomorphism. Consequently Coker $(M \to P) \to$ Coker $(M \to P \otimes Q)$ is a monomorphism. Since $A \approx$ Coker$(M \to P)$ and Coker$(M \to P \otimes Q)$ is divisible, the conclusion follows.

3. INVERSIBLE IDEALS

PROPOSITION 3.1. *In order that a module A be projective it is necessary and sufficient that there exist a family $\{a_\alpha\}$ of elements of A and a family $\{\varphi_\alpha\}$ of homomorphisms φ_α: $A \to \Lambda$ such that for all $a \in A$*

(1) $$a = \Sigma_\alpha(\varphi_\alpha a)a_\alpha$$

where $\varphi_\alpha a$ is zero for all but a finite number of indices α.

PROOF. Let ψ: $F \to A$ be a homomorphism of a free module with base $\{e_\alpha\}$ onto A, and let $a_\alpha = \psi(e_\alpha)$. In order that A be projective it is necessary and sufficient that there exists a homomorphism φ: $A \to F$ such that $\psi\varphi =$ identity. If we write $\varphi a = \Sigma_\alpha(\varphi_\alpha a)e_\alpha$ we obtain homomorphisms φ_α: $A \to \Lambda$ such that for each $a \in A$ we have $\varphi_\alpha a = 0$ for all but a finite number of indices α. The condition $\psi\varphi =$ identity is then equivalent with

$$a = \Sigma_\alpha(\varphi_\alpha a)a_\alpha$$

for all $a \in A$.

The above proof did not utilize the fact that Λ is an integral domain and is therefore valid for modules over any ring.

PROPOSITION 3.2. *In order that a non-zero ideal I of Λ be projective it is necessary and sufficient that I be an inversible ideal, i.e. that there exist $q_1, \dots, q_n \in Q$ and $a_1, \dots, a_n \in I$* with $q_i I \subset \Lambda$, $\Sigma_i\, q_i a_i = 1$, $i = 1, \dots, n$.

PROOF. Assume I inversible and define $\varphi_i x = q_i x$ for $x \in I$. Then φ_i: $I \to \Lambda$ and

$$\Sigma_i(\varphi_i x)a_i = \Sigma_i q_i x a_i = x\Sigma_i q_i a_i = x.$$

Thus, by 3.1, I is projective. Assume now that I is projective and let $\{a_\alpha\}$, $\{\varphi_\alpha\}$ be as in 3.1. Then for each α we have $x(\varphi_\alpha y) = \varphi_\alpha(xy) = y(\varphi_\alpha x)$. Thus $q_\alpha = (\varphi_\alpha x)/x$ for $x \in I$, $x \neq 0$ is an element of Q such that $\varphi_\alpha y = q_\alpha y$ for all $y \in I$. It follows that $q_\alpha I \subset \Lambda$. If $x \neq 0$ then $\varphi_\alpha x = q_\alpha x$ is zero for all except a finite number of indices α. It follows that all q_α except for a finite number are zero. Condition (1) of 3.1 yields

$$x = \Sigma_\alpha(\varphi_\alpha x)a_\alpha = \Sigma(q_\alpha x)a_\alpha = (\Sigma q_\alpha a_\alpha)x$$

which is equivalent with $\Sigma_\alpha q_\alpha a_\alpha = 1$. Thus I is inversible.

PROPOSITION 3.3. *Every inversible ideal is finitely generated.*

PROOF. Let I be an inversible ideal, $b \in I$. With q_i and a_i as above we have $\Sigma q_i b a_i = b$ and $q_i b \in \Lambda$. Thus $a_1, \ldots, a_n$ generate I.

PROPOSITION 3.4. *Let I be an inversible ideal in Λ and A a divisible module. Then the mapping* Hom $(\Lambda,A) \to$ Hom (I,A) *induced by $I \to \Lambda$ is an epimorphism. In other words for each homomorphism $f\colon I \to A$ there is an element $a \in A$ with $f\lambda = \lambda a$ for all $\lambda \in I$.*

PROOF. Since I is inversible there exist $q_1, \ldots, q_n \in Q$ and $\lambda_1, \ldots, \lambda_n \in I$ such that

$$q_i I \subset \Lambda, \qquad \Sigma_i q_i \lambda_i = 1.$$

Since A is divisible, there exist elements $a_i \in A$ with $f\lambda_i = \lambda_i a_i$, $i = 1, \ldots, n$. Then

$$\begin{aligned} f\lambda = f(\Sigma_i q_i \lambda_i \lambda) &= \Sigma_i (q_i \lambda)(f\lambda_i) \\ &= \Sigma_i (q_i \lambda \lambda_i) a_i = \lambda \Sigma_i (q_i \lambda_i) a_i. \end{aligned}$$

Thus setting $a = \Sigma_i (q_i \lambda_i) a_i$ yields $f\lambda = \lambda a$.

From the exact sequence

$$\operatorname{Hom}(\Lambda,A) \to \operatorname{Hom}(I,A) \to \operatorname{Ext}^1(\Lambda/I,A) \to 0$$

it follows that the conclusion of 3.4 may also be stated as $\operatorname{Ext}^1(\Lambda/I,A) = 0$.

4. PRÜFER RINGS

It follows from 3.2 that an integral domain Λ is semi-hereditary if and only if every finitely generated ideal is inversible. Such rings are called *Prüfer rings*. By VI,2.9, we have $\operatorname{Tor}_n^\Lambda = 0$ for $n > 1$ and $\operatorname{Tor}_1^\Lambda$ is left exact for such rings Λ.

PROPOSITION 4.1. *Λ is a Prüfer ring if and only if every finitely generated torsion-free Λ-module is projective.*

PROOF. If every finitely generated torsion-free Λ-module is projective, then every finitely generated ideal in Λ is projective and Λ is a Prüfer ring. If Λ is a Prüfer ring then it is semi-hereditary and by I,6.2 every finitely generated submodule of a free module is projective. Thus the result follows from 2.4.

PROPOSITION 4.2. *If Λ is a Prüfer ring, then A is torsion-free if and only if* $\operatorname{Tor}_1(X,A) = 0$ *for all modules X.*

PROOF. If $\operatorname{Tor}_1(K,A) = 0$ then, by 2.2, $tA = 0$ and A is torsion-free. Conversely assume A torsion-free. Then by 4.1 each finitely generated submodule A_α of A is projective. Thus $\operatorname{Tor}_1(X,A_\alpha) = 0$. Since $A = \varinjlim A_\alpha$, and Tor_1 commutes with direct limits, it follows that $\operatorname{Tor}_1(X,A) = 0$.

COROLLARY 4.3. *If* Λ *is a Prüfer ring and* C *is torsion-free then the functors* $A \otimes C$ *and* $C \otimes A$ *are exact functors of* A.

PROPOSITION 4.4. *If* Λ *is a Prüfer ring then the homomorphisms*

$$\begin{array}{ccc} & \operatorname{Tor}_1(tA,C) & \\ \nearrow & & \searrow \\ \operatorname{Tor}_1(tA,tC) & & \operatorname{Tor}_1(A,C) \\ \searrow & & \nearrow \\ & \operatorname{Tor}_1(A,tC) & \end{array}$$

are isomorphisms.

PROOF. Since the module A/tA is torsion-free, it follows from 4.2 that $\operatorname{Tor}_1(A/tA,C) = 0$. Since also $\operatorname{Tor}_2 = 0$, it follows from exactness that $\operatorname{Tor}_1(tA,C) \to \operatorname{Tor}_1(A,C)$ is an isomorphism. The other isomorphisms are established similarly.

PROPOSITION 4.5. *If* Λ *is a Prüfer ring and* A *and* C *are torsion-free then so is* $A \otimes C$.

PROOF. Since A is torsion-free it follows from 2.1 that the sequence $0 \to A \to Q \otimes A$ is exact. Since C also is torsion-free, 4.3 may be applied to give the exact sequence $0 \to A \otimes C \to Q \otimes A \otimes C$. Thus by 2.1, $A \otimes C$ is torsion-free.

PROPOSITION 4.6. *If* Λ *is a Prüfer ring, then* $\operatorname{Tor}_1(A,C)$ *always is a torsion module.*

This follows directly from 4.4 and 1.7.

5. DEDEKIND RINGS

It follows from 3.2 that an integral domain Λ is hereditary if and only if every ideal is inversible. Such rings are called *Dedekind rings*. It follows from 3.3 that a Dedekind ring is just a Prüfer ring which is Noetherian.

For any functor T defined for Λ-modules over a Dedekind ring, the derived functors R^nT, L_nT and the satellites S^nT, S_nT are zero for $n > 1$. The functor R^1T is right exact and L_1T is left exact.

PROPOSITION 5.1. *For each integral domain* Λ *the following properties are equivalent:*

(a) Λ *is a Dedekind ring.*

(b) *Each divisible* Λ*-module is* Λ*-injective.*

PROOF. (a) $\to$ (b). If Λ is a Dedekind ring then each ideal I of Λ is inversible. Thus 3.4 and I,3.2 imply that each divisible module is injective.

(b) $\to$ (a). Since each quotient of a divisible module is divisible it follows that each quotient of an injective module is injective. Thus, by I,5.4, Λ is hereditary, and therefore a Dedekind ring.

PROPOSITION 5.2. *If* Λ *is a Dedekind ring then the homomorphism*

$$\text{Ext}^1(A,C) \to \text{Ext}^1(A,C/\delta C)$$

is an isomorphism.

This follows from the fact that $\text{Ext}^1(A,\delta C)$ and $\text{Ext}^2(A,\delta C)$ are zero since δC is divisible and thus injective.

PROPOSITION 5.3. *If* Λ *is a Dedekind ring, then the module* A *is torsion-free if and only if* $\text{Ext}^1(A,C)$ *is divisible for all modules* C.

PROOF. If A is torsion free, then since Ext^1 is right exact, it follows from 1.4 that $\text{Ext}^1(A,C)$ is divisible. To prove the converse we use the isomorphism (VI,3.5a)

$$\text{Ext}^1(B, \text{Ext}^1(A,C)) \approx \text{Ext}^1(\text{Tor}_1(B,A),C).$$

If $\text{Ext}^1(A,C)$ is divisible, then by 5.1, it is injective. Therefore the expressions above are zero. Since $\text{Ext}^1(\text{Tor}_1(B,A),C) = 0$ for all C it follows from VI,2.2 that $\text{Tor}_1(B,A)$ is projective, and therefore, by 1.1, is torsion-free. However, by 4.6, $\text{Tor}_1(B,A)$ is a torsion module, so that $\text{Tor}_1(B,A) = 0$. Since this holds for all B it follows from 4.2 that A is torsion-free.

REMARK. 5.1 combined with 2.6 yield a new proof that every Λ-module A (where Λ is a Dedekind ring) admits a monomorphism into an injective Λ-module. In particular, this is valid if $\Lambda = Z$ is the ring of integers (see remark at end of II,6).

6. ABELIAN GROUPS

We shall assume in this section that $\Lambda = Z$ is the ring of rational integers. All modules considered are then simply abelian groups. The results established for Dedekind and Prüfer rings, all apply in this case. In particular "injective" means "divisible." Since a subgroup of a free abelian group is again a free abelian group it follows that "projective" means "free."

Let R denote the additive group of real numbers and let $T = R/Z$ be the group of reals reduced mod 1.

For each abelian group A, the group $\text{Hom}(A,T)$ will be called the dual of A and will be denoted by $D(A)$. Since T is divisible it is injective. Thus $\text{Hom}(A,T)$ is an exact functor of A. Consequently D is a contravariant exact functor. For the moment no topology is imposed on R,T and $D(A)$.

Since T is injective, VI,5.1 gives an isomorphism

$$(1) \qquad \rho^1\colon \text{Ext}^1(A, \text{Hom}(B,T)) \approx \text{Hom}(\text{Tor}_1(A,B),T).$$

This implies

PROPOSITION 6.1. *For any two abelian groups A and B we have*

$$\mathrm{Ext}^1(A,D(B)) \approx D(\mathrm{Tor}_1(A,B)).$$

As an application we prove

COROLLARY 6.2. *If A is a torsion-free abelian group and C is a finite abelian group, then* $\mathrm{Ext}^1(A,C) = 0$.

PROOF. Since C is finite, there is a (finite) group B with $C \approx D(B)$. Then, by 6.1,

$$\mathrm{Ext}^1(A,C) \approx \mathrm{Ext}^1(A,D(B)) \approx D(\mathrm{Tor}_1(A,B)).$$

Since A is torsion-free it follows from 4.2 that the latter term is 0.

Proposition 6.1 acquires more force if we introduce a topology in some of the groups that appear.

So far we have dealt only with categories of Λ-modules and Λ-homomorphisms, where Λ was a ring. Let $\mathscr{C}$ denote the category of compact abelian groups (satisfying the Haussdorf separation axiom) and continuous homomorphisms. In particular, all the results of IV,1 remain valid for compact abelian groups. The continuity of the connecting homomorphisms for homology groups is an easy consequence of compactness. When we pass to graded groups $A = \sum A^n$ (see IV,3) we only require that each A^n be a compact group. We do *not impose any topology* on the direct sum A. With this convention, the definition and basic properties of derived functors remain valid for additive functors whose *values* are in the category $\mathscr{C}$ of compact abelian groups. The same applies to satellites. Also the homomorphisms α and α' of IV,6 are continuous.

An example of such a functor is $T(A) = \mathrm{Hom}(A,C)$, where A is a discrete group and C is a compact abelian group, with the topology in $\mathrm{Hom}(A,C)$ defined as follows. Given a finite subset F of A and a neighborhood of zero V in C, we consider the set $W(F,V)$ of all $f \in \mathrm{Hom}(A,C)$ with $f(F) \subset V$. We consider in $\mathrm{Hom}(A,C)$ the topology in which the sets $W(F,V)$ form a fundamental system of neighborhoods of zero. In this topology $\mathrm{Hom}(A,C)$ is compact (and satisfies Haussdorf's separation axiom) and for each φ: $A \to A'$, $\mathrm{Hom}(\varphi,C)$: $\mathrm{Hom}(A',C) \to \mathrm{Hom}(A,C)$ is continuous.

Let A and B be discrete abelian groups and let C be a compact abelian group. It is immediate that the isomorphism

$$\mathrm{Hom}(A, \mathrm{Hom}(B,C)) \approx \mathrm{Hom}(A \otimes B,C)$$

is a topological isomorphism. In particular, taking $C = T$ with its natural topology derived from the representation $T = R/Z$, we obtain the *topological isomorphism*

$$\mathrm{Hom}(A,D(B)) \approx D(A \otimes B).$$

If we consider the right derived functors of the functor $T(A) = \operatorname{Hom}(A,C)$ with C compact, we find $\operatorname{Ext}^1(A,C)$ has a natural topology and is a compact group. Further all the homeomorphisms used to define ρ^1 in VI,5 being continuous, ρ^1 is continuous and therefore is a homeomorphism (since the groups are compact). It follows that *the isomorphism of* 6.1 *is topological.*

7. A DESCRIPTION OF $\operatorname{Tor}_1(A,C)$

We consider an exact sequence

(S) $$0 \to \Lambda \to R \to T \to 0$$

where Λ is an integral domain and R is a divisible and torsion-free Λ-module; the exact sequence

(S′) $$0 \to \Lambda \to Q \to K \to 0$$

is an example of such a sequence. It is clear that Q may be regarded as a submodule of R. Since, by 1.3, Q is injective we have a direct sum decomposition $R = Q + R'$ where R' again is divisible and torsion-free. Consequently $T = K + R'$ and (S) is obtained as the direct sum of (S′) and a trivial exact sequence $0 \to 0 \to R' \to R' \to 0$.

Since, by 1.3, R is injective we have the exact sequence

$$\operatorname{Hom}(A,T) \xrightarrow{\tau} \operatorname{Ext}^1(A,\Lambda) \longrightarrow 0$$

and since Hom is left exact we obtain an exact sequence

$$0 \longrightarrow \operatorname{Hom}(\operatorname{Ext}^1(A,\Lambda),C) \xrightarrow{\varphi} \operatorname{Hom}(\operatorname{Hom}(A,T),C).$$

Combining φ with the homomorphism

$$\sigma_1\colon \operatorname{Tor}_1(A,C) \to \operatorname{Hom}(\operatorname{Ext}^1(A,\Lambda),C)$$

of VI,5 we obtain the homomorphism

$$\varphi\sigma_1\colon \operatorname{Tor}_1(A,C) \to \operatorname{Hom}(\operatorname{Hom}(A,T),C)$$

which is the object of study of this section.

PROPOSITION 7.1. *If Λ is a Dedekind ring then $\varphi\sigma_1$ is a monomorphism. If further A is a finitely generated torsion module then $\varphi\sigma_1$ is an isomorphism.*

PROOF. It follows from VI,5.5 that σ_1 is a monomorphism. Since φ also is a monomorphism, it follows that $\varphi\sigma_1$ is a monomorphism.

Assume now that A is a finitely generated torsion module. Then σ_1 is an isomorphism by VI,5.4. To show that φ is an isomorphism it suffices to show that $\tau\colon \operatorname{Hom}(A,T) \to \operatorname{Ext}^1(A,\Lambda)$ is an isomorphism. However $T = K + R'$ and we have already shown in 2.3 that $\operatorname{Hom}(A,K) \to \operatorname{Ext}^1(A,\Lambda)$ is an isomorphism (if A is a torsion module). Since A is a torsion module and R' is torsion-free we have $\operatorname{Hom}(A,R') = 0$. Thus τ is an isomorphism.

We now assume that $\Lambda = Z$ is the group of integers, that R is the group of real numbers and that $T = R/Z$. Then 7.1 can be applied and we obtain a monomorphism

$$\varphi\sigma_1\colon \operatorname{Tor}_1(A,C) \to \operatorname{Hom}(D(A),C).$$

In $D(A) = \operatorname{Hom}(A,T)$ we have the (compact) topology defined in the preceding section.

PROPOSITION 7.2. *Let A and C be discrete abelian groups. Then the homomorphism*

$$\varphi\sigma_1\colon \operatorname{Tor}_1(A,C) \to \operatorname{Hom}(D(A),C)$$

maps $\operatorname{Tor}_1(A,C)$ *isomorphically onto the subgroup* $\operatorname{Hom}_c(D(A),C)$ *of* $\operatorname{Hom}(D(A),C)$ *consisting of all continuous homomorphisms* $D(A) \to C$.

PROOF. Let A_α be a finite subgroup of A. We then have a commutative diagram

$$\begin{array}{ccc} \operatorname{Tor}_1(A_\alpha,C) & \xrightarrow{\varphi\sigma_{1\alpha}} & \operatorname{Hom}(D(A_\alpha),C) \\ \downarrow{\scriptstyle i_\alpha} & & \downarrow{\scriptstyle j_\alpha} \\ \operatorname{Tor}_1(A,C) & \xrightarrow[\varphi\sigma_1]{} & \operatorname{Hom}(D(A),C) \end{array}$$

Let $x \in \operatorname{Tor}_1(A_\alpha,C)$. Then $\varphi\sigma_1(i_\alpha x)$ admits a factorization

$$D(A) \to D(A_\alpha) \to C$$

Since $D(A_\alpha)$ is finite it follows that $\varphi\sigma_1(i_\alpha x)$ is continuous. Since each element of $\operatorname{Tor}_1(A,C)$ is in the image of i_α for some finite subgroup A_α, it follows that $\operatorname{Im}\varphi\sigma_1 \subset \operatorname{Hom}_c(D(A),C)$.

Conversely, let $f\colon D(A) \to C$ be a continuous homomorphism. Since C is discrete there is a neighborhood $W(F,V)$ of zero in $D(A)$ such that f maps $W(F,V)$ into zero. In particular, if $\varphi \in D(A)$ is such that $\varphi(a) = 0$ for all $a \in F$, then $\varphi \in W(F,V)$ and $f\varphi = 0$. Let A' be the subgroup of A generated by the finite set F. It follows that f admits a factorization

$$D(A) \longrightarrow D(A') \xrightarrow{g} C$$

where $D(A) \to D(A')$ is induced by the inclusion $A' \to A$ and g is continuous. The group A' being finitely generated, its torsion subgroup $t(A')$ is finite and A' is a direct sum of $t(A')$ and a subgroup E isomorphic with Z^n (= direct sum of n factors isomorphic with Z). Since $D(Z) \approx R$ we have $D(Z^n) \approx T^n$ which is connected. Since $D(A') = D(t(A')) + D(Z^n)$ and g is zero on the connected part $D(Z^n)$, it follows that g admits a factorization $D(A') \to D(t(A')) \to C$. Consequently f admits a factorization

$$D(A) \to D(t(A')) \to C.$$

Taking $A_\alpha = t(A')$ in the diagram above, we find that f is in the image of j_α. Since by 7.1, $\varphi\sigma_{1\alpha}$ is an isomorphism, it follows that f is in the image of $j_\alpha\varphi\sigma_{1\alpha} = \varphi\sigma_1 i_\alpha$ and thus in the image of $\varphi\sigma_1$. This concludes the proof.

If we combine 7.2 with 6.1 we obtain a *natural isomorphism*

$$\text{Ext}(A, D(C)) \approx D[\text{Hom}_c(D(A), C)]$$

of compact groups. This result was established by Eilenberg-MacLane (*Ann. of Math.* 43 (1942), 757–831) using the notion of a "modular trace."

EXERCISES

1. If A and C are finite abelian groups, then

$$\text{Hom}(A,C) \approx A \otimes C \approx \text{Tor}_1(A,C) \approx \text{Ext}^1(A,C).$$

Show that these isomorphisms are *not* natural. Give an example of two infinite torsion groups A, C such that $A \otimes C = 0$, $\text{Tor}_1(A,C) \neq 0$.

2. If Λ is a Dedekind ring, any torsion module has an injective resolution consisting of torsion modules. Using this, prove that the right derived functors $R^nU(A,C)$ of $U(A,C) = A \otimes C$ are zero for any $n \geqq 0$, whenever A or C is a torsion module [Hint: use 1.8].

3. For any integral domain Λ, let $U(A,C)$ denote the functor $A \otimes C$, and let Q denote the field of quotients of Λ. Prove that

$$R^0U(A,C) = Q \otimes A \otimes C$$

when A and C are torsion-free.

[Hint: consider exact consequences

$$0 \to A \to Q \otimes A \to (Q/\Lambda) \otimes A \to 0$$

$$0 \to C \to Q \otimes C \to (Q/\Lambda) \otimes C \to 0$$

$$0 \to (Q/\Lambda) \otimes A \to X_1 \to X_2 \to \cdots$$

$$0 \to (Q/\Lambda) \otimes C \to Y_1 \to Y_2 \to \cdots$$

where X_i and Y_i are injective. Then

$$(X) \qquad Q \otimes A = X_0 \to X_1 \to X_2 \to \cdots$$

$$(Y) \qquad Q \otimes C = Y_0 \to Y_1 \to Y_2 \to \cdots$$

are injective resolutions of A and C. In the complex $X \otimes Y$, d^0 is zero, because $(Q/\Lambda) \otimes A \otimes Y_0$ and $X_0 \otimes (Q/\Lambda) \otimes C$ are zero by 1.8.]

4. Let Λ be a Dedekind ring; then, denoting by $U(A,C)$ the functor $A \otimes C$, prove that

$$R^0U(A,C) = Q \otimes A \otimes C, \qquad R^nU(A,C) = 0 \qquad (n \geqq 1)$$

for all modules A and C.

[Hint: using Exer. 2, prove first that

$$R^nU(A,C) \approx R^nU(A/tA,C/tC).$$

Then, using Exer. 3, prove that

$$R^0U(A,C) = Q \otimes A \otimes C.$$

Finally, observe that $Q \otimes A \otimes C = V(A,C)$ is an exact functor, whenever Λ is a Prüfer ring.]

5. Consider the functor

$$T(A) = A \otimes C,$$

where C is a given torsion module, the ring Λ being an integral domain. Prove that

$$R^nT(A) = 0 \qquad \text{for any } n \geqq 0.$$

Then using Exer. v,4, prove that the right satellite of the functor

$$T_1(A) = \operatorname{Tor}_1(A,C)$$

is $A \otimes C$, whenever C is a torsion module. It follows that, assuming that Λ is a Prüfer ring, the right satellite of $T_1(A) = \operatorname{Tor}_1(A,C)$ is $A \otimes (tC)$ for any module C.

6. Consider the functor

$$T(C) = \operatorname{Hom}_\Lambda(A,C)$$

where A is a torsion module, the ring Λ being an integral domain. Prove that

$$L_nT(C) = 0 \qquad \text{for any } n \geqq 0.$$

Then prove that the left satellite of the functor

$$T^1(C) = \operatorname{Ext}^1_\Lambda(A,C)$$

is $\operatorname{Hom}_\Lambda(A,C)$, whenever A is a torsion module. If Λ is a Dedekind ring and A is finitely generated, the left satellite of $T^1(C) = \operatorname{Ext}^1_\Lambda(A,C)$ is $\operatorname{Hom}_\Lambda(tA,C)$.

7. With the notations of § 7, define a natural map

$$u\colon \operatorname{Tor}_1(A,C) \to \operatorname{Hom}(\operatorname{Hom}(A,T),C)$$

by setting, for $x \in \operatorname{Tor}_1(A,C)$ and $f \in \operatorname{Hom}(A,T)$,

$$(ux)f = v(w_f x),$$

where $w_f\colon \operatorname{Tor}_1(A,C) \to \operatorname{Tor}_1(T,C)$ is induced by f, and $v\colon \operatorname{Tor}_1(T,C) \to \Lambda \otimes C = C$ is the connecting homomorphism induced by the exact sequence (S′) of § 7. Show that, with the notations of § 7,

$$u + \varphi\sigma_1 = 0.$$

[Hint: use Exer. VI,18.]

8. For each Z-complex X, Z-module G and prime p, establish the natural isomorphism

$$H^n(\operatorname{Hom}(X,{}_pG)) \approx \operatorname{Hom}(H_n(X_p),G),$$

where ${}_pG$ denotes $\operatorname{Ker}(p\colon G \to G)$ and X_p denotes X/pX. Derive the isomorphism

$$\operatorname{Ext}^1(A,{}_pG) \approx \operatorname{Hom}({}_pA,G)$$

for each Z-module A. [Hint: note that $\operatorname{Hom}_Z(X,{}_pG) \approx \operatorname{Hom}_{Z_p}(X_p,{}_pG)$ and apply the isomorphism α' over the field Z_p; cf. IV, 7,2.]

Assume that X is torsion-free. The exact sequence $0 \to X \xrightarrow{p} X \to X_p \to 0$ yields a homomorphism

$$H_{n+1}(X_p) \to H_n(X).$$

Combine this with the above to obtain a homomorphism

$$H^n(\operatorname{Hom}(X,G)) \to H^{n+1}(\operatorname{Hom}(X,{}_pG).$$

9. Let Λ be a commutative ring and S a subset of Λ with the following properties: (1°) 0 is not in S; (2°) $1 \in S$; (3°) S is closed under multiplication. For each Λ-module A consider the set of all pairs (a,s), $a \in A$, $s \in S$, and consider the relation

$$(a,s) \approx (a',s')$$

which means: there exists $t \in S$ such that $as't = a'st$. Show that this is an equivalence relation, and that the set of equivalence classes A_S is a Λ-module under the operations

$$(a,s) + (a',s') = (as' + a's, ss'), \qquad (a,s)\lambda = (a\lambda,s).$$

Show that $a \to (a,1)$ yields a Λ-homomorphism $A \to A_S$ whose kernel consists of all those $a \in A$ with $as = 0$ for some $s \in S$.

Convert Λ_S into a ring by setting $(\lambda,s)(\lambda',s') = (\lambda\lambda',ss')$, and show that the mapping $\lambda \to (\lambda,1)$ yields a ring homomorphism $\varphi\colon \Lambda \to \Lambda_S$. Convert A_S into a Λ_S-module by setting

$$(a,s)(\lambda,t) = (a\lambda,st)$$

and show that the mapping $(a,s) \to a \otimes (1,s)$ yields a Λ_S-isomorphism

$$A_S \approx A \otimes_\Lambda \Lambda_S = A_{(\varphi)}.$$

Show that if A is a Λ_S-module, then, if we regard A as a Λ-module we have $A_S = A$.

10. Show that the functor $T(A) = A_S$ as described in Exer. 9 is exact, i.e. that Λ_S is Λ-flat. Apply the results of VI, Exer. 10, and in particular show that

$$\text{w.dim}_\Lambda A_S = \text{w.dim}_{\Lambda_S} A_S.$$

Apply VI, Exer. 11 to obtain the isomorphism

$$(\text{Tor}_n^\Lambda(A,C))_S \approx \text{Tor}_n^{\Lambda_S}(A_S,C_S)$$

and similar homomorphisms for Ext.

11. Let A be a commutative ring and let $\mathbf{M}$ denote the set of all subsets M of Λ such that $\Lambda - M$ is a maximal ideal. Show that for each Λ-module A the relation $A_M = 0$ for all $M \in \mathbf{M}$ implies $A = 0$. Use this and Exer. 10 to show that

$$\text{w.dim}_\Lambda A = \sup_{M \in \mathbf{M}} \text{w.dim}_{\Lambda_M} A_M.$$

If Λ is Noetherian and A is finitely generated, then show that

$$\dim_\Lambda A = \sup_{M \in \mathbf{M}} \dim_{\Lambda_M} A_M$$

[Hint: use VI, Exer. 3.]

12. Let A be a finitely generated abelian group, $A \neq 0$. Show that there exists a prime p such that $A \otimes Z_p \neq 0$. As an application, let A and B be two abelian groups, with A finitely generated and B free; let $f\colon A \to B$ be a homomorphism such that the induced homomorphism $A \otimes Z_p \to B \otimes Z_p$ is a monomorphism for each prime p; show that f is a monomorphism and A is free. [Hint: consider $N = \text{Ker} f$, and observe that N is a direct summand of A.]

CHAPTER VIII

Augmented Rings

Introduction. The homology (and cohomology) theory of augmented rings is the unifying concept of which various more specialized instances will be studied later: homology of associative algebras (Ch. IX), homology of supplemented algebras (Ch. X), homology of groups (X,4) and homology of Lie algebras (Ch. XIII).

Sections 1–3 are devoted to a general exposition, with some examples. The subject matter of § 4–6 is more special; we show how the theorem of "chains of syzygies" of Hilbert ties up with the general notion of the "projective dimension" of a module. This theory is valid either for graded rings or for local rings which are Noetherian.

1. HOMOLOGY AND COHOMOLOGY OF AN AUGMENTED RING

A left *augmented ring* is a triple formed by a ring Λ (always with a unit element), a left Λ-module Q, and a Λ-epimorphism $\varepsilon\colon \Lambda \to Q$. The module Q is called the *augmentation module*, ε is called the *augmentation epimorphism*, the kernel I of ε (a left ideal of Λ) is called the *augmentation ideal*.

We consider the functors

$$T(A) = A \otimes_\Lambda Q, \qquad U(C) = \mathrm{Hom}_\Lambda(Q,C),$$

where A is a right Λ-module and C is a left Λ-module. The groups

$$\mathrm{Tor}_n^\Lambda(A,Q) = S_nT(A), \qquad \mathrm{Ext}_\Lambda^n(Q,C) = S^nU(C)$$

are called the n-th *homology* (resp. *cohomology*) *group of the augmented ring* Λ, with coefficients in A (resp. in C).

Strictly speaking $S_nT(A)$ and $S^nU(C)$ are abelian groups. However if A or C have any additional operators which commute with the operators of Λ, then, following the principles of II,3, these additional operators carry over to $S_nT(A)$ or $S^nU(C)$. In particular, we may always regard these groups as modules over the center of Λ.

To compute $S_nT(A)$ and $S^nU(C)$, we can use a Λ-projective resolution X of A and a Λ-injective resolution Y of C. Then

$$\mathrm{Tor}_n^\Lambda(A,Q) = H_n(X \otimes_\Lambda Q), \qquad \mathrm{Ext}_\Lambda^n(Q,C) = H^n(\mathrm{Hom}_\Lambda(Q,Y)).$$

In the case of $\operatorname{Tor}_n^\Lambda(A,Q)$, this method of computation is due to H. Hopf (*Comment. Math. Helv. 17* (1945), 39–79) who used this process to define homology groups of (discrete) groups.

It is often more convenient to compute using a Λ-projective resolution X of Q. Then

$$\operatorname{Tor}_n^\Lambda(A,Q) = H_n(A \otimes_\Lambda X), \qquad \operatorname{Ext}_\Lambda^n(Q,C) = H^n(\operatorname{Hom}_\Lambda(X,C)).$$

The resolution X consists of a complex

$$\cdots \to X_n \to X_{n-1} \to \cdots \to X_1 \to X_0$$

and an augmentation map $X_0 \to Q$ such that the modules X_n are Λ-projective ($n \geqq 0$) and the sequence

$$\cdots \to X_n \to X_{n-1} \to \cdots \to X_1 \to X_0 \to Q \to 0$$

is exact. Since the sequence

$$0 \longrightarrow I \longrightarrow \Lambda \xrightarrow{\varepsilon} Q \longrightarrow 0 \tag{1}$$

is exact, we can always begin the construction of X by choosing $X_0 = \Lambda$ and letting $X_0 \to Q$ coincide with ε. The remainder of the construction reduces to choosing a projective resolution for the left Λ-module I.

The exact sequence (1) gives rise to connecting homomorphisms for Tor and Ext. Since Λ is Λ-projective we have $\operatorname{Tor}_n^\Lambda(A,\Lambda) = 0 = \operatorname{Ext}_\Lambda^n(\Lambda,C)$ for $n > 0$. This implies the isomorphisms

$$T(A) = A \otimes_\Lambda Q \approx \operatorname{Coker}(A \otimes_\Lambda I \to A) \tag{2}$$

$$U(C) = \operatorname{Hom}_\Lambda(Q,C) \approx \operatorname{Ker}(C \to \operatorname{Hom}_\Lambda(I,C)) \tag{2a}$$

$$S_1T(A) = \operatorname{Tor}_1^\Lambda(A,Q) \approx \operatorname{Ker}(A \otimes_\Lambda I \to A) \tag{3}$$

$$S^1U(C) = \operatorname{Ext}_\Lambda^1(Q,C) \approx \operatorname{Coker}(C \to \operatorname{Hom}_\Lambda(I,C)) \tag{3a}$$

$$S_nT(A) = \operatorname{Tor}_n^\Lambda(A,Q) \approx \operatorname{Tor}_{n-1}^\Lambda(A,I) \qquad (n > 1) \tag{4}$$

$$S^nU(C) = \operatorname{Ext}_\Lambda^n(Q,C) \approx \operatorname{Ext}_\Lambda^{n-1}(I,C) \qquad (n > 1). \tag{4a}$$

The functors $T_n(A) = S_nT(A)$ and $U^n(C) = S^nU(C)$ ($n \geqq 0$) are covariant functors in the variables A and C. Further they form connected sequences of functors: more precisely, given exact sequences

$$0 \to A' \to A \to A'' \to 0 \tag{5}$$

$$0 \to C' \to C \to C'' \to 0, \tag{5a}$$

we have the exact sequences

$$\cdots \to T_n(A') \to T_n(A) \to T_n(A'') \to T_{n-1}(A') \to \cdots \tag{6}$$

$$\cdots \to U^{n-1}(C'') \to U^n(C') \to U^n(C) \to U^n(C'') \to \cdots \tag{6a}$$

In view of v,8 these sequences may be computed as the homology sequences of the following exact sequences of complexes

$$0 \to A' \otimes_\Lambda X \to A \otimes_\Lambda X \to A'' \otimes_\Lambda X \to 0 \tag{7}$$

$$0 \to \operatorname{Hom}_\Lambda (X,C') \to \operatorname{Hom}_\Lambda (X,C) \to \operatorname{Hom}_\Lambda (X,C'') \to 0 \tag{7a}$$

where X is a projective resolution of Q.

THEOREM 1.1. *The connected sequence of covariant functors* $T_n(A) = \operatorname{Tor}_n^\Lambda (A,Q)$ *of the variable A has the following properties:*

(i) *for each exact sequence* (5), *the sequence* (6) *is exact;*

(ii) $T_n(A) = 0$ *if* $n > 0$ *and A is* Λ*-projective;*

(iii) $T_0(A) = A \otimes_\Lambda Q$.

These three properties characterize the connected sequence of functors $T_n(A) = \operatorname{Tor}_n^\Lambda (A,Q)$ *up to an isomorphism.*

THEOREM 1.1a. *The connected sequence of covariant functors* $U^n(C) = \operatorname{Ext}_\Lambda^n (Q,C)$ *of the variable C has the following properties:*

(i) *for each exact sequence* (5a), *the sequence* (6a) *is exact;*

(ii) $U^n(C) = 0$ *if* $n > 0$ *and C is* Λ*-injective;*

(iii) $U^0(C) = \operatorname{Hom}_\Lambda (Q,C)$.

These three properties characterize the connected sequence of functors $U^n(C) = \operatorname{Ext}_\Lambda^n (Q,C)$ *up to an isomorphism.*

The properties listed in 1.1 and 1.1a are special instances of the properties of the functors Tor and Ext. The fact that these properties constitute axiomatic descriptions, follows from III,5.1.

So far we have considered left augmented rings. The definition of right augmented rings and the ensuing discussion are quite similar with $T(A) = Q \otimes_\Lambda A$ and $U(C) = \operatorname{Hom}_\Lambda (Q,C)$.

If the augmentation ideal I is a *two-sided ideal* in Λ, then $Q = \Lambda/I$ is a ring, which may be regarded as a left and as a right Λ-module; thus in this case Λ is simultaneously a left and a right augmented ring. Since Λ operates on the right on Q it follows that for each right Λ-module A, the group $T_n(A) = \operatorname{Tor}_n^\Lambda (A,Q)$ is again a right Λ-module. Similarly for each left Λ-module C, $U^n(C) = \operatorname{Ext}_\Lambda^n (Q,C)$ again is a left Λ-module.

Assuming again that I is a two-sided ideal of Λ, we may replace A by Q in the formulas (2), (3), (4) (Q being considered as a right Λ-module).

If we observe that $Q \otimes_\Lambda I \approx I/I^2$ (I^2 being the image of the homomorphism $I \otimes_\Lambda I \to \Lambda \otimes_\Lambda I = I$) and that the homomorphism $Q \otimes_\Lambda I \to Q$ is zero, we obtain

(8) $$T(Q) = Q \otimes_\Lambda Q \approx Q$$

(9) $$S_1T(Q) = \mathrm{Tor}_1^\Lambda (Q,Q) \approx Q \otimes_\Lambda I \approx I/I^2$$

(10) $$S_2T(Q) = \mathrm{Tor}_2^\Lambda (Q,Q) \approx \mathrm{Tor}_1^\Lambda (Q,I) \approx \mathrm{Ker}\,(I \otimes_\Lambda I \to I).$$

2. EXAMPLES

In the following examples, I will be a two-sided ideal in Λ.

Graded rings. A graded ring is a ring Λ which is graded as an additive group, the grading satisfying the conditions

$$\Lambda^p = 0 \quad \text{for } p < 0, \qquad \Lambda^p\Lambda^q \subset \Lambda^{p+q}.$$

It follows that Λ^0 is a subring which we denote by Q. We define the epimorphism ε: $\Lambda \to Q$ by assigning to each element λ its homogeneous component of degree zero. The augmentation ideal is the two-sided ideal which consists of all elements with a vanishing zero-component.

We list some examples of graded rings. Let K be a ring, and $x_1, \ldots, x_n$ a set of letters. We denote by $\Lambda = F_K(x_1, \ldots, x_n)$ the free left K-module having as a base the elements

$$1, x_i, x_{i_1}x_{i_2}, \ldots, x_{i_1} \ldots x_{i_m}, \ldots$$

where each index i_j assumes any value $1, \ldots, n$. The module Λ is graded by regarding $x_{i_1} \cdots x_{i_m}$ as a homogeneous element of degree m. We define a multiplication in Λ by setting

$$(k_1x_{i_1} \cdots x_{i_m})(k_2x_{j_1} \cdots x_{j_p}) = k_1k_2x_{i_1} \cdots x_{i_m}x_{j_1} \cdots x_{j_p}.$$

The resulting graded ring $F_K(x_1, \ldots, x_n)$ is called the *free* K-ring on the letters $x_1, \ldots, x_n$.

If we divide $F_K(x_1, \ldots, x_n)$ by the two-sided ideal generated by the elements $x_ix_j - x_jx_i$, we obtain the graded K-ring $K[x_1, \ldots, x_n]$, called the *polynomial* K-ring on the letters $x_1, \ldots, x_n$.

If we divide $F_K(x_1, \ldots, x_n)$ by the two-sided ideal generated by the elements x_ix_i and $x_ix_j + x_jx_i$, we obtain the graded K-ring $E_K(x_1, \ldots, x_n)$ called the *exterior* (or Grassmann) K-ring on the letters $x_1, \ldots, x_n$.

For $n = 1$ the exterior ring $E_K(d)$ is easily seen to be the ring of dual numbers $\Lambda = (K,d)$ over K, as defined in IV,2. The elements of Λ are of the form $k_1 + k_2d$, the multiplication being given by

$$(k_1 + k_2d)(k_1' + k_2'd) = k_1k_1' + (k_1k_2' + k_2k_1')d.$$

The augmentation epimorphism ε is defined by

$$\varepsilon(k_1 + k_2d) = k_1.$$

We find a Λ-projective resolution X of K by taking $X_n = \Lambda$ for all $n \geqq 0$, and defining $d_n\colon X_n \to X_{n-1}$ by

$$d_n(k_1 + k_2d) = k_1d \qquad (n > 0).$$

For each right Λ-module A, i.e. for each right K-module A with differentiation operator d, the complex $A \otimes_\Lambda X$ is simply

$$\cdots \longrightarrow A \xrightarrow{d_n} A \longrightarrow \cdots \longrightarrow A \xrightarrow{d_1} A$$

where $d_n = d$. Thus $\mathrm{Tor}_n^\Lambda(A,K)$ as a right K-module, coincides with the homology module $H(A)$ for $n > 0$, and $A \otimes_\Lambda K = \mathrm{Coker}\, d = Z'(A)$.

Similarly, for each left Λ-module C, the complex $\mathrm{Hom}_\Lambda(X,C)$ simply is

$$C \xrightarrow{d^0} C \longrightarrow \cdots \longrightarrow C \xrightarrow{d^n} C \longrightarrow \cdots$$

where $d^n = d$. Thus $\mathrm{Ext}^n_\Lambda(K,C) = H(C)$ for $n > 0$, and $\mathrm{Hom}_\Lambda(K,C) = \mathrm{Ker}\, d = Z(C)$.

The exact sequences (6) and (6a) of § 1 can be seen to coincide with those of IV,1.1.

Another class of augmented rings is the class of *local rings*. A ring Λ is called a *local ring* if it satisfies the following condition:

(*LC*) *The elements of* Λ *which do not have a left inverse form a left ideal* I.

PROPOSITION 2.1. *I is a two-sided ideal and contains all proper left and right ideals of Λ. The elements of I have neither left nor right inverse, while the elements not in I have a two-sided inverse. The factor ring Λ/I is a (not necessarily commutative) field.*

PROOF. If J is a proper left ideal then no element of J has a left inverse. Thus $J \subset I$.

Next we show that no element of I has a right inverse. Indeed, suppose that $x\lambda = 1$ for some $x \in I$. Then $(1 - \lambda x)\lambda = 0$ and since $\lambda x \in I$ it follows that $1 - \lambda x$ is not in I so that $1 - \lambda x$ has a left inverse γ. Then $\lambda = \gamma(1 - \lambda x)\lambda = 0$, a contradiction.

For each $\lambda \in \Lambda$, $I\lambda$ is a left ideal and since $x\lambda \neq 1$ for $x \in I$, $I\lambda$ is a proper left ideal. Thus $I\lambda \subset I$ so that I is a right ideal.

Now consider λ not in I and let γ be a left inverse of λ. Then $\gamma\lambda = 1$

and since I is a right ideal it follows that γ is not in I. Thus γ has a left inverse ζ. Then $\zeta = \zeta\gamma\lambda = \lambda$. Thus $\lambda\gamma = \zeta\gamma = 1$, which shows that λ has a two sided inverse. Since I consists of all the elements which have no right inverses, it follows as above that I contains all proper right ideals.

The conclusion that Λ/I is a field follows from the facts established above.

Because of 2.1, there is no distinction between left and right local rings. The maximal (two sided) ideal I of a local ring Λ defines Λ as an augmented ring with $Q = \Lambda/I$ being a field. Important examples of local rings are the following two:

$K[[x_1, \dots, x_n]]$, the ring of formal power series in the letters $x_1, \dots, x_n$ with coefficients in the field K.

$K\{x_1, \dots, x_n\}$, the ring of convergent power series in the letters $x_1, \dots, x_n$ with coefficients in a commutative field K with a complete non-discrete valuation.

In both cases the ideal I is generated by the elements $x_1, \dots, x_n$.

Another important class of augmented rings is furnished by the *ring of a monoid.* A monoid Π is a multiplicative associative system with a unit element 1. Given a ring K we define the ring $K(\Pi)$ as the free K-module generated by the elements $x \in \Pi$, with multiplication defined by

$$(kx)(k'x') = (kk')(xx'), \qquad k,k' \in K, \qquad x,x' \in \Pi.$$

We observe that if Π is the *free monoid* generated by the elements $x_1, \dots, x_n$, then $K(\Pi)$ may be identified with the free K-ring $F_K(x_1, \dots, x_n)$. If Π is the *free abelian monoid* generated by $x_1, \dots, x_n$, then $K(\Pi)$ may be identified with the polynomial ring $K[x_1, \dots, x_n]$.

Given a ring $K(\Pi)$, there are many possibilities for defining an augmentation $\varepsilon: K(\Pi) \to K$. We shall only consider *multiplicative* augmentations which satisfy the following condition

$$\varepsilon(k1) = k.$$

Such an augmentation is determined by a function $\mu: \Pi \to K$. This function must satisfy the conditions

$$\mu(xx') = \mu(x)\mu(x'), \qquad \mu(1) = 1, \tag{1}$$

and μ must take values *in the center* of K, because of the relation $kx = xk$ in the ring $K(\Pi)$. Conversely, given a function μ satisfying (1) and taking its values in the center of K, we define $\varepsilon: K(\Pi) \to K$ by

$$\varepsilon(kx) = k(\mu x);$$

ε is then a multiplicative augmentation, satisfying $\varepsilon(k1) = k$.

The introduction of the ring $K(\Pi)$ is motivated by the following remark. Let A be a right $K(\Pi)$-module. Then A is a right K-module and further each element $x \in \Pi$ determines a K-endomorphism of A given by $a \to ax$. These endomorphisms satisfy $a1 = a$, $(ax)x' = a(xx')$. Conversely each right K-module with K-endomorphisms $a \to ax$ for $x \in \Pi$ satisfying the conditions above may be regarded as a right $K(\Pi)$-module. Thus $K(\Pi)$ plays the role of an "enveloping ring" for the ring K and the elements $x \in \Pi$. The same applies to left $K(\Pi)$-modules. An analogous example of an "enveloping" ring was the ring of dual numbers $\Lambda = (K,d)$ mentioned earlier. There will be other examples later.

Since a good deal of space will be devoted later to monoids and groups, we shall not pursue the discussion here any further.

3. CHANGE OF RINGS

Given a fixed ring Λ with a left augmentation ε: $\Lambda \to Q$ we have considered the homology groups $\mathrm{Tor}_n^{\Lambda}(A,Q)$ and cohomology groups $\mathrm{Ext}_{\Lambda}^n(Q,C)$ as covariant functors in the variables A and C. We shall now show that in some sense these are also functors of Λ.

Consider two augmented rings Λ and Γ with augmentations

$$\varepsilon_{\Lambda}\colon \Lambda \to Q_{\Lambda}, \qquad \varepsilon_{\Gamma}\colon \Gamma \to Q_{\Gamma}$$

and augmentation ideals I_{Λ} and I_{Γ}. A map φ: $\Lambda \to \Gamma$ of augmented rings is a ring homomorphism such that $\varphi(I_{\Lambda}) \subset I_{\Gamma}$. By passage to quotients we obtain a mapping ψ: $Q_{\Lambda} \to Q_{\Gamma}$ such that the diagram

$$\begin{array}{ccc} \Lambda & \xrightarrow{\varepsilon_{\Lambda}} & Q_{\Lambda} \\ \downarrow{\scriptstyle\varphi} & & \downarrow{\scriptstyle\psi} \\ \Gamma & \xrightarrow[\varepsilon_{\Gamma}]{} & Q_{\Gamma} \end{array}$$

is commutative. It follows that $\psi(\lambda x) = (\varphi\lambda)(\psi x)$ for $\lambda \in \Lambda$, $x \in Q_{\Lambda}$. This shows that ψ is a Λ-homomorphism if we regard Q_{Γ} as a Λ-module by means of φ (see II,6 and VI, 4).

Let A be a right Γ-module and C a left Γ-module. Using φ we may regard A and C also as Λ-modules. We shall define homomorphisms

(1) $$F^{\varphi}\colon \mathrm{Tor}^{\Lambda}(A,Q_{\Lambda}) \to \mathrm{Tor}^{\Gamma}(A,Q_{\Gamma})$$

(1a) $$F_{\varphi}\colon \mathrm{Ext}_{\Gamma}(Q_{\Gamma},C) \to \mathrm{Ext}_{\Lambda}(Q_{\Lambda},C).$$

To this end let X_{Λ} be a Λ-projective resolution of Q_{Λ} and X_{Γ} a Γ-projective resolution of Q_{Γ}. Further let

$$g\colon {}_{(\varphi)}Q_{\Lambda} = \Gamma \otimes_{\Lambda} Q_{\Lambda} \to Q_{\Gamma}$$

be defined by $g(\gamma \otimes x) = \gamma(\psi x)$. By II,6.1, $\Gamma \otimes_\Lambda X_\Lambda = {}_{(\varphi)}X_\Lambda$ is a Γ-projective complex over $\Gamma \otimes_\Lambda Q_\Lambda$. Following V,1.1 there is a map

$$G\colon \Gamma \otimes_\Lambda X_\Lambda \to X_\Gamma$$

over g and this map G is unique up to a homotopy. This yields homomorphisms

$$(2) \qquad H(A \otimes_\Lambda X_\Lambda) = H(A \otimes_\Gamma (\Gamma \otimes_\Lambda X_\Lambda)) \to H(A \otimes_\Gamma X_\Gamma)$$

$$(2a) \quad H(\mathrm{Hom}_\Gamma (X_\Gamma,C)) \to H(\mathrm{Hom}_\Gamma (\Gamma \otimes_\Lambda X_\Lambda,C)) = H(\mathrm{Hom}_\Lambda (X_\Lambda,C)$$

which are the desired homomorphisms F^φ and F_φ.

THEOREM 3.1 (*Mapping theorem*). *In order that F^φ be an isomorphism for all right Γ-modules A it is necessary and sufficient that*

$$\text{(i)} \qquad g\colon \Gamma \otimes_\Lambda Q_\Lambda \approx Q_\Gamma$$

$$\text{(ii)} \qquad \mathrm{Tor}_n^\Lambda (\Gamma,Q_\Lambda) = 0 \qquad \textit{for } n > 0.$$

If these conditions are satisfied then F_φ also is an isomorphism for all left Γ-modules C; and, for any Λ-projective resolution X_Λ of Q_Λ, the complex $\Gamma \otimes_\Lambda X_\Lambda$ with the augmentation $\Gamma \otimes_\Lambda X_\Lambda \longrightarrow \Gamma \otimes_\Lambda Q_\Lambda \overset{g}{\approx} Q_\Gamma$ is a Γ-projective resolution of Q_Γ.

PROOF. Assume F^φ is an isomorphism. In particular, taking $A = \Gamma$, we obtain that $F^\varphi\colon \mathrm{Tor}^\Lambda (\Gamma,Q_\Lambda) \approx \mathrm{Tor}^\Gamma (\Gamma,Q_\Gamma)$. This yields precisely (i) and (ii).

Assume that (i) and (ii) hold and let X_Λ be a Λ-projective resolution of Q_Λ. Then $H_n(\Gamma \otimes_\Lambda X_\Lambda) = \mathrm{Tor}_n^\Lambda (\Gamma,Q_\Lambda) = 0$ for $n > 0$. Thus (i) and (ii), precisely express the fact that $\Gamma \otimes_\Lambda X_\Lambda$ (with the augmentation as above) is a Γ-projective resolution of Q_Γ. Taking $X_\Gamma = \Gamma \otimes_\Lambda X_\Lambda$ the map G may be taken to be the identity map. Then (2) and (2a) become isomorphisms.

The "Mapping theorem" will have many applications.

4. DIMENSION

Let Λ be a left augmented ring with $\varepsilon\colon \Lambda \to Q$ and $I = \mathrm{Ker}\, \varepsilon$. We shall be interested in the projective dimension of Q as a left Λ-module (see VI, 2). Clearly $\mathrm{l.dim}_\Lambda Q \leq \mathrm{l.gl.dim}\, \Lambda$.

PROPOSITION 4.1. *If Q is not projective, then* $1 + \mathrm{l.dim}_\Lambda I = \mathrm{l.dim}_\Lambda Q$.

This is an immediate consequence of VI, 2.3.

THEOREM 4.2. *Suppose that I is generated (as a left ideal) by elements $x_1, \ldots, x_n$ which commute with each other. Let $I_k (0 \leq k \leq n)$ denote the left ideal generated by $x_i (i \leq k)$. If*

$$(i) \qquad (\lambda \in \Lambda \text{ and } \lambda x_k \in I_{k-1}) \Rightarrow (\lambda \in I_{k-1}), \quad k = 1, \ldots, n,$$

then $\mathrm{l.dim.}_\Lambda Q = n$, *provided* $Q \neq 0$.

Before we proceed with the proof, we list the three most important examples to which 4.2 applies. These are:

$\Lambda = K[x_1, \dots, x_n]$, the graded ring of polynomials in the letters $x_1, \dots, x_n$ with coefficients in the (not necessarily commutative) ring K.

$\Lambda = K[[x_1, \dots, x_n]]$, the ring of formal power series in the letters $x_1, \dots, x_n$ with coefficients in the ring K. If K is a (not necessarily commutative) field, then Λ is a local ring.

$\Lambda = K\{x_1, \dots, x_n\}$, the ring of convergent power series in the letters $x_1, \dots, x_n$ with coefficients in the commutative field K with a complete non-discrete valuation. This is also a local ring.

In all three cases the augmentation ideal I is the two-sided ideal I generated by $x_1, \dots, x_n$ and $\Lambda/I = K$. Condition (i) of 4.2 is verified. Therefore, by 4.2, $\text{l.dim}_\Lambda K = n$. (For the case of $\Lambda = K[[x_1, \dots, x_n]]$, the fact that the dimension of K is finite has been proved by F. Recillas (unpublished).)

PROOF of 4.2. Since $x_1, \dots, x_n$ commute, we may regard Λ as a (right) module over the ring $\Gamma = Z[x_1, \dots, x_n]$. Let $J_k (0 \leqq k \leqq n)$ be the ideal of Γ generated by $x_i (i \leqq k)$. Then $I_k = \Lambda J_k$.

More generally, for each Γ-module M we shall define a left complex over the module M/MJ_n (in the sense of v,1). Consider the exterior algebra $E(y_1, \dots, y_k)$ on n letters $y_1, \dots, y_n$ with integral coefficients, as defined in §2. The tensor product (over Z)

$$X = M \otimes E(y_1, \dots, y_n)$$

is graded by the modules

$$X_i = M \otimes E_i(y_1, \dots, y_n),$$

where $E_i(y_1, \dots, y_n)$ denotes the group of elements of degree i in $E(y_1, \dots, y_n)$. We define an augmentation $\varepsilon\colon X_0 \to M/MJ_n$ as the natural map of $X_0 = M$ onto M/MJ_n. We define a differentiation $d_i\colon X_i \to X_{i-1} \, (i > 0)$ by

$$d_i(m \otimes y_{p_1} \cdots y_{p_i}) = \sum_{1 \leq j \leq i} (-1)^{j+1} (m x_{p_j}) \otimes y_{p_1} \cdots \hat{y}_{p_j} \cdots y_{p_i}$$

where $\hat{y}_{p_j}$ indicates that y_{p_j} is to be omitted. Using the fact that $x_1, \dots, x_n$ commute with each other, it is easy to verify that $d_{i-1} d_i = 0$ for $i > 1$ and $\varepsilon d_1 = 0$.

Before proving 4.2 we establish

PROPOSITION 4.3. *If the Γ-module M satisfies*

(i′) $\qquad (m \in M, m x_k \in MJ_{k-1}) \Rightarrow (m \in MJ_{k-1}), \qquad k = 1, \dots, n,$

then the complex X is acyclic.

PROOF of 4.3. Condition (i') expresses the fact that the mapping $M/MJ_{k-1} \to MJ_k/MJ_{k-1}$ induced by $m \to mx_k$ is an isomorphism.

We introduce the left complexes $X^{(k)} = M \otimes E(y_1, \ldots, y_k)$ over MJ_k with differentiation and augmentation defined as above. We want to show by induction that $X^{(k)}$ is acyclic. For $k = 0$ this is clear because $J_0 = 0$ and $X^{(0)} = M$. Suppose now that we already know that $X^{(k-1)}$ is acyclic over MJ_{k-1} $(k > 0)$. Consider the complex

$$Y^{(k)}: \cdots \to 0 \to X_k^{(k)} \to \cdots \to X_1^{(k)} \to MJ_k \to 0.$$

Since the augmentation $X_0^{(k)} = M \to M/MJ_k$ is an epimorphism with MJ_k as kernel, the acyclicity of the complex $X^{(k)}$ is equivalent to $H(Y^{(k)}) = 0$. Since $Y^{(k-1)}$ may be regarded as a subcomplex of $Y^{(k)}$ and since $H(Y^{(k-1)}) = 0$, it follows from the exactness of

$$H(Y^{(k-1)}) \to H(Y^{(k)}) \to H(Y^{(k)}/Y^{(k-1)})$$

that it suffices to show that $H(Y^{(k)}/Y^{(k-1)}) = 0$. To this end we consider the diagram

$$\begin{array}{ccccccccc} 0 \to X_{k-1}^{(k-1)} \to & X_{k-2}^{(k-1)} & \to \cdots \to & X_0^{(k-1)} & \to M/MJ_{k-1} & \to 0 \\ \downarrow & \downarrow & & \downarrow & \downarrow & \\ 0 \to X_k^{(k)} & \to X_{k-1}^{(k)}/X_{k-1}^{(k-1)} & \to \cdots \to & X_1^{(k)}/X_1^{(k-1)} & \to MJ_k/MJ_{k-1} & \to 0 \end{array}$$

where $X_{i-1}^{(k-1)} \to X_i^{(k)}/X_i^{(k-1)}$ is defined by right multiplication by y_k. The mapping $M/MJ_{k-1} \to MJ_k/MJ_{k-1}$ is the isomorphism induced by $m \to mx_k$. This diagram is commutative; the vertical maps are all isomorphisms, and the upper row is exact by the inductive assumption; it follows that the lower row is exact, i.e. that $H(Y^{(k)}/Y^{(k-1)}) = 0$.

We now return to the proof of 4.2. If we replace M by the ring Λ of 4.2, we find that X is a Λ-complex. Since the products $y_{p_1} \cdots y_{p_i}$ $(p_1 < \cdots < p_i)$ form a base for the free abelian group $E_i(y_1, \ldots, y_n)$, it follows that X is Λ-free and has dimension n. Thus X is a projective resolution of Q of dimension n, so that $\text{l.dim}_\Lambda Q \leqq n$.

The preceding resolution X of $Q = \Lambda/I$ can be used to express the homology and cohomology modules of the augmented ring Λ with any coefficient module. For a right Λ-module A, we have

(1) $$\text{Tor}_i^\Lambda (A,Q) = H_i(A \otimes E(y_1, \ldots, y_n)).$$

Indeed $\text{Tor}_i^\Lambda (A,Q)$ are the homology modules of the complex

$$A \otimes_\Lambda X = A \otimes_\Lambda (\Lambda \otimes E(y_1, \ldots, y_n)) = A \otimes E(y_1, \ldots, y_n),$$

with the differentiation

(2) $$d_i(a \otimes y_{p_1} \cdots y_{p_i}) = \sum_{1 \leq j \leq i} (-1)^{j+1}(ax_{p_j}) \otimes y_{p_1} \cdots \hat{y}_{p_j} \cdots y_{p_i}.$$

Similarly for cohomology, we have for any left Λ-module C:

$$\operatorname{Ext}^i_\Lambda(Q,C) = H^i(\operatorname{Hom}(E(y_1,\dots,y_n),C)).$$

Indeed

$$\operatorname{Hom}_\Lambda(X,C) = \operatorname{Hom}_\Lambda(\Lambda \otimes E(y_1,\dots,y_n),C) = \operatorname{Hom}(E(y_1,\dots,y_n),C).$$

An element of degree i in $\operatorname{Hom}(E(y_1,\dots,y_n),C)$ may be identified with a function $f(p_1,\dots,p_i)$ defined for integers satisfying $1 \leqq p_1 < \cdots < p_i \leqq n$; the differentiation is then given by the formula

$$\text{(2a)} \qquad (\delta f)(p_1,\dots,p_{i+1}) = \sum_{1\leqq j \leqq i+1} x_{p_j} f(p_1,\dots,\hat{p}_j,\dots,p_{i+1}).$$

We observe now that in the above method of computation the ring Λ has almost completely disappeared; the only thing that needs to be known is how the elements $x_1,\dots,x_n$ operate on A (resp. C) and that these endomorphisms commute with each other.

In particular, we find

$$\begin{aligned}\operatorname{Ext}^n_\Lambda(Q,Q) = H^n(\operatorname{Hom}(E(y_1,\dots,y_n),Q) &= \operatorname{Hom}(E_n(y_1,\dots,y_n),Q)\\ &= \operatorname{Hom}(Z,Q) = Q.\end{aligned}$$

Thus if $Q \neq 0$ then $\operatorname{l.dim}_\Lambda Q = n$. This completes the proof of 4.2.

The complex $A \otimes E(y_1,\dots,y_n)$ was first found by J. L. Koszul (*Colloque de topologie*, Bruxelles, 1950), in connection with cohomology theory of Lie groups.

REMARK 1. If we apply (1) to calculate $\operatorname{Tor}^\Gamma_i(M,Z)$ with M as in 4.3, the acyclicity of the complex $M \otimes E(y_1,\dots,y_n)$ is equivalent to

$$\text{(3)} \qquad \operatorname{Tor}^\Gamma_i(M,Z) = 0, \qquad i > 0.$$

The fact that hypothesis (i') for M implies (3) can be established directly: indeed we can show by induction on k that

$$\operatorname{Tor}^\Gamma_i(M, \Gamma/J_k) = 0,$$

using exact sequences $0 \longrightarrow \Gamma/J_{k-1} \xrightarrow{x_k} \Gamma/J_{k-1} \longrightarrow \Gamma/J_k \longrightarrow 0$.

REMARK 2. Consider the algebra $\Lambda = K[x_\alpha]$ of polynomials in an arbitrary set $\{x_\alpha\}$ of variables. We may regard Λ as the direct limit of algebras $\Lambda_J = K(J)$ where J runs through the finite subsets of the collection $\{x_\alpha\}$. The complexes X_J constructed above form then a direct system with a complex X as limit. It is immediately clear that X is a Λ-projective resolution of K and that X is the tensor product $\Lambda \otimes E(y_\alpha)$, where $E(y_\alpha)$ is the exterior algebra on the letters $\{y_\alpha\}$. The differentiation in X is given by the same formulae as above.

5. FAITHFUL SYSTEMS

We shall assume here that Λ is an augmented ring with a proper two sided augmentation ideal I. Then $Q = \Lambda/I$ may be regarded as a ring and $\varepsilon\colon \Lambda \to Q$ is a ring homomorphism.

Let A be a right Λ-module. Then $A \otimes_\Lambda Q$ is a right Q-module. From the exact sequence $0 \to I \to \Lambda \to Q \to 0$ we deduce

$$(1) \qquad A \otimes_\Lambda Q \approx A/AI.$$

A right Λ-module A will be called *proper* if either $A = 0$ or $A \otimes_\Lambda Q \neq 0$ (i.e. $A \neq AI$). A free module clearly is proper.

Given a subset N of a module A we consider the free module F_N generated by the elements of N. We have a natural homomorphism $F_N \to A$ which leads to the exact sequence

$$(2) \qquad 0 \to R_N \to F_N \to A \to L_N \to 0.$$

with $R_N = \operatorname{Ker}(F_N \to A)$, $L_N = \operatorname{Coker}(F_N \to A)$.

DEFINITION. A subset M of a right Λ-module A is called *faithful* if for each $N \subset M$ the modules R_N and L_N are proper. A family $\mathscr{D}$ of right Λ-modules is called *allowable* if for each $A \in \mathscr{D}$ there is a faithful set M generating A and such that in the exact sequence $0 \to R_M \to F_M \to A \to 0$ we have $R_M \in \mathscr{D}$.

If A has a faithful subset M, then taking $N = 0$ we have $L_N = A$ and therefore A is proper. Thus all the modules in an allowable family are proper.

To illustrate the notions just introduced we consider two important special cases.

First we take up the case when Λ is a graded ring (see § 2), $Q = \Lambda_0$, and I is the ideal of elements with a vanishing component of degree zero. A right Λ-module A is *graded* if a grading

$$A = A^0 + A^1 + \cdots + A^n + \cdots$$

of A as an abelian group is given, with $A^p\Lambda^q \subset A^{p+q}$. Right Λ-modules for which such a grading exists will be called *gradable*. Clearly a free Λ-module is gradable.

PROPOSITION 5.1. *Let Λ be a graded ring. Every graded Λ-module A is proper and every set of homogeneous elements is faithful. The family of all gradable Λ-modules is allowable.*

PROOF. Let A be a graded Λ-module with $A \neq 0$ and let $a \in A^m$ be a non-zero, homogeneous element of lowest possible degree. If $A = AI$ then $a \in AI$, i.e. $a = \sum a_i\lambda_i$ with homogeneous elements $a_i \in A$ and $\lambda_i \in I$.

Since each a_i has degree at least m and each λ_i has degree at least 1, it follows that a has degree at least $m + 1$. This contradiction shows that A is proper.

If N is any set of homogeneous elements of A then the module F_N may be graded by the requirement that the map $F_N \to A$ be homogeneous of degree zero. Then R_N and L_N are graded modules and therefore proper. It follows that any subset of A composed of homogeneous elements is faithful. In particular, since each graded module is generated by its homogeneous elements it follows that the family of gradable modules is allowable.

As a second illustration, we take up the case when Λ is a local ring and I its maximal ideal.

PROPOSITION 5.1′. *Let Λ be a local ring. Every finitely generated right Λ-module is proper. If Λ is right Noetherian then every finite subset of a finitely generated right Λ-module is faithful and the class of all finitely generated right Λ-modules is allowable.*

PROOF. Let $A \neq 0$ be a finitely generated right Λ-module and let $(a_1, \ldots, a_n)$ be a minimal system of generators of A. Let B be the submodule generated by $a_2, \ldots, a_n$. If $a_1 \in AI$ then $a_1 = a_1\lambda_1 + \cdots + a_n\lambda_n$ for some $\lambda_1, \ldots, \lambda_n \in I$. Thus $a_1(1 - \lambda_1) \in B$. Since $1 - \lambda_1$ is not in I we may choose $\lambda \in \Lambda$ with $(1 - \lambda_1)\lambda = 1$. Then $a_1 = a_1(1 - \lambda_1)\lambda \in B$. This contradiction shows that a_1 is not in AI. Thus $AI \neq A$ and A is proper.

If A is finitely generated and N is a finite subset of A, then F_N and L_N are finitely generated. If further Λ is right Noetherian then R_N also is finitely generated. Thus every finite subset of A is faithful. This proves the second half of the proposition.

With the notions of "proper," "faithful," and "allowable" thus illustrated, we return to the abstract treatment.

PROPOSITION 5.2. *Let A be a right Λ-module and M a faithful subset of A. If the image of M in $A \otimes_\Lambda Q = A/AI$ generates $A \otimes_\Lambda Q$ as a right Q-module, then M generates A. If further* $\operatorname{Tor}_1^\Lambda(A,Q) = 0$ *and the images of the elements of M in $A \otimes_\Lambda Q$ form a Q-base for $A \otimes_\Lambda Q$ then M is a Λ-base for A.*

PROOF. We consider the exact sequence

$$0 \to R_M \to F_M \to A \to L_M \to 0$$

By assumption the map $F_M \otimes_\Lambda Q \to A \otimes_\Lambda Q$ is an epimorphism. Therefore, by the right exactness of the tensor product, $L_M \otimes_\Lambda Q = 0$. Since L_M is proper, it follows that $L_M = 0$, i.e. $F_M \to A$ is an epimorphism. Thus M generates A.

Since $L_M = 0$, we obtain an exact sequence

$$\mathrm{Tor}_1^\Lambda(A,Q) \longrightarrow R_M \otimes_\Lambda Q \longrightarrow F_M \otimes_\Lambda Q \xrightarrow{\varphi} A \otimes_\Lambda Q.$$

The condition that the images of the elements of M form a Q-base of $A \otimes_\Lambda Q$ means precisely that φ is an isomorphism. Since $\mathrm{Tor}_1^\Lambda(A,Q)=0$, it follows that $R_M \otimes_\Lambda Q = 0$. However R_M is proper so that $R_M = 0$. This implies that $F_M \to A$ is an isomorphism, i.e. M is a Λ-base for A.

THEOREM 5.3. *Assume that Q is a (not necessarily commutative) field, and that* $\mathrm{Tor}_1^\Lambda(A,Q) = 0$. *Then every faithful subset Λ-generating A contains a Λ-base for A. In particular, if A is generated by a faithful set, then A is Λ-free.*

PROOF. Let M be a faithful subset generating A. Then the image of M in $A \otimes_\Lambda Q$ generates $A \otimes_\Lambda Q$ as a right Q-module. Since Q is a field M contains a subset N such that the images of the elements of N in $A \otimes_\Lambda Q$ are a Q-base for $A \otimes_\Lambda Q$. Since N also is faithful, it follows from 5.2 that N is a Λ-base for A.

THEOREM 5.4. *Assume that $Q = \Lambda/I$ is a field, and let $\mathscr{D}$ be an allowable family of right Λ-modules. Then for every module A of $\mathscr{D}$*

$$\mathrm{r.dim}_\Lambda A \leqq \mathrm{l.dim}_\Lambda Q.$$

PROOF. Let M be a faithful subset generating A. Then $0 \to R_M \to F_M \to A \to 0$ is exact. Since R_M is again in $\mathscr{D}$, this process may be repeated with A replaced by R_M. Thus by iteration we obtain an exact sequence

$$0 \to X_n \to X_{n-1} \to \cdots \to X_0 \to A \to 0$$

of right Λ-modules, with $X_0, \ldots, X_{n-1}$ Λ-free and with $X_n \in \mathscr{D}$. It then follows from v,7.2 that the iterated connecting homomorphism yields an isomorphism

$$\mathrm{Tor}_1^\Lambda(X_n,Q) \approx \mathrm{Tor}_{n+1}^\Lambda(A,Q).$$

Assume now that $\dim Q = n$. Then $\mathrm{Tor}_{n+1}^\Lambda(A,Q) = 0$ so that $\mathrm{Tor}_1^\Lambda(X_n,Q) = 0$. Since $X_n \in \mathscr{D}$, X_n is generated by a faithful subset and therefore, by 5.3, X_n is Λ-free. Thus $\dim A \leqq n$.

6. APPLICATIONS TO GRADED AND LOCAL RINGS

THEOREM 6.1. *Let Λ be a graded ring with $\Lambda_0 = Q$ a field. If A is a graded right Λ-module with* $\mathrm{Tor}_1^\Lambda(A,Q) = 0$ *then A is free; every homogeneous system of generators of such a module A contains a base for A.*

This follows from 5.1 and 5.3.

THEOREM 6.2. *Let Λ be a graded ring with $\Lambda_0 = Q$ a field. For each graded right Λ-module A,*

$$\text{r.dim}_\Lambda A \leqq \text{l.dim}_\Lambda Q.$$

This follows from 5.1 and 5.4.

COROLLARY 6.3. *Under the assumptions of* 6.2, *let J be a homogeneous right ideal in Λ. Then*

$$1 + \text{r.dim}_\Lambda J \leqq \text{l.dim}_\Lambda Q$$

unless $\text{l.dim}_\Lambda Q = 0$, *in which case* $\text{r.dim}_\Lambda J \leqq 0$.

Indeed, consider the graded module Λ/J. Then by VI,2.3 we have $1 + \dim J = \dim \Lambda/J$, unless $\dim \Lambda/J \leqq 0$ in which case $\dim J \leqq 0$. Since $\dim \Lambda/J \leqq \dim Q$, the conclusion follows.

Note that Q itself is a graded right Λ-module. Therefore

COROLLARY 6.4. $\text{l.dim}_\Lambda Q = \text{r.dim}_\Lambda Q$. *The conclusions of* 6.2 *and* 6.3 *apply equally well to left Λ-modules and left ideals.*

Now consider the particular case when $\Lambda = K[x_1, \ldots, x_n]$ is the ring of polynomials in the letters $x_1, \ldots, x_n$ with coefficients in the (not necessarily commutative) field K. Then $Q = K$ and by 4.2, $\dim_\Lambda Q = n$. We thus obtain

THEOREM 6.5. *Let K be a field and let $\Lambda = K[x_1, \ldots, x_n]$, $n \geqq 1$. For each graded* (*right or left*) *Λ-module A we have*

$$\dim_\Lambda A \leqq n.$$

For each homogeneous (*right or left*) *ideal J in Λ we have*

$$\dim_\Lambda J \leqq n - 1.$$

This theorem contains Hilbert's theorem on "chains of syzygies" (see W. Gröbner, *Monatshefte für Mathematik* 53 (1949), 1–16). The method used here is an extension of the one indicated by J. L. Koszul (*Colloque de topologie*, Bruxelles, 1950). We shall see later (IX,7.11) that if K is a commutative semi-simple ring, then $\text{gl.dim}\, \Lambda = n$.

We now pass to the case of a local ring Λ with a maximal ideal I and with $Q = \Lambda/I$. We know then by 2.1 that Q is a field.

THEOREM 6.1′. *Let Λ be a right Noetherian local ring with maximal ideal I and with $Q = \Lambda/I$. Every finitely generated right Λ-module A such that* $\text{Tor}_1^\Lambda (A,Q) = 0$ *is then free; every finite set of generators of such a module A contains a base.*

This follows from 5.1′ and 5.3.

THEOREM 6.2′. *Let Λ be a right Noetherian local ring with maximal ideal I and with $Q = \Lambda/I$. For each finitely generated right Λ-module A*

$$\text{r.dim}_\Lambda A \leqq \text{l.dim}_\Lambda Q.$$

This follows from 5.1′ and 5.4.

COROLLARY 6.3′. *Under the assumptions of* 6.2′ *let J be any right ideal in Λ. Then*

$$1 + \text{r.dim}_\Lambda J \leqq \text{l.dim}_\Lambda Q,$$

unless $\text{l.dim}_\Lambda Q = 0$ *in which case* $\text{r.dim}_\Lambda J \leqq 0$.

The proof is the same as for 6.3.

COROLLARY 6.4′. *Under the conditions of* 6.2′ *assume that Λ is also left Noetherian. Then* $\text{l.dim}_\Lambda Q = \text{r.dim}_\Lambda Q$. *The conclusions of* 6.2′ *and* 6.3′ *apply equally well to left as to right Λ-modules and left ideals.*

We now consider two particular rings; the ring $\Lambda = K[[x_1, \ldots, x_n]]$ of formal power series in $x_1, \ldots, x_n$ with coefficients in the field K, and the ring $\Lambda = K\{x_1, \ldots, x_n\}$ of convergent power series in $x_1, \ldots, x_n$ with coefficients in a commutative field K with a complete non-discrete valuation. Then Λ is a local ring with $Q = K$ and, by 4.2, $\dim_\Lambda Q = n$.

Further Λ is (both left and right) Noetherian; for $K[[x_1, \ldots, x_n]]$ see W. Krull (*Crelle* 179 (1938), p. 204–226); for $K\{x_1, \ldots, x_n\}$ see Bochner and Martin (*Several Complex Variables*, Princeton, 1948; Ch. x, th. 1). We thus obtain

THEOREM 6.5′. *Let K be a field and let* $\Lambda = K[[x_1, \ldots, x_n]]$, $n \geqq 1$, *or let K be a commutative field with a complete non-discrete valuation and let* $\Lambda = K\{x_1, \ldots, x_n\}$, $n \geqq 1$. *For each finitely generated* (*right or left*) *Λ-module A we have*

$$\dim_\Lambda A \leqq n.$$

For each (*right or left*) *ideal J in Λ we have*

$$\dim_\Lambda J \leqq n - 1.$$

EXERCISES

1. In the situation treated in § 3 establish the commutativity of the following diagrams

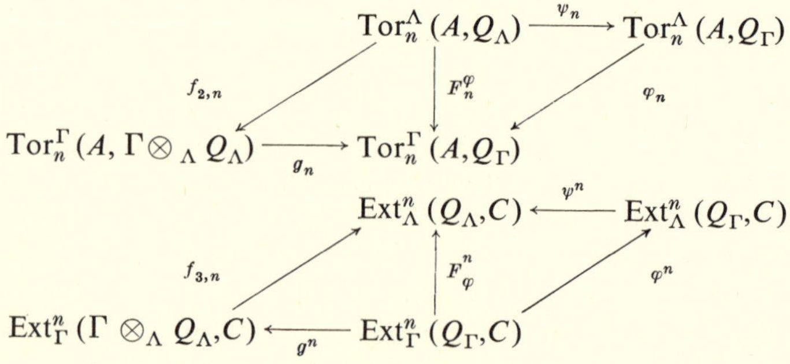

where g_n and g^n are induced by g, ψ_n and ψ^n are induced by ψ, and the maps $f_{2,n}, f_{3,n}, \varphi_n$ and φ^n are defined in VI,4.

2. Show that under the conditions of 5.4 and using the notion of weak dimension of VI, Exer. 3, the conclusion of 5.4 may be strengthened as follows

$$\text{w.dim } A = \dim A \leqq \text{w.dim } Q$$

for each $A \in \mathscr{D}$. Apply this result to the considerations of § 6.

3. Let Λ be an augmented ring with a two sided augmentation ideal I such that $\bigcap_p I^p = 0$ where I^p is defined by recursion as $I^p = I^{p-1}I$, $I^1 = I$. Show that each submodule of a free Λ-module is proper.

4. Let Λ be a local ring with maximal ideal I such that $\bigcap_p I^p = 0$, and let $Q = \Lambda/I$. Show that if A is a finitely generated right (resp. left) Λ-module such that $\mathrm{Tor}_1^\Lambda(A,Q) = 0$ (resp. $\mathrm{Tor}_1^\Lambda(Q,A) = 0$), then A is free; each system of generators of such a module A contains a base.

5. Consider the example of the ring Λ given at the end of I,7, and show that it can be put in the form

$$\Lambda = Z[x] + Z[x]$$

with multiplication given for $a,a',b,b' \in Z[x]$,

$$(a,b)(a',b') = (aa', ab' + (\varepsilon a')b).$$

Replace $Z[x]$ by a commutative ring Γ with a ring endomorphism $\varepsilon\colon \Gamma \to \Gamma$. Prove that the result is a ring Λ with (1,0) as unit element. Show that in Λ an element (a,b) has a right (or left) inverse if and only if a has an inverse in Γ. If Γ is a (commutative) local ring then Λ also is a local ring with the same field of augmentation as Γ.

Taking $\Gamma = K[[x]]$ where K is a commutative field and ε is the augmentation of Γ, prove that the local ring Λ is left Noetherian without being right Noetherian.

6. Let $\Lambda = K[[x]]$ be the ring of formal power series in one variable x with coefficients in a commutative ring K. Let $\varepsilon\colon \Lambda \to K$ assign to each series its "constant term." Then an element of Λ has an inverse if and only if its image under ε has an inverse in K. Hence, by recursion on n: the ring $K[[x_1, \ldots, x_n]]$ is a local ring if and only if K is a local ring.

7. Let $\Lambda = K[x_1, \ldots, x_n]$ be the ring of polynomials with coefficients in a commutative ring K. For any Λ-module A, establish natural isomorphisms

$$\mathrm{Tor}_q^\Lambda(K,A) \approx \mathrm{Ext}_\Lambda^{n-q}(K,A) \qquad 0 \leqq q \leqq n.$$

[Hint: let X be the complex $\Lambda \otimes E(y_1, \ldots, y_n)$ as defined in § 4. Define isomorphisms

$$X_q \to \mathrm{Hom}_\Lambda(X_{n-q}, \Lambda)$$

in order to induce an isomorphism of complexes

$$X \to \mathrm{Hom}_\Lambda(X,\Lambda)$$

raising the degrees by n. Observe that

$$\mathrm{Hom}_\Lambda(X,\Lambda) \otimes_\Lambda A \approx \mathrm{Hom}_\Lambda(X,A).]$$

8. Let A be a left Λ-module with Λ-endomorphisms x_k $(1 \leqq k \leqq n)$ which we shall write as $a \to ax_k$. Let I_k $(0 \leqq k \leqq n)$ denote the submodule of all elements of the type $a_1x_1 + \cdots + a_kx_k$ (in particular, $I_0 = 0$). Assume that

(ii) $\qquad (a \in A \text{ and } ax_k \in I_{k-1}) \Longleftrightarrow (a \in I_{k-1}), \qquad k = 1, \ldots, n.$

Show that

$$\mathrm{l.dim}_\Lambda A/I_k \leqq k + \mathrm{l.dim}_\Lambda A, \qquad k = 0, \ldots, n.$$

9. Let Λ be an augmented ring. Assume that the augmentation ideal I is such that $1 + x$ has a right inverse for any $x \in I$. Show that the conclusions of 5.1′ remain valid.

10. In the ring $K[[x]]$, where K is a (not necessarily commutative) field, consider the subring Λ consisting of all power series without terms of degree 1. Show that Λ is a local ring.

Let A be the right Λ-module consisting of all series without a constant term. Show that A is not Λ-free and thus is not Λ-projective. Establish the exact sequence

$$0 \longrightarrow A \xrightarrow{\psi} \Lambda + \Lambda \xrightarrow{\varphi} A \longrightarrow 0$$

with $\varphi(\lambda_1,\lambda_2) = \lambda_1x^2 - \lambda_2x^3$, $\psi\lambda = (\lambda x,\lambda)$. As an application derive that $\mathrm{r.dim}_\Lambda A = \infty$.

11. In the ring $K[[x,y]]$, where K is a (not necessarily commutative) field, consider the subring Λ consisting of all power series in which all terms are of even total degree. Show that Λ is a local ring.

Let A be the right Λ-module consisting of all series in $K[[x,y]]$ in which all the terms have odd degree. Show that A is not Λ-free and thus is not Λ-projective. Establish the exact sequence

$$0 \longrightarrow A \xrightarrow{\psi} \Lambda + \Lambda \xrightarrow{\varphi} A \longrightarrow 0$$

with $\varphi(\lambda_1,\lambda_2) = \lambda_1y - \lambda_2x$, $\psi\lambda = (\lambda x,\lambda y)$. As an application derive that $\mathrm{r.dim}_\Lambda A = \infty$.

12. In the ring $K[[x]] + K[[y]]$, where K is a (not necessarily commutative) field, consider the subring consisting of all pairs $(f(x),g(y))$ with $f(0) = g(0)$. Show that Λ is a local ring.

Let A and B be the right modules consisting respectively of all pairs $(f(x)x,0)$ and $(0,g(y)y)$. Show that A and B are not Λ-free and thus are not Λ-projective. Establish the exact sequences

$$0 \longrightarrow A \xrightarrow{\psi} \Lambda \xrightarrow{\varphi} B \longrightarrow 0, \quad 0 \longrightarrow B \xrightarrow{\psi'} \Lambda \xrightarrow{\varphi'} A \longrightarrow 0,$$

where ψ and ψ' are inclusions while $\varphi(f(x),g(y)) = (0,g(y)y)$ and $\varphi'(f(x),g(y)) = (f(x)x,0)$. As an application derive that $\text{r.dim}_\Lambda A = \text{r.dim}_\Lambda B = \infty$.

CHAPTER IX

Associative Algebras

Introduction. The homology (and cohomology) theory of an associative K-algebra Λ is that of an augmented ring, namely the ring $\Lambda^e = \Lambda \otimes_K \Lambda^*$ (where Λ^* is the "opposite" algebra of Λ) with the augmentation $\rho\colon \Lambda^e \to \Lambda$, $\rho(\lambda \otimes \mu^*) = \lambda\mu$. The homology groups $H_n(\Lambda,A)$ and the cohomology groups $H^n(\Lambda,A)$ are then defined for any two sided Λ-module A (§ 4). If the algebra Λ is K-projective, the homology and cohomology groups may be calculated using a "standard complex" (§ 6); we thus retrieve the initial definition of Hochschild (*Ann. of Math.* 46 (1945), 58–67).

The last section (§ 7) is devoted to the study of "dimension" of algebras from the homological point of view. The method utilized leads to new connections with the theory of algebras, and deserves further study.

1. ALGEBRAS AND THEIR TENSOR PRODUCTS

Let K be a *commutative* ring (with a unit element denoted by 1). A *K-algebra* is a ring Λ (with a unit element also denoted by 1) which is also a K-module such that

$$(k_1\lambda_1)(k_2\lambda_2) = (k_1k_2)(\lambda_1\lambda_2)$$

for $k_1,k_2 \in K$, $\lambda_1,\lambda_2 \in \Lambda$. Setting $\eta k = k1$ yields a ring homomorphism $\eta\colon K \to \Lambda$ whose image is in the center of Λ. Clearly K itself is a K-algebra.

Clearly every ring Λ may be regarded as a Z-algebra, where Z is the ring of integers, with $n\lambda$, $n \in Z$, $\lambda \in \Lambda$ defined in the obvious way.

Let Λ and Γ be K-algebras. A K-algebra homomorphism $\Lambda \to \Gamma$ is a ring homomorphism which also is a K-homomorphism.

Given two K-algebras Λ and Γ, the tensor product $\Lambda \otimes_K \Gamma$ is a K-module, and the multiplication

$$(\lambda_1 \otimes \gamma_1)(\lambda_2 \otimes \gamma_2) = (\lambda_1\lambda_2) \otimes (\gamma_1\gamma_2)$$

converts $\Lambda \otimes_K \Gamma$ into a K-algebra. As long as the ring K is fixed we shall frequently write $\Lambda \otimes \Gamma$ for $\Lambda \otimes_K \Gamma$. We have the natural K-algebra homomorphisms

$$\Lambda \to \Lambda \otimes \Gamma, \qquad \Gamma \to \Lambda \otimes \Gamma$$

given by $\lambda \to \lambda \otimes 1$, $\gamma \to 1 \otimes \gamma$.

The tensor product of K-algebras has the usual associative property of the tensor product over a commutative ring.

If Λ is a K-algebra, every (left or right) Λ-module A may then also be regarded as a K-module. We shall frequently have to consider the situation where A is also a Γ-module for another K-algebra Γ. In this case we shall always assume that (1°) the operators of Λ and Γ on A commute; (2°) the structure of A as a K-module induced by Λ is the same as that induced by Γ. In particular, suppose A is a left Λ-module and a left Γ-module (situation ${}_{\Lambda-\Gamma}A$). Then setting

$$(\lambda \otimes \gamma)a = \lambda(\gamma a) = \gamma(\lambda a)$$

converts A into a left $\Lambda \otimes_K \Gamma$-module. The converse is also clear: every left $\Lambda \otimes_K \Gamma$-module is obtained in this way from a unique Λ-Γ-module. The same applies with "left" replaced by "right."

Suppose now that A is a left Λ-module and a right Γ-module (situation ${}_\Lambda A_\Gamma$). We first use the opposite algebra Γ^* (see VI,1) to convert A into a left Γ^*-module: $\gamma^* a = a\gamma$, and then, using the definition above, A becomes a left $\Lambda \otimes \Gamma^*$-module with

$$(\lambda \otimes \gamma^*)a = (\lambda a)\gamma = \lambda(a\gamma).$$

Similarly A may be regarded as a right $\Lambda^* \otimes \Gamma$-module.

These considerations generalize in an obvious way to a module A over any finite number of K-algebras.

The concept of the tensor product discussed above, can be applied to define "the extension of the ring of operators" in an augmented ring. Let Λ be an augmented ring with augmentation $\varepsilon\colon \Lambda \to Q$. Assume that Λ is a K-algebra and ε is a K-homomorphism and let L be another K-algebra. Then clearly

$$L \otimes \varepsilon\colon L \otimes_K \Lambda \to L \otimes_K Q$$

is an epimorphism which defines an augmentation for the ring $L \otimes_K \Lambda$. We obtain a commutative diagram

$$\begin{array}{ccc} \Lambda & \xrightarrow{\varepsilon} & Q \\ \downarrow{\scriptstyle\varphi} & & \downarrow{\scriptstyle\psi} \\ L \otimes \Lambda & \xrightarrow[L \otimes \varepsilon]{} & L \otimes Q \end{array}$$

where $\varphi\lambda = 1 \otimes \lambda$, $\psi q = 1 \otimes q$. Thus, by VIII,3, we obtain homomorphisms

$$F^{\varphi}\colon \operatorname{Tor}^{\Lambda}(A,Q) \to \operatorname{Tor}^{L\otimes\Lambda}(A, L \otimes Q), \qquad A_{L-\Lambda}$$

$$F_{\varphi}\colon \operatorname{Ext}_{L\otimes\Lambda}(L \otimes Q, C) \to \operatorname{Ext}_{\Lambda}(Q,C), \qquad {}_{\Lambda-\Gamma}C.$$

PROPOSITION 1.1. *If the algebra* Λ *is K-projective and if* $\operatorname{Tor}_n^K(L,Q)=0$ *for* $n > 0$, *then* F^{φ} *and* F_{φ} *are isomorphisms. If, furthermore, X is a* Λ*-projective resolution of Q, then* $L \otimes_K X$ *is a* $L \otimes_K \Lambda$*-projective resolution of* $L \otimes_K Q$.

PROOF. The result will follow from the "mapping theorem" VIII,3.1 provided we verify its hypotheses which in this case become:

(i) $(L \otimes_K \Lambda) \otimes_{\Lambda} Q \to L \otimes_K Q$ is an isomorphism;

(ii) $\operatorname{Tor}_n^{\Lambda}(L \otimes_K \Lambda, Q) = 0$ for $n > 0$.

Condition (i) follows directly from associativity. To verify (ii) we consider a Λ-projective resolution X of Q. Then

$$\operatorname{Tor}_n^{\Lambda}(L \otimes_K \Lambda, Q) = H_n(L \otimes_K \Lambda \otimes_{\Lambda} X) = H_n(L \otimes_K X).$$

Since Λ is K-projective, it follows from II,6.2 that X is also a K-projective resolution of Q. Thus $H_n(L \otimes_K X) = \operatorname{Tor}_n^K(L,Q)$ which was assumed to be zero for $n > 0$.

We shall see useful applications of 1.1.

The definition of the tensor product $\Lambda \otimes_K \Gamma$ of K-algebras has an important variant when Λ and Γ are graded K-algebras. A *graded K-algebra* Λ is a graded ring (VIII,2) which is also a K-algebra such that $K\Lambda^q \subset \Lambda^q$. If Γ is another graded K-algebra then $\Lambda \otimes_K \Gamma$ is a doubly graded K-module which is converted into a (singly) graded module in the usual fashion. To define the product $(\lambda_1 \otimes \gamma_1)(\lambda_2 \otimes \gamma_2)$, $\lambda_1 \in \Lambda^p$, $\lambda_2 \in \Lambda^m$, $\gamma_1 \in \Gamma^q$, $\gamma_2 \in \Gamma^n$ we denote by $f\colon \Lambda \to \Lambda$, $g\colon \Gamma \to \Gamma$ the endomorphisms given by left multiplication by λ_1 and γ_1 respectively. Clearly f and g have degrees p and q. Then $f \otimes g$ should be left multiplication by $\lambda_1 \otimes \gamma_1$. However, by IV,5

$$(f \otimes g)(\lambda_2 \otimes \gamma_2) = (-1)^{mq} f\lambda_2 \otimes g\gamma_2$$

and therefore we obtain the multiplication rule

$$(\lambda_1 \otimes \gamma_1)(\lambda_2 \otimes \gamma_2) = (-1)^{mq} \lambda_1\lambda_2 \otimes \gamma_1\gamma_2$$

where m is the degree of λ_2 and q is the degree of γ_1. With this multiplication $\Lambda \otimes_K \Gamma$ becomes a graded K-algebra, called the *skew* tensor product of the graded K-algebras Λ and Γ.

PROPOSITION 1.2. *Let* Λ *and* Γ *be graded K-algebras. The map* $\varphi\colon \Lambda \otimes_K \Gamma \to \Gamma \otimes_K \Lambda$ *defined by setting*

$$\varphi(\lambda \otimes \gamma) = (-1)^{pq} \gamma \otimes \lambda, \qquad \lambda \in \Lambda^p, \qquad \gamma \in \Gamma^q$$

is a K-algebra isomorphism.

The proof is left to the reader.

REMARK. The definition of the tensor product $\Lambda \otimes \Gamma$ of non-graded algebras may be considered as a special case of the tensor product of graded algebras, by defining on Λ and Γ the "trivial" grading which assigns to each element the degree zero. With this convention, we may apply 1.2 to the case when no grading is given on Λ and Γ.

2. ASSOCIATIVITY FORMULAE

We shall consider three K-algebras Λ, Γ and Σ. In the situation $(A_{\Lambda-\Gamma}, {}_\Lambda B_\Sigma)$ we convert $A \otimes_\Lambda B$ into a right $\Gamma \otimes \Sigma$-module by setting

$$(a \otimes b)(\gamma \otimes \sigma) = a\gamma \otimes b\sigma.$$

Similarly in the situation $({}_\Lambda B_\Sigma, C_{\Gamma-\Sigma})$ we convert $\operatorname{Hom}_\Sigma(B,C)$ into a right $\Lambda \otimes \Gamma$-module by setting

$$(f(\lambda \otimes \gamma))b = (f(\lambda b))\gamma$$

for $f \in \operatorname{Hom}_\Sigma(B,C)$.

We leave it to the reader to give similar definitions for other situations.

We can now state associativity formulae which generalize those of II,5.

PROPOSITION 2.1. *Let Λ, Γ, Σ be K-algebras. In the situation $(A_{\Lambda-\Gamma}, {}_\Lambda B_\Sigma, {}_{\Gamma-\Sigma}C)$, there is a unique homomorphism*

$$r\colon (A \otimes_\Lambda B) \otimes_{\Gamma\otimes\Sigma} C \to A \otimes_{\Lambda\otimes\Gamma} (B \otimes_\Sigma C)$$

such that $r((a \otimes b) \otimes c) = a \otimes (b \otimes c)$. This homomorphism is an isomorphism and establishes a natural equivalence of functors.

PROPOSITION 2.2. *Let Λ, Γ, Σ be K-algebras. In the situation $(A_{\Lambda-\Gamma}, {}_\Lambda B_\Sigma, C_{\Gamma-\Sigma})$, there is a unique homomorphism*

$$s\colon \operatorname{Hom}_{\Lambda\otimes\Gamma}(A, \operatorname{Hom}_\Sigma(B,C)) \to \operatorname{Hom}_{\Gamma\otimes\Sigma}(A \otimes_\Lambda B, C)$$

such that for each $\Lambda \otimes \Gamma$-homomorphism $f\colon A \to \operatorname{Hom}_\Sigma(B,C)$ we have $(sf)(a \otimes b) = (fa)b$. This homomorphism is an isomorphism and establishes a natural equivalence of functors.

We leave the proofs to the reader. Proposition 2.2 has an analogue with left and right operators interchanged and with $A \otimes_\Lambda B$ replaced by $B \otimes_\Lambda A$.

PROPOSITION 2.3. *In the situation $(A_{\Lambda-\Gamma}, {}_\Lambda B_\Sigma)$ assume that A is $\Lambda \otimes \Gamma$-projective and B is Σ-projective. Then $A \otimes_\Lambda B$ is projective as a right $\Gamma \otimes \Sigma$-module.*

PROOF. It suffices to prove that $\operatorname{Hom}_{\Gamma\otimes\Sigma}(A \otimes_\Lambda B, C)$ is an exact functor of the right $\Gamma \otimes \Sigma$-module C. In view of 2.2 this last functor is equivalent with the composition of $\operatorname{Hom}_{\Lambda\otimes\Gamma}(A,D)$ with $D = \operatorname{Hom}_\Sigma(B,C)$.

Since B is Σ-projective, $\operatorname{Hom}_\Sigma(B,C)$ is an exact functor of C. Since A is $\Lambda \otimes \Gamma$-projective, $\operatorname{Hom}_{\Lambda\otimes\Gamma}(A,D)$ is an exact functor of D. Thus the composite functor is an exact functor of C.

Replacing $\Lambda, \Gamma, \Sigma, A, B$ by $\Lambda^*, \Sigma, K, B, \Lambda$ we obtain

COROLLARY 2.4. *If B is a projective right $\Lambda^* \otimes \Sigma$-module and if Λ is K-projective then B is Σ-projective.*

COROLLARY 2.5. *In the situation (A_Γ, B_Σ), if A is Γ-projective and B is Σ-projective, then $A \otimes_K B$ is $\Gamma \otimes \Sigma$-projective.*

Quite analogously we prove

PROPOSITION 2.3a. *In the situation $({}_\Lambda B_\Sigma, C_{\Gamma-\Sigma})$ assume that B is Λ-projective and C is $\Gamma \otimes \Sigma$-injective. Then $\operatorname{Hom}_\Sigma(B,C)$ is injective as a right $\Lambda \otimes \Gamma$-module.*

COROLLARY 2.4a. *If C is an injective right $\Gamma \otimes \Sigma$-module and if Γ is K-projective, then C is Σ-injective.*

COROLLARY 2.5a. *In the situation $({}_\Lambda B, C_\Gamma)$ if B is Λ-projective and C is Γ-injective then $\operatorname{Hom}_K(B,C)$ is $\Lambda \otimes \Gamma$-injective.*

PROPOSITION 2.6. *Let Λ, Γ, Σ be K-projective K-algebras. In the situation $(A_{\Lambda-\Gamma}, {}_\Lambda B_\Sigma)$ let X be a $\Lambda \otimes \Gamma$-projective resolution of A, and Y a $\Lambda^* \otimes \Sigma$-projective resolution of B. If $\operatorname{Tor}_n^\Lambda(A,B) = 0$ for $n > 0$ then $X \otimes_\Lambda Y$ is a $\Gamma \otimes \Sigma$-projective resolution of $A \otimes_\Lambda B$.*

PROOF. First we note that following IV,5, $X \otimes_\Lambda Y$ is to be regarded first as a double complex and then converted into a complex. Since Λ is K-projective it follows from 2.4 that Y is Σ-projective. Consequently by 2.3, $X \otimes_\Lambda Y$ is $\Gamma \otimes \Sigma$-projective. There remains to be shown that $X \otimes_\Lambda Y$ is an acyclic complex over $A \otimes_\Lambda B$. Since the tensor product is right exact it follows from II,4.3 that the sequence

$$X_1 \otimes_\Lambda Y_0 + X_0 \otimes_\Lambda Y_1 \longrightarrow X_0 \otimes_\Lambda Y_0 \longrightarrow A \otimes_\Lambda B \longrightarrow 0$$

is exact. Thus it suffices to show that $H_n(X \otimes_\Lambda Y) = 0$ for $n > 0$. Since Σ is K-projective, it follows from 2.4 that Y is Λ-projective; similarly, X is Λ-projective since Γ is K-projective. Thus X and Y are Λ-projective resolutions of A and B. Consequently $H_n(X \otimes_\Lambda Y) = \operatorname{Tor}_n^\Lambda(A,B) = 0$ for $n > 0$ by hypothesis.

COROLLARY 2.7. *Let Γ and Σ be K-projective K-algebras. In the situation (A_Γ, B_Σ) let X be a Γ-projective resolution of A and Y a Σ-projective resolution of B. If $\operatorname{Tor}_n^K(A,B) = 0$ for $n > 0$ then $X \otimes_K Y$ is a $\Gamma \otimes \Sigma$-projective resolution of $A \otimes_K B$.*

We similarly prove

PROPOSITION 2.6a. *Let Λ, Γ, Σ be K-projective K-algebras. In the situation $({}_\Lambda B_\Sigma, C_{\Gamma-\Sigma})$ let X be a $\Lambda^* \otimes \Sigma$-projective resolution of B, and Y a $\Gamma \otimes \Sigma$-injective resolution of C. If $\operatorname{Ext}_\Sigma^n(B,C) = 0$ for $n > 0$ then $\operatorname{Hom}_\Sigma(X,Y)$ is a $\Lambda \otimes \Gamma$-injective resolution of $\operatorname{Hom}_\Sigma(B,C)$.*

COROLLARY 2.7a. *Let Λ,Γ be K-projective K-algebras. In the situation $({}_\Lambda B,C_\Gamma)$ let X be a Λ-projective resolution of B and Y a Γ-injective resolution of C. If* $\operatorname{Ext}^n_K(B,C)=0$ *for* $n>0$ *then* $\operatorname{Hom}_K(X,Y)$ *is a* $\Lambda\otimes\Gamma$*-injective resolution of* $\operatorname{Hom}_K(B,C)$.

THEOREM 2.8. *Let Λ, Γ, Σ be K-projective K-algebras. In the situation* $(A_{\Lambda-\Gamma},{}_\Lambda B_{\Sigma},{}_{\Gamma-\Sigma}C)$ *assume*

$$\text{(1)}\qquad \operatorname{Tor}^\Lambda_n(A,B)=0=\operatorname{Tor}^\Sigma_n(B,C)\qquad \text{for } n>0.$$

Then there is an isomorphism

$$\operatorname{Tor}^{\Gamma\otimes\Sigma}(A\otimes_\Lambda B,C)\approx\operatorname{Tor}^{\Lambda\otimes\Gamma}(A,B\otimes_\Sigma C)$$

which, in degree zero, reduces to the isomorphism of 2.1.

PROOF. Let X be a $\Lambda\otimes\Gamma$-projective resolution of A, Y a $\Lambda^*\otimes\Sigma$-projective resolution of B and Z a $\Gamma\otimes\Sigma$-projective resolution of C. In view of (1), it follows from 2.6 that $X\otimes_\Lambda Y$ is a $\Gamma\otimes\Sigma$-projective resolution of $A\otimes_\Lambda B$ and $Y\otimes_\Sigma Z$ is a $\Lambda\otimes\Gamma$-projective resolution of $B\otimes_\Sigma C$. Therefore

$$\begin{aligned}\operatorname{Tor}^{\Gamma\otimes\Sigma}(A\otimes_\Lambda B,C)=H((X\otimes_\Lambda Y)\otimes_{\Gamma\otimes\Sigma}Z)&\approx H(X\otimes_{\Lambda\otimes\Gamma}(Y\otimes_\Sigma Z))\\&=\operatorname{Tor}^{\Lambda\otimes\Gamma}(A,B\otimes_\Sigma C).\end{aligned}$$

Quite analogously we prove

THEOREM 2.8a. *Let Λ, Γ, Σ be K-projective K-algebras. In the situation* $(A_{\Lambda-\Gamma},{}_\Lambda B_\Sigma,C_{\Gamma-\Sigma})$ *assume*

$$\operatorname{Tor}^\Lambda_n(A,B)=0=\operatorname{Ext}^n_\Sigma(B,C)\qquad\text{for } n>0.$$

Then there is an isomorphism

$$\operatorname{Ext}_{\Gamma\otimes\Sigma}(A\otimes_\Lambda B,C)\approx\operatorname{Ext}_{\Lambda\otimes\Gamma}(A,\operatorname{Hom}_\Sigma(B,C))$$

which, in degree zero, reduces to the isomorphism of 2.2.

3. THE ENVELOPING ALGEBRA Λ^e

Let Λ be a K-algebra. A two-sided Λ-module is an abelian group A on which Λ operates on the left and on the right in such a way that $(\lambda a)\mu=\lambda(a\mu)$ and $ka=ak$ for $a\in A$, $\lambda,\mu\in\Lambda$, $k\in K$. With the notations of the preceding section we are thus in the situation $({}_\Lambda A_\Lambda)$.

A two-sided Λ-module may be regarded as a left module over the algebra $\Lambda\otimes_K\Lambda^*$, by setting

$$(\lambda\otimes\mu^*)a=\lambda a\mu.$$

The algebra $\Lambda\otimes_K\Lambda^*$ will be called the *enveloping algebra* of Λ and will be denoted by Λ^e. We may also regard A as a right Λ^e-module by setting $a(\lambda \otimes \mu^*) = \mu a\lambda$. In particular, Λ is a two-sided Λ-module; we shall therefore always regard Λ as a left Λ^e-module with operators

$$(\mu \otimes \gamma^*)\lambda = \mu\lambda\gamma \qquad \lambda,\mu,\gamma \in \Lambda.$$

In particular, taking $\lambda = 1$, we obtain a mapping

$$\rho: \Lambda^e \to \Lambda$$

given by $\rho(\mu \otimes \gamma^*) = \mu\gamma$. This mapping ρ is an epimorphism of left Λ^e-modules, and thus defines on Λ^e the structure of an augmented ring (VIII,1). We shall denote by J the augmentation ideal i.e. the kernel of ρ.

PROPOSITION 3.1. *As a left Λ-module, J is generated by the elements* $\lambda \otimes 1 - 1 \otimes \lambda^*$.

PROOF. Let $\sum \mu_i \otimes \gamma_i^* \in J$. Then $\sum \mu_i\gamma_i = 0$ and

$$\sum \mu_i \otimes \gamma_i^* = \sum (\mu_i \otimes 1)(1 \otimes \gamma_i^* - \gamma_i \otimes 1). \tag{1}$$

We define a K-homomorphism

$$j: \Lambda \to J$$

by setting $j\lambda = \lambda \otimes 1 - 1 \otimes \lambda^*$. We verify the identity

$$j(\lambda\mu) = \lambda(j\mu) + (j\lambda)\mu.$$

In general, given any left Λ^e-module A (i.e. a two-sided Λ-module A) we define a *crossed homomorphism* (also called *derivation*) $f: \Lambda \to A$ as a K-homomorphism such that

$$f(\lambda\mu) = \lambda(f\mu) + (f\lambda)\mu.$$

Each crossed homomorphism satisfies $f1 = 0$; therefore we may regard as defined on $\Lambda' = \text{Coker}\,(K \to \Lambda)$.

PROPOSITION 3.2. *If with each $h \in \text{Hom}_{\Lambda^e}(J,A)$ we associate the mapping hj, we obtain an isomorphism of the K-module* $\text{Hom}_{\Lambda^e}(J,A)$ *with the K-module of all crossed homomorphisms of Λ into A.*

PROOF. The essential part of the proof consists in showing that each crossed homomorphism $f: \Lambda \longrightarrow A$ admits a factorization $\Lambda \xrightarrow{j} J \xrightarrow{h} A$ where h is a Λ^e-homomorphism. Let $x = \sum \mu_i \otimes \gamma_i^* \in J$. Guided by (1) we define

$$hx = \sum \mu_i f(\gamma_i).$$

Clearly h is a K-homomorphism. To show that h is a Λ^e-homomorphism we compute

$$\begin{aligned} h((\lambda \otimes \tau^*)x) &= h(\sum \lambda\mu_i \otimes (\gamma_i\tau)^*) \\ &= \sum \lambda\mu_i f(\gamma_i\tau) = \sum \lambda\mu_i(f\gamma_i)\tau + \sum \lambda\mu_i\gamma_i(f\tau). \end{aligned}$$

The last term is zero because $\sum \mu_i \gamma_i = 0$. The term before last yields $(\lambda \otimes \tau^*)hx$ as desired.

A Λ^e-homomorphism $h: J \to A$ is extendable to a Λ^e-homomorphism $\Lambda^e \to A$ if and only if there is an element $a \in A$ such that $h(\sum \mu_i \otimes \gamma_i^*) = \sum \mu_i a \gamma_i$. It follows that the associated crossed homomorphism $f = hj: \Lambda \to A$ is given by $f\lambda = \lambda a - a\lambda$. Such a crossed homomorphism is called *principal* (or *inner derivation*). The K-module of all principal crossed homomorphisms corresponds to the image of $A \approx \mathrm{Hom}_{\Lambda^e}(\Lambda^e, A) \to \mathrm{Hom}_{\Lambda^e}(J, A)$.

4. HOMOLOGY AND COHOMOLOGY OF ALGEBRAS

Let Λ be a K-algebra and A a two-sided Λ-module. Using the augmented algebra Λ^e (with augmentation $\rho: \Lambda^e \to \Lambda$) we shall now define the homology and cohomology groups of Λ with coefficients in A.

First we regard A as a right Λ^e-module and define the n-th *homology group* as

$$H_n(\Lambda, A) = \mathrm{Tor}_n^{\Lambda^e}(A, \Lambda).$$

Then we regard A as a left Λ^e-module and define the n-th *cohomology group* as

$$H^n(\Lambda, A) = \mathrm{Ext}_{\Lambda^e}^n(\Lambda, A).$$

Both the homology and cohomology groups are K-modules. We shall see in § 6 that the cohomology groups $H^n(\Lambda, A)$ coincide with those defined by Hochschild (*Ann. of Math. 46* (1945), 58–67) in the case when K is a field. For this reason we shall frequently refer to the groups above as the Hochschild homology and cohomology groups of the algebra Λ.

REMARK. The notation $H_n(\Lambda, A)$ and $H^n(\Lambda, A)$ is contrary to our general conventions concerning graded modules; indeed with the notation as is we cannot use the symbol $H(\Lambda, A)$ to denote either the graded homology or the graded cohomology module.

To compute the homology and cohomology groups of Λ, a projective resolution X of Λ as a left Λ^e-module may be used:

$$H_n(\Lambda, A) = H_n(A \otimes_{\Lambda^e} X)$$
$$H^n(\Lambda, A) = H^n(\mathrm{Hom}_{\Lambda^e}(X, A)).$$

The functors $H_n(\Lambda, A)$ and $H^n(\Lambda, A)$ are connected sequences of covariant functors of A. If $0 \to A' \to A \to A'' \to 0$ is an exact sequence of two-sided Λ-modules, we obtain the usual exact sequences

$$\cdots \to H_n(\Lambda, A') \to H_n(\Lambda, A) \to H_n(\Lambda, A'') \to H_{n-1}(\Lambda, A') \to \cdots$$
$$\cdots \to H^{n-1}(\Lambda, A'') \to H^n(\Lambda, A') \to H^n(\Lambda, A) \to H^n(\Lambda, A'') \to \cdots$$

These exact sequences are the homology sequences of the exact sequences of complexes

$$0 \to A' \otimes_{\Lambda^e} X \to A \otimes_{\Lambda^e} X \to A'' \otimes_{\Lambda^e} X \to 0$$

$$0 \to \operatorname{Hom}_{\Lambda^e}(X,A') \to \operatorname{Hom}_{\Lambda^e}(X,A) \to \operatorname{Hom}_{\Lambda^e}(X,A'') \to 0.$$

The formulae (2)–(4a) of VIII,1 are applicable. In particular we have

(2) $$H_0(\Lambda,A) = \operatorname{Coker}(A \otimes_{\Lambda^e} J \to A) = A/AJ$$

(2a) $$H^0(\Lambda,A) = \operatorname{Ker}(A \to \operatorname{Hom}_{\Lambda^e}(J,A))$$

(3a) $$H^1(\Lambda,A) = \operatorname{Coker}(A \to \operatorname{Hom}_{\Lambda^e}(J,A)).$$

The submodule AJ of A is, by 3.1, generated by the elements of the form $a\lambda - \lambda a$ ($a \in A$, $\lambda \in \Lambda$). Let $a \in A$ and let $f \in \operatorname{Hom}_{\Lambda^e}(J,A)$ be the corresponding homomorphism. In order that $f = 0$ it is necessary and sufficient that f be zero on elements $\lambda \otimes 1 - 1 \otimes \lambda^*$, i.e. that $\lambda a = a\lambda$ for all $\lambda \in \Lambda$. We call such elements of A, *invariant* elements. As for the terms in (4a) an interpretation is given in 3.2 and the subsequent remark. Summarizing we obtain

PROPOSITION 4.1. *The homology group $H_0(\Lambda,A)$ may be identified with the quotient of A by the submodule generated by the elements $a\lambda - \lambda a$, $a \in A$, $\lambda \in \Lambda$. The cohomology group $H^0(\Lambda,A)$ may be identified with the subgroup of the invariant elements of A. The cohomology group $H^1(\Lambda,A)$ may be identified with the group of all crossed homomorphisms $\Lambda \to A$ factored by the subgroup of principal crossed homomorphisms.*

Let us now apply the associativity theorems 2.8 and 2.8a. In 2.8 we replace (Γ,C) by (Σ^*,Σ), then $\operatorname{Tor}_n^{\Sigma}(B,\Sigma) = 0$ for $n > 0$, and if we assume that Λ is semi-simple then also $\operatorname{Tor}_n^{\Lambda}(A,B) = 0$ for $n > 0$. Thus 2.8 yields

PROPOSITION 4.2. *Let Λ and Σ be K-projective K-algebras with Λ semi-simple. In the situation $({}_{\Sigma}A_{\Lambda},{}_{\Lambda}B_{\Sigma})$ we have isomorphisms*

$$H_n(\Sigma,A \otimes_{\Lambda} B) \approx \operatorname{Tor}_n^{\Lambda\otimes\Sigma^*}(A,B).$$

Similarly, in 2.8a, we replace (Γ,A) by (Λ^*,Λ). We obtain

PROPOSITION 4.3. *Let Λ and Σ be K-projective K-algebras with Σ semi-simple. In the situation $({}_{\Lambda}B_{\Sigma},{}_{\Lambda}C_{\Sigma})$ we have isomorphisms*

$$H^n(\Lambda,\operatorname{Hom}_{\Sigma}(B,C)) \approx \operatorname{Ext}^n_{\Lambda^*\otimes\Sigma}(B,C).$$

Replacing (Λ,Σ) by (K,Λ) in 4.2, and Σ by K in 4.3 we obtain

COROLLARY 4.4. *If Λ is a K-algebra with K semi-simple, we have the isomorphisms*

$$H_n(\Lambda, A \otimes_K B) \approx \operatorname{Tor}_n^{\Lambda}(B,A) \qquad ({}_{\Lambda}A,B_{\Lambda})$$

$$H^n(\Lambda,\operatorname{Hom}_K(B,C) \approx \operatorname{Ext}^n_{\Lambda}(B,C) \qquad ({}_{\Lambda}B,{}_{\Lambda}C).$$

REMARK 1. Throughout this discussion we have treated Λ as a left Λ^e-module and thus regarded Λ^e as a left augmented ring. We could regard Λ^e as a right augmented ring with augmentation $\rho'(\mu \otimes \gamma^*) = \gamma\mu$, and accordingly regard Λ as a right Λ^e-module. One then obtains the same homology and cohomology groups $H_n(\Lambda,A)$ and $H^n(\Lambda,A)$, because these are the satellites of $H_0(\Lambda,A)$ and $H^0(\Lambda,A)$ whose description is independent of the choice of ρ or ρ' as augmentation. If Λ is commutative then $\rho = \rho'$.

REMARK 2. The assumption that K is semi-simple may be replaced without any loss of generality by the assumption that K is a field. Indeed, a commutative semi-simple ring K is a direct sum $K_1 + \cdots + K_n$ of fields. This induces a decomposition $\Lambda = \Lambda_1 + \cdots + \Lambda_n$ of any K-algebra Λ into a direct sum of K_i-algebras $\Lambda_i = K_i\Lambda$. For the algebra Λ^e we then have a decomposition $\Lambda^e = \Lambda_1^e + \cdots + \Lambda_n^e$, where $\Lambda_i^e = \Lambda_i \otimes_{K_i} \Lambda_i^* = K_i\Lambda^e$. These direct sum decompositions induce similar direct sum decompositions for Ext_Λ, Ext_{Λ^e}, etc.

5. THE HOCHSCHILD GROUPS AS FUNCTORS OF Λ

Let Λ and Γ be K-algebras and $\varphi\colon \Lambda \to \Gamma$ a K-algebra homomorphism. Then φ induces a homomorphism $\varphi^e\colon \Lambda^e \to \Gamma^e$. More generally let Λ be a K-algebra and Γ an L-algebra. Consider a pair of ring homomorphisms

$$\varphi\colon \Lambda \to \Gamma, \qquad \psi\colon K \to L$$

such that $\varphi(k\lambda) = \psi(k)\varphi(\lambda)$, $k \in K$, $\lambda \in \Lambda$. Then φ induces a homomorphism φ^e of $\Lambda^e = \Lambda \otimes_K \Lambda^*$ into $\Gamma^e = \Gamma \otimes_L \Gamma^*$ such that the diagram

$$\begin{array}{ccc} \Lambda^e & \xrightarrow{\rho_\Lambda} & \Lambda \\ \downarrow{\scriptstyle \varphi^e} & & \downarrow{\scriptstyle \varphi} \\ \Gamma^e & \xrightarrow[\rho_\Gamma]{} & \Gamma \end{array}$$

is commutative. We are thus in the situation described in VIII,3 for augmented rings. Therefore for each two-sided Γ-module A (which using φ may also be regarded as a two-sided Λ-module) we have the homomorphisms

(1) $$F_n^\varphi\colon H_n(\Lambda,A) \to H_n(\Gamma,A)$$

(2) $$F_\varphi^n\colon H^n(\Gamma,A) \to H^n(\Lambda,A).$$

In this sense, $H_n(\Lambda,A)$ is a covariant functor of Λ while $H^n(\Lambda,A)$ is contravariant in Λ.

The mapping theorem VIII,3.1 gives necessary and sufficient conditions in order that F_n^φ and F_φ^n be isomorphisms. We shall apply this theorem to the case when Γ is obtained from Λ by an extension of the ground ring from K to L. Thus we suppose that Λ and L are K-algebras, L is commutative, $\Gamma = L \otimes_K \Lambda$, $\varphi(\lambda) = 1 \otimes \lambda$; and $\psi(k) = k1 \in L$. We have

$$(L \otimes_K \Lambda)^e = (L \otimes_K \Lambda) \otimes_L (L \otimes_K \Lambda)^* \approx (L \otimes_K \Lambda) \otimes_L (L \otimes_K \Lambda^*)$$
$$\approx L \otimes_K \Lambda \otimes_K \Lambda^* = L \otimes_K \Lambda^e.$$

We may therefore apply 1.1 with Λ, Q replaced by Λ^e, Λ. If Λ is K-projective then $\mathrm{Tor}_n^K(L,\Lambda) = 0$ for $n > 0$, so that we obtain

PROPOSITION 5.1. *If the K-algebra Λ is K-projective and L is a commutative K-algebra, then for each two-sided $L \otimes_K \Lambda$-module A we have the isomorphisms*

$$F_n\colon\ H_n(\Lambda, A) \approx H_n(L \otimes_K \Lambda, A),$$
$$F^n\colon\ H^n(L \otimes_K \Lambda, A) \approx H^n(\Lambda, A).$$

Further, if X is a Λ^e-projective resolution of Λ then $L \otimes_K X$ is an $(L \otimes_K \Lambda)^e$-projective resolution of $L \otimes_K \Lambda$.

We should remark here that since $L \otimes_K \Lambda$ is regarded as an L-algebra, the left operators of L on A must coincide with the right operators of L on A.

Proposition 5.1 may be applied when K is a field because then Λ is always K-projective. If L is a field extension of K, we can then say that the homology and cohomology groups remain unchanged under an extension of the ground field.

Let Λ and Γ be two K-algebras. The direct sum $\Lambda + \Gamma$ with multiplication and operators defined by

$$(\lambda_1,\gamma_1)(\lambda_2,\gamma_2) = (\lambda_1\lambda_2,\gamma_1\gamma_2), \qquad k(\lambda,\gamma) = (k\lambda,k\gamma)$$

is then again a K-algebra Σ, called the *direct product* of Λ and Γ. If e_Λ and e_Γ are the unit elements of Λ and Γ, then (e_Λ,e_Γ) is the unit element of Σ.

We consider the homomorphisms $\varphi\colon \Sigma \to \Lambda$, $\psi\colon \Sigma \to \Gamma$ given by $\varphi(\lambda,\gamma) = \lambda$, $\psi(\lambda,\gamma) = \gamma$. These are K-algebra homomorphisms which induce homomorphisms

$$\varphi^e\colon\ \Sigma^e \to \Lambda^e, \qquad \psi^e\colon\ \Sigma^e \to \Gamma^e.$$

Therefore every two-sided Λ-module A may be regarded also as a two-sided Σ-module. Similarly every two-sided Γ-module A' may be regarded as a two-sided Σ-module. Consequently $A + A'$ is a two-sided Σ-module.

Let A be a two-sided Σ-module. We introduce the module $e_\Lambda A e_\Lambda = \Lambda A \Lambda$ which is a two-sided Λ-module. It is easy to see that $\Lambda A \Lambda$ may be identified with $\operatorname{Hom}_{\Sigma^e}(\Lambda^e, A)$ and with $A \otimes_{\Sigma^e} \Lambda^e$. We further note the following identities

$$C \otimes_{\Sigma^e} (A+A') = (\Lambda C \Lambda + \Gamma C \Gamma) \otimes_{\Sigma^e} (A+A') = \Lambda C \Lambda \otimes_{\Lambda^e} A + \Gamma C \Gamma \otimes_{\Gamma^e} A'$$

$$\begin{aligned}\operatorname{Hom}_{\Sigma^e}(A+A',C) &= \operatorname{Hom}_{\Sigma^e}(A+A',\Lambda C \Lambda + \Gamma C \Gamma)\\ &= \operatorname{Hom}_{\Lambda^e}(A,\Lambda C \Lambda) + \operatorname{Hom}_{\Gamma^e}(A',\Gamma C \Gamma)\end{aligned}$$

in the situation $({}_\Lambda A_\Lambda, {}_\Gamma A'_\Gamma, {}_\Sigma C_\Sigma)$.

PROPOSITION 5.2. *A two-sided Λ-module A is Λ^e-projective if and only if it is Σ^e-projective.*

PROOF. For any Σ^e-module C we have $\operatorname{Hom}_{\Sigma^e}(A,C) = \operatorname{Hom}_{\Lambda^e}(A,\Lambda C \Lambda)$. Assume A is Λ^e-projective. Since $\Lambda C \Lambda$ is an exact functor of C it follows that $\operatorname{Hom}_{\Sigma^e}(A,C)$ is an exact functor of C, so that A is Σ^e-projective. Assume now that A is Σ^e-projective. If C is any Λ^e-module then $\operatorname{Hom}_{\Sigma^e}(A,C) = \operatorname{Hom}_{\Lambda^e}(A,C)$. Thus $\operatorname{Hom}_{\Lambda^e}(A,C)$ is an exact functor of C, so that A is Λ^e-projective.

THEOREM 5.3. (*Additivity theorem*). *If X is a Λ^e-projective resolution of Λ and Y is a Γ^e-projective resolution of Γ, then $X + Y$ is a Σ^e-projective resolution of $\Sigma = \Lambda + \Gamma$. Further for any two-sided Σ-module A*

$$H_n(\Sigma,A) \approx H_n(\Sigma,\Lambda A \Lambda + \Gamma A \Gamma) \approx H_n(\Lambda,\Lambda A \Lambda) + H_n(\Gamma,\Gamma A \Gamma), \tag{3}$$

$$H^n(\Sigma,A) \approx H^n(\Sigma,\Lambda A \Lambda + \Gamma A \Gamma) \approx H^n(\Lambda,\Lambda A \Lambda) + H^n(\Gamma,\Gamma A \Gamma). \tag{4}$$

PROOF. By 5.2, X and Y are Σ^e-projective, thus $X + Y$ also is Σ^e-projective. Since $H(X + Y) = H(X) + H(Y)$ it follows that $X + Y$ is a Σ^e-projective resolution of Σ. We have

$$\begin{aligned}A \otimes_{\Sigma^e}(X+Y) &= (\Lambda A \Lambda + \Gamma A \Gamma) \otimes_{\Sigma^e}(X+Y)\\ &= \Lambda A \Lambda \otimes_{\Lambda^e} X + \Gamma A \Gamma \otimes_{\Gamma^e} Y,\end{aligned}$$

$$\begin{aligned}\operatorname{Hom}_{\Sigma^e}(X+Y,A) &= \operatorname{Hom}_{\Sigma^e}(X+Y,\Lambda A \Lambda + \Gamma A \Gamma)\\ &= \operatorname{Hom}_{\Lambda^e}(X,\Lambda A \Lambda) + \operatorname{Hom}_{\Gamma^e}(Y,\Gamma A \Gamma)\end{aligned}$$

Thus passing to homology we obtain the desired isomorphisms (3) and (4).

COROLLARY 5.4. *If A is a two-sided Λ-module then $\varphi\colon \Sigma \to \Lambda$ induces isomorphisms*

$$H_n(\Sigma,A) \approx H_n(\Lambda,A)$$

$$H^n(\Lambda,A) \approx H^n(\Sigma,A).$$

6. STANDARD COMPLEXES

As was said in § 4 the homology and cohomology groups of a K-algebra Λ are usually computed using a Λ^e-projective resolution of Λ. The existence of such resolutions and their uniqueness up to a homotopy equivalence are guaranteed by the results of v,1. We shall describe here a construction which to each K-algebra Λ assigns an acyclic left complex $S(\Lambda)$ over Λ as a left Λ^e-module. If Λ is K-projective then $S(\Lambda)$ will be a Λ^e-projective resolution of Λ. In addition, the complex $S(\Lambda)$ will be a functor of Λ.

For each integer $n \geqq -1$, let $S_n(\Lambda)$ denote the $(n+2)$-fold tensor product (over K) of Λ with itself. Thus $S_{-1}(\Lambda) = \Lambda$, $S_{n+1}(\Lambda) = \Lambda \otimes_K S_n(\Lambda)$. We convert $S_n(\Lambda)$ into a two-sided Λ-module by setting

$$(\mu \otimes \gamma^*)(\lambda_0 \otimes \lambda_1 \otimes \cdots \otimes \lambda_n \otimes \lambda_{n+1}) = (\mu\lambda_0) \otimes \lambda_1 \otimes \cdots \otimes \lambda_n \otimes (\lambda_{n+1}\gamma).$$

We define a K-homomorphism

$$s_n\colon\ S_n(\Lambda) \to S_{n+1}(\Lambda)$$

by the formula $s_n a = 1 \otimes a$, $a \in S_n(\Lambda)$. Clearly s_n is a Λ-homomorphism for the right operators. Further, setting $t_n(\lambda \otimes a) = \lambda a$ we obtain a map $t_n\colon\ S_{n+1}(\Lambda) \to S_n(\Lambda)$ such that $t_n s_n = \text{identity}$. Thus s_n is a monomorphism.

We shall now define for each $n \geqq 0$ a left Λ-homomorphism

$$d_n\colon\ S_n(\Lambda) \to S_{n-1}(\Lambda)$$

such that

$$d_0(\lambda \otimes \mu) = \lambda\mu \in \Lambda \tag{1}$$

$$d_{n+1}s_n x + s_{n-1}d_n x = x \qquad \text{for } x \in S_n(\Lambda),\ n \geqq 0. \tag{2}$$

It is immediate that these conditions determine d_n by induction; given d_n, the homomorphism d_{n+1} is determined by (2) on the image of s_n; since the image of s_n generates $S_{n+1}(\Lambda)$ as a left Λ-module, d_{n+1} is unique. The following closed formula for d_n can easily be seen to verify (1) and (2)

$$d_n(\lambda_0 \otimes \cdots \otimes \lambda_{n+1}) = \sum_{0 \leq i \leq n} (-1)^i \lambda_0 \otimes \cdots \otimes (\lambda_i\lambda_{i+1}) \otimes \cdots \otimes \lambda_{n+1}. \tag{3}$$

We further see from this formula that d_n also is a right Λ-homomorphism. Thus d_n is a Λ^e-homomorphism.

We now prove that $d_{n-1}d_n = 0$ for $n > 0$. For $n = 1$ this follows from the associativity relation $(\lambda_0\lambda_1)\lambda_2 = \lambda_0(\lambda_1\lambda_2)$ in Λ. For $n > 1$ we argue by induction on n using (2). Using (2) we compute

$$d_n d_{n+1} s_n = d_n - d_n s_{n-1} d_n = s_{n-2} d_{n-1} d_n.$$

Consequently $d_n d_{n+1} s_n = 0$. Since the image of s_n generates $S_{n+1}(\Lambda)$ as a left Λ-module, it follows that $d_n d_{n+1} = 0$.

We observe that $S_0(\Lambda) = \Lambda \otimes \Lambda$ coincides with $\Lambda \otimes \Lambda^* = \Lambda^e$ as a two-sided Λ-module; further the map $d_0\colon S_0(\Lambda) \to S_{-1}(\Lambda)$ is precisely the augmentation $\rho\colon \Lambda^e \to \Lambda$. It follows that $S(\Lambda) = \sum_{n \geqq 0} S_n(\Lambda)$ with the differentiation d_n and the augmentation $d_0 = \rho$ is a left complex over the Λ^e-module Λ. Relations (2) prove that this complex is *acyclic*.

It is frequently convenient to write $S_n(\Lambda)$ in the form

$$S_n(\Lambda) = \Lambda \otimes_K \tilde{S}_n(\Lambda) \otimes_K \Lambda = \Lambda^e \otimes_K \tilde{S}_n(\Lambda)$$

where $\tilde{S}_0(\Lambda) = K$, and $\tilde{S}_n(\Lambda)$ for $n > 0$ is the K-module obtained by taking the n-fold tensor product of Λ over K. This form shows more explicitly the operators of Λ^e on $S_n(\Lambda)$.

If Λ is K-projective, then by 2.5, $\tilde{S}_n(\Lambda)$ is K-projective and, again by 2.5, $S_n(\Lambda)$ is Λ^e-projective. Thus in this case $S(\Lambda)$ is a Λ^e-projective resolution of Λ. This is the *standard complex* of Λ. It is clear how a map $\varphi\colon \Lambda \to \Gamma$ induces a map $S(\varphi)\colon S(\Lambda) \to S(\Gamma)$.

In computing the homology groups we use the identification

$$A \otimes_{\Lambda^e} S_n(\Lambda) = A \otimes_{\Lambda^e} \Lambda^e \otimes_K \tilde{S}_n(\Lambda) = A \otimes_K \tilde{S}_n(\Lambda).$$

Thus $H_n(\Lambda,A)$ are the homology groups of the complex $A \otimes_K \tilde{S}(\Lambda)$ with differentiation

$$\begin{aligned} d_n(a \otimes \lambda_1 \otimes \cdots \otimes \lambda_n) = a\lambda_1 \otimes \lambda_2 \otimes \cdots \otimes \lambda_n \\ + \sum_{0<i<n} (-1)^i a \otimes \lambda_1 \otimes \cdots \otimes \lambda_i\lambda_{i+1} \otimes \cdots \otimes \lambda_n \\ + (-1)^n \lambda_n a \otimes \lambda_1 \otimes \cdots \otimes \lambda_{n-1}. \end{aligned}$$

In computing the cohomology groups we use

$$\operatorname{Hom}_{\Lambda^e}(S_n(\Lambda),A) = \operatorname{Hom}_{\Lambda^e}(\Lambda^e \otimes \tilde{S}_n(\Lambda), A) = \operatorname{Hom}_K(\tilde{S}_n(\Lambda),A).$$

The elements of the latter group are called *n-dimensional cochains*, and are K-linear functions of n variables in Λ with values in A. The "coboundary" δf of an n-cochain f is

$$\begin{aligned} (\delta f)(\lambda_1, \ldots, \lambda_{n+1}) = \lambda_1 f(\lambda_2, \ldots, \lambda_{n+1}) \\ + \sum_{0<i<n+1} (-1)^i f(\lambda_1, \ldots, \lambda_i\lambda_{i+1}, \ldots, \lambda_{n+1}) \\ + (-1)^{n+1} f(\lambda_1, \ldots, \lambda_n)\lambda_{n+1}. \end{aligned}$$

This formula shows that the cohomology groups coincide with those defined by Hochschild.

There is a very useful variant of the standard complex $S(\Lambda)$ called the *normalized* standard complex $N(\Lambda)$. We define $N_n(\Lambda) = \Lambda^e \otimes_K \tilde{N}_n(\Lambda)$ where $\tilde{N}_0(\Lambda) = K$ and $\tilde{N}_n(\Lambda)$ for $n > 0$ is the n-fold tensor product over K of the K-module $\Lambda' = \text{Coker}\,(K \to \Lambda)$ with itself. The natural K-epimorphisms $\Lambda \to \Lambda'$ induces K-epimorphisms $\tilde{S}_n(\Lambda) \to \tilde{N}_n(\Lambda)$ and Λ^e-epimorphisms $S_n(\Lambda) \to N_n(\Lambda)$. The operators s_n and d_n pass to the quotients and yield similar operators in $N(\Lambda) = \sum N_n(\Lambda)$ with (1) and (2) still satisfied. Thus $N(\Lambda)$ also is an acyclic left complex over Λ. If Λ' is K-projective, then $N(\Lambda)$ is a Λ^e-projective resolution of Λ.

If $\lambda_0 \otimes \cdots \otimes \lambda_{n+1} \in S_n(\Lambda)$ we denote by $\lambda_0[\lambda_1, \ldots, \lambda_n]\lambda_{n+1}$ the corresponding element of $N_n(\Lambda)$. For $\lambda_0 = 1$ (resp. $\lambda_{n+1} = 1$) we write simply $[\lambda_1, \ldots, \lambda_n]\lambda_{n+1}$ (resp. $\lambda_0[\lambda_1, \ldots, \lambda_n]$). We thus have the boundary formula

$$d_n[\lambda_1, \ldots, \lambda_n] = \lambda_1[\lambda_2, \ldots, \lambda_n] + \sum_{0<i<n} (-1)^i[\lambda_1, \ldots, \lambda_i\lambda_{i+1}, \ldots, \lambda_n]$$
$$+ (-1)^n[\lambda_1, \ldots, \lambda_{n-1}]\lambda_n.$$

The above convention applies also in the case $n = 0$. The symbol $[\,]$ stands then for the unit element of $\tilde{N}_0(\Lambda) = K$. Thus the element $\lambda_0 \otimes \lambda_1 \in S_0(\Lambda) = N_0(\Lambda)$ will be written as $\lambda_0[\,]\lambda_1$. With this convention the boundary formula above yields

$$d_1[\lambda] = \lambda[\,] - [\,]\lambda.$$

The notation just introduced will also be used for the non-normalized complex $S(\Lambda)$. Thus the symbol $[\lambda_1, \ldots, \lambda_n]$ will be ambiguously regarded as representing elements of either $S(\Lambda)$ or $N(\Lambda)$. However it must be remembered that $[\lambda_1, \ldots, \lambda_n]$ regarded as an element of $N(\Lambda)$ is zero whenever one of its coordinates λ_i is in the image of $K \to \Lambda$.

7. DIMENSION

Let Λ be a K-algebra, Λ^e its enveloping algebra. We shall be concerned with the projective dimension of Λ as a left Λ^e-module. According to the conventions introduced in VI,2 this integer (or $+\infty$) is denoted by $\dim_{\Lambda^e} \Lambda$ or simply by $\dim \Lambda$. This coincides also with the projective dimension of Λ as a right Λ^e-module. The relation $\dim \Lambda \leqq n$ means, by definition, that there is a Λ^e-projective resolution X of Λ such that $X_k = 0$ for $k > n$ (i.e. X is a complex of dimension $\leqq n$). It follows from VI,2.1 that this is equivalent to

$$H^{n+1}(\Lambda,A) = \text{Ext}^{n+1}_{\Lambda^e}(\Lambda,A) = 0$$

for all two-sided Λ-modules A.

PROPOSITION 7.1. *Let Λ be a K-projective K-algebra and L a commutative K-algebra. Then*

$$\dim (L \otimes_K \Lambda) \leqq \dim \Lambda.$$

If further the natural mapping $K \to L$ is a monomorphism of K onto a direct factor of L (as a K-module) then

$$\dim (L \otimes_K \Lambda) = \dim \Lambda.$$

PROOF. The first inequality follows directly from 5.1. To prove the second part, consider a K-homomorphism $\sigma\colon L \to K$ such that the composition $K \to L \to K$ is the identity. Let A be any two-sided Λ-module. Then $L \otimes_K A$ may be regarded as a two-sided $L \otimes_K \Lambda$-module, and by 5.1

$$H^n(L \otimes_K \Lambda, L \otimes_K A) \approx H^n(\Lambda, L \otimes_K A).$$

Since the composition of the homomorphisms

$$H^n(\Lambda,A) \to H^n(\Lambda, L \otimes_K A) \to H^n(\Lambda,A)$$

is the identity it follows that the relation $H^n(L \otimes_K \Lambda, L \otimes_K A) = 0$ implies $H^n(\Lambda,A) = 0$. Thus $\dim \Lambda \leqq \dim (L \otimes_K \Lambda)$.

COROLLARY 7.2. *If Λ is an algebra over a commutative field K, and L is a commutative field containing K, then*

$$\dim (L \otimes_K \Lambda) = \dim \Lambda.$$

PROPOSITION 7.3. *Let Λ and Γ be K-algebras and $\Lambda + \Gamma$ their direct product. Then*

$$\dim (\Lambda + \Gamma) = \max (\dim \Lambda, \dim \Gamma).$$

This follows directly from 5.3 and 5.4.

PROPOSITION 7.4. *Let Λ and Γ be K-projective K-algebras. Then*

$$\dim (\Lambda \otimes_K \Gamma) \leqq \dim \Lambda + \dim \Gamma.$$

If further K is a field and Λ and Γ are finitely K-generated then

$$\dim (\Lambda \otimes_K \Gamma) = \dim \Lambda + \dim \Gamma.$$

PROOF. Let X be a Λ^e-projective resolution of Λ, of dimension $\leqq p$ and let Y be a Γ^e-projective resolution of Γ, of dimension $\leqq q$. Since $\operatorname{Tor}_n^K (\Lambda,\Gamma) = 0$ for $n > 0$ we may apply 2.7, to deduce that $X \otimes_K Y$ is a $\Lambda^e \otimes_K \Gamma^e$-projective resolution of $\Lambda \otimes_K \Gamma$. Since $\Lambda^e \otimes_K \Gamma^e \approx (\Lambda \otimes_K \Gamma)^e$ and since $X \otimes_K Y$ has dimension $\leqq p + q$, the first inequality follows.

Now assume that K is a field and that Λ and Γ are finitely K-generated. Then Λ^e and Γ^e are also finitely K-generated and therefore are Noetherian. The projective resolutions X and Y may then be chosen so that each X_n (resp. Y_n) is Λ^e-free (resp. Γ^e-free) on a finite base. Then for any two-sided Λ-module A and any two-sided Γ-module A' we have the natural isomorphisms

$$\operatorname{Hom}_{\Lambda^e}(X,A) \otimes_K \operatorname{Hom}_{\Gamma^e}(Y,A') \approx \operatorname{Hom}_{(\Lambda\otimes\Gamma)^e}(X \otimes_K Y, A \otimes_K A').$$

Passing to homology this yields an isomorphism

$$H^n(\operatorname{Hom}_{\Lambda^e}(X,A) \otimes_K \operatorname{Hom}_{\Gamma^e}(Y,A')) \approx H^n(\Lambda \otimes_K \Gamma, A \otimes_K A').$$

Since K is a field, it follows from IV,7.2 that the mapping α yields an isomorphism of the left hand side with

$$\sum_{p+q=n} H^p(\operatorname{Hom}_{\Lambda^e}(X,A)) \otimes_K H^q(\operatorname{Hom}_{\Gamma^e}(Y,A')).$$

Thus, finally, we obtain an isomorphism

$$\sum_{p+q=n} H^p(\Lambda,A) \otimes_K H^q(\Gamma,A') \approx H^n(\Lambda \otimes_K \Gamma, A \otimes_K A').$$

Therefore, if $H^p(\Lambda,A) \neq 0$ and $H^q(\Gamma,A') \neq 0$ then since K is a field $H^p(\Lambda,A) \otimes_K H^q(\Gamma,A') \neq 0$ and consequently $H^{p+q}(\Lambda \otimes_K \Gamma, A \otimes_K A') \neq 0$. Thus $\dim(\Lambda \otimes_K \Gamma) \geqq p + q$.

Proposition 7.4 includes a theorem by Rose (*Amer. Jour. of Math. 74* (1952), 531–546).

We now propose to compare $\dim \Lambda$ with the various other dimensions, namely: l.gl.dim Λ, r.gl.dim Λ, l.gl.dim Λ^e, r.gl.dim Λ^e. Since Λ^e is isomorphic with its opposite ring $(\Lambda^e)^*$, the last two numbers are equal and will be denoted simply by gl.dim Λ^e.

PROPOSITION 7.5. *For any K-algebra* Λ

$$\dim \Lambda \leqq \text{gl.dim } \Lambda^e.$$

If further Λ *is semi-simple and K-projective then*

$$\dim \Lambda = \text{gl. dim } \Lambda^e.$$

PROOF. The first part follows directly from the definition of the global dimension. To prove the second part we use 4.3 with $\Sigma = \Lambda$. We obtain an isomorphism

$$H^n(\Lambda, \operatorname{Hom}_\Lambda(B,C)) \approx \operatorname{Ext}^n_{\Lambda^e}(B,C)$$

for any two-sided Λ-modules B and C, where $\operatorname{Hom}_\Lambda(B,C)$ is the group of right Λ-homomorphisms $B \to C$. This implies gl. dim $\Lambda^e \leqq \dim \Lambda$.

PROPOSITION 7.6. *If* Λ *is a K-algebra with K semi-simple then*

$$\text{l.gl.dim}\ \Lambda \leqq \dim \Lambda, \qquad \text{r.gl.dim}\ \Lambda \leqq \dim \Lambda.$$

PROOF. We apply 4.3 with $\Sigma = K$. This yields

$$H^n(\Lambda, \operatorname{Hom}_K(B,C)) \approx \operatorname{Ext}^n_\Lambda(B,C)$$

for any left Λ-modules B and C. This implies l.gl.dim $\Lambda \leqq \dim \Lambda$.

We shall see in x,6.2 numerous examples where the inequalities of 7.6 are replaced by equalities.

We now proceed to discuss in greater detail algebras for which $\dim \Lambda = 0$, i.e. algebras Λ which are Λ^e-projective.

PROPOSITION 7.7. *In order that* $\dim \Lambda = 0$ *it is necessary and sufficient that there exist an element e of the two-sided* Λ*-module* $\Lambda \otimes \Lambda$ *(isomorphic with* $\Lambda^e = \Lambda \otimes \Lambda^*$*) such that* $\lambda e = e\lambda$ *(i.e. e is invariant) and that under the mapping* $x \otimes y \to xy$ *the image of e in* Λ *is* 1.

PROOF. If Λ is Λ^e-projective, there is a Λ^e-homomorphism $f\colon \Lambda \to \Lambda^e$ such that the composition $\Lambda \xrightarrow{f} \Lambda^e \xrightarrow{\rho} \Lambda$ is the identity. Then $e = f1$ has the desired properties. Conversely, given an element e with the properties listed above, the map $f\lambda = \lambda e$ is a Λ^e-homomorphism $\Lambda \to \Lambda^e$ such that ρf is the identity. Thus Λ is Λ^e-projective.

PROPOSITION 7.8. *Let* $M_n(K)$ *be the algebra of square matrices of order n with coefficients in K. Then* $\dim M_n(K) = 0$.

PROOF. We shall apply the criterion of 7.7. Let e_{ij} be the matrix with 1 at the intersection of the i-th row and j-th column, and with zero everywhere else. Then $\sum_i e_{ii}$ is the unit matrix. Clearly the elements e_{ij} constitute a K-base of $\Lambda = M_n(K)$. Consider the element $e = \sum_i e_{i1} \otimes e_{1i} \in \Lambda \otimes_K \Lambda$. Then

$$e_{rs}e = e_{r1} \otimes e_{1s} = ee_{rs}$$

$$\sum_i e_{i1}e_{1i} = \sum_i e_{ii} = \text{unit matrix.}$$

Thus the conditions of 7.7 are fulfilled and $\dim \Lambda = 0$.

THEOREM 7.9. *Let* Λ *be a K-algebra with K semi-simple. Then* $\dim \Lambda = 0$ *if and only if* Λ^e *is semi-simple.*

PROOF. If $\dim \Lambda = 0$ then, by 7.6, Λ is semi-simple. Therefore by 7.5, gl.dim $\Lambda^e = 0$, i.e. Λ^e is semi-simple. Conversely, if Λ^e is semi-simple then, by 7.5, $\dim \Lambda = 0$.

THEOREM 7.10. *Let K be a commutative field and* Λ *a K-algebra, finitely K-generated. In order that* $\dim \Lambda = 0$ *it is necessary and sufficient that* Λ *be separable (i.e. that* $L \otimes_K \Lambda$ *be semi-simple for every commutative field L containing K).*

PROOF. Assume $\dim \Lambda = 0$. Then, by 7.1, $\dim (L \otimes_K \Lambda) = 0$ and therefore, by 7.6, $L \otimes_K \Lambda$ is semi-simple. Thus Λ is separable.

Conversely assume that Λ is separable. Then it is well known (see Albert, *Structure of Algebras*, New York, 1939, p. 45) that there exists a commutative field L containing K which is a splitting field for Λ, i.e. such that $L \otimes_K \Lambda$ is isomorphic to a direct product $\Lambda_1 + \cdots + \Lambda_r$ of full matrix algebras over L. By 7.8 we have $\dim \Lambda_i = 0$ so that 7.3 implies $\dim (\Lambda_1 + \cdots + \Lambda_r) = 0$. Thus $\dim (L \otimes_K \Lambda) = 0$ which, by 7.2, implies $\dim \Lambda = 0$.

To prove the existence of algebras of dimension n for an arbitrary integer n we consider the algebra $\Lambda = K[x_1, \ldots, x_n]$ of polynomials in the letters $x_1, \ldots, x_n$. We have shown in VIII, 4.2 that $\dim_\Lambda K = n$ provided K is converted into a Λ-module by means of $\varepsilon_0 \colon \Lambda \to K$, $\varepsilon_0 x_i = 0$. If $\varepsilon \colon \Lambda \to K$ is any K-algebra homomorphism, then the substitution $x_i \to x_i - \varepsilon x_i$ yields an automorphism φ of Λ such that $\varepsilon\varphi = \varepsilon_0$. Therefore $\dim_\Lambda K = n$ also with respect to ε.

Now we have

$\Lambda^e = \Lambda \otimes_K \Lambda \approx K[x_1, \ldots, x_n, y_1, \ldots, y_n] = \Lambda[y_1, \ldots, y_n]$. The map $\eta \colon \Lambda^e \to \Lambda$ yields a Λ-algebra homomorphism $\eta \colon \Lambda[y_1, \ldots, y_n] \to \Lambda$ (actually $\eta y_i = x_i$). Thus $\dim_{\Lambda^e} \Lambda = n$, i.e. $\dim \Lambda = n$.

Since $\dim_\Lambda K = n$, we have $\text{gl.dim}\, \Lambda \geqq n$; if further K is semi-simple, then, by 7.6, $\text{gl.dim}\, \Lambda \leqq \dim \Lambda = n$. Thus we have

THEOREM 7.11. *Let K be a commutative ring and $\Lambda = K[x_1, \ldots, x_n]$. Then*

$$\dim \Lambda = \dim_\Lambda K = n.$$

If K is semi-simple then

$$\text{gl.dim}\, \Lambda = n.$$

This supplements theorem VIII,6.5 in which K was assumed to be a (not necessarily commutative) field and only graded modules were considered.

EXERCISES

1. In the situation $({}_{\Lambda\text{-}\Gamma}A, B_{\Lambda\text{-}\Sigma}, {}_\Gamma C_\Sigma)$ where Λ, Γ, Σ are K-algebras, establish the isomorphism

$$\text{Hom}_{\Lambda\otimes\Gamma} (A, \text{Hom}_\Sigma (B,C)) \approx \text{Hom}_{\Lambda\otimes\Sigma} (B, \text{Hom}_\Gamma (A,C)).$$

Prove that if Λ, Γ, Σ are K-projective and

$$\text{Ext}^n_\Gamma (A,C) = 0 = \text{Ext}^n_\Sigma (B,C) \qquad \text{for } n > 0$$

then

$$\text{Ext}_{\Lambda\otimes\Gamma} (A, \text{Hom}_\Sigma (B,C)) \approx \text{Ext}_{\Lambda\otimes\Sigma} (B, \text{Hom}_\Gamma (A,C)).$$

2. Let $\Lambda = F_K(x_1, \ldots, x_n)$ be the free K-algebra generated by $x_1, \ldots, x_n$. Show that any crossed homomorphism $f\colon \Lambda \to A$ is determined by its values on $x_1, \ldots, x_n$ and that these may be chosen arbitrarily. Deduce from this that J is Λ^e-free with the elements $x_i \otimes 1 - 1 \otimes x_i^*$, $i = 1, \ldots, n$, as base. Show that dim $\Lambda = 1$ for $n > 0$.

3. Show that in the normalized standard complex $N(\Lambda)$ the "contracting" homotopy s has the form

$$s(\lambda[\lambda_1, \ldots, \lambda_n]\lambda') = [\lambda, \lambda_1, \ldots, \lambda_n]\lambda'$$

and that the sequence

$$0 \longrightarrow N_{-1}(\Lambda) \xrightarrow{s_{-1}} N_0(\Lambda) \xrightarrow{s_0} \cdots \longrightarrow N_n(\Lambda) \xrightarrow{s_n} N_{n+1}(\Lambda) \longrightarrow \cdots$$

is exact. As a consequence show that d_n $(n > 0)$ maps Ker s_n isomorphically onto Im d_n.

4. Given a K-algebra Λ consider the K-algebra $\Lambda^+ = K + \Lambda$ with multiplication and operators given by

$$(k_1,\lambda_1)(k_2,\lambda_2) = (k_1k_2, k_1\lambda_2 + k_2\lambda_1 + \lambda_1\lambda_2), \qquad k'(k,\lambda) = (k'k, k'\lambda).$$

Show that each two-sided Λ-module may be regarded as a two-sided Λ^+-module. Compare the complexes $S(\Lambda)$ and $N(\Lambda^+)$. Prove that if Λ is K-projective and A is a two-sided Λ-module, then

$$H_n(\Lambda^+,A) \approx H_n(\Lambda,A), \qquad H^n(\Lambda^+,A) \approx H^n(\Lambda,A).$$

5. Let Λ and Γ be K-algebras. Show that if Λ is K-free and dim $\Lambda = 0$ then dim $(\Lambda \otimes_K \Gamma) =$ dim Γ. [Hint: assuming $H^n(\Gamma, C) \neq 0$ show that $H^n(\Lambda \otimes_K \Gamma, \mathrm{Hom}_K(\Lambda^e, C)) \neq 0$.]

6. Let Λ be a K-algebra. Show that dim $\Lambda = 0$ if and only if $H^1(\Lambda,J) = 0$.

7. Consider the K-algebra with the basis $1, a, r$ with multiplication $aa = a$, $rr = 0$, $ar = r$, $ra = 0$. Show that dim $\Lambda = 1$.

8. Using the results of VI,5 establish the homomorphisms

$$(1) \qquad H^n(\Lambda, \mathrm{Hom}_K(A,K)) \to \mathrm{Hom}_K(H_n(\Lambda,A),K)$$

$$(2) \qquad H_n(\Lambda, \mathrm{Hom}_K(A,K)) \to \mathrm{Hom}_K(H^n(\Lambda,A),K).$$

Show that if K is a field then (1) is an isomorphism. If K is a field and Λ is finitely K-generated then (2) also is an isomorphism.

CHAPTER X

Supplemented Algebras

Introduction. The notion of a supplemented algebra is a very special but very important case of an augmented ring. The homology theory of supplemented algebras includes both the homology theory of groups (or more generally, of monoids) and of Lie algebras (the .latter will be treated in Ch. XIII).

The homology groups $H_n(\Lambda,A)$ of a supplemented algebra Λ are defined for each right Λ-module A; the cohomology groups $H^n(\Lambda,C)$ are defined for each left Λ-module C. In the most interesting case when the algebra Λ is K-projective, these homology and cohomology groups may be included in the Hochschild theory of Ch. IX. Theorem 2.1 shows precisely how a complex which is constructed to be used for the computation of the Hochschild groups of Λ, may be used to compute the homology and cohomology groups of Λ, as a supplemented algebra. This procedure is applied to the standard complex; in the case Λ is the algebra $K(\Pi)$ of a group Π, we find the "non-homogeneous" complex introduced by Eilenberg-MacLane (*Proc. Nat. Acad. Sci. U.S.A.* 29 (1943), 155–158).

For some particular monoids and groups, it is more convenient to use complexes especially constructed rather than the standard complex. Some such examples are given in § 5; the cyclic groups will be discussed in XII,7.

In § 7 we study some relations between algebras and subalgebras, as well as between groups and subgroups.

1. HOMOLOGY OF SUPPLEMENTED ALGEBRAS

A K-algebra Λ together with a K-algebra homomorphism ε: $\Lambda \to K$ is called a *supplemented algebra.* Clearly the kernel I of ε is a two-sided ideal, and Λ is a left and right augmented ring (with ε as augmentation). If η: $K \to \Lambda$ is the map defining the K-algebra structure in Λ then $\varepsilon\eta =$ identity. It follows that K may be regarded as a subalgebra of Λ with η as inclusion map. Then Λ (as a right or left K-module) is the direct sum $K + I$.

A supplemented algebra being a special case of an augmented ring the definitions introduced in VIII,1 apply. In particular, the homology groups

$\operatorname{Tor}_n^\Lambda(A,K)$ and cohomology groups $\operatorname{Ext}_\Lambda^n(K,C)$ are defined for any right Λ-module A and any left Λ-module C. These may be computed as $H_n(A \otimes_\Lambda X)$ and $H^n(\operatorname{Hom}_\Lambda(X,C))$ using any Λ-projective resolution X of K as a left Λ-module. All the facts listed in VIII,1 apply with Q replaced by K.

Using the augmentation map ε: $\Lambda \to K$ we may convert any K-module A into a left (or right) Λ-module ${}_\varepsilon A$ (or A_ε) by setting $\lambda a = (\varepsilon\lambda)a$ (or $a\lambda = (\varepsilon\lambda)a$). We then say that the operators of Λ on ${}_\varepsilon A$ are *trivial*. If A is already a right (or left) Λ-module then ${}_\varepsilon A$ (or A_ε) is a two-sided Λ-module, i.e. a left Λ^e-module. In particular K has trivial Λ-operators.

Using this process, several definitions made for two-sided modules may be translated for Λ-modules. For instance, let C be a left Λ-module. An element $c \in C$ will be called *invariant* if c as an element of C_ε is invariant in the sense of IX,4, i.e. if $\lambda c = c\lambda$. Since $c\lambda = c(\varepsilon\lambda) = (\varepsilon\lambda)c$, it follows that C is invariant if and only if $(\lambda - \varepsilon\lambda)c = 0$ for all $\lambda \in \Lambda$ or equivalently if $Ic = 0$. The invariant elements of c form a Λ-submodule C^Λ; this is the largest submodule of C with trivial Λ-operators. Formula (2a) of VIII,1 now can be interpreted as $\operatorname{Hom}_\Lambda(K,C) = C^\Lambda$. On the other hand it follows from IX,4 that the 0-th Hochschild cohomology group $H^0(\Lambda,C_\varepsilon)$ also coincides with C^{Λ}. We thus obtain isomorphisms

(1) $$\operatorname{Hom}_\Lambda(K,C) \approx H^0(\Lambda,C_\varepsilon) \approx C^\Lambda.$$

Similarly if A is a right Λ-module, then we observe that $AI = ({}_\varepsilon A)J$ where J is the kernel of the augmentation μ: $\Lambda^e \to \Lambda$. We therefore obtain isomorphisms

(2) $$A \otimes_\Lambda K \approx H_0(\Lambda, {}_\varepsilon A) \approx A_\Lambda$$

where $A_\Lambda = A/AI$. Clearly AI is the largest Λ-submodule in A such that the operators of Λ on the quotient module are trivial.

We return to the consideration of a left Λ-module C. Following the definition made in IX,3 a crossed homomorphism f: $\Lambda \to C$ (or rather f: $\Lambda \to C_\varepsilon$) is a K-homomorphism satisfying

$$f(\lambda_1\lambda_2) = \lambda_1 f(\lambda_2) + (f\lambda_1)(\varepsilon\lambda_2).$$

It is easy to see that each such crossed homomorphism admits a unique factorization

$$\Lambda \xrightarrow{p} I \xrightarrow{g} C$$

where g is a Λ-homomorphism and p is the projection operator $p\lambda = \lambda - \varepsilon\lambda$. The crossed homomorphism f is principal if and only if the homomorphism g admits an extension to a Λ-homomorphism $\Lambda \to C$.

Combining this with IX,3.2 we obtain an isomorphism

$$\mathrm{Hom}_\Lambda(I,C) \approx \mathrm{Hom}_{\Lambda^e}(J,C_\varepsilon).$$

Further, VIII,1 (3a) and IX,4.1 combine to give

$$\mathrm{Ext}^1_\Lambda(K,C) \approx H^1(\Lambda,C_\varepsilon). \tag{3}$$

Both groups are isomorphic with the group of all crossed homomorphisms $\Lambda \to C$ reduced modulo principal crossed homomorphisms.

The 0-th homology group $A \otimes_\Lambda K$ and the 0-th cohomology group $\mathrm{Hom}_\Lambda(K,A)$ both reduce to the module A, if A has trivial Λ-operators. The earlier discussion of the group $\mathrm{Ext}^1_\Lambda(K,A)$ shows that this is the group of all crossed homomorphisms $f\colon \Lambda \to A$ (i.e. K-homomorphisms satisfying $f(\lambda_1\lambda_2) = (\varepsilon\lambda_1)(f\lambda_2) + (f\lambda_1)(\varepsilon\lambda_2)$), all the principal crossed homomorphisms being zero. For the 1-dimensional homology group we have by (3) of VIII,1

$$\mathrm{Tor}^\Lambda_1(A,K) \approx \mathrm{Ker}(A \otimes_\Lambda I \to A) \approx \mathrm{Ker}(A \otimes_K (K \otimes_\Lambda I) \to A \otimes_K K).$$

Since the homomorphism $K \otimes_\Lambda I \to K$ is zero we have $\mathrm{Tor}^\Lambda_1(A,K) \approx A \otimes_K (K \otimes_\Lambda I)$. By (9) of VIII,1 we have $K \otimes_\Lambda I \approx I/I^2$. Thus we obtain

$$\mathrm{Tor}^\Lambda_1(A,K) \approx A \otimes_K I/I^2 \tag{4}$$

if A has trivial Λ-operators.

Consider supplemented algebras

$$\Lambda \xrightarrow{\varepsilon} K, \qquad \Gamma \xrightarrow{\varepsilon'} L.$$

A map of the first algebra into the second is a pair of ring homomorphisms $\varphi\colon \Lambda \to \Gamma$, $\psi\colon K \to L$ such that $\varepsilon'\varphi = \psi\varepsilon$ and $\varphi(k\lambda) = (\psi k)(\varphi\lambda)$. This places us in the situation discussed in VIII,3, and we obtain homomorphisms

$$F^\varphi\colon \mathrm{Tor}^\Lambda_n(A,K) \to \mathrm{Tor}^\Gamma_n(A,L)$$

$$F_\varphi\colon \mathrm{Ext}^n_\Gamma(L,C) \to \mathrm{Ext}^n_\Lambda(K,C)$$

defined for any right Γ-module A and any left Γ-module C. The case that most commonly applies is that of $K = L$, $\psi =$ identity.

A somewhat different case is the following one. Let $\Lambda \xrightarrow{\varepsilon} K$ be a supplemented K-algebra and let L be a (not necessarily commutative) K-algebra. By extending the ground ring we obtain an augmented ring

$$L \otimes_K \varepsilon\colon L \otimes_K \Lambda \to L.$$

This places us in the situation discussed in IX,1.1 and we obtain homomorphisms

(5) $$\operatorname{Tor}_n^{\Lambda}(A,K) \to \operatorname{Tor}_n^{L\otimes\Lambda}(A,L)$$

(5a) $$\operatorname{Ext}_{L\otimes\Lambda}^n(L,C) \to \operatorname{Ext}_{\Lambda}^n(K,C)$$

for any right $L \otimes_K \Lambda$-module A and any left $L \otimes_K \Lambda$-module C. Applying IX,1.1 we obtain

PROPOSITION 1.1. *If the supplemented K-algebra Λ is K-projective then* (5) *and* (5a) *are isomorphisms. Further if X is a Λ-projective resolution of K then $L \otimes_K X$ is an $L \otimes_K \Lambda$-projective resolution of L.*

If the ring L is commutative, then $L \otimes_K \Lambda$ is a supplemented L-algebra (obtained from Λ by covariant extension of the ground ring). Proposition 1.1 then asserts the invariance of the homology and cohomology groups under such extensions.

2. COMPARISON WITH HOCHSCHILD GROUPS

Formulae (1)–(3) of the preceding section show that in low dimensions the homology and cohomology of a supplemented algebra Λ coincide with the Hochschild homology and cohomology groups of Λ. To carry out this comparison more systematically we consider the diagram

$$\begin{array}{ccc} \Lambda^e & \xrightarrow{\rho} & \Lambda \\ \downarrow{\scriptstyle\varphi} & & \downarrow{\scriptstyle\varepsilon} \\ \Lambda & \xrightarrow[\varepsilon]{} & K \end{array}$$

where $\varphi(\lambda \otimes \gamma^*) = \lambda(\varepsilon\gamma)$. Since $\varepsilon\varphi(\lambda \otimes \gamma^*) = \varepsilon(\lambda\gamma) = \varepsilon\rho(\lambda \otimes \gamma^*)$, the diagram is commutative, and thus the pair (φ,ε) is a map of the augmented ring Λ^e into the augmented ring Λ. We are thus in the situation treated in VIII,3, and we find homomorphisms

$$F^{\varphi}\colon\ H_n(\Lambda, {}_{\varepsilon}A) = \operatorname{Tor}_n^{\Lambda^e}({}_{\varepsilon}A,\Lambda) \to \operatorname{Tor}_n^{\Lambda}(A,K)$$

$$F_{\varphi}\colon\ \operatorname{Ext}_{\Lambda}^n(K,C) \to \operatorname{Ext}_{\Lambda^e}^n(\Lambda,C_{\varepsilon}) = H^n(\Lambda,C_{\varepsilon})$$

for a right Λ-module A and a left Λ-module C.

THEOREM 2.1. *If the supplemented K-algebra Λ is K-projective, then F^{φ} and F_{φ} are isomorphisms, and for each Λ^e-projective resolution X of Λ, the complex $X \otimes_{\Lambda} K$ is a Λ-projective resolution of $K = \Lambda \otimes_{\Lambda} K$ as a left Λ-module.*

We begin with

LEMMA 2.2. *Let B be a two-sided Λ-module. Then the homomorphism*

$$\tau: {}_\varepsilon\Lambda \otimes_{\Lambda^e} B \to B \otimes_\Lambda K$$

given by $\tau(\lambda \otimes b) = \lambda b \otimes 1$ is an isomorphism.

Indeed, define a homomorphism σ: $B \otimes_\Lambda K \to {}_\varepsilon\Lambda \otimes_{\Lambda^e} B$ by $\sigma(b \otimes k) = k \otimes b$. Then

$$\sigma\tau(\lambda \otimes_{\Lambda^e} b) = \sigma(\lambda b \otimes_\Lambda 1) = 1 \otimes_{\Lambda^e} \lambda b = \lambda \otimes_{\Lambda^e} b$$

$$\tau\sigma(b \otimes_\Lambda k) = \tau(k \otimes_{\Lambda^e} b) = kb \otimes_\Lambda 1 = b \otimes_\Lambda k.$$

Thus τ is an isomorphism.

We now return to the proof of 2.1. It suffices to verify conditions (i) and (ii) of the mapping theorem VIII,3.1. Applying the lemma with $B = \Lambda$ we find that condition (i) holds. Next we take $B = X$ where X is a Λ^e-projective resolution of Λ. Then

$$\operatorname{Tor}_n^{\Lambda^e}({}_\varepsilon\Lambda,\Lambda) = H_n({}_\varepsilon\Lambda \otimes_{\Lambda^e} X) \approx H_n(X \otimes_\Lambda K).$$

Since Λ is K-projective, it follows from IX,2.4 that X is Λ^*-projective, i.e. X is Λ-projective as a right Λ-module. Therefore $H_n(X \otimes_\Lambda K) = \operatorname{Tor}_n^\Lambda(\Lambda,K)$ which is zero for $n > 0$. This proves condition (ii) of the mapping theorem and thus completes the proof of theorem 2.1.

Theorem 2.1 reduces completely the homology and cohomology theory of a K-projective supplemented K-algebra to the Hochschild theory. This will allow us to replace the notation $\operatorname{Tor}_n^\Lambda(A,K)$ by the homological notation $H_n(\Lambda,{}_\varepsilon A)$ for a right Λ-module A. To further simplify the notation we shall write $H_n(\Lambda,A)$ omitting ε. Similarly for cohomology.

We may apply 2.1 to the standard complex $S(\Lambda)$ or the normalized standard complex $N(\Lambda)$ of IX,6 (note that the normalized standard complex may be used because $\Lambda' = \Lambda/K \approx I$ is K-projective as a K-direct summand of Λ). We denote the complexes $S(\Lambda) \otimes_\Lambda K$ and $N(\Lambda) \otimes_\Lambda K$ by $S(\Lambda,\varepsilon)$ and $N(\Lambda,\varepsilon)$ respectively. To give an explicit description of $N(\Lambda,\varepsilon)$ note that $N_n(\Lambda) = \Lambda \otimes_K \tilde{N}_n(\Lambda) \otimes_K \Lambda$ so that $N_n(\Lambda,\varepsilon) = N_n(\Lambda) \otimes_\Lambda K = \Lambda \otimes_K \tilde{N}_n(\Lambda)$. The differentiation operator in

$$N(\Lambda,\varepsilon) = \sum_{n \geqq 0} \Lambda \otimes_K \tilde{N}_n(\Lambda)$$

is

$$d_1[\lambda] = \lambda - \varepsilon\lambda,$$

$$d_n[\lambda_1, \ldots, \lambda_n] = \lambda_1[\lambda_2, \ldots, \lambda_n] + \sum_{0<i<n} (-1)^i[\lambda_1, \ldots, \lambda_i\lambda_{i+1}, \ldots, \lambda_n]$$

$$+ (-1)^n[\lambda_1, \ldots, \lambda_{n-1}](\varepsilon\lambda_n).$$

We recall that the symbol $[\lambda_1, \ldots, \lambda_n](n \geqq 0)$ is K-multilinear and is zero whenever $\lambda_i = 1$ for some $i = 1, \ldots, n$. If this last condition is dropped, we obtain the (unnormalized) complex $S(\Lambda,\varepsilon)$.

The whole discussion of this section could be repeated by regarding K as a right Λ-module and thus treating the supplemented algebra Λ as a right augmented ring. If Λ is K-projective then $S(\varepsilon,\Lambda) = K \otimes_\Lambda S(\Lambda)$ and $N(\varepsilon,\Lambda) = K \otimes_\Lambda N(\Lambda)$ are Λ-projective resolutions of K as a right Λ-module.

We conclude this section by discussing the case when the coefficient module A (for homology and cohomology) has trivial Λ-operators, i.e. is simply a K-module.

PROPOSITION 2.3. *Let Λ be a supplemented K-algebra and A a K-module. Given a projective resolution X of K as a left Λ-module, define $\bar{X} = K \otimes_\Lambda X$. We then have natural isomorphisms*

$$\operatorname{Tor}_n^\Lambda (A,K) \approx H_n(A \otimes_K \bar{X})$$
$$\operatorname{Ext}_\Lambda^n (K,A) \approx H^n(\operatorname{Hom}_K (\bar{X},A)).$$

PROOF. We have

$$\operatorname{Tor}_n^\Lambda (A,K) = H_n(A \otimes_\Lambda X) = H_n((A \otimes_K K) \otimes_\Lambda X) \approx H_n(A \otimes_K (K \otimes_\Lambda X)) = H_n(A \otimes_K \bar{X}).$$

A similar proof applies to cohomology.

Applying the above result to the normalized standard complex $N(\Lambda,\varepsilon)$ (under the assumption that Λ is K-projective) we find the complex $N(\varepsilon,\Lambda,\varepsilon) = K \otimes_\Lambda N(\Lambda,\varepsilon)$ composed of modules $N_n(\varepsilon,\Lambda,\varepsilon) = K \otimes_\Lambda N_n(\Lambda,\varepsilon) = K \otimes_\Lambda \Lambda \otimes_K \tilde{N}_n(\Lambda) = \tilde{N}_n(\Lambda)$. The differentiation operator in the complex

$$N(\varepsilon,\Lambda,\varepsilon) = \sum_{n \geqq 0} \tilde{N}_n(\Lambda)$$

is

$$d_1[\lambda] = 0,$$

$$d_n[\lambda_1, \ldots, \lambda_n] = (\varepsilon\lambda_1)[\lambda_2, \ldots, \lambda_n] + \sum_{0<i<n} (-1)^i[\lambda_1, \ldots, \lambda_i\lambda_{i+1}, \ldots, \lambda_n] + (-1)^n[\lambda_1, \ldots, \lambda_{n-1}](\varepsilon\lambda_n).$$

3. AUGMENTED MONOIDS

We return to the discussion of monoids initiated in VIII,2. Let Π be a monoid, L a ring (not necessarily commutative) and ε: $L(\Pi) \to L$ an augmentation of the ring $L(\Pi)$. As we have seen in VIII,2, the augmentation is uniquely determined by a function $\Pi \to L$ also denoted by ε, satisfying

$$\varepsilon(xx') = (\varepsilon x)(\varepsilon x'), \qquad \varepsilon 1 = 1$$

and whose values are in the center of L. The monoid Π together with the augmentation function $\varepsilon\colon \Pi \to L$ is called an *augmented monoid.*

We shall show that for the purposes of homology theory, one may always assume that L is commutative. Indeed, consider a factorization

$$\Pi \xrightarrow{\varepsilon'} K \xrightarrow{\mu} L$$

of the augmentation ε, where K is a commutative ring, μ is a ring homomorphism with values in the center of L, and ε' is an augmentation function. Such a factorization always exists; it suffices to take $K =$ center of L. The map μ defines on L the structure of a K-algebra. It is then clear that $L(\Pi)$ may be identified with the tensor product of K-algebras $L \otimes_K K(\Pi)$. The augmentation $\varepsilon\colon L(\Pi) \to L$ then becomes $L \otimes \varepsilon'\colon L \otimes_K K(\Pi) \to L$. Since the elements $x \in \Pi$ form a K-base for $K(\Pi)$, it follows that $K(\Pi)$ is K-projective. We are thus in a position to apply 1.1, obtaining isomorphisms

(1) $$\operatorname{Tor}_n^{K(\Pi)}(A,K) \approx \operatorname{Tor}_n^{L(\Pi)}(A,L),$$

(1a) $$\operatorname{Ext}_{K(\Pi)}^n(K,C) \approx \operatorname{Ext}_{L(\Pi)}^n(L,C),$$

for any right $L(\Pi)$-module A and any left $L(\Pi)$-module C.

Relations (1), (1a) place us squarely in the theory of supplemented algebras. Since $K(\Pi)$ is K-projective, we may (in view of 2.1) regard the left sides of (1) and (1a) also as Hochschild homology and cohomology groups. We shall use the notation $H_n(\Pi,A)$ and $H^n(\Pi,C)$ to denote the left sides of (1) and (1a). This notation does not exhibit the ring K and the augmentation $\varepsilon\colon \Pi \to K$ and will be used only if it is clear what these are. The group A is assumed to be a right $K(\Pi)$-module, while C is a left $K(\Pi)$-module. The augmentation ideal, i.e. the kernel of $\varepsilon\colon K(\Pi) \to K$ will usually be denoted by $I(\Pi)$.

We further minimize the role of K by using the expression "Π-module" instead of "$K(\Pi)$-module". Similarly we use the notation $\otimes_\Pi$, Hom_Π, Tor_n^Π, Ext_Π^n instead of $\otimes_{K(\Pi)}$, $\operatorname{Hom}_{K(\Pi)}$, $\operatorname{Tor}_n^{K(\Pi)}$, $\operatorname{Ext}_{K(\Pi)}^n$.

The most important examples of augmentations in a monoid Π are the following two: (1°) the *unit augmentation* $\varepsilon x = 1$ for all $x \in \Pi$; (2°) the *zero augmentation* $\varepsilon x = 0$ for $x \in \Pi$, $x \neq 1$ and $\varepsilon 1 = 1$; this augmentation may be used only if the relation $xx' = 1$ in Π implies $x = 1 = x'$ (i.e. if no inverses exist in Π). In the case of either of these two augmentations the ring K may be taken to be the ring Z of integers.

The standard complexes $S(\Lambda,\varepsilon)$ and $N(\Lambda,\varepsilon)$ for $\Lambda = K(\Pi)$ will be denoted simply by $S(\Pi,\varepsilon)$ and $N(\Pi,\varepsilon)$. The description given in § 2 need not be repeated. We only observe that the elements $[x_1, \cdots, x_n]$,

$x_i \in \Pi$, form a $K(\Pi)$-basis for $S_n(\Pi,\varepsilon)$. The same applies to $N_n(\Pi,\varepsilon)$, except that in $N_n(\Pi,\varepsilon)$ we set $[x_1, \dots, x_n] = 0$ whenever one of the coordinates x_i is 1.

The functorial properties relative to the variable Π may be easily derived. Let Π, Π' be monoids with augmentations $\varepsilon\colon \Pi \to K$, $\varepsilon'\colon \Pi' \to K$. A map $\varphi\colon \Pi' \to \Pi$ is a multiplicative map such that $\varphi 1 = 1$ and $\varepsilon\varphi = \varepsilon'$. This clearly induces a map $\varphi\colon K(\Pi') \to K(\Pi)$ of supplemented algebras, and thus yields homomorphisms

$$F^{\varphi}\colon H_n(\Pi',A) \to H_n(\Pi,A)$$

$$F_{\varphi}\colon H^n(\Pi,C) \to H^n(\Pi',C)$$

where A is a right Π-module and C is a left Π-module.

As an application of the mapping theorem we obtain

PROPOSITION 3.1. *In order that F^{φ} be an isomorphism for all right Π-modules A it is necessary and sufficient that*

(i) $K(\Pi) \otimes_{\Pi'} K \approx K$ *under the mapping* $x \otimes k \to \varepsilon(x)k$,

(ii) $H_n(\Pi',K(\Pi)) = 0$ *for* $n > 0$.

If these conditions are satisfied then F_{φ} also is an isomorphism for all left Π-modules C. Further, for any Π'-projective resolution X of K, the complex $K(\Pi) \otimes_{\Pi'} X$ is a Π-projective resolution of K.

It should be noted that condition (i) is always satisfied when $\varphi\colon \Pi' \to \Pi$ is an epimorphism.

4. GROUPS

We assume here that Π is a group. We first show that no generality is lost by assuming that the augmentation is the unit augmentation. Indeed, let $\varepsilon\colon \Pi \to K$ be any augmentation function. We denote by $K(\Pi,\varepsilon)$ the K-algebra $K(\Pi)$ with the supplementation given by ε and by $K(\Pi,i)$ the same algebra with the supplementation given by the unit augmentation i. The mapping $\varphi\colon K(\Pi,\varepsilon) \to K(\Pi,i)$ given by $x \to (\varepsilon x)x$ for $x \in \Pi$ is then an isomorphism of supplemented algebras.

As a consequence we shall always assume that the augmentation is the unit augmentation. As a further consequence we shall always assume that $K = Z$ is the ring of integers. We shall therefore deal with the algebra $Z(\Pi)$ supplemented by $\varepsilon(\sum z_i x_i) = \sum z_i$, $z_i \in Z$, $x_i \in \Pi$. The augmentation ideal $I(\Pi)$ is then a free abelian group with the elements $x - 1$ as basis ($x \in \Pi$).

The standard complexes $S(\Pi)$ and $N(\Pi)$ may be put in a somewhat different form by a change of basis. We introduce the symbols

$$(1) \qquad (x_0, \dots, x_n) = x_0[x_0^{-1}x_1, \dots, x_{n-1}^{-1}x_n];$$

then

$$x(x_0, \ldots, x_n) = (xx_0, \ldots, xx_n) \tag{2}$$

$$[x_1, \ldots, x_n] = (1, x_1, x_1x_2, \ldots, x_1 \cdots x_n) \tag{3}$$

$$d_n(x_0, \ldots, x_n) = \sum_{i=0}^{n} (-1)^i (x_0, \ldots, \hat{x}_i, \ldots, x_n). \tag{4}$$

It follows that $S_n(\Pi)$ is the free group generated by the elements $(x_0, \ldots, x_n)$, $x_i \in \Pi$ with Π-operators defined by (2) and with differentiation given by (4). This form of the standard complex is known as the *homogeneous form*. The same applies to the normalized standard complex $N(\Pi)$ provided we set $(x_0, \ldots, x_n) = 0$ whenever $x_{i-1} = x_i$ for some $i = 1, \ldots, n$. We observe that the differentiation operator does not involve the operators of Π and has the standard form encountered in the homology theory of simplicial complexes.

For low dimensions the homology and cohomology groups may be described as follows:

$H_0(\Pi,A) = A/AI = A_\Pi$ is the factor group of A by the subgroup generated by the elements $a(x-1)$, $a \in A$, $x \in \Pi$.

$H^0(\Pi,C) = C^\Pi$ is the subgroup of *invariant* elements of C, i.e. elements c with $xc = c$ for all $x \in \Pi$.

$H^1(\Pi,C)$ is the group of all crossed homomorphisms $f\colon \Pi \to C$ (i.e. all functions satisfying $f(xy) = x(fy) + fx$, for $x,y \in \Pi$) reduced modulo principal crossed homomorphisms (i.e. functions of the form $fx = xc - c$ for a fixed $c \in C$).

If Π operates trivially on A (i.e. $A^\Pi = A$) these results simplify as follows

$$H_0(\Pi,A) = A = H^0(\Pi,A) \tag{5}$$

$$H^1(\Pi,A) \approx \operatorname{Hom}(\Pi,A). \tag{6}$$

Clearly $\operatorname{Hom}(\Pi,A)$ may be replaced by $\operatorname{Hom}(\Pi/[\Pi,\Pi],A)$ where $[\Pi,\Pi]$ is the commutator subgroup of Π. In the case of trivial operators we can also calculate the group $H_1(\Pi,A)$. We have from formula (4) of § 1 that $H_1(\Pi,A) \approx A \otimes I/I^2$ where $I = I(\Pi)$. We establish maps

$$\varphi\colon I/I^2 \to \Pi/[\Pi,\Pi], \qquad \psi\colon \Pi/[\Pi,\Pi] \to I/I^2$$

by setting $\varphi(x-1) = x$, $\psi x = x - 1$. Simple calculations show that we obtain an isomorphism

$$I/I^2 \approx \Pi/[\Pi,\Pi]. \tag{7}$$

Thus we have

$$(8)\qquad H_1(\Pi,A) \approx A \otimes \Pi/[\Pi,\Pi].$$

We shall now prove a proposition which shows that in some cases the homology and cohomology groups of a group are the same as those of a monoid contained in the group.

PROPOSITION 4.1. *Let* Π *be a group and* Π' *a monoid contained in* Π *such that each element of* Π *has the form* $x^{-1}y$, $x \in \Pi'$, $y \in \Pi'$. *Then the homomorphisms*

$$H_n(\Pi',A) \to H_n(\Pi,A), \qquad H^n(\Pi,C) \to H^n(\Pi',C)$$

induced by the inclusion map $\Pi' \to \Pi$ *are isomorphisms. Further, if* X *is a* Π'*-projective resolution of* Z, *then* $Z(\Pi) \otimes_{\Pi'} X$ *is a* Π*-projective resolution of* Z.

PROOF. We apply 3.1. To verify that condition (i) of 3.1 is satisfied we must show that the relation $z \otimes_{\Pi'} 1 = 1 \otimes_{\Pi'} 1$ is valid in $Z(\Pi) \otimes_{\Pi'} Z$ for each $z \in \Pi$. Let then $z = x^{-1}y$, $x \in \Pi'$, $y \in \Pi'$. Then, writing $\otimes$ for $\otimes_{\Pi'}$, we have

$$z \otimes 1 = x^{-1}y \otimes 1 = x^{-1} \otimes y1 = x^{-1} \otimes 1 = x^{-1} \otimes x1 = 1 \otimes 1.$$

To verify condition (ii) of 3.1 we must show that $\operatorname{Tor}_n^{\Pi'}(Z(\Pi),Z) = 0$ for $n > 0$. Since $\operatorname{Tor}_n^{\Pi'}$ commutes with direct limits (VI,1.3) it suffices to prove that $Z(\Pi)$ is a direct limit of Π'-projective right modules. Indeed, we shall show that $Z(\Pi)$ is the union of a directed family of submodules M_s each of which is isomorphic with $Z(\Pi')$.

For each $s \in \Pi$ consider the map $f_s\colon \Pi' \to \Pi$ given by $f_s x = sx$. Since Π is a group, f_s induces a Π'-isomorphism of $Z(\Pi')$ with a right Π'-submodule M_s of $Z(\Pi)$. Since $s \in M_s$, $Z(\Pi)$ is the union of the submodules M_s. There remains to be shown that the family $\{M_s\}$ is directed. Given $s,t \in \Pi$ we have $s^{-1}t = v^{-1}w$ with $v,w \in \Pi'$. Thus $sv^{-1} = tw^{-1}$. Setting $u = sv^{-1} = tw^{-1}$ we have

$$sx = u(vx) \in M_u, \qquad tx = u(wx) \in M_u, \qquad x \in \Pi'.$$

Consequently $M_s \subset M_u$ and $M_t \subset M_u$ so that the family $\{M_s\}$ is directed, as required.

COROLLARY 4.2. *Let* Π *be an abelian group and* Π' *a submonoid of* Π *generating* Π. *Then the conclusions of* 4.1 *hold.*

One fundamental difference between groups and monoids is the existence of the transformation $\omega x = x^{-1}$ which yields an isomorphism

$$\omega\colon Z(\Pi) \approx (Z(\Pi))^*.$$

This "antipodism" allows us to convert any right Π-module A into a left Π-module by setting

$$xa = ax^{-1}.$$

It is because of this antipodism, that it is possible to build the whole homology and cohomology theory of groups using left Π-modules exclusively.

5. EXAMPLES OF RESOLUTIONS

The standard complex $S(\Pi)$ has the advantage of being defined for all augmented monoids Π and being a covariant functor of the variable Π. However, for each individual monoid Π there are usually simpler Π-projective resolutions of K which lead more quickly to the computation of the homology and cohomology groups of Π. We shall discuss here a number of such examples.

As our first example we shall treat simultaneously the following cases:

Π is the *free monoid* generated by a (finite or infinite) set of letters $\{x_\alpha\}$ with any augmentation $\varepsilon\colon \Pi \to K$.

Π is the *free group* generated by a set of letters $\{x_\alpha\}$ with the unit augmentation $\Pi \to Z$.

Let C be any left $K(\Pi)$-module. It is easy to verify that each crossed homomorphism $f\colon \Pi \to C$ is uniquely determined by its values on the elements x_α and that these may be arbitrarily prescribed in advance. In view of the 1–1-correspondence between crossed homomorphisms and $K(\Pi)$-homomorphisms $g\colon I(\Pi) \to C$ it follows that each such homomorphism g is uniquely determined by its values on $x_\alpha - \varepsilon x_\alpha$ and that these values may be arbitrarily prescribed. It follows that the elements $\{x_\alpha - \varepsilon x_\alpha\}$ form a $K(\Pi)$-base of $I(\Pi)$ as a left $K(\Pi)$-module. Thus $I(\Pi)$ is $K(\Pi)$-free. Therefore the exact sequence

$$0 \longrightarrow I(\Pi) \longrightarrow K(\Pi) \xrightarrow{\varepsilon} K \longrightarrow 0$$

yields a projective $K(\Pi)$-resolution of K. This implies the well known result:

$$H_n(\Pi,A) = 0 = H^n(\Pi,C) \qquad \text{for } n > 1.$$

Following formula (9) of VIII,1 we have $H_1(\Pi,K) = \mathrm{Tor}_1^\Pi(K,K) \approx I/I^2$. The identity $xy - 1 = (x-1) + (y-1) + (x-1)(y-1)$ implies that in the module I/I^2 the images of the elements $x_\alpha - 1$ form a K-base. Thus if $\{x_\alpha\}$ is not empty this module is non-zero. This shows that

$$\text{l.dim}_{K(\Pi)} K = 1$$

if Π is the free monoid (or group) on a non-empty base $\{x_\alpha\}$. Similarly $\text{r.dim}_{K(\Pi)} K = 1$.

Incidentally we have shown that if Π is a free group generated by the letters $\{x_\alpha\}$ then for each $x \in \Pi$ we have

$$x - 1 = \sum a_\alpha(x)(x_\alpha - 1)$$

with $a_\alpha(x) \in Z(\Pi)$ uniquely determined. The functions a_α are crossed homomorphisms $\Pi \to Z(\Pi)$ uniquely determined by the conditions $a_\alpha(x_\alpha) = 1$ and $a_\alpha(x_\beta) = 0$ for $\alpha \neq \beta$. The elements $a_\alpha(x)$ are called the partial derivatives of x and are written as $\dfrac{\partial x}{\partial x_\alpha}$.

Our next example is that of the *free abelian monoid* Π generated by a finite set of letters $x_1, \ldots, x_n$ with an arbitrary augmentation $\varepsilon\colon \Pi \to K$. We introduce the elements $x_i' = x_i - \varepsilon x_i$ of $K(\Pi)$. It is then clear that the ring $K(\Pi)$ may be identified with the ring $K[x_1', \ldots, x_n']$ of polynomials in $x_1', \ldots, x_n'$ with the augmentation given by $\varepsilon(x_i') = 0$. We thus fall into the case treated in VIII,4 and obtain the complex

$$K(\Pi) \otimes_K E(y_1, \ldots, y_n)$$

where $E(y_1, \ldots, y_n)$ is the exterior K-algebra on the letters $y_1, \ldots, y_n$, with differentiation

$$d_i(x \otimes y_{p_1} \cdots y_{p_i}) = \sum_{1 \leqq j \leqq i} (-1)^{j+1} x(x_{p_j} - \varepsilon x_{p_j}) \otimes y_{p_1} \cdots \hat{y}_{p_j} \cdots y_{p_i}.$$

In particular, the discussion carried out at the end of VIII,4 applies.

The last case treated here will be that of the *free abelian group* Π generated by a finite set of letters $x_1, \ldots, x_n$ (with the unit augmentation $\Pi \to Z$). In this group we have the free monoid Π' generated by $x_1, \ldots, x_n$ and 4.1 may be applied. Using the complex constructed above for the free monoid, we find the complex

$$Z(\Pi) \otimes E(y_1, \ldots, y_n)$$

with

$$d_i(x \otimes y_{p_1} \cdots y_{p_i}) = \sum_{1 \leqq j \leqq i} (-1)^{j+1} x(x_{p_j} - 1) \otimes y_{p_1} \cdots \hat{y}_{p_j} \cdots y_{p_i}.$$

Further examples will be given in Ch. XII for finite groups.

6. THE INVERSE PROCESS

We have seen in 2.1 that the homology and cohomology theory of a K-projective supplemented K-algebra is expressible in terms of the Hochschild theory. We shall see here that in some very important cases, the converse also holds: the Hochschild homology theory may be derived from the homology theory of supplemented algebras.

Let Λ be a supplemented K-algebra and assume that a K-algebra homomorphism

$$E\colon \Lambda \to \Lambda^e$$

is given, such that the diagram

$$\begin{array}{ccc} \Lambda & \xrightarrow{\varepsilon} & K \\ {\scriptstyle E}\downarrow & & \downarrow{\scriptstyle \eta} \\ \Lambda^e & \xrightarrow[\rho]{} & \Lambda \end{array}$$

is commutative. In this diagram ε is the augmentation of Λ, η is determined by the K-algebra structure of Λ and ρ is the augmentation of Λ^e. This commutativity relation is equivalent with the inclusion

$$EI \subset J. \tag{1}$$

We are now in the situation covered by VIII,3 and we obtain homomorphisms

$$F^E\colon \operatorname{Tor}_n^\Lambda(A_E,K) \to H_n(\Lambda,A)$$

$$F_E\colon H^n(\Lambda,A) \to \operatorname{Ext}_\Lambda^n(K,{}_EA)$$

where A is a two-sided Λ-module, ${}_EA$ (or A_E) is the left (or right) Λ-module obtained by regarding A as a left (or right) Λ^e-module and then defining the Λ-module structure using E. In particular, we shall denote by Λ_E^e the algebra Λ^e regarded (1°) as a left Λ^e-module, (2°) as a right Λ-module by means of the map E.

THEOREM 6.1. *Assume that the following conditions hold*

(E.1) $$J = \Lambda_E^e I$$

(E.2) Λ_E^e *is a projective right* Λ-*module.*

Then the maps F^E *and* F_E *are isomorphisms, and for each projective resolution* X *of* K *as a left* Λ-*module,* $\Lambda_E^e \otimes_\Lambda X$ *is a projective resolution of* Λ *as a left* Λ^e-*module.*

PROOF. We only need to verify conditions (i) and (ii) of the mapping theorem VIII,3.1.

From the exact sequence $0 \to I \to \Lambda \to K \to 0$ we deduce the exact sequence

$$\Lambda_E^e \otimes_\Lambda I \to \Lambda_E^e \to \Lambda_E^e \otimes_\Lambda K \to 0$$

which implies

$$\Lambda_E^e \otimes_\Lambda K \approx \operatorname{Coker}(\Lambda_E^e \otimes_\Lambda I \to \Lambda_E^e) = \Lambda_E^e/\Lambda_E^e I = \Lambda_E^e/J = \Lambda.$$

The isomorphism $\Lambda^e_E \otimes_\Lambda K \approx \Lambda$ is given by $\gamma \otimes_\Lambda 1 \to \rho\gamma$ and this proves condition (i) of the mapping theorem.

Condition (ii) of the mapping theorem is $\mathrm{Tor}^\Lambda_n(\Lambda^e_E, K) = 0$ for $n > 0$. This is a direct consequence of (E.2).

We return for a moment to the condition (E.1) above. The inclusion $\Lambda^e_E I \subset J$ is equivalent with the inclusion (1). The other inclusion $J \subset \Lambda^e_E I$ expresses the fact that J is contained in the left ideal of Λ^e generated by EI. Since J as a left ideal is generated by elements of the form $\lambda \otimes 1 - 1 \otimes \lambda^*$, $\lambda \in \Lambda$ (see IX,3.1), we find that in the presence of (1), condition (E.1) is equivalent with

($E.1'$). *For each* $\lambda \in \Lambda$, *the element* $\lambda \otimes 1 - 1 \otimes \lambda^*$ *is in the left ideal of* Λ^e *generated by* EI.

THEOREM 6.2. *Assume the conditions of* 6.1 *and that* Λ *is K-projective. Then*

$$\dim \Lambda = \mathrm{l.dim}_\Lambda K = \mathrm{r.dim}_\Lambda K.$$

If further K is semi-simple, then

$$\dim \Lambda = \mathrm{l.gl.dim}\, \Lambda = \mathrm{r.gl.dim}\, \Lambda.$$

PROOF. The equality $\dim \Lambda = \mathrm{l.dim}_\Lambda K$ follows directly from 2.1 and 6.1. To prove $\dim \Lambda = \mathrm{r.dim}_\Lambda K$ it suffices to prove $\dim \Lambda = \mathrm{l.dim}_{\Lambda^*} K$. The map E: $\Lambda \to \Lambda^e$ induces a map E^*: $\Lambda^* \to (\Lambda^e)^* = (\Lambda^*)^e$ and it is easy to see that conditions (E.1) and (E.2) still hold. Thus by the part of the theorem already established, $\dim \Lambda^* = \mathrm{l.dim}_{\Lambda^*} K$. Since $\dim \Lambda = \dim \Lambda^*$, the conclusion follows.

If K is semi-simple then, by IX,7.6, we have $\mathrm{l.gl.dim}\, \Lambda \leqq \dim \Lambda$. Since $\dim \Lambda = \mathrm{l.dim}_\Lambda K \leqq \mathrm{l.gl.dim}\, \Lambda$ we obtain $\dim \Lambda = \mathrm{l.gl.dim}\, \Lambda$. Similarly for $\mathrm{r.gl.dim}\, \Lambda$.

As an illustration of the inverse process consider the ring $\Lambda = Z(\Pi)$ (or more generally $\Lambda = K(\Pi)$) where Π is a group with unit augmentation. Define E: $\Lambda \to \Lambda^e$ by setting $Ex = x \otimes (x^{-1})^*$ for $x \in \Pi$. Then $\rho Ex = 1$ so that (1) holds. Since

$$x \otimes 1 - 1 \otimes x^* = (x \otimes 1)(1 \otimes 1 - x^{-1} \otimes x^*) = (x \otimes 1)E(1 - x^{-1})$$

it follows that ($E.1'$) holds. To verify (E.2) observe that the elements $1 \otimes x^*$ ($x \in \Pi$) form a base of Λ^e as a right Λ-module. Thus theorems 6.1 and 6.2 may be applied. Note that if A is a two-sided Λ-module then ${}_EA$ is the left Π-module with operators $a \to xax^{-1}$ and A_E is the right Π-module with operators $a \to x^{-1}ax$.

In XIII,5 we shall show that the inverse process can also be applied to the homology theory of Lie algebras.

7. SUBALGEBRAS AND SUBGROUPS

Let Λ and Γ be K-algebras and let $\varphi\colon \Lambda \to \Gamma$ be a K-algebra homomorphism. Given an augmentation $\varepsilon\colon \Gamma \to K$ which converts Γ into a supplemented algebra, we define Λ as a supplemented algebra using the augmentation $\varepsilon\varphi\colon \Lambda \to K$. In most cases Λ will be a subalgebra of Γ and φ will be the inclusion map. We shall use here the notations of II.6.

PROPOSITION 7.1. *If Γ, as a left (right) Λ-module, is projective, then for any left (right) Γ-module A, any Γ-projective or Γ-injective resolution X of A is also a Λ-projective or Λ-injective resolution of A.*

This follows directly from II,6.2 and II,6.2a.

PROPOSITION 7.2. *If Γ is projective as a left Λ-module, then we have isomorphisms*

$$\operatorname{Tor}_n^\Lambda(A,K) \approx \operatorname{Tor}_n^\Gamma(A_{(\varphi)},K), \qquad \operatorname{Ext}_\Gamma^n(K,{}^{(\varphi)}C) \approx \operatorname{Ext}_\Lambda^n(K,C)$$

for each right Λ-module A and each left Λ-module C.

This follows directly from VI,4.1.1 and VI,4.1.4. Similarly applying VI,4.1.2 and VI,4.1.3 we find

PROPOSITION 7.3. *If Γ is projective as a right Λ-module, then we have isomorphisms*

$$\operatorname{Tor}_n^\Lambda(A,K) \approx \operatorname{Tor}_n^\Gamma(A,\Gamma \otimes_\Lambda K), \qquad \operatorname{Ext}_\Gamma^n(\Gamma \otimes_\Lambda K,C) \approx \operatorname{Ext}_\Lambda^n(K,C)$$

for each right Γ-module A and each left Γ-module C.

We apply these results to groups. Let π be a subgroup of Π and let $\varphi\colon Z(\pi) \to Z(\Pi)$ be induced by the inclusion $\pi \to \Pi$. If $\{x_\alpha\}$ is a system of representatives of right cosets of π in Π, then it is clear that $\{x_\alpha\}$ is a base of $Z(\Pi)$ regarded as a left $Z(\pi)$-module. Similarly $Z(\Pi)$ is free as a right $Z(\pi)$-module. We may thus apply 7.2, replacing $A_{(\varphi)}$ and ${}^{(\varphi)}C$ by their definitions.

PROPOSITION 7.4. *Let π be a subgroup of Π. Then*

(1) $$H_n(\pi,A) \approx H_n(\Pi,A \otimes_\pi Z(\Pi))$$

(1a) $$H^n(\pi,C) \approx H^n(\Pi, \operatorname{Hom}_\pi(Z(\Pi),C))$$

for each right π-module A and each left π-module C.

Before we apply 7.3 we reinterpret the module $Z(\Pi) \otimes_\pi Z$ as follows. Let $Z(\Pi/\pi)$ be the free abelian group generated by the left cosets $x\pi$ of π in Π. Then Π operates on $Z(\Pi/\pi)$ on the left and we may identify $Z(\Pi/\pi)$ with $Z(\Pi) \otimes_\pi Z$.

PROPOSITION 7.5. *Let π be a subgroup of Π. Then*

(2) $$H_n(\pi,A) \approx \operatorname{Tor}_n^\Pi(A,Z(\Pi/\pi)),$$

(2a) $$H^n(\pi,C) \approx \operatorname{Ext}_\Pi^n(Z(\Pi/\pi),C),$$

for each right Π-module A and each left Π-module C.

Let π be a subgroup of Π and let $x \in \Pi$.
In the situation $(A_{\Pi}, {}_{\Pi}C)$ we define

$$c_x\colon\ A \otimes_{\pi} C \to A \otimes_{x\pi x^{-1}} C$$

by setting $c_x(a \otimes_{\pi} c) = ax^{-1} \otimes_{x\pi x^{-1}} xc$. Replacing C by a Π-projective resolution of Z, and passing to homology we obtain isomorphisms

$$c_x\colon\ H_n(\pi,A) \approx H_n(x\pi x^{-1},A)$$

for any right Π-module A.

Similarly, in the situation $({}_{\Pi}A, {}_{\Pi}C)$ we define

$$c_x\colon\ \mathrm{Hom}_{\pi}\,(A,C) \to \mathrm{Hom}_{x\pi x^{-1}}\,(A,C)$$

by setting $(c_x f)a = x(f(x^{-1}a))$. Replacing A by a Π-projective resolution of Z and passing to homology we obtain isomorphisms

$$c_x\colon\ H^n(\pi,C) \approx H^n(x\pi x^{-1},C)$$

for any left Π-module C.

The following properties of c_x are directly verified

(3) $$c_x c_y = c_{xy},$$

(4) If $x \in \pi$ then c_x is the identity.

Assume now that π is an *invariant* subgroup of Π. Then $\pi = x\pi x^{-1}$ and it is clear from (3) and (4) that c_x defines left operators of Π/π on $H_n(\pi,A)$ and $H^n(\pi,C)$ for each right Π-module A and each left Π-module C.

There is another way in which these operators may be arrived at. The Π-module $Z(\Pi/\pi)$ appearing in (2) and (2a) is the group algebra of the group Π/π regarded as a left Π-module. The structure of $Z(\Pi/\pi)$ as a right Π/π-module can then be used to define right Π/π-operators of the groups appearing in (2) and left Π/π-operators of the groups appearing in (2a). The verification that these operators agree with the operators defined using c_x, is left to the reader.

In Ch. XIII we shall carry out a similar discussion for Lie algebras. In Ch. XVI we shall obtain further results using spectral sequences.

8. WEAKLY INJECTIVE AND PROJECTIVE MODULES

Let Λ be a K-algebra and $\eta\colon\ K \to \Lambda$ the natural map. A right Λ-module A is said to be *weakly projective* if it is η-projective in the sense of II,6. This means that the kernel of the map $g\colon\ A \otimes_K \Lambda \to A$ given by

$a \otimes \lambda \to a\lambda$ is a direct summand of $A \otimes_K \Lambda$ regarded as a right Λ-module using the right operators of Λ on Λ (not on A!). A similar definition applies to left Λ-modules using $\Lambda \otimes_K A \to A$.

Similarly a left Λ module C is said to be *weakly injective* if it is η-injective in the sense of II,6. This means that the image of the homomorphism h: $C \to \operatorname{Hom}_K(\Lambda,C)$ which to each c assigns the homomorphism $\lambda \to \lambda c$ is a direct summand of $\operatorname{Hom}_K(\Lambda,C)$ regarded as a left Λ-module using the right operators of Λ on Λ. Similarly for right Λ-modules.

PROPOSITION 8.1. *Let M be a K-module, and A a weakly projective right Λ-module. Then $M \otimes_K A$ is weakly projective and* $\operatorname{Hom}_K(A,M)$ *is weakly injective.*

PROOF. We shall only prove the second part. Consider the commutative diagram

$$\begin{array}{ccc} \operatorname{Hom}_K(A \otimes_K \Lambda, M) & \xleftarrow{\;s\;} & \operatorname{Hom}_K(\Lambda, \operatorname{Hom}_K(A,M)) \\ {\scriptstyle \operatorname{Hom}_K(g,M)}\nwarrow & & \nearrow {\scriptstyle h} \\ & \operatorname{Hom}_K(A,M) & \end{array}$$

where s is the Λ-isomorphism of II,5.2. Since A is weakly projective, $\operatorname{Ker} g$ is a direct Λ-summand of $A \otimes_K \Lambda$. It follows that the image of $\operatorname{Hom}_K(g,M)$ is a Λ-direct summand. Thus the same holds for h, so that $\operatorname{Hom}_K(A,M)$ is weakly injective.

PROPOSITION 8.2. *Let Λ be a K-projective K-algebra. In the situation $(A_\Lambda, {}_\Lambda C)$, if C is K-projective and A is weakly projective, then*

$$\operatorname{Tor}_n^\Lambda(A,C) = 0 \qquad \text{for } n > 0.$$

PROPOSITION 8.2a. *Let Λ be a K-projective K-algebra. In the situation $({}_\Lambda A, {}_\Lambda C)$, if A is K-projective and C is weakly injective then*

$$\operatorname{Ext}_\Lambda^n(A,C) = 0 \qquad \text{for } n > 0.$$

These are immediate consequences of VI,4.2.1 and VI,4.2.4. As a consequence we obtain

COROLLARY 8.3. *Let Λ be a K-projective supplemented K-algebra. Then*

$$H_n(\Lambda,A) = 0 = H^n(\Lambda,C) \qquad \text{for } n > 0$$

for any weakly projective right Λ-module A and any weakly injective left Λ-module C.

We now turn to the discussion of weakly projective and injective modules in the case $\Lambda = Z(\Pi)$ where Π is a group.

Let A be a right Π-module. On the group $A \otimes Z(\Pi)$ we consider two right Π-module structures given for $a \in A$, $x,y \in \Pi$ by

(1) $$(a \otimes x)y = a \otimes xy$$

(1′) $$(a \otimes x)y = ay \otimes xy$$

respectively. The mapping φ: $a \otimes x \to ax \otimes x$ maps structure (1) isomorphically onto structure (1′). In defining weakly projective modules we used the map g: $A \otimes Z(\Pi) \to A$ given by $g(a \otimes x) = ax$ which was a Π-homomorphism on the structure (1). The map $g' = g\varphi^{-1}$ is given by $g'(a \otimes x) = a$, and is a Π-homomorphism on the structure (1′). We thus obtain

Proposition 8.4. *A right Π-module A is weakly projective if and only if there exists a Π-homomorphism ν: $A \to A \otimes Z(\Pi)$ (rel. to structure (1′)) such that $g'\nu =$ identity.*

Let C be a left Π-module. On the group Hom $(Z(\Pi),C)$ we consider two left Π-module structures given for $f \in$ Hom $(Z(\Pi),C)$, $x,y \in \Pi$ by

(2) $$(yf)x = f(xy)$$

(2′) $$(yf)x = yf(y^{-1}x)$$

respectively. The mapping $f \to \psi f$ given by $(\psi f)x = x(fx^{-1})$ maps structure (2) isomorphically onto structure (2′). In defining weakly injective modules we used the map h: $C \to$ Hom $(Z(\Pi),C)$ given by $(hc)x = xc$. The map $h' = \psi h$: $C \to$ Hom $(Z(\Pi),C)$ is given by $(h'c)x = c$, and is a Π-homomorphism on structure (2′). Thus we obtain

Proposition 8.4a. *A left Π-module C is weakly injective if and only if there exists a Π-homomorphism μ:* Hom $(Z(\Pi),C) \to C$ *(rel. to structure (2′)) such that $\mu h' =$ identity, i.e. such that if fx is constant and has value c then $\mu f = c$. Such a function μ will be called a* mean.

As an application of the above criteria we prove the following two propositions that will be used in Ch. XVI.

Proposition 8.5. *In the situation $({}_\Pi A, {}_\Pi C)$ assume that A is weakly projective. Then $A \otimes C$ with left operators*

$$x(a \otimes c) = xa \otimes xc$$

is weakly projective. Similarly Hom (A,C) *with left operators*

$$(xf)a = x(f(x^{-1}a))$$

is weakly injective.

The proof is similar to that of 8.1 and uses the diagrams

$$\begin{array}{ccc} (\Lambda \otimes A) \otimes C & \xrightarrow{r} & \Lambda \otimes (A \otimes C) \\ & {\scriptstyle g' \otimes C} \searrow \quad \swarrow {\scriptstyle g'} & \\ & A \otimes C & \end{array}$$

$$\begin{array}{ccc} \operatorname{Hom}(\Lambda \otimes A, C) & \xleftarrow{s'} & \operatorname{Hom}(\Lambda, \operatorname{Hom}(A,C)) \\ & {\scriptstyle \operatorname{Hom}(g',C)} \nwarrow \quad \nearrow {\scriptstyle h'} & \\ & \operatorname{Hom}(A,C) & \end{array}$$

where $\Lambda = Z(\Pi)$ is treated as a left Π-module. The horizontal maps r and s' are given by II,5.1 and II,5.2′ and are Π-isomorphisms.

PROPOSITION 8.6. *A left Π-module A is weakly projective if and only if there exists a Z-endomorphism $\rho\colon A \to A$ such that*

(i) *for each $a \in A$, $\rho(x^{-1}a) = 0$ for all but a finite number of elements $x \in \Pi$,*

(ii) $a = \sum_{x \in \Pi} x\rho(x^{-1}a)$ *for all $a \in A$.*

PROOF. Let $g'\colon Z(\Pi) \otimes A \to A$ be given by $g'(x \otimes a) = a$. If A is weakly projective then, by 8.4, there exists a Π-homomorphism $\nu\colon A \to Z(\Pi) \otimes A$ such that

$$g'\nu a = a \qquad \text{for all } a \in A.$$

Here $Z(\Pi) \otimes A$ has operators $y(x \otimes a) = yx \otimes ya$. Since the elements $x \in \Pi$ form a Z-base for $Z(\Pi)$ each element νa can be written as a finite linear combination

$$\nu a = \sum x \otimes g(x,a), \qquad x \in \Pi. \tag{3}$$

Since for $y \in \Pi$

$$\nu(ya) = \sum_x x \otimes g(x,ya)$$

$$y(\nu a) = \sum_x yx \otimes yg(x,a) = \sum_x x \otimes yg(y^{-1}x,a)$$

it follows that the condition $\nu(ya) = y(\nu a)$ is equivalent to

$$g(x,ya) = yg(y^{-1}x,a) \qquad \text{for all } x,y \in \Pi. \tag{4}$$

This in turn is equivalent to

$$g(x,a) = xg(1,x^{-1}a) \qquad \text{for all } x \in \Pi. \tag{5}$$

Therefore setting $\rho a = g(1,a)$ we have

$$g(x,a) = x\rho(x^{-1}a). \tag{6}$$

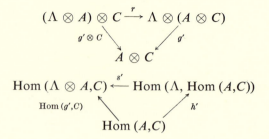

Since for a fixed a, $g(x,a) = 0$ for all but a finite number of elements $x \in \Pi$, condition (i) follows. Finally

$$a = g'\nu a = g'(\sum_x x \otimes g(x,a)) = \sum_x g(x,a) = \sum_x x\rho(x^{-1}a).$$

Conversely, given $\rho\colon A \to A$ satisfying (i) and (ii), we may define $g(x,a)$ using (6) and ν using (3). Then ν is a Π-homomorphism and $g'\nu$=identity.

In the discussion concerning groups the ground ring Z may be replaced by any commutative ring K.

EXERCISES

1. Let Λ be a supplemented K-algebra, and A a K-module. Show that $\Lambda \otimes_K A$ is Λ-projective if and only if A is K-projective. Assume further that K is such that K-projective modules are K-free. Then show that if $\Lambda \otimes_K A$ is Λ-projective then it is Λ-free.

2. Show that the inverse process of § 6 can be applied to the algebras $\Lambda = F_K(x_1, \dots, x_n)$ and $\Lambda = K[x_1, \dots, x_n]$ supplemented by $\varepsilon x_i = 0$ using the map $E\colon \Lambda \to \Lambda^e$ given by $Ex_i = x_i \otimes 1 - 1 \otimes x_i^*$.

3. Let Λ be a K-projective supplemented K-algebra with K a hereditary (commutative) ring. Establish the exact splitting sequences

$$0 \to H_n(\Lambda,K) \otimes_K A \to H_n(\Lambda,A) \to \mathrm{Tor}_1^K (H_{n-1}(\Lambda,K),A) \to 0,$$

$$0 \to \mathrm{Ext}_K^1 (H^{n-1}(\Lambda,K),A) \to H^n(\Lambda,A) \to \mathrm{Hom}_K (H^n(\Lambda,K),A) \to 0,$$

for any module A with trivial Λ-operators.

4. Let Π be a group (or a monoid with an augmentation $\Pi \to Z$). Establish the splitting exact sequences

$$0 \to H_n(\Pi,Z) \otimes A \to H_n(\Pi,A) \to \mathrm{Tor}_1 (H_{n-1}(\Pi,Z),A) \to 0$$

$$0 \to \mathrm{Ext}^1 (H^{n-1}(\Pi,Z),A) \to H^n(\Pi,A) \to \mathrm{Hom}\, (H^n(\Pi,Z),A) \to 0$$

for any abelian group A with trivial Π-operators.

5. Let Λ be a K-projective supplemented K-algebra. Then $\Lambda' = \mathrm{Coker}\,(K \to \Lambda)$ may be identified with the augmentation ideal I. Using this remark give a description of the complex $T(\Lambda,\varepsilon)$ using symbols $[\lambda_1, \dots, \lambda_n]$ with $\lambda_i \in I$.

6. Let Λ be a supplemented K-algebra and let z be an element in the center of Λ. For any left Λ-module C, multiplication by z defines an endomorphism $C \to C$ which induces an endomorphism $H^n(\Lambda,C) \to H^n(\Lambda,C)$. Show that the latter is given by multiplication by the element εz of K. State a similar result for homology.

CHAPTER XI

Products

Introduction. The functors Tor and Ext may be combined with each other using four product operations $\top, \bot, \vee, \wedge$. Each of these products involves three algebras, more precisely two algebras Λ and Γ and their tensor product $\Lambda \otimes \Gamma$. They satisfy a number of associative, anticommutative and other rules.

There are also internal products involving only one algebra. To obtain internal products $\cap, \cup$ corresponding to $\top, \bot$, we need an algebra homomorphism $\Lambda \otimes \Lambda \to \Lambda$; such a homomorphism always exists if Λ is commutative. For the internal products $\cup, \cap$ corresponding to $\vee, \wedge$, we need an algebra homomorphism $\Lambda \to \Lambda \otimes \Lambda$; such a homomorphism (usually called a "diagonal map") will be exhibited in a number of interesting cases.

The external and internal products may be computed using suitable multiplication formulae in complexes (§ 5).

The situation outlined above closely resembles that encountered in algebraic topology.

The general products for Tor and Ext are applied (§§ 6, 7) to the homology theories of Chs. IX and X. The internal products $\cup, \cap$ will be modified in § 8, using an "antipodism" $\Lambda \to \Lambda^*$; this leads to reduction theorems (§ 9) which generalize the "cup product reduction theorem" of Eilenberg-MacLane (*Ann. of Math. 48* (1947), 51–78, Ch. III).

1. EXTERNAL PRODUCTS

In this and the following sections we shall consider K-algebras over the same commutative ring K. We shall therefore simplify the notation by writing $\otimes$ and Hom instead of $\otimes_K$ and Hom_K.

Given complexes X and Y composed of K-modules we obtain new complexes $X \otimes Y$ and Hom (X, Y) and homomorphisms

(1)
$$\alpha: H(X) \otimes H(Y) \to H(X \otimes Y)$$

(1')
$$\alpha': H(\mathrm{Hom}\,(X, Y)) \to \mathrm{Hom}\,(H(X), H(Y))$$

defined in IV,6. The properties of these homomorphisms relative to the degrees imply homomorphisms

$$\alpha\colon\ H^p(X) \otimes H^q(Y) \to H^{p+q}(X \otimes Y)$$

$$\alpha'\colon\ H^{p+q}(\operatorname{Hom}(X,Y)) \to \operatorname{Hom}(H_p(X), H^q(Y)).$$

Since these homomorphisms are fundamental for the theory of products that we are about to develop, we give a brief survey of their definitions.

Let $h_1 \in H^p(X)$, $h_2 \in H^q(Y)$ and let $z_1 \in Z^p(X)$, $z_2 \in Z^q(Y)$ be representatives of h_1 and h_2. Regard z_1 and z_2 as elements of X^p and Y^q. Then $z_1 \otimes z_2 \in X^p \otimes Y^q$ and $d(z_1 \otimes z_2) = 0$, thus $z_1 \otimes z_2 \in Z^{p+q}(X \otimes Y)$. The element $\alpha(h_1 \otimes h_2) \in Z^{p+q}(X \otimes Y)$ is the class of $z_1 \otimes z_2$.

Let $h_1 \in H^{p+q}(\operatorname{Hom}(X,Y))$, $h_2 \in H_p(X)$ and let $f \in Z^{p+q}(\operatorname{Hom}(X,Y))$, $z_2 \in Z_p(X)$ be representatives of h_1 and h_2. Then $fz_2 \in Y^q$ and $d(fz_2) = 0$. Thus fz_2 determines an element of $H^q(Y)$ which is precisely $(\alpha' h_1)h_2$.

Let Λ and Γ be two K-algebras and consider the K-algebra

$$\Omega = \Lambda \otimes \Gamma$$

where as usual $\otimes$ stands for $\otimes_K$.

If A is a left Λ module and A' is a left Γ-module then $A \otimes A'$ is a left Ω-module.

Let X be a Λ-projective resolution of A and X' a Γ-projective resolution of A'. Then, by IX,2.5, $X \otimes X'$ is an Ω-projective left complex over $A \otimes A'$. Thus we have the homomorphisms (of degree zero)

(2) $$H(B \otimes_\Omega (X \otimes X')) \to \operatorname{Tor}^\Omega(B, A \otimes A') \qquad B_\Omega$$

(2′) $$\operatorname{Ext}_\Omega(A \otimes A', C) \to H(\operatorname{Hom}_\Omega(X \otimes X', C)). \qquad {}_\Omega C$$

PROPOSITION 1.1. (cf. IX,2.7). *If Λ and Γ are K-projective and* $\operatorname{Tor}_n^K(A,A') = 0$ *for $n > 0$ then $X \otimes X'$ is an Ω-projective resolution of $A \otimes A'$. In particular,* (2) *and* (2′) *are then isomorphisms.*

PROOF. Since Λ and Γ are K-projective, it follows from II,6.2 that X and X' are K-projective resolutions of A and A'. Thus $H_n(X \otimes X') = \operatorname{Tor}_n^K(A,A') = 0$ for $n > 0$.

We now place ourselves in the situation described by the symbol $({}_\Lambda A, C_\Lambda, {}_\Gamma A', C'_\Gamma)$ and define the homomorphism

$$\varphi_1\colon\ (C \otimes_\Lambda A) \otimes (C' \otimes_\Gamma A') \to (C \otimes C') \otimes_\Omega (A \otimes A')$$

given by $\varphi((c \otimes a) \otimes (c' \otimes a')) = (c \otimes c') \otimes (a \otimes a')$. Replacing A and A' by X and X' we obtain

$$\Phi_1\colon\ (C \otimes_\Lambda X) \otimes (C' \otimes_\Gamma X') \to (C \otimes C') \otimes_\Omega (X \otimes X').$$

Passing to homology and applying α we obtain the homomorphism

$$\operatorname{Tor}^{\Lambda}(C,A) \otimes \operatorname{Tor}^{\Gamma}(C',A') \to H((C \otimes C') \otimes_{\Omega} (X \otimes X')).$$

Composing this with (2) we obtain the $\top$-product

$$\top\colon \operatorname{Tor}^{\Lambda}(C,A) \otimes \operatorname{Tor}^{\Gamma}(C',A') \to \operatorname{Tor}^{\Omega}(C \otimes C'.A \otimes A').$$

Since $\top$ has degree zero, it yields maps

$$\boxed{\top\colon \operatorname{Tor}_p^{\Lambda}(C,A) \otimes \operatorname{Tor}_q^{\Gamma}(C',A') \to \operatorname{Tor}_{p+q}^{\Omega}(C \otimes C',A \otimes A').}$$

PROPOSITION 1.2.1. *For $p = q = 0$ the $\top$-product reduces to the map φ_1.*

This follows readily from the definition of $\top$ by applying the augmentation maps $X \to A$ and $X' \to A'$.

Next we consider the situation described by the symbol $({}_{\Lambda}A, C_{\Lambda}, {}_{\Gamma}A', {}_{\Gamma}C')$. We define $\operatorname{Hom}(C,C')$ as a left Ω-module by setting

$$((\lambda \otimes \gamma)f)c = \gamma(f(c\lambda)) \qquad c \in C, f \in \operatorname{Hom}(C,C')$$

and define the homomorphism

$$\varphi_2\colon \operatorname{Hom}_{\Omega}(A \otimes A', \operatorname{Hom}(C,C')) \to \operatorname{Hom}(C \otimes_{\Lambda} A, \operatorname{Hom}_{\Gamma}(A',C'))$$

by setting

$$((\varphi_2 f)(c \otimes a))a' = (f(a \otimes a'))c.$$

Replacing A and A' by X and X' we obtain

$$\Phi_2\colon \operatorname{Hom}_{\Omega}(X \otimes X', \operatorname{Hom}(C,C')) \to \operatorname{Hom}(C \otimes_{\Lambda} X, \operatorname{Hom}_{\Gamma}(X',C')).$$

Passing to homology and applying α' we obtain the homomorphisms

$$H(\operatorname{Hom}_{\Omega}(X \otimes X', \operatorname{Hom}(C,C')) \to \operatorname{Hom}(\operatorname{Tor}^{\Lambda}(C,A), \operatorname{Ext}_{\Gamma}(A',C')).$$

Composing this with (2') we obtain the $\perp$-product

$$\perp\colon \operatorname{Ext}_{\Omega}(A \otimes A', \operatorname{Hom}(C,C')) \to \operatorname{Hom}(\operatorname{Tor}^{\Lambda}(C,A), \operatorname{Ext}_{\Gamma}(A',C')).$$

Since $\perp$ has degree zero, it yields maps

$$\boxed{\perp\colon \operatorname{Ext}_{\Omega}^{p+q}(A \otimes A', \operatorname{Hom}(C,C')) \to \operatorname{Hom}(\operatorname{Tor}_p^{\Lambda}(C,A), \operatorname{Ext}_{\Gamma}^{q}(A',C')).}$$

PROPOSITION 1.2.2. *For $p = q = 0$, the $\perp$-product reduces to the map φ_2.*

For the remaining two products $\vee$ and $\wedge$ we make the following two assumptions:

(i) Λ and Γ are K-projective

(ii) $\operatorname{Tor}_n^K(A,A') = 0$ for $n > 0$.

It follows from 1.1 that the homomorphisms (2) and (2′) are isomorphisms.

We now place ourselves in the situation described by the symbol $({}_\Lambda A, {}_\Lambda C, {}_\Gamma A', {}_\Gamma C')$. We define the homomorphism

$$\varphi_3\colon \operatorname{Hom}_\Lambda(A,C) \otimes \operatorname{Hom}_\Gamma(A',C') \to \operatorname{Hom}_\Omega(A \otimes A', C \otimes C')$$

by setting

$$(\varphi_3(f \otimes f'))(a \otimes a') = fa \otimes f'a'.$$

Replacing A and A' by X and X', we obtain

$$\Phi_3\colon \operatorname{Hom}_\Lambda(X,C) \otimes \operatorname{Hom}_\Gamma(X',C') \to \operatorname{Hom}_\Omega(X \otimes X', C \otimes C').$$

Passing to homology and applying α we obtain the homomorphism

$$\operatorname{Ext}_\Lambda(A,C) \otimes \operatorname{Ext}_\Gamma(A',C') \to H(\operatorname{Hom}_\Omega(X \otimes X', C \otimes C'))$$

Combining this with the inverse of (2′) we obtain the $\vee$-product

$$\vee\colon \operatorname{Ext}_\Lambda(A,C) \otimes \operatorname{Ext}_\Gamma(A',C') \to \operatorname{Ext}_\Omega(A \otimes A', C \otimes C').$$

Since $\vee$ is of degree zero, it yields maps

$$\boxed{\vee\colon \operatorname{Ext}^p_\Lambda(A,C) \otimes \operatorname{Ext}^q_\Gamma(A',C') \to \operatorname{Ext}^{p+q}_\Omega(A \otimes A', C \otimes C').}$$

PROPOSITION 1.2.3. *For $p = q = 0$ the $\vee$-product reduces to the homomorphism φ_3.*

Finally we place ourselves in the situation described by the symbol $({}_\Lambda A, {}_\Lambda C, {}_\Gamma A', C'_\Gamma)$. We define $\operatorname{Hom}(C,C')$ as a right Ω-module by setting

$$(f(\lambda \otimes \gamma))c = (f(\lambda c))\gamma \qquad c \in C, f \in \operatorname{Hom}(C,C')$$

and define the homomorphism

$$\varphi_4\colon \operatorname{Hom}(C,C') \otimes_\Omega (A \otimes A') \to \operatorname{Hom}(\operatorname{Hom}_\Lambda(A,C), C' \otimes_\Gamma A')$$

by setting

$$(\varphi_4(f \otimes a \otimes a'))g = f(ga) \otimes a'$$

for $f \in \operatorname{Hom}(C,C')$, $g \in \operatorname{Hom}_\Lambda(A,C)$. Replacing A and A' by X and X' we obtain

$$\Phi_4\colon \operatorname{Hom}(C,C') \otimes_\Omega (X \otimes X') \to \operatorname{Hom}(\operatorname{Hom}_\Lambda(X,C), C' \otimes_\Gamma X').$$

Passing to homology and applying α' we obtain the homomorphism

$$H(\operatorname{Hom}(C,C') \otimes_\Omega (X \otimes X')) \to \operatorname{Hom}(\operatorname{Ext}_\Lambda(A,C), \operatorname{Tor}^\Gamma(C',A')).$$

Combining this with the inverse of (2) we obtain the $\wedge$ product

$$\wedge: \operatorname{Tor}^{\Omega}(\operatorname{Hom}(C,C'),A \otimes A') \to \operatorname{Hom}(\operatorname{Ext}_{\Lambda}(A,C), \operatorname{Tor}^{\Gamma}(C',A')).$$

Since $\wedge$ has degree zero, it yields maps

$$\boxed{\wedge: \operatorname{Tor}^{\Omega}_{p+q}(\operatorname{Hom}(C,C'),A \otimes A') \to \operatorname{Hom}(\operatorname{Ext}^{p}_{\Lambda}(A,C), \operatorname{Tor}^{\Gamma}_{q}(C',A')).}$$

PROPOSITION 1.2.4. *If $p = q = 0$, the $\wedge$ product reduces to the map φ_4.*

The notation used above for the four products is convenient as long as we do not exhibit individual elements of the groups Tor and Ext involved. For formulas involving elements, it is more convenient to adopt the following notation

$$\top(a \otimes b) = a \top b, \qquad (\perp a)b = a \perp b,$$

$$\vee(a \otimes b) = a \vee b, \qquad (\wedge a)b = a \wedge b.$$

We recall that $\vee$ and $\wedge$ are defined only if conditions (i) and (ii) are satisfied.

2. FORMAL PROPERTIES OF THE PRODUCTS

The formal properties of the four products are too numerous to be listed in detail. We shall therefore be satisfied with an informal discussion omitting most proofs. It should be remembered that whenever the products $\vee$ and $\wedge$ occur, suitable assumptions should be made in order that these products be defined.

First, for fixed K-algebras Λ and Γ we may consider maps $A \to A_1$, $C \to C_1$, $A' \to A'_1$ and $C' \to C'_1$. For the product $\top$ we then obtain a commutative diagram

$$\begin{array}{ccc} \operatorname{Tor}^{\Lambda}(C,A) \otimes \operatorname{Tor}^{\Gamma}(C',A') & \xrightarrow{\top} & \operatorname{Tor}^{\Omega}(C \otimes C',A \otimes A') \\ \downarrow & & \downarrow \\ \operatorname{Tor}^{\Lambda}(C_1,A_1) \otimes \operatorname{Tor}^{\Gamma}(C'_1,A'_1) & \xrightarrow{\top} & \operatorname{Tor}^{\Omega}(C_1 \otimes C'_1,A_1 \otimes A'_1). \end{array}$$

Next we can consider homomorphisms φ: $\Lambda' \to \Lambda$, ψ: $\Gamma' \to \Gamma$ of K-algebras. Every Λ-module (resp. Γ-module) may then be regarded as a Λ'-module (resp. Γ' module). In the situation $({}_{\Lambda}A,C_{\Lambda},{}_{\Gamma}A',C'_{\Gamma})$ we then obtain a commutative diagram

$$\begin{array}{ccc} \operatorname{Tor}^{\Lambda'}(C,A) \otimes \operatorname{Tor}^{\Gamma'}(C',A') & \xrightarrow{\top} & \operatorname{Tor}^{\Omega'}(C \otimes C',A \otimes A') \\ \downarrow{\scriptstyle \varphi_* \otimes \psi_*} & & \downarrow{\scriptstyle (\varphi \otimes \psi)_*} \\ \operatorname{Tor}^{\Lambda}(C,A) \otimes \operatorname{Tor}^{\Gamma}(C',A') & \xrightarrow{\top} & \operatorname{Tor}^{\Omega}(C \otimes C',A \otimes A') \end{array}$$

where φ_*: $\operatorname{Tor}^{\Lambda'} \to \operatorname{Tor}^{\Lambda}$ and ψ_*: $\operatorname{Tor}^{\Gamma'} \to \operatorname{Tor}^{\Gamma}$ are induced by φ and ψ.

Keeping the K-algebras Λ and Γ fixed we may consider a ring homomorphism ζ: $L \to K$ where L is a commutative ring. Λ and Γ may then be regarded as L algebras. This leads to a commutative diagram

$$\begin{array}{ccc} \mathrm{Tor}^{\Lambda}(C,A) \otimes_L \mathrm{Tor}^{\Gamma}(C',A') & \xrightarrow{\top} & \mathrm{Tor}^{\Lambda\otimes_L\Gamma}(C \otimes_L C', A \otimes_L A') \\ \downarrow & & \downarrow \\ \mathrm{Tor}^{\Lambda}(C,A) \otimes_K \mathrm{Tor}^{\Gamma}(C',A') & \xrightarrow{\top} & \mathrm{Tor}^{\Lambda\otimes_K\Gamma}(C \otimes_K C', A \otimes_K A'). \end{array}$$

Similar diagrams to the above hold for the remaining three products. One can also consider more complicated situations in which the modules $A,C, \ldots$, the algebras Λ, Γ and the ground ring K are all mapped simultaneously.

Next we turn to commutativity rules. To formulate these rules we must identify $\Omega = \Lambda \otimes \Gamma$ with $\Gamma \otimes \Lambda$ (cf. IX, 1.2).

PROPOSITION 2.1. *The following diagram is commutative*

$$\begin{array}{ccc} \mathrm{Tor}_p^{\Lambda}(A,C) \otimes \mathrm{Tor}_q^{\Gamma}(A',C') & \xrightarrow{\top} & \mathrm{Tor}_{p+q}^{\Omega}(A \otimes A', C \otimes C') \\ \downarrow{\scriptstyle (-1)^{pq}f} & & \downarrow{\scriptstyle g} \\ \mathrm{Tor}_q^{\Gamma}(A',C') \otimes \mathrm{Tor}_p^{\Lambda}(A,C) & \xrightarrow{\top} & \mathrm{Tor}_{p+q}^{\Omega}(A' \otimes A, C' \otimes C) \end{array}$$

where f is the map establishing the commutativity of the tensor products, while g is induced by similar maps f_1: $A \otimes A' \to A' \otimes A$, f_2: $C \otimes C' \to C' \otimes C$.

PROOF. The proof is an easy consequence of the fact that a map F_1: $X \otimes X' \to X' \otimes X$ over f_1 is obtained by setting $F_1(x \otimes x') = (-1)^{pq}$ $(x' \otimes x)$ for $x \in X_p$, $x' \in X'_q$.

A similar commutativity rule for the $\vee$ product is obtained by simply replacing Tor by Ext in the diagram above.

We now come to the associativity rules, of which there are six. These will be stated without proofs. We consider three K-algebras Λ, Γ, Σ and define $\Omega = (\Lambda \otimes \Gamma) \otimes \Sigma = \Lambda \otimes (\Gamma \otimes \Sigma)$. In general, we shall regard the tensor product as an associative operation.

PROPOSITION 2.2. *In the situation* $({}_\Lambda A, C_\Lambda, {}_\Gamma A', C'_\Gamma, {}_\Sigma A'', C''_\Sigma)$ *consider* $a \in \mathrm{Tor}_p^{\Lambda}(C,A)$, $b \in \mathrm{Tor}_q^{\Gamma}(C',A')$, $c \in \mathrm{Tor}_r^{\Sigma}(C'',A'')$. *Then* $(a \top b) \top c = a \top (b \top c)$.

PROPOSITION 2.2a. *In the situation* $({}_\Lambda A, {}_\Lambda C, {}_\Gamma A', {}_\Gamma C', {}_\Sigma A'', {}_\Sigma C'')$ *consider* $a \in \mathrm{Ext}_\Lambda^p(A,C)$, $b \in \mathrm{Ext}_\Gamma^q(A',C')$, $c \in \mathrm{Ext}_\Sigma^r(A'',C'')$. *Then* $(a \vee b) \vee c = a \vee (b \vee c)$.

To state the next two associative laws we use the identification $\mathrm{Hom}(C, \mathrm{Hom}(C',C'')) = \mathrm{Hom}(C \otimes C', C'')$. It should be observed that in the situations $(C_\Lambda, C'_\Gamma, {}_\Sigma C'')$ or $({}_\Lambda C, {}_\Gamma C', C''_\Sigma)$ this identification is compatible with the operators of Ω on both groups.

PROPOSITION 2.3. *In the situation* $({}_\Lambda A, C_\Lambda, {}_\Gamma A', C'_\Gamma, {}_\Sigma A'', {}_\Sigma C'')$, *consider* $a \in \operatorname{Ext}^\Omega_{p+q+r}(A \otimes A' \otimes A'', \operatorname{Hom}(C \otimes C', C''))$, $b \in \operatorname{Tor}^\Lambda_p(C,A)$ *and* $c \in \operatorname{Tor}^\Gamma_q(C',A')$. *Then* $a \perp (b \top c) = (a \perp b) \perp c$.

PROPOSITION 2.3a. *In the situation* $({}_\Lambda A, {}_\Lambda C, {}_\Gamma A', {}_\Gamma C', {}_\Sigma A'', C''_\Sigma)$ *consider* $a \in \operatorname{Tor}^\Omega_{p+q+r}(\operatorname{Hom}(C \otimes C', C''), A \otimes A' \otimes A'')$, $b \in \operatorname{Ext}^p_\Lambda(A,C)$ *and* $c \in \operatorname{Ext}^q_\Gamma(A',C')$. *Then* $a \wedge (b \vee c) = (a \wedge b) \wedge c$.

To formulate the last pair of associative laws we consider the natural homomorphism

$$\xi\colon \operatorname{Hom}(C,C') \otimes C'' \to \operatorname{Hom}(C, C' \otimes C'')$$

given by

$$[\xi(f \otimes c'')]c = (fc) \otimes c''.$$

We observe that in the situation $({}_\Lambda C, C'_\Gamma, C''_\Sigma)$ and $(C_\Lambda, {}_\Gamma C', {}_\Sigma C'')$, ξ is an Ω-homomorphism.

PROPOSITION 2.4. *In the situation* $({}_\Lambda A, {}_\Lambda C, {}_\Gamma A', C'_\Gamma, {}_\Sigma A'', C''_\Sigma)$ *consider* $a \in \operatorname{Tor}^{\Lambda\otimes\Gamma}_{p+q}(\operatorname{Hom}(C,C'), A \otimes A')$, $b \in \operatorname{Ext}^p_\Lambda(A,C)$, $c \in \operatorname{Tor}^\Sigma_r(C'',A'')$. *Then* $(a \wedge b) \top c = [\xi_*(a \top c)] \wedge b$ *where*

$$\xi_*\colon \operatorname{Tor}^\Omega_n(\operatorname{Hom}(C,C') \otimes C'', D) \to \operatorname{Tor}^\Omega_n(\operatorname{Hom}(C, C' \otimes C''), D)$$

is induced by ξ.

PROPOSITION 2.4a. *In the situation* $({}_\Lambda A, C_\Lambda, {}_\Gamma A', {}_\Gamma C', {}_\Sigma A'', {}_\Sigma C'')$ *consider* $a \in \operatorname{Ext}^{\Lambda\otimes\Gamma}_{p+q}(A \otimes A', \operatorname{Hom}(C,C'))$, $b \in \operatorname{Tor}^\Lambda_p(C,A)$, $c \in \operatorname{Ext}^r_\Sigma(A'',C'')$. *Then* $(a \perp b) \vee c = [\xi^*(a \vee c)] \perp b$ *where*

$$\xi^*\colon \operatorname{Ext}^n_\Omega(D, \operatorname{Hom}(C,C') \otimes C'') \to \operatorname{Ext}^n_\Omega(D, \operatorname{Hom}(C, C' \otimes C''))$$

is induced by ξ.

We now pass to the discussion of connecting homomorphisms. Since there are four products and each of them involves four variables, there is a total of sixteen commutativity rules with connecting homomorphisms. We shall only state two of these (concerning the variables C, C' in the $\top$-product), the others being quite similar.

PROPOSITION 2.5. *Let*

$$(1) \qquad 0 \to A_1 \to A \to A_2 \to 0$$

be an exact sequence such that the sequence

$$(2) \qquad 0 \to A_1 \otimes A' \to A \otimes A' \to A_2 \otimes A' \to 0$$

is exact. Then the diagram

$$\begin{array}{ccc}
\operatorname{Tor}^\Lambda(C,A_2) \otimes \operatorname{Tor}^\Gamma(C',A') & \xrightarrow{\ \top\ } & \operatorname{Tor}^\Omega(C \otimes C', A_2 \otimes A') \\
\Big\downarrow{\scriptstyle \delta\otimes i} & & \Big\downarrow{\scriptstyle \Delta} \\
\operatorname{Tor}^\Lambda(C,A_1) \otimes \operatorname{Tor}^\Gamma(C',A') & \xrightarrow{\ \top\ } & \operatorname{Tor}^\Omega(C \otimes C', A_1 \otimes A')
\end{array}$$

is commutative. Here δ is the connecting homomorphism relative to (1), *Δ is the connecting homomorphism relative to* (2) *and i is the appropriate identity map.*

PROOF. Let $0 \to X_1 \to X \to X_2 \to 0$ be a projective resolution of $0 \to A_1 \to A \to A_2 \to 0$ and let X' be a projective resolution of A'. In view of the definition of $\top$, the required relation follows from the following commutativity relations

$$\begin{array}{ccc}
H_p(C \otimes_\Lambda X_2) \otimes H_q(C' \otimes_\Gamma X') & \longrightarrow & H_{p-1}(C \otimes_\Lambda X_1) \otimes H_q(C' \otimes_\Gamma X') \\
\downarrow \alpha & & \downarrow \alpha \\
H_{p+q}((C \otimes_\Lambda X_2) \otimes (C' \otimes_\Gamma X')) & \longrightarrow & H_{p+q-1}((C \otimes_\Lambda X_1) \otimes (C' \otimes_\Gamma X')) \\
\downarrow \Phi_1 & & \downarrow \Phi_1 \\
H_{p+q}((C \otimes C') \otimes_\Omega (X_2 \otimes X')) & \longrightarrow & H_{p+q-1}((C \otimes C') \otimes_\Omega (X_1 \otimes X')) \\
\downarrow & & \downarrow \\
\operatorname{Tor}^\Omega_{p+q}(C \otimes C', A_2 \otimes A') & \longrightarrow & \operatorname{Tor}^\Omega_{p+q-1}(C \otimes C'), A_1 \otimes A')
\end{array}$$

The first of these relations follows from IV,7.1 (restated for left exact functors), the second one follows from the naturality of the map Φ_1, the third one follows from V,4.3.

PROPOSITION 2.5′. *Let*

$$(1') \qquad 0 \to A_1' \to A' \to A_2' \to 0$$

be an exact sequence such that the sequence

$$(2') \qquad 0 \to A \otimes A_1' \to A \otimes A' \to A \otimes A_2' \to 0$$

is exact. Then the diagram

$$\begin{array}{ccc}
\operatorname{Tor}^\Lambda (C,A) \otimes \operatorname{Tor}^\Gamma (C',A_2') & \longrightarrow & \operatorname{Tor}^\Omega (C \otimes C', A \otimes A_2') \\
\downarrow i \otimes \delta & & \downarrow \Delta \\
\operatorname{Tor}^\Lambda (C,A) \otimes \operatorname{Tor}^\Gamma (C',A_1') & \longrightarrow & \operatorname{Tor}^\Omega (C \otimes C', A \otimes A_1')
\end{array}$$

is commutative.

It should be noted that the definition of the map $i \otimes \delta$ includes a sign, since δ has degree $+1$. Therefore, for $a \in \operatorname{Tor}^\Lambda_p (C,A)$, $b \in \operatorname{Tor}^\Gamma_q (C',A')$, we have $\Delta(a \top b) = (-1)^p a \top \delta b$.

3. ISOMORPHISMS

THEOREM 3.1. *If K is semi-simple then the maps $\top$ and $\perp$ are isomorphisms. If further Λ and Γ are left Noetherian, A is finitely Λ-generated and A' is finitely Γ-generated then $\vee$ and $\wedge$ also are isomorphisms.*

PROOF. First we observe that since K is semi-simple, condition (i) and (ii) of § 1 are satisfied and therefore the maps (2) and (2′) of § 1 are isomorphisms. Further the functors $\otimes_K$ and Hom_K are exact and therefore, by IV,7.2, the maps α and α' employed in § 1 are isomorphisms.

Next we consider the maps

$$\varphi_1: (C \otimes_\Lambda A) \otimes (C' \otimes_\Gamma A') \to (C \otimes C') \otimes_\Omega (A \otimes A')$$

$$\varphi_2: \mathrm{Hom}_\Omega (A \otimes A', \mathrm{Hom}(C,C')) \to \mathrm{Hom}(C \otimes_\Lambda A, \mathrm{Hom}_\Gamma (A',C'))$$

$$\varphi_3: \mathrm{Hom}_\Lambda (A,C) \otimes \mathrm{Hom}_\Gamma (A',C') \to \mathrm{Hom}_\Omega (A \otimes A', C \otimes C')$$

$$\varphi_4: \mathrm{Hom}(C,C') \otimes_\Omega (A \otimes A') \quad \to \mathrm{Hom}(\mathrm{Hom}_\Lambda (A,C), C' \otimes_\Gamma A')$$

as defined in § 1. It is easy to see that φ_1 and φ_2 are isomorphisms. This implies that the maps Φ_1 and Φ_2 obtained by replacing the modules A and A' by their projective resolutions X and X' also are isomorphisms. This proves that $\top$ and $\perp$ are isomorphisms.

As for the maps φ_3 and φ_4, they are isomorphisms if $A = \Lambda$ and $A' = \Gamma$. Therefore, by direct sum properties, it follows that φ_3 and φ_4 are isomorphisms if A is projective and finitely Λ-generated and A' is projective and finitely Γ-generated. Now, if Λ and Γ are left Noetherian and A and A' are finitely generated then, by V,1.3, the resolutions X and X' may be chosen so that each module X_p is projective and finitely Λ-generated while each module X'_q is projective and finitely Γ-generated. Thus in this case the maps Φ_3 and Φ_4 also are isomorphisms. This concludes the proof.

THEOREM 3.2. *If the algebras Λ and Γ are K-projective, A is a left Λ-module and A' is a left Γ-module such that* $\mathrm{Tor}_n^K (A,A') = 0$ *for* $n > 0$, *then*

$$\dim_{\Lambda \otimes \Gamma} (A \otimes A') \leqq \dim_\Lambda A + \dim_\Gamma A'.$$

If, further, K is a field, Λ and Γ are left Noetherian, A is finitely Λ-generated and A' is finitely Γ-generated, then the above inequality is an equality.

PROOF. Let $\dim_\Lambda A \leqq m$, $\dim_\Gamma A' \leqq n$. There exist then projective resolutions X of A and X' of A' such that $X_p = 0$ for $p > m$ and $X_q = 0$ for $q > n$. Then, by 1.1, $X \otimes X'$ is a projective $\Lambda \otimes \Gamma$-resolution of $A \otimes A'$. Since $X \otimes X'$ is zero in degrees $> n + m$, it follows that $\dim_{\Lambda\otimes\Gamma} (A \otimes A') \leqq m + n$.

Assume now that the second set of hypotheses is satisfied. Suppose $\dim_\Lambda A \geqq m$, $\dim_\Gamma A' \geqq n$. Then there exist modules C and C' such that $\mathrm{Ext}^m_\Lambda (A,C) \neq 0$ and $\mathrm{Ext}^n_\Gamma (A',C') \neq 0$. Since K is a field, we have $\mathrm{Ext}^m_\Lambda (A,C) \otimes \mathrm{Ext}^n_\Gamma (A',C') \neq 0$. Now, by 3.1, $\vee$ is an isomorphism so that $\mathrm{Ext}^{m+n}_{\Lambda\otimes\Gamma} (A \otimes A', C \otimes C') \neq 0$. This implies $\dim_{\Lambda\otimes\Gamma}(A \otimes A') \geqq m+n$.

We next consider the products $\perp$ and $\wedge$ in the case $\Gamma = K = A'$. Then $\Omega = \Lambda$, $\operatorname{Ext}_\Gamma(A',C') = \operatorname{Hom}(A',C') = C'$, $\operatorname{Tor}^\Gamma(C',A') = C'$. The products thus become

$$\perp: \operatorname{Ext}_\Lambda(A, \operatorname{Hom}(C,C')) \to \operatorname{Hom}(\operatorname{Tor}^\Lambda(C,A),C') \tag{1}$$

$$\wedge: \operatorname{Tor}^\Lambda(\operatorname{Hom}(C,C'),A) \to \operatorname{Hom}(\operatorname{Ext}_\Lambda(A,C),C') \tag{2}$$

defined in the situations $({}_\Lambda A, C_{\Lambda}, {}_K C')$, $({}_\Lambda A, {}_\Lambda C, {}_K C')$ respectively. If we inspect the definition of (1) and (2) using a Λ-projective resolution of A (and using K as a K-projective resolution of K) we find that (2) is defined without the condition that Λ be K-projective. Further from this direct definition it becomes clear that (1) is a special case of the homomorphism ρ of VI,5 while (2) is a special case of σ of VI,5. Consequently VI,5.1 and VI,5.3 imply

PROPOSITION 3.3. *If C' is K-injective then* (1) *is an isomorphism. If further Λ is left Noetherian and A is finitely Λ-generated then* (2) *is also an isomorphism.*

4. INTERNAL PRODUCTS

Let Λ be a *commutative* ring. If we regard Λ as a Λ-algebra and observe that $\Lambda \otimes_\Lambda \Lambda = \Lambda$, the products $\top$ and $\perp$ yield the following *internal products*

$$\pitchfork: \operatorname{Tor}^\Lambda(C,A) \otimes_\Lambda \operatorname{Tor}^\Lambda(C',A') \to \operatorname{Tor}^\Lambda(C \otimes_\Lambda C', A \otimes_\Lambda A')$$

$$\Cup: \operatorname{Ext}_\Lambda(A \otimes A', \operatorname{Hom}_\Lambda(C,C')) \to \operatorname{Hom}_\Lambda(\operatorname{Tor}^\Lambda(C,A), \operatorname{Ext}_\Lambda(A',C'))$$

defined for any Λ-modules A, C, A', C'. Both products are Λ-homomorphisms.

We recall that Λ operates on $C \otimes_\Lambda C'$ and $\operatorname{Hom}_\Lambda(C,C')$ as follows

$$\lambda(c \otimes c') = \lambda c \otimes c' = c \otimes \lambda c', \qquad (\lambda f)c = \lambda(fc) = f(\lambda c).$$

These internal products being special instances of the products of § 1, all the formal properties stated in § 2 remain valid.

There is another kind of internal products that can be obtained for a K-algebra Λ (Λ no longer assumed commutative) provided we are given a ring homomorphism D: $\Lambda \to \Lambda \otimes \Lambda$ (tensor product over K) which we shall call a *diagonal map*. The map D induces homomorphisms $\operatorname{Tor}_n^\Lambda \to \operatorname{Tor}_n^{\Lambda\otimes\Lambda}$ and $\operatorname{Ext}^n_{\Lambda\otimes\Lambda} \to \operatorname{Ext}^n_\Lambda$. Composing these homomorphisms with the $\vee$- and $\wedge$-products we obtain the following two products (called the *cup-product* and the *cap-product*):

$$\cup: \operatorname{Ext}_\Lambda(A,C) \otimes \operatorname{Ext}_\Lambda(A',C') \to \operatorname{Ext}_\Lambda(A \otimes A', C \otimes C'),$$

$$\cap: \operatorname{Tor}^\Lambda(\operatorname{Hom}(C,C'), A \otimes A') \to \operatorname{Hom}(\operatorname{Ext}_\Lambda(A,C), \operatorname{Tor}^\Lambda(C',A')).$$

Both products are defined only under the condition that Λ is K-projective and $\mathrm{Tor}_n^K(A,A') = 0$ for $n > 0$. The $\cup$-product is defined in the situation $({}_\Lambda A, {}_\Lambda C, {}_\Lambda A', {}_\Lambda C')$ while the $\cap$-product is defined for $({}_\Lambda A, {}_\Lambda C, {}_\Lambda A', C'_\Lambda)$. The operators of Λ on $A \otimes A'$, $C \otimes C'$ and $\mathrm{Hom}(C,C')$ are obtained by composing the operators of $\Lambda \otimes \Lambda$ with the diagonal map D.

The rules for connecting homomorphisms are the same as for the $\vee$- and $\wedge$-products. For the remaining formal rules, conditions must be imposed on the diagonal map D: $\Lambda \to \Lambda \otimes \Lambda$. Specifically, we shall say that D is *commutative*, if the diagram

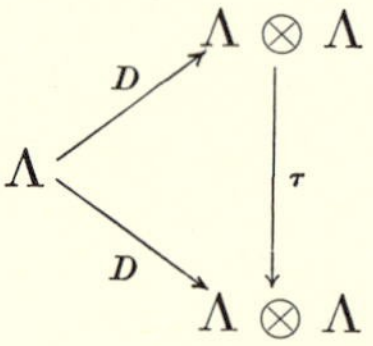

where $\tau(\lambda \otimes \lambda') = \lambda' \otimes \lambda$, is commutative. If D is commutative then we have the commutation rule

(1) $$a \cup b = (-1)^{pq}\, b \cup a$$

for $a \in \mathrm{Ext}_\Lambda^p(A,C)$, $b \in \mathrm{Ext}_\Lambda^q(A',C')$.

The diagonal map D will be called *associative* if the diagram

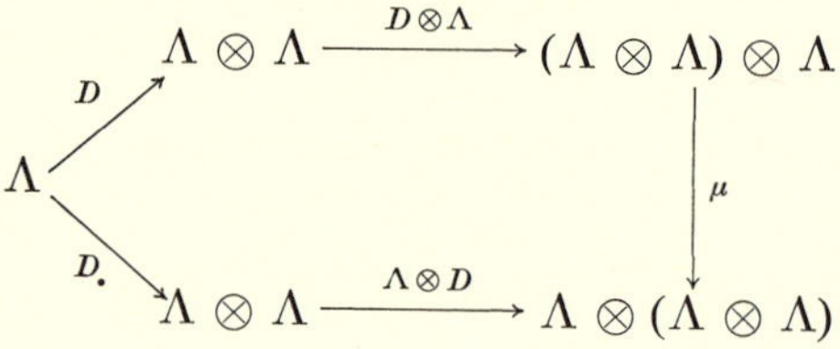

where $\mu((\lambda \otimes \lambda') \otimes \lambda'') = \lambda \otimes (\lambda' \otimes \lambda'')$, is commutative. Under this condition we obtain the associative rules $(a \cup b) \cup c = a \cup (b \cup c)$ as in 2.2a, and $a \cap (b \cup c) = (a \cap b) \cap c$ as in 2.3a.

The precise statements and proofs of these formal rules is left to the reader.

A $\cup$-*pairing* is a Λ-homomorphism

$$C \otimes C' \to B \qquad ({}_\Lambda B, {}_\Lambda C, {}_\Lambda C')$$

Using this homomorphism, the $\cup$-product yields

$$\cup:\ \mathrm{Ext}_\Lambda(A,C) \otimes \mathrm{Ext}_\Lambda(A',C') \to \mathrm{Ext}_\Lambda(A \otimes A',B).$$

A $\cap$-*pairing* is a Λ-homomorphism

$$B \to \mathrm{Hom}(C,C') \qquad (B_\Lambda, {}_\Lambda C, C'_\Lambda)$$

Using this homomorphism, the $\cap$-product yields

$$\cap: \operatorname{Tor}^\Lambda (B, A \otimes A') \to \operatorname{Hom} (\operatorname{Ext}_\Lambda (A,C), \operatorname{Tor}^\Lambda (C',A')).$$

5. COMPUTATION OF PRODUCTS

We shall discuss here the question of computation of the products using specific projective resolutions of A, A' and $A \otimes A'$. The procedures derived in this section will be used later to obtain "product formulae" in various specific situations.

Let X be a Λ-projective resolution of A, X' a Γ-projective resolution of A' and Y an Ω-projective resolution of $A \otimes A'$.

We begin with the products $\top$ and $\perp$. The definitions of $\top$ and $\perp$ show that these maps admit factorizations

$$H(C \otimes_\Lambda X) \otimes H(C' \otimes_\Gamma X') \xrightarrow{\Phi_1 \alpha} H((C \otimes C') \otimes_\Omega (X \otimes X')) \longrightarrow H((C \otimes C' \otimes_\Omega Y)$$

$$H(\operatorname{Hom}_\Omega (Y,D)) \longrightarrow H(\operatorname{Hom}_\Omega (X \otimes X',D)) \xrightarrow{\alpha' \Phi_2} \operatorname{Hom} (H(C \otimes_\Lambda X), H(\operatorname{Hom}_\Gamma (X',C)))$$

where $D = \operatorname{Hom}(C,C')$. The maps $\Phi_1 \alpha$ and $\alpha' \Phi_2$ are "explicit." The remaining two maps were obtained using the fact that $X \otimes X'$ is an Ω-projective left complex over $A \otimes A'$. Thus to render the remaining two maps explicit we need an Ω-map

(1) $$f\colon X \otimes X' \to Y$$

over the identity map of $A \otimes A'$.

The situation concerning the products $\vee$ and $\wedge$ is similar. Again looking at the definitions of these products we find that they admit factorizations

$$H(\operatorname{Hom}_\Lambda(X,C)) \otimes H(\operatorname{Hom}_\Gamma(X',C')) \xrightarrow{\Phi_3 \alpha} H(\operatorname{Hom}_\Omega(X \otimes X', C \otimes C')) \longrightarrow H(\operatorname{Hom}_\Omega (Y, C \otimes C'))$$

$$H(D \otimes_\Omega Y) \longrightarrow H(D \otimes_\Omega (X \otimes X')) \xrightarrow{\alpha' \Phi_4} \operatorname{Hom}(H(\operatorname{Hom}_\Lambda (X,C)), H(C' \otimes_\Gamma X'))$$

where $D = \operatorname{Hom}(C,C')$. Therefore the products above obtain explicit definition if we have an Ω-map

(2) $$g\colon Y \to X \otimes X'$$

over the identity map of $A \otimes A'$.

In order to give general procedures for finding the maps f and g we must enter into a more detailed discussion of the resolutions X, X' and Y.

In practice, the resolution X of A will be not only Λ-projective but Λ-free. This implies that each X_n may be written in the form $\Lambda \otimes \tilde{X}_n$ where $\tilde{X}_n$ is a K-module. Actually all the resolutions encountered so far and all those encountered later, are directly given in the form $X = \Lambda \otimes \tilde{X}$ where $\tilde{X}$ is a graded K-module. Clearly $\tilde{X}$ may be regarded as a K-submodule of X, but not as a subcomplex. Under these circumstances we say that X is given in *split* form. Similar remarks apply to X' and Y. It should further be noted that if $X = \Lambda \otimes \tilde{X}$ and $X' = \Gamma \otimes \tilde{X}'$ are given in split form then $X \otimes X' = \Omega \otimes \tilde{X} \otimes \tilde{X}'$ also has the split form.

Another notion that we need is that of a *contracting homotopy.* A contracting homotopy $\{s_n, \sigma\}$ for a left complex X over A is a family of K-homomorphisms

$$\sigma\colon A \to X_0, \qquad s_n\colon X_n \to X_{n+1},$$

such that

$$d_{n+1}s_n x + s_{n-1}d_n x = x - \sigma\varepsilon x \qquad \text{for } x \in X_n,$$

$$\varepsilon\sigma x = x \qquad \text{for } x \in A,$$

where $\varepsilon\colon X_0 \to A$ is the augmentation map. Of course the existence of a contracting homotopy implies that X is an acyclic left complex over A. Conversely assume that X is an acyclic left complex over A and that A and X are K-projective. Then the sequence

$$(3) \qquad \cdots \to X_n \to \cdots \to X_0 \to A \to 0$$

is exact and may be regarded as a K-projective resolution of the zero module. It follows from v,1.2 that the identity map of (3) is homotopic with the zero map. This yields a contracting homotopy for X.

PROPOSITION 5.1. *If Λ and A are K-projective then every Λ-projective resolution of A has a contracting homotopy.*

Indeed, let X be a Λ-projective resolution of A. Since Λ is K-projective, it follows from II,6.2 that X is K-projective. Thus the existence of a contracting homotopy for X follows from what was said above.

We further note that if $\{s_n, \sigma\}$ is a contracting homotopy for X while $\{s'_n, \sigma'\}$ is a contracting homotopy for X' then a contracting homotopy $\{t_n, \tau\}$ for $X \otimes X'$ may be obtained by setting

$$\tau = \sigma \otimes \sigma',$$

$$t = s \otimes i' + (\sigma\varepsilon) \otimes s'$$

where i' is the identity map of X'.

The usefulness of the above notions is shown by the following proposition.

PROPOSITION 5.2. *Let C be a left Ω-module. Let $Z = \Omega \otimes \tilde{Z}$ be a left complex over C given in split form and let Z' be a left complex over C with a contracting homotopy $\{t_n, \tau\}$. Then the inductive formulae*

$$f(\omega \otimes z) = \omega\tau\varepsilon z \qquad z \in \tilde{Z}_0$$

$$f(\omega \otimes z) = \omega t_{q-1} f dz \qquad z \in \tilde{Z}_q,\ q > 0$$

yield a map

$$f\colon Z \to Z'$$

over the identity map of C. This map is uniquely characterized by the condition

$$dtfz = 0 \qquad z \in \tilde{Z}.$$

PROOF. Clearly f is an Ω-homomorphism. For $z \in \tilde{Z}_0$ we have

$$\varepsilon' f(\omega \otimes z) = \omega\varepsilon'\tau\varepsilon z = \omega\varepsilon z = \varepsilon(\omega \otimes z)$$

so that $\varepsilon' f = \varepsilon$. For $z \in \tilde{Z}_1$, we have

$$df(\omega \otimes z) = \omega dtfdz = \omega fdz - \omega\tau\varepsilon' fdz$$
$$= fd(\omega \otimes z) - \omega\tau\varepsilon dz = fd(\omega \otimes z).$$

For $z \in \tilde{Z}_q$, $q > 1$ we have inductively
$df(\omega \otimes z) = \omega dtfdz = \omega fdz - \omega tdfdz = fd(\omega \otimes z) - \omega tfddz = fd(\omega \otimes z)$.
This shows that f is a map as required.

For any map $f\colon Z \to Z'$ over the identity map we have

$$fz = \tau\varepsilon z + dtfz \qquad z \in \tilde{Z}_0$$

$$fz = tdfz + dtfz = tfdz + dtfz \qquad z \in \tilde{Z}_q, q > 0.$$

Therefore f satisfies the inductive definition if and only if $dtfz = 0$ for $z \in \tilde{Z}$.

We now return to the question of finding maps (1) and (2) above. To find $f\colon X \otimes X' \to Y$ we assume that X and X' are given in split form while in Y we are given a contracting homotopy. Then $X \otimes X'$ also is given in split form and 5.2 may be used to define f.

To find $g\colon Y \to X \otimes X'$ we assume that Y is given in split form while in X and X' we are given contracting homotopies. Then as shown above we may construct a contracting homotopy in $X \otimes X'$ and then use 5.2 to define g.

Next we assume that Λ is commutative and consider the internal products $\frown$ and $\smile$. The computation requires a Λ-map

(3) $$h\colon X \otimes_\Lambda X' \to Y$$

over the identity map of $A \otimes_\Lambda A'$, where X, X' and Y are Λ-projective resolutions of A, A' and $A \otimes_\Lambda A'$. We shall assume that X and X' are given in split form $X = \Lambda \otimes \tilde{X}$, $X' = \Lambda \otimes \tilde{X}$, while in Y a contracting homotopy is given. Then $X \otimes_\Lambda X' = \Lambda \otimes \tilde{X} \otimes \tilde{X}'$ also is given in split form so that 5.2 may be applied to find h.

Finally we consider the $\cup$- and $\cap$-products corresponding to a diagonal map $D: \Lambda \to \Lambda \otimes \Lambda$. The computation requires a map

$$j\colon\ Y \to X \otimes X' \tag{4}$$

where X, X' and Y are Λ-projective resolutions of A, A' and $A \otimes A'$, and where Λ operates on $A \otimes A'$ and $X \otimes X'$ using the diagonal map D. Again if Y is given in split form, while in X and X' we are given contracting homotopies, then a contracting homotopy in $X \otimes X'$ also is given and j may be found using 5.2.

6. PRODUCTS IN THE HOCHSCHILD THEORY

Let Λ and Γ be K-algebras and $\Lambda \otimes_K \Gamma$ their tensor product. We have

$$(\Lambda \otimes \Gamma)^e = (\Lambda \otimes \Gamma) \otimes (\Lambda \otimes \Gamma)^* \approx \Lambda \otimes \Gamma \otimes \Lambda^* \otimes \Gamma^* \approx \Lambda^e \otimes \Gamma^e$$

where all tensor products are over K. Henceforth we shall identify $(\Lambda \otimes \Gamma)^e$ with $\Lambda^e \otimes \Gamma^e$. Having made this identification we apply the products of § 1 replacing the symbol $(\Lambda,\Gamma,A,A',C,C')$ by $(\Lambda^e,\Gamma^e,\Lambda,\Gamma,A,A')$ and replacing Ω by $(\Lambda \otimes \Gamma)^e$. We obtain products

$$\top\colon\ H_p(\Lambda,A) \otimes H_q(\Gamma,A') \to H_{p+q}(\Lambda \otimes \Gamma, A \otimes A'),$$

$$\perp\colon\ H^{p+q}(\Lambda \otimes \Gamma, \operatorname{Hom}(A,A')) \to \operatorname{Hom}(H_p(\Lambda,A), H^q(\Gamma,A')),$$

$$\vee\colon\ H^p(\Lambda,A) \otimes H^q(\Gamma,A') \to H^{p+q}(\Lambda \otimes \Gamma, A \otimes A'),$$

$$\wedge\colon\ H_{p+q}(\Lambda \otimes \Gamma, \operatorname{Hom}(A,A')) \to \operatorname{Hom}(H^p(\Lambda,A), H_q(\Gamma,A')),$$

where $\otimes$ and Hom stand for $\otimes_K$ and Hom_K, A is a two-sided Λ-module, A' is a two-sided Γ-module and $A \otimes A'$ and $\operatorname{Hom}(A,A')$ are converted into two-sided $\Lambda \otimes \Gamma$ modules as follows

$$(\lambda_1 \otimes \gamma_1)(a \otimes a')(\lambda_2 \otimes \gamma_2) = \lambda_1 a \lambda_2 \otimes \gamma_1 a' \gamma_2,$$

$$[(\lambda_1 \otimes \gamma_1) f (\lambda_2 \otimes \gamma_2)]a = \gamma_1[f(\lambda_2 a \lambda_1)]\gamma_2.$$

The products themselves are K-homomorphisms. The products $\top$ and $\perp$ are defined without any restrictions. For the products $\vee$ and $\wedge$ to be defined we must assume that conditions (i) and (ii) of § 1 are satisfied. These read: (i) Λ^e and Γ^e are K-projective; (ii) $\operatorname{Tor}_n^K(\Lambda,\Gamma) = 0$ for $n > 0$.

Clearly the assumption that Λ and Γ are K-projective, suffices for both (i) and (ii). Henceforth we shall always assume that Λ and Γ are K-projective, whenever the products $\vee$ or $\wedge$ are involved.

The formal properties established in § 2 all apply here and will not be restated. The same holds for the results of § 3. We shall only observe that proposition IX,7.4 is an easy corollary of 3.2 and actually the proof of IX,7.4 used implicitly the $\vee$-product and repeated the arguments of 3.1 and 3.2.

We now pass to internal products. Assume that Λ is a commutative K-algebra. Then the same holds for Λ^e and we have $\Lambda \otimes_{\Lambda^e} \Lambda = \Lambda$. The products $\pitchfork$ and $\curlyvee$ of § 4 are defined

$$\pitchfork: \; H_p(\Lambda,A) \otimes_{\Lambda^e} H_q(\Lambda,A') \rightarrow H_{p+q}(\Lambda,A \otimes_{\Lambda^e} A')$$

$$\curlyvee: \; H^{p+q}(\Lambda, \operatorname{Hom}_{\Lambda^e}(A,A')) \rightarrow \operatorname{Hom}_{\Lambda^e}(H_p(\Lambda,A),H^q(\Lambda,A')).$$

Here both A and A' are two-sided Λ-modules and Λ operates two-sidedly on $A \otimes_{\Lambda^e} A'$ and $\operatorname{Hom}_{\Lambda^e}(A,A')$ as follows

$$\lambda(a \otimes a')\lambda' = \lambda a \otimes a'\lambda',$$

$$(\lambda f \lambda')a = \lambda[f(a\lambda')].$$

Both products $\pitchfork$ and $\curlyvee$ are Λ^e-homomorphisms.

To discuss the cup- and cap-products we consider a K-projective K-algebra Λ together with a K-algebra homomorphism D: $\Lambda \rightarrow \Lambda \otimes_K \Lambda$ called the diagonal map. Clearly D induces a K-algebra homomorphism D^e: $\Lambda^e \rightarrow (\Lambda \otimes_K \Lambda)^e$. We thus obtain the products

$$\cup: \; H^p(\Lambda,A) \otimes H^q(\Lambda,A') \rightarrow H^{p+q}(\Lambda,A \otimes A')$$

$$\cap: \; H_{p+q}(\Lambda, \operatorname{Hom}(A,A')) \rightarrow \operatorname{Hom}(H^p(\Lambda,A),H_q(\Lambda,A'))$$

where A and A' are two-sided Λ-modules. $A \otimes A'$ and $\operatorname{Hom}(A,A')$ are first regarded as two-sided $\Lambda \otimes \Lambda$-modules and then converted into two-sided Λ-modules using D. As in the case of $\vee$ and $\wedge$, the symbols $\otimes$ and Hom stand for $\otimes_K$ and Hom_K, the products $\cup$ and $\cap$ themselves are K-homomorphisms.

We now come to the question of the computation of the products using complexes. We begin by the consideration of the normalized standard complexes $N(\Lambda)$, $N(\Gamma)$ and $N(\Lambda \otimes \Gamma)$. We assume that Λ, Γ, Coker $(K \rightarrow \Lambda)$, Coker $(K \rightarrow \Gamma)$ are K-projective. Then Coker $(K \rightarrow \Lambda \otimes \Gamma)$ also is K-projective, and therefore the normalized standard complexes above are appropriate projective resolutions.

To compute the $\top$ and $\perp$ products we need a map

$$f\colon N(\Lambda) \otimes N(\Gamma) \to N(\Lambda \otimes \Gamma)$$

over the identity map of $\Lambda \otimes \Gamma$. Since the standard complexes are always given in split form, and since their definition is always accompanied by a contracting homotopy, the procedure given in § 5 may be applied. We shall give a closed formula for a map f and then verify that it satisfies the inductive definition given in 5.2. Consider elements

$$a = [\lambda_1, \ldots, \lambda_p], \qquad b = [\gamma_1, \ldots, \gamma_q]$$

of $N_p(\Lambda)$ and $N_q(\Gamma)$. If $p = 0$ then $a = [\lambda_1, \ldots, \lambda_p]$ is the unit element of $N_0(\Lambda) = \Lambda^e$ and similarly if $q = 0$. We define

$$f(a \otimes b) = \sum \pm [\zeta_1, \ldots, \zeta_{p+q}] \tag{1}$$

where $\zeta_1, \ldots, \zeta_{p+q}$ ranges over all permutations of the sequence $\lambda_1 \otimes 1, \ldots, \lambda_p \otimes 1, 1 \otimes \gamma_1, \ldots, 1 \otimes \gamma_q$ for which $\lambda_1 \otimes 1, \ldots, \lambda_p \otimes 1$ remain in order and $1 \otimes \gamma_1, \ldots, 1 \otimes \gamma_q$ remain in order. The sign is $+$ for even permutations and $-$ for odd ones. The mapping f is extended to $N_p(\Lambda) \otimes N_q(\Gamma)$ by $(\Lambda \otimes \Gamma)^e$-linearity (see Eilenberg-MacLane, *Ann. of Math.*, 58, (1953)). To convert this into an inductive definition, we recall that $[\zeta_1, \ldots, \zeta_{p+q}] = s(\zeta_1[\zeta_2, \ldots, \zeta_{p+q}])$ where s is the contracting homotopy of the complex $N(\Lambda \otimes \Gamma)$. Now ζ_1 is either $\lambda_1 \otimes 1$ or $1 \otimes \gamma_1$ and considering these two cases separately we obtain for $p > 0$, $q > 0$

$$f(a \otimes b) = s\{(\lambda_1 \otimes 1) f(a_1 \otimes b) + (-1)^p (1 \otimes \gamma_1) f(a \otimes b_1)\} \tag{2}$$

where $a_1 = [\lambda_2, \ldots, \lambda_p]$ and $b_1 = [\gamma_2, \ldots, \gamma_q]$. For $p = 0$ we have

$$f(1 \otimes b) = [1 \otimes \gamma_1, \ldots, 1 \otimes \gamma_q]$$

while for $q = 0$ we have

$$f(a \otimes 1) = [\lambda_1 \otimes 1, \ldots, \lambda_p \otimes 1].$$

To prove that the above definition of f checks with the inductive definition of 5.2, we must show that $f(a \otimes b) = sfd(a \otimes b)$. We recall that in the complex $N(\Lambda \otimes \Gamma)$ the operator s is zero on all elements of form $[\omega_1, \ldots, \omega_k]\omega$, which do not have an operator in front. Since f introduces no operators, $d(a \otimes b)$ may be replaced by $\lambda_1 a_1 \otimes b + (-1)^p a \otimes \gamma_1 b_1$ assuming that $p > 0$, $q > 0$. Thus $sfd(a \otimes b)$ gives precisely formula (2) above. The cases $p = 0$ or $q = 0$ are similarly verified by an easier argument.

We now pass to the products $\vee$ and $\wedge$. Here we need a map

$$g\colon N(\Lambda \otimes \Gamma) \to N(\Lambda) \otimes N(\Gamma)$$

over the identity map of $\Lambda \otimes \Gamma$. Such a map is given by the formula

(3) $$g[\lambda_1 \otimes \gamma_1, \ldots, \lambda_n \otimes \gamma_n] = \sum_{0 \leq p \leq n} [\lambda_1, \ldots, \lambda_p]\lambda_{p+1} \cdots \lambda_n \otimes \gamma_1 \cdots \gamma_p[\gamma_{p+1}, \ldots, \gamma_n].$$

The verification that this formula is precisely the one given by the inductive procedure of 5.2 using the standard splitting and contracting homotopies, is analogous to the one just discussed and is left to the reader.

If Λ is commutative, then to compute the products $\pitchfork$ and $\Cup$ we need a map

$$h: N(\Lambda) \otimes_{\Lambda^e} N(\Lambda) \to N(\Lambda).$$

Such a map is given for

$$a = [\lambda_1, \ldots, \lambda_p], \qquad b = [\lambda'_1, \ldots, \lambda'_q]$$

by the formula

$$h(a \otimes_\Lambda b) = \sum \pm (\zeta_1, \ldots, \zeta_{p+q})$$

where $\zeta_1, \ldots, \zeta_{p+q}$ ranges over all permutations of the sequence $\lambda_1, \ldots, \lambda_p, \lambda'_1, \ldots, \lambda'_q$ for which $\lambda_1, \ldots, \lambda_p$ remain in order and $\lambda'_1, \ldots, \lambda'_q$ remain in order. The sign is $+$ for even permutations and $-$ for odd ones.

For the $\cup$ and $\cap$ products depending on a diagonal map, we have no explicit formulae except when the diagonal map D is explicitly given.

So far we have used the normalized standard complexes. For the unnormalized standard complexes, the inductive procedures of § 5 can equally well be applied to obtain maps f, g, h. These will be considerably more complicated. It turns out, however, that the same formulae used to define f, g, h for the normalized complexes preserve their meaning and commute with the differentiation even if the complexes are taken unnormalized. They no longer are the maps given by the inductive procedure of 5.2. We have no rational explanation of this phenomenon.

7. PRODUCTS FOR SUPPLEMENTED ALGEBRAS

Let Λ and Γ be supplemented K-algebras with augmentations $\varepsilon_\Lambda: \Lambda \to K$, $\varepsilon_\Gamma: \Gamma \to K$. The tensor product $\Lambda \otimes_K \Gamma$ is supplemented by setting

$$\varepsilon(\lambda \otimes \gamma) = (\varepsilon_\Lambda \lambda)(\varepsilon_\Gamma \gamma).$$

In the products of § 1 we replace A and A' by K, thus obtaining products

$$\top: \operatorname{Tor}_p^\Lambda(A,K) \otimes \operatorname{Tor}_q^\Gamma(A',K) \to \operatorname{Tor}_{p+q}^{\Lambda\otimes\Gamma}(A \otimes A',K)$$

$$\perp: \operatorname{Ext}_{\Lambda\otimes\Gamma}^{p+q}(K, \operatorname{Hom}(A,C')) \to \operatorname{Hom}(\operatorname{Tor}_p^\Lambda(A,K), \operatorname{Ext}_\Gamma^q(K,C'))$$

for $(A_\Lambda, A'_\Gamma, {}_\Gamma C')$. If Λ and Γ are K-projective then so is $\Lambda \otimes \Gamma$ and we may pass to homology notation

$$\top\colon\ H_p(\Lambda,A) \otimes H_q(\Gamma,A') \to H_{p+q}(\Lambda \otimes \Gamma, A \otimes A')$$

$$\perp\colon\ H^{p+q}(\Lambda \otimes \Gamma, \operatorname{Hom}(A,C')) \to \operatorname{Hom}(H_p(\Lambda,A),H^q(\Gamma,C')).$$

For the products $\vee$ and $\wedge$ we must assume that Λ and Γ are K-projective. We obtain

$$\vee\colon\ H^p(\Lambda,C) \otimes H^q(\Gamma,C') \to H^{p+q}(\Lambda \otimes \Gamma, C \otimes C')$$

$$\wedge\colon\ H_{p+q}(\Lambda \otimes \Gamma, \operatorname{Hom}(C,A')) \to \operatorname{Hom}(H^p(\Lambda,C),H_q(\Gamma,A'))$$

for $({}_\Lambda C, A'_\Gamma, {}_\Gamma C')$.

If Λ is commutative we have the internal products $\Cap$ and $\Cup$ which, if Λ is K-projective, are Λ-homomorphisms

$$\Cap\colon\ H_p(\Lambda,A) \otimes_\Lambda H_q(\Lambda,A') \to H_{p+q}(\Lambda, A \otimes_\Lambda A')$$

$$\Cup\colon\ H^{p+q}(\Lambda, \operatorname{Hom}_\Lambda(A,A')) \to \operatorname{Hom}_\Lambda(H_p(\Lambda,A),H^q(\Lambda,A')).$$

Finally we come to the products $\cup$ and $\cap$ given by a diagonal map $D\colon \Lambda \to \Lambda \otimes \Lambda$. Here we postulate that D is compatible with the augmentation ε. This is expressed by requiring that the diagrams

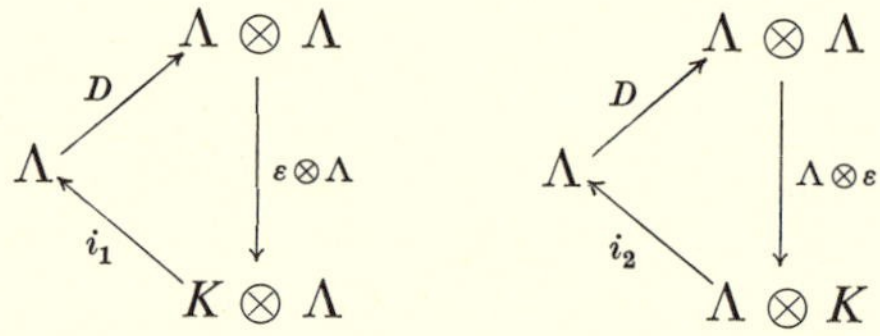

be commutative, where $i_1\lambda = 1 \otimes \lambda$ and $i_2\lambda = \lambda \otimes 1$. If we identify Λ with $K \otimes \Lambda$ and $\Lambda \otimes K$ and regard i_1 and i_2 as identity maps, the conditions become

$$(\varepsilon \otimes \Lambda)D = \text{identity} = (\Lambda \otimes \varepsilon)D.$$

The conditions imposed on D imply for each (left or right) Λ-module A, that the usual identifications of A with $K \otimes A$ and with $A \otimes K$ are compatible with the Λ-operators on the modules $K \otimes A$ and $A \otimes K$ induced by D. In particular we may identify K with $K \otimes K$. This leads to products

$$\cup\colon\ H^p(\Lambda,C) \otimes H^q(\Lambda,C') \to H^{p+q}(\Lambda, C \otimes C')$$

$$\cup\colon\ H_{p+q}(\Lambda, \operatorname{Hom}(C,A')) \to \operatorname{Hom}(H^p(\Lambda,C),H_q(\Lambda,A')).$$

The formulae for these products in terms of the standard complexes differ only in that the complexes $N(\Lambda)$ must be replaced by $N(\Lambda,\varepsilon)$. The formulae for the maps

$$f\colon N(\Lambda,\varepsilon_\Lambda) \otimes N(\Gamma,\varepsilon_\Gamma) \to N(\Lambda \otimes \Gamma,\varepsilon)$$

$$h\colon N(\Lambda,\varepsilon) \otimes_\Lambda N(\Lambda,\varepsilon) \to N(\Lambda,\varepsilon) \qquad \Lambda \text{ commutative}$$

are unchanged. The formula for

$$g\colon N(\Lambda \otimes \Gamma,\varepsilon) \to N(\Lambda,\varepsilon_\Lambda) \otimes N(\Gamma,\varepsilon_\Gamma)$$

reads

$$g[\lambda_1\otimes \gamma_1,\ldots,\lambda_n\otimes \gamma_n] = \sum_{0\leq p\leq n} [\lambda_1,\ldots,\lambda_p]\varepsilon_\Lambda(\lambda_{p+1}\cdots\lambda_n)\otimes \gamma_1\cdots\gamma_p[\gamma_{p+1},\ldots,\gamma_n]$$

The same formulae apply to the unnormalized standard complexes.

We now pass to monoids. Given monoids Π and Π' with augmentations $\varepsilon\colon \Pi \to K$, $\varepsilon'\colon \Pi' \to K$ we define the direct product $\Pi \times \Pi'$ as the monoid with elements (x,x'), $x \in \Pi$, $x' \in \Pi'$, multiplication $(x,x')(y,y') = (xy,x'y')$ and augmentation $\bar{\varepsilon}(x,x') = \varepsilon(x)\varepsilon'(x')$. It is then easy to see that $K(\Pi \times \Pi')$ may be identified with $K(\Pi) \otimes_K K(\Pi')$ under the identification $(x,x') = x \otimes x'$. Thus all the considerations concerning the $\top$, $\perp$, $\pitchfork$ and $\curlyvee$ products apply with Λ, Γ, $\Lambda \otimes \Gamma$ replaced by Π, Π', $\Pi \times \Pi'$. We shall not duplicate the various formulae.

To introduce the $\cup$ and $\cap$ products we consider the diagonal map $D\colon K(\Pi) \to K(\Pi \times \Pi)$ induced by $x \to (x,x)$. In order that this map be compatible with the augmentation we must assume that $\varepsilon(x) = \varepsilon(x)\varepsilon(x)$ i.e. that the augmentation $\varepsilon\colon \Pi \to K$ is *idempotent*. The unit augmentation and the zero augmentation clearly are idempotent. Thus assuming an idempotent augmentation, we obtain the products

$$\cup\colon H^p(\Pi,C) \otimes H^q(\Pi,C') \to H^{p+q}(\Pi,C \otimes C')$$

$$\cap\colon H_{p+q}(\Pi, \operatorname{Hom}(C,A)) \to \operatorname{Hom}(H^p(\Pi,C),H_q(\Pi,A)),$$

in the situation $(A_\Pi,{}_\Pi C,{}_\Pi C')$. The diagonal map D is commutative and associative in the sense of § 4, thus the commutative and associative rules stated in § 4 apply to the products $\cup$ and $\cap$ defined above.

To obtain formulae for the computation of the products, we define a map

$$j\colon N(\Pi) \to N(\Pi) \otimes N(\Pi)$$

over the identity map of K, as the composition of the map $N(\Pi) \to N(\Pi \times \Pi)$ induced by D with the map $g\colon N(\Pi \times \Pi) \to N(\Pi) \otimes N(\Pi)$ defined above. We obtain

$$j[x_1,\ldots,x_n] = \sum_{0\leq p\leq n} [x_1,\ldots,x_p]\varepsilon(x_{p+1}\cdots x_n) \otimes x_1\cdots x_p[x_{p+1},\ldots,x_n].$$

This formula also is valid for the unnormalized standard complexes.

8. ASSOCIATIVITY FORMULAE

Let Λ be a supplemented K-algebra with augmentation $\varepsilon\colon \Lambda \to K$. We shall assume that a diagonal map

$$D\colon \Lambda \to \Lambda \otimes \Lambda$$

and an "antipodism"

$$\omega\colon \Lambda \to \Lambda^*$$

are given subject to the following conditions:

(i) D and ω are homomorphisms of K-algebras.

(ii) D and ω are compatible with ε, i.e.

$$\varepsilon = \varepsilon^*\omega, \qquad (\varepsilon \otimes \Lambda)D = \text{identity} = (\Lambda \otimes \varepsilon)D.$$

(iii) $\omega^*\omega = $ identity, where $\omega^*\colon \Lambda^* \to \Lambda$ is induced by ω.

(iv) D is associative, i.e. $(\Lambda \otimes D)D = (D \otimes \Lambda)D$.

(v) D and ω commute, i.e. $D^*\omega = (\omega \otimes \omega)D$.

(vi) The map $E\colon \Lambda \to \Lambda^e$ defined by $E = (\Lambda \otimes \omega)D$ satisfies condition (E.1) of x,6.

It follows from (iii) that $\omega\omega^*$ also is the identity and therefore ω is an isomorphism with $\omega^{-1} = \omega^*$.

Consider the situation $(A_\Lambda, {}_\Lambda C)$. We may regard $A \otimes C$ as a right $\Lambda \otimes \Lambda^*$-module, and, using the map E, also as a right Λ-module. It follows from condition (vi) that $(A \otimes C)I = (A \otimes C)J$ where I is the kernel of $\varepsilon\colon \Lambda \to K$ while J is the kernel of $\rho\colon \Lambda^e \to \Lambda$. Since J is generated (as a left Λ^e-module) by the elements $\lambda \otimes 1 - 1 \otimes \lambda^*$ and since $(a \otimes c)(\lambda \otimes 1 - 1 \otimes \lambda^*) = a\lambda \otimes c - a \otimes \lambda c$ it follows that $(A \otimes C)I$ is the kernel of the natural map $A \otimes C \to A \otimes_\Lambda C$. We thus obtain

$$(1) \qquad A \otimes_\Lambda C \approx (A \otimes C)/(A \otimes C)I = (A \otimes C)_\Lambda \approx (A \otimes C) \otimes_\Lambda K.$$

Next consider the situation $({}_\Lambda A, {}_\Lambda C)$. We consider $\operatorname{Hom}(A,C)$ as a left Λ^e-module with operators $[(\lambda \otimes \gamma^*)f]a = \lambda f(\gamma a)$. Using the map E we may regard $\operatorname{Hom}(A,C)$ also as a left Λ-module. It then follows from (vi) that the elements of $\operatorname{Hom}(A,C)$ invariant under the operators of Λ^e (i.e. the annihilators of J) coincide with the elements of $\operatorname{Hom}(A,C)$ invariant under the operators of Λ (i.e. the annihilators of I). Since $[(\lambda \otimes 1 - 1 \otimes \lambda^*)f]a = \lambda(fa) - f(\lambda a)$, it follows that the invariant elements of $\operatorname{Hom}(A,C)$ are precisely those of the subgroup $\operatorname{Hom}_\Lambda(A,C)$. We thus obtain

$$(2) \qquad \operatorname{Hom}_\Lambda(A,C) = [\operatorname{Hom}(A,C)]^\Lambda \approx \operatorname{Hom}_\Lambda(K, \operatorname{Hom}(A,C)).$$

PROPOSITION 8.1. *In the situation* $({}_\Lambda A, {}_\Lambda B, {}_\Lambda C)$ *the isomorphism*

$$\text{Hom}(A, \text{Hom}(B,C)) \approx \text{Hom}(A \otimes B, C) \tag{3}$$

of II,5.2′ *induces an isomorphism*

$$\text{Hom}_\Lambda(A, \text{Hom}(B,C)) \approx \text{Hom}_\Lambda(A \otimes B, C) \tag{4}$$

where Hom (B,C) *and* $A \otimes B$ *are regarded as left* Λ*-modules using the maps* E *and* D *respectively.*

We first regard both sides of (3) as left $(\Lambda \otimes \Lambda^* \otimes \Lambda^*)$-modules by setting

$$\{[(\lambda \otimes \gamma^* \otimes \mu^*)f]a\}b = \lambda\{[f(\mu a)]\gamma b\}$$

for the left side of (3) and

$$[(\lambda \otimes \gamma^* \otimes \mu^*)g](a \otimes b) = \lambda[g(\mu a \otimes \gamma b)]$$

for the right side of (3). Then (3) becomes an operator isomorphism. Then we convert both sides of (3) into left Λ-modules using the map

$$\varphi\colon \Lambda \to \Lambda \otimes \Lambda^* \otimes \Lambda^*$$

given by

$$\begin{aligned}\varphi &= (\Lambda \otimes \omega \otimes \omega)(\Lambda \otimes D)D\\ &= (\Lambda \otimes \omega \otimes \omega)(D \otimes \Lambda)D.\end{aligned}$$

Consequently the invariant elements of both sides of (3) correspond to each other under the isomorphism (3). Since

$$\varphi = (E \otimes \Lambda^*)E$$

the operators of Λ on Hom $(A$, Hom $(B,C))$ may be arrived at as follows. First regard Hom (B,C) as a left Λ-module using the map E, then regard Hom $(A$, Hom $(B,C))$ as a left Λ-module again using the map E. Thus by (2) the invariant elements are $\text{Hom}_\Lambda(A, \text{Hom}(B,C))$. Now examine the right hand side of (3). Since

$$\varphi = (\Lambda \otimes D^*)E$$

we may regard first $A \otimes B$ as a left Λ-module using the map D and then regard Hom $(A \otimes B,C)$ as a left Λ-module using the map E. Thus by (2), the invariant elements are $\text{Hom}_\Lambda(A \otimes B, C)$.

Quite analogously we prove

PROPOSITION 8.1a. *In the situation* $(A_\Lambda, {}_\Lambda B, {}_\Lambda C)$ *the isomorphism*

$$(A \otimes B) \otimes C \approx A \otimes (B \otimes C) \tag{3a}$$

of II,5.1 *induces an isomorphism*

$$(A \otimes B) \otimes_\Lambda C \approx A \otimes_\Lambda (B \otimes C). \tag{4a}$$

where $A \otimes B$ is regarded as a right Λ-module using the map E, and $B \otimes C$ is regarded as a left Λ-module using the map D.

As an application of 8.1 we prove

PROPOSITION 8.2. *In the situation $({}_\Lambda A, {}_\Lambda B, {}_\Lambda C)$, if A is Λ-projective and B is K-projective then $A \otimes B$ is Λ-projective. If B is K-projective and C is Λ-injective then* Hom (B,C) *is Λ-injective.*

PROOF. Assume A is Λ-projective and B is K-projective. Then $\text{Hom}_\Lambda(A,D)$ and $\text{Hom}(B,C)$ are exact functors of D and C. Thus $\text{Hom}_\Lambda(A, \text{Hom}(B,C))$ is an exact functor of C. Consequently $\text{Hom}_\Lambda(A \otimes B, C)$ is an exact functor of C, which implies that $A \otimes B$ is Λ-projective. The second half is proved similarly.

Now consider the situation

$$({}_\Lambda A, {}_\Lambda A', {}_\Lambda B, {}_\Lambda C, C'_\Lambda).$$

Applying 8.1 to the triples $(\text{Hom}(B,C), B, C)$ and $(C', B, C' \otimes B)$ (with C' regarded as a left Λ-module using ω) we obtain isomorphisms

$$\text{Hom}_\Lambda(\text{Hom}(B,C), \text{Hom}(B,C)) \approx \text{Hom}_\Lambda(\text{Hom}(B,C) \otimes B, C)$$

$$\text{Hom}_\Lambda(C' \otimes B, C' \otimes B) \approx \text{Hom}_\Lambda(C', \text{Hom}(B, C' \otimes B)).$$

Substituting on the left the identity maps $\text{Hom}(B,C) \to \text{Hom}(B,C)$ and $C' \otimes B \to C' \otimes B$ we obtain the Λ-homomorphisms

$$\varphi\colon \text{Hom}(B,C) \otimes B \to C \tag{5}$$

$$\psi\colon C' \to \text{Hom}(B, C' \otimes B) \tag{5a}$$

given by

$$\varphi(f \otimes b) = fb, \qquad (\psi c')b = c' \otimes b.$$

Now assume that Λ is K-projective and that

$$\text{Tor}_n^K(A', A) = 0 \qquad \text{for } n > 0. \tag{*}$$

We then have the products

$$\cup\colon \text{Ext}_\Lambda(A', \text{Hom}(B,C)) \otimes \text{Ext}_\Lambda(A,B) \to \text{Ext}_\Lambda(A' \otimes A, \text{Hom}(B,C) \otimes B)$$

$$\cap\colon \text{Tor}^\Lambda(\text{Hom}(B, C' \otimes B), A \otimes A') \to \text{Hom}(\text{Ext}_\Lambda(A,B), \text{Tor}^\Lambda(C' \otimes B, A')).$$

Combining these with the pairings φ and ψ we obtain the *modified products*

$$\cup\colon \text{Ext}_\Lambda(A', \text{Hom}(B,C)) \otimes \text{Ext}_\Lambda(A,B) \to \text{Ext}_\Lambda(A' \otimes A, C) \tag{6}$$

$$\cap\colon \text{Tor}^\Lambda(C', A \otimes A') \to \text{Hom}(\text{Ext}_\Lambda(A,B), \text{Tor}^\Lambda(C' \otimes B, A')). \tag{6a}$$

Taking $A' = K$ we find $A \otimes A' = A$ and (*) is automatically satisfied. We thus obtain

(7) $\quad \cup\colon H^p(\Lambda, \operatorname{Hom}(B,C)) \otimes \operatorname{Ext}^q_\Lambda(A,B) \to \operatorname{Ext}^{p+q}_\Lambda(A,C)$

(7a) $\quad \cap\colon \operatorname{Tor}^\Lambda_{p+q}(C',A) \to \operatorname{Hom}(\operatorname{Ext}^p_\Lambda(A,B), H_q(\Lambda, C' \otimes B)).$

Further, taking $A = K$ we have

(8) $\quad \cup\colon H^p(\Lambda, \operatorname{Hom}(B,C)) \otimes H^q(\Lambda,B) \to H^{p+q}(\Lambda,C) \qquad ({}_\Lambda B, {}_\Lambda C),$

(8a) $\quad \cap\colon H_{p+q}(\Lambda,C') \to \operatorname{Hom}(H^p(\Lambda,B), H_q(\Lambda,C' \otimes B)) \qquad ({}_\Lambda B, C'_\Lambda).$

We recall that $\operatorname{Hom}(B,C)$ is regarded as a left Λ-module and $C' \otimes B$ as a right Λ-module using the map E.

PROPOSITION 8.3. *For $p = q = 0$ the maps*

$$\cup\colon \operatorname{Hom}_\Lambda(A', \operatorname{Hom}(B,C)) \otimes \operatorname{Hom}_\Lambda(A,B) \to \operatorname{Hom}_\Lambda(A' \otimes A, C)$$

$$\cap\colon C' \otimes_\Lambda (A \otimes A') \to \operatorname{Hom}(\operatorname{Hom}_\Lambda(A,B), (C' \otimes B) \otimes_\Lambda A'))$$

are given by

$$(f \cup g)(a' \otimes a) = (fa')(ga)$$

$$[c' \otimes (a \otimes a')] \cap f = (c' \otimes fa) \otimes a'.$$

This is a direct consequence of 1.2.3, 1.2.4 and the definition of the modified products.

In the case $\Lambda = Z(\Pi)$, where Π is a group with unit augmentation, we define

$$Dx = (x,x) = x \otimes x, \qquad \omega x = (x^{-1})^*.$$

Axioms (i)–(vi) are then satisfied. Thus all the above considerations apply in this case. We note that the operators of Π on $\operatorname{Hom}(B,C)$ and $C' \otimes B$ are given by

$$(xf)b = x[f(x^{-1}b)], \qquad (c' \otimes b)x = c'x \otimes x^{-1}b.$$

We shall see in Ch. XIII that the discussion of this section applies also to Lie algebras.

9. REDUCTION THEOREMS

We continue with the assumption that Λ is a K-projective supplemented K-algebra. We assume that $D\colon \Lambda \to \Lambda \otimes \Lambda$ and $\omega\colon \Lambda \to \Lambda^*$ satisfying the conditions (i)–(vi) of § 8 are given.

Taking $A = B$ in the products (7), (7a) of § 8 we obtain

$$\cup\colon H^p(\Lambda, \operatorname{Hom}(A,C)) \otimes \operatorname{Ext}^q_\Lambda(A,A) \to \operatorname{Ext}^{p+q}_\Lambda(A,C)$$

$$\cap\colon \operatorname{Tor}^\Lambda_{p+q}(C',A) \to \operatorname{Hom}(\operatorname{Ext}^p_\Lambda(A,A), H_q(\Lambda,C' \otimes A))$$

in the situation $({}_\Lambda A, {}_\Lambda C, C'_\Lambda)$. We concentrate our attention on the element

$$j \in \operatorname{Hom}_\Lambda(A,A) = \operatorname{Ext}^0_\Lambda(A,A)$$

which is the identity mapping of A. We thus obtain maps

(1) $$\cup j\colon\ H^p(\Lambda, \operatorname{Hom}(A,C)) \to \operatorname{Ext}^p_\Lambda(A,C)$$

(1a) $$\cap j\colon\ \operatorname{Tor}^\Lambda_p(C',A) \to H_p(\Lambda, C' \otimes A)$$

which we propose to investigate.

PROPOSITION 9.1. *For $p = 0$, the maps* (1) *and* (1*a*) *reduce to the isomorphisms*

$$\operatorname{Hom}_\Lambda(K, \operatorname{Hom}(A,C)) \approx \operatorname{Hom}_\Lambda(A,C)$$

$$C' \otimes_\Lambda A \approx (C' \otimes A) \otimes_\Lambda K$$

given by 8.1 *and* 8.1a.

This is an immediate consequence of 1.2.3, 1.2.4 and the definition of the modified products.

PROPOSITION 9.2. *If A is K-projective then $\cup j$ and $\cap j$ are isomorphisms.*

PROOF. Assuming that $\cup j$ and $\cap j$ are isomorphisms in degree p $(p \geq 0)$ we shall prove the same for $p + 1$.

Consider an exact sequence $0 \to C \to Q \to N \to 0$ with Q Λ-injective. Since A is K-projective, the sequence

$$0 \to \operatorname{Hom}(A,C) \to \operatorname{Hom}(A,Q) \to \operatorname{Hom}(A,N) \to 0$$

is exact, and by 8.2, $\operatorname{Hom}(A,Q)$ is Λ-injective. Using the rules for commutation of $\cap$ with connecting homomorphisms, we obtain a commutative diagram

$$\begin{array}{ccccccc}
H^p(\Lambda, \operatorname{Hom}(A,Q)) & \longrightarrow & H^p(\Lambda, \operatorname{Hom}(A,N)) & \longrightarrow & H^{p+1}(\Lambda, \operatorname{Hom}(A,C)) & \longrightarrow & 0 \\
\downarrow{\scriptstyle \cup j} & & \downarrow{\scriptstyle \cup j} & & \downarrow{\scriptstyle \cup j} & & \\
\operatorname{Ext}^p_\Lambda(A,Q) & \longrightarrow & \operatorname{Ext}^p_\Lambda(A,N) & \longrightarrow & \operatorname{Ext}^{p+1}_\Lambda(A,C) & \longrightarrow & 0
\end{array}$$

with exact rows. Since the first two vertical maps are isomorphisms the same follows for the third vertical map.

The proof for $\cap j$ is similar using an exact sequence $0 \to M \to P \to C' \to 0$ with P Λ-projective.

Consider an exact sequence

(S) $$0 \to A \to F_{q-1} \to \cdots \to F_0 \to F \to 0$$

of left Λ-modules $(q > 0)$. The iterated connecting homomorphism

$$\delta_S\colon\ \operatorname{Hom}_\Lambda(A,A) \to \operatorname{Ext}^q_\Lambda(F,A)$$

is then defined. The image $\delta_S j \in \mathrm{Ext}^q_\Lambda(F,A)$ of the element j is called the *characteristic element* of (S).

The products (7) and (7a) with A and B replaced by F and A yield maps

$$\gamma\colon H^p(\Lambda, \mathrm{Hom}(A,C)) \to \mathrm{Ext}^{p+q}_\Lambda(F,C) \qquad {}_\Lambda C$$

$$\vartheta\colon \mathrm{Tor}^\Lambda_{p+q}(C',F) \to H_p(\Lambda, C' \otimes A) \qquad C'_\Lambda$$

given by

$$\gamma h = (-1)^{pq} h \cup \delta_S j, \qquad h \in H^p(\Lambda, \mathrm{Hom}(A,C)),$$

$$\vartheta h' = h' \cap \delta_S j, \qquad h' \in \mathrm{Tor}^\Lambda_{p+q}(C',F).$$

PROPOSITION 9.3. *The homomorphisms γ and ϑ admit factorizations*

$$H^p(\Lambda, \mathrm{Hom}(A,C)) \xrightarrow{\cup j} \mathrm{Ext}^p_\Lambda(A,C) \xrightarrow{\Delta} \mathrm{Ext}^{p+q}_\Lambda(F,C)$$

$$\mathrm{Tor}^\Lambda_{p+q}(C',F) \xrightarrow{\Delta'} \mathrm{Tor}^\Lambda_p(C',A) \xrightarrow{\cap j} H_p(\Lambda, C' \otimes A)$$

where Δ and Δ' are the iterated connecting homomorphisms relative to the sequence (S).

PROOF. This clearly follows from the commutativity of the diagrams

$$\begin{array}{ccc}
\mathrm{Ext}_\Lambda(K, \mathrm{Hom}(A,C)) \otimes \mathrm{Ext}_\Lambda(A,A) & \xrightarrow{\cup} & \mathrm{Ext}_\Lambda(A,C) \\
\downarrow{\scriptstyle I \otimes \delta_S} & & \downarrow{\scriptstyle \Delta} \\
\mathrm{Ext}_\Lambda(K, \mathrm{Hom}(A,C)) \otimes \mathrm{Ext}_\Lambda(F,A) & \xrightarrow{\cup} & \mathrm{Ext}_\Lambda(F,C)
\end{array}$$

$$\begin{array}{ccc}
\mathrm{Tor}^\Lambda(C',F) & \to & \mathrm{Hom}(\mathrm{Ext}_\Lambda(F,A), \mathrm{Tor}^\Lambda(C' \otimes A,K)) \\
\downarrow{\scriptstyle \Delta'} & & \downarrow{\scriptstyle \mathrm{Hom}(\delta_S, I')} \\
\mathrm{Tor}^\Lambda(C',A) & \to & \mathrm{Hom}(\mathrm{Ext}_\Lambda(A,A), \mathrm{Tor}^\Lambda(C' \otimes A,K))
\end{array}$$

where I and I' stand for appropriate identity maps.

THEOREM 9.4. (*Reduction theorem.*) *If $F_0, \dots, F_{q-1}$ are Λ-projective and A is K-projective, then the maps γ and ϑ are isomorphisms for $p > 0$. For $p = 0$ we have the exact sequences*

$$\mathrm{Hom}_\Lambda(F_{q-1},C) \to \mathrm{Hom}_\Lambda(A,C) \to \mathrm{Ext}^q_\Lambda(F,C) \to 0$$

$$0 \to \mathrm{Tor}^\Lambda_q(C',F) \to C' \otimes_\Lambda A \to C' \otimes_\Lambda F_{q-1}.$$

PROOF. We apply the factorizations given in 9.3 and observe that since A is K-projective, the maps $\cup j$ and $\cap j$ are isomorphisms by 9.2. This reduces 9.4 to a statement about the iterated connecting homomorphisms Δ and Δ', which is a consequence of v,7.2.

COROLLARY 9.5. *Let*

$$(S) \qquad 0 \to A \to F_{q-1} \to \cdots \to F_0 \to K \to 0 \qquad q > 0$$

be an exact sequence of left Λ*-modules with* $F_{q-1}, \ldots, F_0$ Λ*-projective. Then for* $p > 0$ *we have isomorphisms*

$$H^p(\Lambda, \operatorname{Hom}(A,C)) \approx H^{p+q}(\Lambda,C) \qquad {}_{\Lambda}C$$

$$H_{p+q}(\Lambda,C') \approx H_p(\Lambda,C' \otimes A) \qquad C'_{\Lambda}$$

given by $h \to h \cup \delta_S j$ *and* $h' \to h' \cap \delta_S j$. *For* $p = 0$ *we have the exact sequences*

$$\operatorname{Hom}_{\Lambda}(F_{q-1},C) \to \operatorname{Hom}_{\Lambda}(A,C) \to H^q(\Lambda,C) \to 0$$

$$0 \to H_q(\Lambda,C') \to C' \otimes_{\Lambda} A \to C' \otimes_{\Lambda} F_{q-1}.$$

This follows directly from 9.4 provided we show that A is K-projective. This is immediately seen by decomposing (S) into short exact sequences.

In particular, if we consider the exact sequence

$$0 \to I \to \Lambda \to K \to 0$$

we obtain isomorphisms $(p > 0)$

$$H^p(\Lambda, \operatorname{Hom}(I,C)) \approx H^{p+1}(\Lambda,C), \qquad {}_{\Lambda}C$$

$$H_{p+1}(\Lambda,C') \approx H_p(\Lambda,C' \otimes I), \qquad C'_{\Lambda}$$

given by appropriate products.

EXERCISES

1. Let Λ, Γ, Σ be K-algebras with Σ assumed to be K-projective. Define the products

$$\top : \operatorname{Tor}^{\Lambda \otimes \Sigma^*}(C,A) \otimes \operatorname{Tor}^{\Sigma \otimes \Gamma^*}(C',A') \to \operatorname{Tor}^{\Lambda \otimes \Gamma^*}(C' \otimes_{\Sigma} C, A \otimes_{\Sigma} A')$$

in the situation $({}_{\Lambda}A_{\Sigma}, {}_{\Sigma}C_{\Lambda}, {}_{\Sigma}A'_{\Gamma}, {}_{\Gamma}C'_{\Sigma})$, and the product

$$\perp : \operatorname{Ext}_{\Lambda \otimes \Gamma^*}(A \otimes_{\Sigma} A', \operatorname{Hom}_{\Sigma}(C,C')) \to \operatorname{Hom}(\operatorname{Tor}^{\Lambda \otimes \Sigma^*}(C,A), \operatorname{Ext}_{\Sigma \otimes \Gamma^*}(A',C'))$$

in the situation $({}_{\Lambda}A_{\Sigma}, {}_{\Sigma}C_{\Lambda}, {}_{\Sigma}A'_{\Gamma}, {}_{\Sigma}C'_{\Gamma})$.

Assume further that Λ, Γ, Σ are K-projective and that $\operatorname{Tor}_n^{\Sigma}(A,A') = 0$ for $n > 0$. Define the product

$$\vee : \operatorname{Ext}_{\Lambda \otimes \Sigma^*}(A,C) \otimes \operatorname{Ext}_{\Sigma \otimes \Gamma^*}(A',C') \to \operatorname{Ext}_{\Lambda \otimes \Gamma^*}(A \otimes_{\Sigma} A', C \otimes_{\Sigma} C')$$

in the situation $({}_{\Lambda}A_{\Sigma}, {}_{\Lambda}C_{\Sigma}, {}_{\Sigma}A'_{\Gamma}, {}_{\Sigma}C'_{\Gamma})$, and the product

$$\wedge : \operatorname{Tor}^{\Lambda \otimes \Gamma^*}(\operatorname{Hom}_{\Sigma}(C,C'), A \otimes_{\Sigma} A') \to \operatorname{Hom}(\operatorname{Ext}_{\Lambda \otimes \Sigma^*}(A,C), \operatorname{Tor}^{\Sigma \otimes \Gamma^*}(C',A')$$

in the situation $({}_{\Lambda}A_{\Sigma}, {}_{\Lambda}C_{\Sigma}, {}_{\Sigma}A'_{\Gamma}, {}_{\Gamma}C'_{\Sigma})$.

Show that replacing the triple (Λ,Γ,Σ) by (Λ,Γ^*,K) gives the products defined in § 1. Establish the formal properties of the generalized products.

2. Let Λ be a K-projective K-algebra. Taking $\Lambda = \Gamma = \Sigma$ in Exer. 1 derive the "products of the second kind" for the Hochschild groups:

$$\begin{aligned}
\top:&\ H_p(\Lambda,A) \otimes H_q(\Lambda,A') &&\to H_{p+q}(\Lambda,A \otimes_\Lambda A')\\
\perp:&\ H^{p+q}(\Lambda, \mathrm{Hom}_\Lambda(A,A')) &&\to \mathrm{Hom}(H_p(\Lambda,A),H^q(\Lambda,A'))\\
\vee:&\ H^p(\Lambda,A) \otimes H^q(\Lambda,A') &&\to H^{p+q}(\Lambda, A \otimes_\Lambda A')\\
\wedge:&\ H_{p+q}(\Lambda, \mathrm{Hom}_\Lambda(A,A')) &&\to \mathrm{Hom}(H^p(\Lambda,A),H_q(\Lambda,A'))
\end{aligned}$$

defined for any two-sided Λ-modules A and A'.

Find maps

$$f'\colon N(\Lambda) \otimes_\Lambda N(\Lambda) \to N(\Lambda)$$
$$g'\colon N(\Lambda) \to N(\Lambda) \otimes_\Lambda N(\Lambda)$$

analogous to the maps f and g of § 6, and show that the products of the second kind may be computed using f' and g'.

3. Show that the composition

$$N(\Lambda) \otimes N(\Gamma) \xrightarrow{f} N(\Lambda \otimes \Gamma) \xrightarrow{g} N(\Lambda) \otimes N(\Gamma)$$

is the identity. Similarly for the maps f', g' of Exer. 2.

4. Show that in the normalized standard complex $N(\Lambda)$ the contracting homotopy s has the form

$$s(\lambda[\lambda_1, \ldots, \lambda_n]\lambda') = [\lambda,\lambda_1, \ldots, \lambda_n]\lambda'$$

and that the sequence

$$0 \longrightarrow N_{-1}(\Lambda) \xrightarrow{s_{-1}} N_0(\Lambda) \xrightarrow{s_0} \cdots \longrightarrow N_n(\Lambda) \xrightarrow{s_n} N_{n+1}(\Lambda) \longrightarrow \cdots$$

is exact. As a consequence show that d_n $(n \geqq 0)$ maps $\mathrm{Ker}\, s_n$ isomorphically onto $\mathrm{Im}\, d_n$.

5. Show that the map $f\colon N(\Lambda) \otimes N(\Gamma) \to N(\Lambda \otimes \Gamma)$ constructed in § 6 satisfies $sf(\tilde{N}(\Lambda) \otimes \tilde{N}(\Gamma)) = 0$. Show that this property characterizes the map f in a unique fashion. Using this method establish commutativity and associativity properties of the map f. Apply a similar discussion to $f\colon N(\Lambda,\varepsilon_\Lambda) \otimes N(\Gamma,\varepsilon_\Gamma) \to N(\Lambda \otimes \Gamma,\varepsilon)$ for supplemented algebras.

6. Show that the map $g\colon N(\Lambda \otimes \Gamma) \to N(\Lambda) \otimes N(\Gamma)$ constructed in § 6 satisfies $ng(\tilde{N}(\Lambda \otimes \Gamma)) = 0$ where n is the contracting homotopy for $N(\Lambda) \otimes N(\Gamma)$ constructed, as in § 5, using the contracting homotopies s in $N(\Lambda)$ and $N(\Gamma)$. Show that the above property of g characterizes that map in a unique fashion. Derive an associativity property of g by this method. Apply a similar discussion to the map $N(\Lambda \otimes \Gamma,\varepsilon) \to N(\Lambda,\varepsilon_\Lambda) \otimes N(\Gamma,\varepsilon_\Gamma)$ for supplemented algebras.

7. Let $\varphi\colon \Lambda \to \Gamma$ be an epimorphism of commutative K-algebras. Let X be a projective resolution of Γ as a Λ-module, given in split form $X = \Lambda \otimes \tilde{X}$ with $\tilde{X}_0 = K$. Let further (s_n,σ) be a contracting homotopy for X. Then $\Gamma \otimes_\Lambda \Gamma = \Gamma \otimes_\Gamma \Gamma = \Gamma$ and the complex $X \otimes_\Lambda X = \Lambda \otimes \tilde{X} \otimes \tilde{X}$ is a left complex over Γ also given in split form. Construct a map

$$h\colon X \otimes_\Lambda X \to X$$

over the identity map of Γ using the inductive procedure of 5.2. Establish the following properties of h:

(i) $h(x \otimes y) = (-1)^{pq} h(y \otimes x)$ for $x \in X_p$, $y \in X_q$.

(ii) If $s(\tilde{X}) = 0$ then the element $1 \otimes \Lambda = X_0$ is a unit element for h, i.e. $h(1 \otimes x) = x = h(x \otimes 1)$.

(iii) If $h(\tilde{X} \otimes \tilde{X}) \subset \tilde{X}$ then h is associative, i.e. $h(x \otimes h(y \otimes z)) = h(h(x \otimes y) \otimes z)$.

If all these conditions hold then h converts X into a graded algebra with differentiation.

Apply the above to the case $\rho\colon \Lambda^e \to \Lambda$ and $X = N(\Lambda)$, where Λ is a commutative K-algebra. Conclude that $N(\Lambda)$ is a graded algebra with differentiation under the map h given in § 6.

Apply the above to the case $\varepsilon\colon \Lambda \to K$ where Λ is a supplemented commutative K-algebra. Conclude that $N(\Lambda,\varepsilon)$ is a graded algebra with differentiation, under the map h of § 7. In particular, for a commutative augmented monoid $\Pi \to K$, $N(\Pi)$ is a graded algebra with differentiation.

8. Let Λ be a K-algebra, A a two-sided Λ-module and M a K-module. Using the homomorphisms at the end of § 3 (also those of VI,5) establish homomorphisms

$$\rho^n\colon H^n(\Lambda, \operatorname{Hom}(A,M)) \to \operatorname{Hom}(H_n(\Lambda,A),M),$$

$$\sigma_n\colon H_n(\Lambda, \operatorname{Hom}(A,M)) \to \operatorname{Hom}(H^n(\Lambda,A),M).$$

Show that ρ^n is an isomorphism if M is K-injective, and that σ_n is an isomorphism if M is K-injective and Λ^e is Noetherian. In particular ρ^n and σ_n are isomorphisms if K is a field, $M = K$ and Λ is finitely K-generated.

9. Let Π be a monoid with an augmentation $\Pi \to Z$. Let A be a right Π-module and $T = R/Z$ the group of reals reduced mod 1. Then with the topology defined in VII,6, the group $D(A) = \operatorname{Hom}(A,T)$ is a compact abelian group with continuous left Π-operators. Then ρ^n of Exer. 8 becomes

$$\rho^n\colon H^n(\Pi,D(A)) \approx D(H_n(\Pi,A)).$$

Assign a natural topology to the group $H^n(\Pi, D(A))$ using a Π-projective resolution of Z and show that the isomorphism ρ^n is topological. Carry out a similar discussion for

$$\sigma_n:\ H_n(\Pi, D(C)) \approx D(H^n(\Pi, C))$$

where C is a left Π-module and the monoid Π is *finite*.

10. Let $\Lambda = (K,d)$ be the algebra of dual numbers over a commutative ring K, as defined in IV,2. Show that there is no diagonal map D: $\Lambda \to \Lambda \otimes \Lambda$ satisfying conditions (i)–(vi) of § 8 (with $\omega =$ identity).

CHAPTER XII

Finite Groups

Introduction. If Π is a finite group it is convenient to consider the homology groups $H_n(\Pi,A)$ and the cohomology groups $H^n(\Pi,A)$ using a left Π-module A in both cases. The norm homomorphism N: $A \to A$ induces a homomorphism N^*: $H_0(\Pi,A) \to H^0(\Pi,A)$. Using the method introduced in v,10, this allows us to combine the homology and cohomology groups into a single sequence $\hat{H}^q(\Pi,A)$ $(-\infty < q < \infty)$ called the *complete derived sequence* of Π. The interesting fact is that for this complete derived sequence a product theory may be established (§ 4–6) which generalizes the $\cup$ and $\cap$ products. In § 8–10 we study the relations between $\hat{H}(\Pi,A)$ and $\hat{H}(\pi,A)$ where π is a subgroup of Π. The last section (§ 11) is devoted to the study of groups Π for which $\hat{H}^q(\Pi,A)$ has a periodicity with respect to q.

The theory presented here has been developed by Tate (unpublished) with a view to applications in class-field theory. The results of § 11 are due to Artin and Tate (unpublished). The writing of this chapter was made possible only through the generous help of G. P. Hochschild and J. Tate.

1. NORMS

We shall be concerned with finite groups Π. The ground ring for the construction of the group algebra $\Lambda = Z(\Pi)$ will always be the ring Z of integers; the augmentation $Z(\Pi) \to Z$ will always be the unit augmentation. Unless otherwise stated all Π-modules will be assumed to be *left* Π-modules.

In the group ring $Z(\Pi)$ we distinguish a particular element

$$N = \Sigma x, \qquad x \in \Pi.$$

Rather than deal with the element N directly we shall consider the *norm homomorphism*

$$N\colon\ A \to A$$

defined for each Π-module A by

$$Na = \Sigma xa, \qquad x \in \Pi.$$

Since $N(x-1)=0$ and $xN=N$ it follows that

$$IA \subset \operatorname{Ker} N, \qquad \operatorname{Im} N \subset A^{\Pi}.$$

Consequently N induces a homomorphism

$$N^*\colon A_{\Pi} \to A^{\Pi}$$

where as usual $A_{\Pi} = A/IA$, and A^{Π} is the set of invariant elements of A.

The image of the norm homomorphism $N\colon A \to A$ will be denoted by $N(A)$ and will be regarded as a covariant functor of the Π-module A. Clearly we have a commutative diagram

(1)
$$\begin{array}{ccc} & N(A) & \\ {}^{g}\nearrow & & \searrow^{h} \\ A_{\Pi} & \xrightarrow[N^*]{} & A^{\Pi} \end{array}$$

where g is defined by N and is an epimorphism, while h is an inclusion map.

The kernel of the homomorphism $N\colon A \to A$ will be denoted by ${}_NA$.

If A and C are Π-modules, we convert Hom (A,C) into a Π-module by setting

$$(xf)a = x(f(x^{-1}a)).$$

We thus obtain a norm homomorphism

(2) $$N\colon \operatorname{Hom}(A,C) \to \operatorname{Hom}(A,C)$$

defined as $(Nf)a = \sum xf(x^{-1}a)$, $x \in \Pi$. The image of (2) is in the subgroup $\operatorname{Hom}_{\Pi}(A,C)$. If $f\colon A \to C$ is a Π-homomorphism then $Nf = (\Pi : 1)f$, where $(\Pi : 1)$ is the order of the group Π.

Consider homomorphisms

$$A \xrightarrow{f} B \xrightarrow{g} C \xrightarrow{h} D$$

where f and h are Π-homomorphisms and g is only a Z-homomorphism. Then

$$N(hgf) = h(Ng)f.$$

Proposition 1.1. *For each Π-module A the following properties are equivalent:*

(a) *The identity map $A \to A$ is the norm of some Z-endomorphism $\rho\colon A \to A$.*

(b) *A is weakly projective.*

(c) *A is weakly injective.*

Proof. The equivalence of (a) and (b) is stated in x,8.6.

(a) $\Rightarrow$ (c). Let $\rho\colon A \to A$ be such that $N\rho =$ identity. For any $f \in \operatorname{Hom}(Z(\Pi),A)$ define

$$\mu f = \sum x\rho(x^{-1}(fx)), \qquad x \in \Pi.$$

Then for $s \in \Pi$ we have

$$\mu(sf) = \sum x\rho(x^{-1}sf(s^{-1}x)) = s(\mu f).$$

If f is a constant with value a then

$$\mu f = \sum x\rho(x^{-1}a) = (N\rho)a = a.$$

Thus μ is a "mean" in the sense of x,8.4a and A is weakly injective.

(c) $\Rightarrow$ (a). Suppose A is weakly injective and let μ: Hom $(Z(\Pi),A) \to A$ be a mean in the sense of x,8.4a. For each $a \in A$ define $f_a \in$ Hom $(Z(\Pi),A)$ by setting $f_a 1 = a$, $f_a x = 0$ for $x \neq 1$. Then $\sum x f_{x^{-1}a}$ is a function constant on Π with value a. Define $\rho a = \mu f_a$. Then

$$(N\rho)a = \sum x\rho(x^{-1}a) = \sum x\mu(f_{x^{-1}a}) = \mu(\sum x f_{x^{-1}a}) = a.$$

Thus $N\rho$ is the identity.

PROPOSITION 1.2. *In order that $f \in \text{Hom}_\Pi(A,C)$ be the norm of an element $h \in$ Hom (A,C) it is necessary and sufficient that f admit a factorization*

$$A \xrightarrow{g} Z(\Pi) \otimes C \xrightarrow{h} C$$

where g and h are Π-homomorphisms and Π operates on $Z(\Pi) \otimes C$ as $y(x \otimes c) = yx \otimes c$.

PROOF. Assume $f = hg$. Since $Z(\Pi) \otimes C$ is weakly projective (x,8.1) there is a Z-endomorphism ρ of $Z(\Pi) \otimes C$ such that $N\rho =$ identity. Then $f = hg = h(N\rho)g = N(h\rho g)$.

Conversely assume $f = Nk$ for some $k \in$ Hom (A,C). Define $ga = \sum x \otimes k(x^{-1}a)$ and $h(x \otimes c) = xc$. Then $hga = \sum xk(x^{-1}a) = (Nk)a = fa$.

PROPOSITION 1.3. *If A is weakly projective, then in the diagram* (1), *the maps N^*, g, h are isomorphisms, i.e.*

$$\text{Ker } N = I.A, \qquad \text{Im } N = A^\Pi.$$

PROOF. By 1.1 there exists a Z-endomorphism ρ: $A \to A$ such that $a = \sum_{x \in \Pi} x\rho(x^{-1}a)$ for all $a \in A$. Suppose $Na = 0$. Then

$$a = \sum x\rho(x^{-1}a) = \sum x\rho(x^{-1}a) - \sum \rho(x^{-1}a)$$
$$= \sum(x - 1)\rho(x^{-1}a) \in I.A$$

so that Ker $g = 0$ and g is an isomorphism. Suppose $a \in A^\Pi$. Then $x^{-1}a = a$ for all $x \in \Pi$ and

$$a = \sum x\rho(x^{-1}a) = \sum x\rho(a) = N\rho a.$$

Thus Im $N = A^\Pi$ and h is an isomorphism.

2. THE COMPLETE DERIVED SEQUENCE

We shall consider the homology and cohomology groups of Π with coefficients in the same left Π-module A. Thus for $n \geqq 0$

$$H_n(\Pi,A) = \mathrm{Tor}_n^{\Pi}(Z,A), \qquad H^n(\Pi,A) = \mathrm{Ext}_{\Pi}^n(Z,A)$$

where on the left Z is regarded as a right Π-module while on the right Z is regarded as a left Π-module.

In addition to these functors we also consider the covariant functor N which to each module A assigns the image $N(A)$ of the norm homomorphism $N\colon A \to A$.

Diagram (1) of the preceding section may thus be rewritten

(1)
$$\begin{array}{ccc} & N & \\ {}^{g}\nearrow & & \searrow^{h} \\ H_0 & \xrightarrow[N^*]{} & H^0 \end{array}$$

where N^*, g, h are natural transformations of functors.

The three functors and the three maps in diagram (1), each give rise to a derived sequence in the sense of v,10. We shall denote these by DH_0, DH^0, DN, DN^*, Dh, Dg. Between these six derived sequences we have the following maps

(2)
$$\begin{array}{ccccc} Dg & \to & DN & \to & Dh \\ \uparrow & & & & \downarrow \\ DH_0 & \to & DN^* & \to & DH^0 \end{array}$$

which form a commutative diagram.

For instance the map $DN \to Dh$ is defined by the diagram

$$\begin{array}{ccc} N & \longrightarrow & N \\ \downarrow & & \downarrow h \\ N & \xrightarrow[h]{} & H^0 \end{array}$$

PROPOSITION 2.1. *All the maps in diagram* (2) *are isomorphisms.*

PROOF. We consider for example the map $DN \to Dh$, induced by the diagram above. Since by 1.3, $h\colon N(A) \to H^0(\Pi,A)$ is an isomorphism whenever A is projective, it follows from v,10.3 that $DN \to Dh$ is an isomorphism.

In view of 2.1 we shall identify the six derived sequences above into one sequence called the *complete derived sequence* of Π. We shall use either the notation

$$\ldots, H_n, \ldots, H_1, \tilde{H}_0, \tilde{H}^0, H^1, \ldots, H^n, \ldots$$

or

$$\ldots, \hat{H}^{-n}, \ldots, \hat{H}^{-1}, \hat{H}^0, \hat{H}^1, \ldots, \hat{H}^n, \ldots$$

Thus we have

$$\hat{H}^n(\Pi,A) = H^n(\Pi,A) = \operatorname{Ext}^n_\Pi(Z,A) \qquad n > 0$$

$$\hat{H}^0(\Pi,A) = \tilde{H}^0(\Pi,A) = \operatorname{Coker}(H_0 \to H^0) = \operatorname{Coker}(N \to H^0) = A^\Pi/NA.$$

$$\hat{H}^{-1}(\Pi,A) = \tilde{H}_0(\Pi,A) = \operatorname{Ker}(H_0 \to H^0) = \operatorname{Ker}(H_0 \to N) = {}_NA/I.A.$$

$$\hat{H}^n(\Pi,A) = H_{-n-1}(\Pi,A) = \operatorname{Tor}^\Pi_{-n-1}(Z,A). \qquad n < -1$$

The reason for renumbering the groups and introducing the symbol $\hat{H}^n$ is to enable us to consider the graded module $\hat{H}(\Pi,A)$ including all the terms of the complete derived sequence. The graded functor $\hat{H}$ is an exact connected sequence of functors, i.e. for each exact sequence $0 \to A' \to A \to A'' \to 0$ of Π-modules we have an exact sequence

$$\cdots \to \hat{H}^{n-1}(\Pi,A'') \to \hat{H}^n(\Pi,A') \to \hat{H}^n(\Pi,A) \to \hat{H}^n(\Pi,A'') \to \hat{H}^{n+1}(\Pi,A') \to \cdots$$

PROPOSITION 2.2. *If A is weakly projective (= weakly injective) then* $\hat{H}(\Pi,A) = 0$.

PROOF. Since A is weakly projective, it follows from x,8.2 that $H_n(\Pi,A) = 0$ for $n > 0$. Since A also is weakly injective, it follows from x,8.2a that $H^n(\Pi,A) = 0$ for $n > 0$. Finally it follows from 1.3 that $\tilde{H}^0(\Pi,A) = 0 = \tilde{H}_0(\Pi,A)$.

PROPOSITION 2.3. *In the complete derived sequence of* Π, *each functor is the satellite S_1 of the following one and the satellite S^1 of the preceding one*:

$$\hat{H}^{n-1} = S_1\hat{H}^n, \qquad \hat{H}^{n+1} = S^1\hat{H}^n.$$

This follows directly from the axiomatic description of satellites given by III,5.1.

Given a Π-homomorphism $f\colon A \to C$ of Π-modules we shall denote by $\hat{f}$ the induced homomorphism $\hat{H}(\Pi,A) \to \hat{H}(\Pi,C)$.

PROPOSITION 2.4. *If $f\colon A \to C$ is the norm of an element of* Hom (A,C) *then* $\hat{f} = 0$.

PROOF. It follows from 1.2 that f admits a factorization $A \to Z(\Pi) \otimes C \to C$. Since $Z(\Pi) \otimes C$ is weakly projective (x,8.1), we have $\hat{H}(\Pi,Z(\Pi) \otimes C) = 0$. Thus $\hat{f} = 0$.

PROPOSITION 2.5. *If* Π *is of order* $r = (\Pi : 1)$, *then* $r\hat{H}(\Pi,A) = 0$.

PROOF. The map $f\colon A \to A$ given by $a \to ra$ is the norm of the identity map $A \to A$. Thus $\hat{f} = 0$ and $r\hat{H}(\Pi,A) = 0$.

COROLLARY 2.6. *If* $\Pi = 1$ *is the trivial group, then* $\hat{H}(\Pi,A) = 0$.

COROLLARY 2.7. *If* Π *is of order* $r = (\Pi : 1)$ *and* $nA = 0$ *for some* n *relatively prime to* r, *then* $\hat{H}(\Pi,A) = 0$.

It should be noted that in contrast with $H^n(\Pi,A)$ and $H_n(\Pi,A)$ the functors $\hat{H}(\Pi,A)$ are not functors in the variable Π. We shall see substitute concepts in § 8 when we discuss the relations between a group Π and a subgroup π.

Using the group Z with trivial Π-operators as coefficient group, we have

$$(4)\qquad \begin{cases} \hat{H}^1(\Pi,Z) = 0, & \\ \hat{H}^0(\Pi,Z) = Z_r = Z/rZ, & r = (\Pi : 1), \\ \hat{H}^{-1}(\Pi,Z) = 0, & \\ \hat{H}^{-2}(\Pi,Z) = \Pi/[\Pi,\Pi]. & \end{cases}$$

The first result follows from the fact that each crossed homomorphism $\Pi \to Z$ is zero. The second and third follow from the fact that $N\colon Z \to Z$ consists in multiplication by r. The last formula follows from x,4, (8).

It will also be useful to determine the connecting homomorphism

$$(5)\qquad \delta\colon \hat{H}^{-1}(\Pi,A'') \to \hat{H}^0(\Pi,A')$$

corresponding to an exact sequence $0 \longrightarrow A' \xrightarrow{\psi} A \xrightarrow{\varphi} A'' \longrightarrow 0$. Replacing $\hat{H}^{-1}$ and $\hat{H}^0$ by their definitions, we have

$$(5')\qquad \delta\colon {}_NA''/I.A'' \to A'^{\Pi}/NA'.$$

This homomorphism may be explicitly described as follows. Given $a'' \in {}_NA''$ choose $a \in A$ with $\varphi a = a''$. Then $\varphi Na = N\varphi a = Na'' = 0$ so that there is an element $a' \in A'$ with $\psi a' = Na$. Since $\psi xa' = xNa = Na = \psi a'$ it follows that $a' \in A'^{\Pi}$ and determines an element of A'^{Π}/NA'. This is $\delta a''$. This description is in agreement with the description obtained from the diagrams of v,10.

3. COMPLETE RESOLUTIONS

We shall introduce a new type of resolutions which will allow us to compute the complete derived sequence of a finite group using a single complex.

We need some preliminary considerations. For each Z-module C, we denote by C^0 the Z-module Hom (C,Z). Clearly, if C has a finite Z-base, then C^0 also has one. For any Z-module A, the homomorphism

$$\sigma\colon C \otimes A \to \mathrm{Hom}\,(C^0,A)$$

given by (cf. VI,5)

$$[\sigma(c \otimes a)]f = (fc)a, \qquad c \in C, \qquad a \in A, \qquad f \in C^0,$$

is an isomorphism whenever C has a finite Z-base. In particular, taking $A = Z$, we have an isomorphism $C \approx C^{00}$ if C has a finite Z-base.

PROPOSITION 3.1. *Given an exact sequence of Z-free modules*

$$(X) \qquad \cdots \longrightarrow X_{n+1} \xrightarrow{d_{n+1}} X_n \xrightarrow{d_n} X_{n-1} \longrightarrow \cdots$$

there exists a "contracting homotopy," i.e. a sequence of Z-homomorphisms $s_n\colon X_n \to X_{n+1}$ *such that*

$$(1) \qquad d_{n+1}s_n + s_{n-1}d_n = identity.$$

PROOF. Let $U_n = \operatorname{Ker} d_n = \operatorname{Im} d_{n+1}$. Then U_n is Z-free (cf. I,5.3), and we have an exact sequence

$$0 \longrightarrow U_n \xrightarrow{i_n} X_n \xrightarrow{\partial_n} U_{n-1} \longrightarrow 0$$

where i_n is the inclusion map and ∂_n is induced by d_n. Since U_{n-1} is projective, this sequence splits: there exist homomorphisms

$$U_n \xleftarrow{\varphi_n} X_n \xleftarrow{\psi_n} U_{n-1}$$

which together with i_n and ∂_n yield a representation of X_n as a direct sum. Set $s_n = \psi_{n+1}\varphi_n$; then (1) follows immediately.

COROLLARY 3.2. *Given an exact sequence (X) of Z-free modules, the corresponding sequence*

$$(X^0) \qquad \cdots \longrightarrow X^0_{n-1} \xrightarrow{d^0_{n-1}} X^0_n \xrightarrow{d^0_n} X^0_{n+1} \longrightarrow \cdots$$

where $X^0_n = \operatorname{Hom}(X_n, Z)$ *and* $d^0_n = \operatorname{Hom}(d_{n+1}, Z)$, *is exact.*

PROOF. Let $s^0_n = \operatorname{Hom}(s_{n-1}, Z)$. Then

$$d^0_{n-1}s^0_n + s^0_{n+1}d^0_n = \text{identity},$$

showing that the complex (X^0) is acyclic.

We observe that if the Z-free modules X_n have a finite base, then the modules X^0_n also are Z-free with a finite base, and the sequence (X^{00}) is simply (X).

We now consider $Z(\Pi)$-modules, Π being a finite group. If C is a left Π-module, then $C^0 = \operatorname{Hom}(C, Z)$ is a left Π-module by setting

$$(1) \qquad (xf)c = f(x^{-1}c);$$

this definition agrees with that given in § 1, if we consider Z as a Π-module on which Π operates trivially. If A is another left Π-module, then the homomorphism σ becomes a Π-homomorphism. Moreover, if C is $Z(\Pi)$-free with a finite base, then σ is an isomorphism, because C has a finite Z-base. We shall always identify $C \otimes A$ and $\mathrm{Hom}\,(C^0,A)$ as (left) Π-modules, when C is $Z(\Pi)$-free with a finite base. In particular taking $A = Z$ with trivial operators, we shall identify C and C^{00} as Π-modules.

If C is $Z(\Pi)$-free with a finite base, then C^0 also has one: it suffices to give the proof for the case $C = Z(\Pi)$. In fact, we have a natural isomorphism

$$\mathrm{Hom}\,(Z(\Pi),Z) \approx Z(\Pi) \tag{2}$$

defined in the following way: let (e_i) be the finite Z-base of $C = Z(\Pi)$ consisting of all elements of Π, and let (e_i^*) be the "dual base" of C^0, defined by

$$e_j^*(e_i) = \begin{cases} 0 & \text{if } i \neq j \\ 1 & \text{if } i = j. \end{cases}$$

There is a Z-isomorphism φ: $C \to C^0$ such that $\varphi(e_i) = e_i^*$, and it is immediately seen that φ is a $Z(\Pi)$-isomorphism, which proves (2).

This result together with 3.1 and 3.2 implies

PROPOSITION 3.3. *Given an exact sequence of (left)* $Z(\Pi)$*-modules and* $Z(\Pi)$*-homomorphisms*

$$\cdots \longrightarrow X_{n+1} \xrightarrow{d_{n+1}} X_n \xrightarrow{d_n} X_{n-1} \longrightarrow \cdots \tag{X}$$

such that each X_n *is* $Z(\Pi)$*-free with a finite base, then the sequence*

$$\cdots \longrightarrow X^0_{n-1} \xrightarrow{\overset{0}{d}_{n-1}} X^0_n \xrightarrow{\overset{0}{d}_n} X^0_{n+1} \longrightarrow \cdots \tag{X^0}$$

is exact and each X^0_n *is* $Z(\Pi)$*-free with a finite base.*

Consider now two (left) Π-modules C and A. In the tensor product $C \otimes_\Pi A$ it is understood that C is considered as a right Π-module by setting

$$cx = x^{-1}c, \qquad c \in C, x \in \Pi.$$

It follows that, considering $C \otimes A$ as a left Π-module by setting

$$x(c \otimes a) = (xc) \otimes (xa), \qquad c \in C, a \in A, x \in \Pi,$$

we have

$$(C \otimes A)_\Pi = C \otimes_\Pi A. \tag{3}$$

Moreover

$$(\mathrm{Hom}\,(C,A))^\Pi = \mathrm{Hom}_\Pi\,(C,A). \tag{3a}$$

Assuming now that C has a finite $Z(\Pi)$-base, we use the isomorphism σ

$$C \otimes A \approx \operatorname{Hom}(C^0, A),$$

or, replacing C by C^0,

$$C^0 \otimes A \approx \operatorname{Hom}(C, A). \tag{4}$$

Since, by x,8.5, $C^0 \otimes A$ is weakly projective (= weakly injective), it follows from 1.3 that

$$N^*\colon (C^0 \otimes A)_\Pi \to (C^0 \otimes A)^\Pi$$

is an isomorphism. By using (3), (3a), and (4) we have finally an isomorphism

$$\tau\colon C^0 \otimes_\Pi A \approx \operatorname{Hom}_\Pi(C, A)$$

when C and A are (left) Π-modules, and C has a finite $Z(\Pi)$-base. This isomorphism can be made explicit by the formula

$$[\tau(f \otimes a)]c = \sum_{x \in \Pi} f(x^{-1}c)xa, \qquad f \in C^0, c \in C, a \in A. \tag{5}$$

In particular, taking $A = Z(\Pi)$, we have, for each $Z(\Pi)$-free module C with a finite base, an isomorphism

$$\tau\colon C^0 \approx \operatorname{Hom}_\Pi(C, Z(\Pi)),$$

defined by

$$(\tau f)c = \sum_{x \in \Pi} f(x^{-1}c)x, \qquad f \in C^0, c \in C.$$

After these preliminaries we return to the main objective of this section. A *complete resolution* X for a finite group Π is an exact sequence

$$(X) \quad \cdots \longrightarrow X_n \xrightarrow{d_n} X_{n-1} \longrightarrow \cdots \longrightarrow X_0 \xrightarrow{d_0} X_{-1} \longrightarrow \cdots \longrightarrow X_{-n} \longrightarrow \cdots$$

of finitely generated free (left) Π-modules, together with an element $e \in (X_{-1})^\Pi$ such that the image of d_0 is generated by e.

Since $xe = e$ for each $x \in \Pi$ it follows that Im d_0 is the sub-Z-module generated by e. Further, since X_{-1} is Z-free, we have $ne \neq 0$ for $n \in Z$, $n \neq 0$. Therefore the mapping d_0 admits a factorization

$$X_0 \xrightarrow{\varepsilon} Z \xrightarrow{\mu} X_{-1} \tag{6}$$

where ε is a Π-epimorphism while μ is a Π-monomorphism given by $\mu 1 = e$. We consider the exact sequences

$$(X_L) \qquad \cdots \longrightarrow X_n \longrightarrow X_{n-1} \longrightarrow \cdots \longrightarrow X_0 \xrightarrow{\varepsilon} Z \longrightarrow 0$$

$$(X_R) \qquad 0 \longrightarrow Z \xrightarrow{\mu} X_{-1} \longrightarrow \cdots \longrightarrow X_{-n+1} \longrightarrow X_{-n} \longrightarrow \cdots$$

The sequence (X_L) provides a projective resolution of Z by means of finitely generated Π-free modules; by 3.3, the "dual" of (X_R):

$$(X_R^0) \qquad \cdots \to X^0_{-n} \to X^0_{-n+1} \to \cdots \to X^0_{-1} \to Z \to 0$$

provides also a projective resolution of Z by means of finitely generated Π-free modules.

Conversely, given two resolutions (X_L) and (X_L') of Z by finitely generated Π-free modules, we can construct a complete resolution X by "splicing" (X_L) with the sequence $(X_L'^0)$ suitably renumbered.

Given a complete resolution X and a (left) Π-module A, consider the complex

$$\operatorname{Hom}_\Pi(X,A).$$

For $n \geqq 0$ we leave the group $\operatorname{Hom}_\Pi(X_n,A)$ as it is. For $n < 0$ we replace $\operatorname{Hom}_\Pi(X_n,A)$ by the isomorphic group $X_n^0 \otimes_\Pi A$, using the isomorphism τ. We must examine in detail the map

$$(7) \qquad X^0_{-1} \otimes_\Pi A \to \operatorname{Hom}_\Pi(X_0,A)$$

induced by d_0: $X_0 \to X_{-1}$. In view of the factorization (6) of d_0, we obtain a commutative diagram

$$\begin{array}{ccccc} X^0_{-1} \otimes_\Pi A & \xrightarrow{\sigma} & \operatorname{Hom}_\Pi(X_{-1},A) & \longrightarrow & \operatorname{Hom}_\Pi(X_0,A) \\ \downarrow & & \downarrow & \nearrow & \\ Z \otimes_\Pi A & \xrightarrow[\sigma]{} & \operatorname{Hom}_\Pi(Z,A) & & \end{array}$$

Thus (7) admits the factorization

$$X^0_{-1} \otimes_\Pi A \longrightarrow H_0(\Pi,A) \xrightarrow{N^*} H^0(\Pi,A) \longrightarrow \operatorname{Hom}_\Pi(X_0,A).$$

Using the notation of v,10, we obtain

$$\operatorname{Hom}_\Pi(X,A) = (X_R^0 \otimes_\Pi A, N^*, \operatorname{Hom}_\Pi(X_L,A)).$$

This applying v,10.4 we obtain

THEOREM 3.2. *For any left* Π*-module* A*, the group* $\hat{H}^n(\Pi,A)$ *may be computed as* $H^n(\operatorname{Hom}_\Pi(X,A))$ *where* X *is any complete resolution of* Π. *If* f: $A \to A'$ *is a homomorphism then* $\hat{f}$ *may be computed from* $\operatorname{Hom}_\Pi(X,A) \to \operatorname{Hom}_\Pi(X,A')$. *If* $0 \to A' \to A \to A'' \to 0$ *is an exact sequence of left* Π*-modules, then the connecting homomorphisms* $\hat{H}(\Pi,A'') \to \hat{H}(\Pi,A')$ *may be computed from the exact sequence*

$$0 \to \operatorname{Hom}_\Pi(X,A') \to \operatorname{Hom}_\Pi(X,A) \to \operatorname{Hom}_\Pi(X,A'') \to 0.$$

REMARK. In the definition of a complete resolution it would be possible to use finitely generated *projective* Π-modules, instead of finitely generated *free* Π-modules. Actually, the first category of modules is strictly greater than the second (cf. Dock Sang Rim, *Ann. of Math.*, 69, 1959, pp. 700–712).

4. PRODUCTS FOR FINITE GROUPS

Given two (left) Π-modules A and A' we consider the tensor product (over Z) $A \otimes A'$ with the diagonal operators $x(a \otimes a') = xa \otimes xa'$. If $a \in A^{\Pi}$ and $a' \in A'^{\Pi}$ then $x(a \otimes a') = a \otimes a'$ and therefore $a \otimes a' \in (A \otimes A')^{\Pi}$. If $a \in A^{\Pi}$ and $a' = Nb', b' \in A'$ then $a \otimes a' = a \otimes \sum xb' = \sum x(a \otimes b') = N(a \otimes b')$ so that $a \otimes a' \in N(A \otimes A')$. Similarly if $a \in NA$, $a' \in A'^{\Pi}$. There results a homomorphism

(1) $$\xi\colon A^{\Pi}/NA \otimes A'^{\Pi}/NA' \to (A \otimes A')^{\Pi}/N(A \otimes A')$$

or

$$\xi\colon \hat{H}^0(\Pi,A) \otimes \hat{H}^0(\Pi,A') \to \hat{H}^0(\Pi,A \otimes A').$$

THEOREM 4.1. *There is a unique family of homomorphisms*

$$\xi^{p,q}\colon \hat{H}^p(\Pi,A) \otimes \hat{H}^q(\Pi,A') \to \hat{H}^{p+q}(\Pi,A \otimes A')$$

defined for each pair of Π*-modules* A *and* A' *and all integers* p,q *such that* $\xi^{0,0}$ *coincides with* ξ *and* $\xi^{p,q}$ *commutes with the connecting homomorphisms with respect to the variables* A *and* A' *as stated in* XI,2.5 *and* XI,2.5′.

We shall only be concerned in this section with the existence of the products $\xi^{p,q}$. Uniqueness is postponed to the next section.

Let X be a complete resolution for Π with selected element $e \in X_{-1}$. We consider the double complex $X \otimes X$ with differentiations $d' = d \otimes X$ and $d'' = X \otimes d$ (the definition of d'' involves the usual sign). A *mapping*

$$\Phi\colon X \to X \otimes X$$

is a family of Π-homomorphisms

$$\Phi_{p,q}\colon X_{p+q} \to X_p \otimes X_q$$

satisfying the following conditions:

(i) $$\Phi_{p,q}d = d'\Phi_{p+1,q} + d''\Phi_{p,q+1}$$

(ii) if $x \in X_0$ and $dx = e$ then $(d \otimes d)\Phi_{0,0}x = e \otimes e$.

The last condition may be rephrased as follows: if $\varepsilon\colon X_0 \to Z$ is the augmentation map obtained from the factorization of $d\colon X_0 \to X_{-1}$, then $(\varepsilon \otimes \varepsilon)\Phi_{0,0} = \varepsilon$.

Now given two cochains $f \in \operatorname{Hom}_{\Pi}(X_p,A)$, $g \in \operatorname{Hom}_{\Pi}(X_q,A')$ we define the product cochain $f.g \in \operatorname{Hom}_{\Pi}(X_{p+q},A \otimes A')$ as $f.g = (f \otimes g)\Phi_{p,q}$. Then $d(f.g) = (df).g + (-1)^p f.(dg)$, so that passing to cohomology we obtain a *bilinear* map $\hat{H}(\Pi,A) \otimes \hat{H}(\Pi,A') \to \hat{H}(\Pi,A \otimes A')$. The verification that this map verifies the condition of the theorem is immediate.

We now proceed to show that there exist mappings $\Phi\colon\ X \to X \otimes X$. In this construction we shall utilize

1° a contracting homotopy s for X. This is given by 3.1.

2° a Z-endomorphism $\rho\colon\ X \to X$ (of degree zero) such that $N\rho = I$, where I is the identity map of X. The existence of such a ρ is proved by 1.1 since each X_n is projective.

Next we introduce

$$s' = s \otimes I, \qquad s'' = I \otimes s,$$

$$\rho' = \rho \otimes I \qquad \rho'' = I \otimes \rho.$$

To define $\Phi_{0,0}$ we consider the diagram

$$\begin{array}{ccccc} & & X_0 & & \\ & & \downarrow{\scriptstyle \varepsilon} & & \\ X_0 \otimes X_0 & \xrightarrow{\varepsilon \otimes \varepsilon} & Z & \longrightarrow & 0 \end{array}$$

in which the row is exact. Since X_0 is projective, there is a map $\Phi_{0,0} = X_0 \to X_0 \otimes X_0$ such that $(\varepsilon \otimes \varepsilon)\Phi_{0,0} = \varepsilon$. This implies $d''d'\Phi_{0,0}d = 0$. We first define $\Phi_{p,q}$ with $p + q = 0$ as follows

$$\Phi_{p,-p} = -N(\rho''s'd''\Phi_{p-1,1-p}) \qquad p > 0$$

$$\Phi_{-p,p} = -N(\rho's''d'\Phi_{1-p,p-1}) \qquad p > 0$$

and verify by induction that

$$(d'\Phi_{p,-p} + d''\Phi_{p-1,1-p})d = 0 \qquad \text{all } p.$$

We now suppose that $\Phi_{p,q}$ is already defined for $|p + q| < t$ where t is a positive integer, that it satisfies (i) for $-t < p + q < t - 1$ and that

(iii) $$(d'\Phi_{p+1,q} + d''\Phi_{p,q+1})d = 0 \qquad \text{for } p + q = -t.$$

Now for $p + q = -t$ define

$$\Psi_{p,q} = d'\Phi_{p+1,q} + d''\Phi_{p,q+1}$$

$$\Phi_{p,q} = N(\rho'\Psi_{p,q}s).$$

Then

$$\begin{aligned} \Phi_{p,q}d = N(\rho'\Psi_{p,q}sd) &= N(\rho'\Psi_{p,q}(I - ds)) = N(\rho'\Psi_{p,q}) \\ &= \Psi_{p,q} = d'\Phi_{p+1,q} + d''\Phi_{p,q+1}. \end{aligned}$$

Further

$$\begin{aligned} (d'\Phi_{p,q} + d''\Phi_{p-1,q+1})d &= d'\Psi_{p,q} + d''\Psi_{p-1,q+1} \\ &= d'd''\Phi_{p,q+1} + d''d'\Phi_{p,q+1} = 0, \end{aligned}$$

which verifies condition (iii) at the next stage.

Next define

$$\Phi_{p,q} = N(s'\Phi_{p-1,q}d\rho) \qquad p+q=t.$$

Then

$$\begin{aligned} d'\Phi_{p,q} &= N((I - s'd')\Phi_{p-1,q}d\rho) \\ &= \Phi_{p-1,q}d - N(s'd'\Phi_{p-1,q}d\rho) \\ &= \Phi_{p-1,q}d + N(s'd''\Phi_{p-2,q+1}d\rho) \\ &= \Phi_{p-1,q}d - d''N(s'\Phi_{p-2,q+1}d\rho) \\ &= \Phi_{p-1,q}d - d''\Phi_{p-1,q+1}. \end{aligned}$$

Thus (i) holds as desired.

This concludes the existence proof.

5. THE UNIQUENESS THEOREM

The argument that will be used in proving the uniqueness of the products reappears many times in subsequent considerations. Therefore we shall give it an abstract formulation applicable in various situations.

Let Π be a fixed finite group; letters $A,A_1, \dots, A_n,B,C$ etc. will all be used to denote left Π-modules. Let $U_1, \dots, U_k,V$ each represent an exact connected sequence of covariant functors of A. A *map*

$$F\colon\ U_1 \otimes \cdots \otimes U_k \to V$$

is a family of homomorphisms

$$F\colon\ U_1^{i_1}(A_1) \otimes \cdots \otimes U_k^{i_k}(A_k) \to V^{i_1+\cdots+i_k}(A_1 \otimes \cdots \otimes A_k)$$

which is natural relative to Π-homomorphisms of the variables $A_1, \dots, A_k$ and which commutes with connecting homomorphisms in the following sense: If $0 \to A_j' \to A_j \to A_j'' \to 0$ is an exact sequence of Π-modules which splits over Z, then the diagram

$$\begin{array}{ccc} U_1(A_1) \otimes \cdots \otimes U_j(A_j'') \otimes \cdots \otimes U_k(A_k) & \to & V(A_1 \otimes \cdots \otimes A_j'' \otimes \cdots \otimes A_k) \\ \downarrow & & \downarrow \\ U_1(A_1) \otimes \cdots \otimes U_j(A_j') \otimes \cdots \otimes U_k(A_k) & \to & V(A_1 \otimes \cdots \otimes A_j' \otimes \cdots \otimes A_k) \end{array}$$

is commutative.

THEOREM 5.1. (*Uniqueness theorem.*) *Assume that the functors* $U_1, \dots, U_k,V$ *satisfy*

$$U_i(\operatorname{Hom}(Z(\Pi),A)) = 0$$

$$V(Z(\Pi) \otimes A) = 0$$

for any Π*-module* A. *If*

$$F,G\colon\ U_1 \otimes \cdots \otimes U_k \to V$$

are two maps such that F and G coincide on $U_1^0 \otimes \cdots \otimes U_k^0$, then $F = G$.

$Z(\Pi) \otimes A$ and $\operatorname{Hom}(Z(\Pi),A)$ are treated as Π-modules with operators

$$y(x \otimes a) = yx \otimes a, \qquad (yf)x = f(xy).$$

PROOF. Since $F - G$ also is a map $U_1 \otimes \cdots \otimes U_k \to V$ we may assume that $G = 0$ and prove that $F = 0$. To simplify the notation we shall limit our attention to the case $k = 2$.

Suppose that we already know that the map $F^{p,q}\colon U_1^p(A_1) \otimes U_2^q(A_2) \to V^{p+q}(A_1 \otimes A_2)$ is zero. Consider the exact sequence

$$0 \longrightarrow B \longrightarrow Z(\Pi) \otimes A_1 \xrightarrow{\varphi} A_1 \longrightarrow 0 \tag{1}$$

where $\varphi(x \otimes a) = a$ and $B = \operatorname{Ker} \varphi$. The Z-homomorphism $\varphi'\colon A_1 \to Z(\Pi) \otimes A_1$ given by $\varphi' a = 1 \otimes a$ shows that the exact sequence splits over the ring Z. It follows that the sequence

$$0 \to B \otimes A_2 \to Z(\Pi) \otimes A_1 \otimes A_2 \to A_1 \otimes A_2 \to 0$$

is exact, and we obtain the commutative diagram

$$\begin{array}{ccc} U_1^{p-1}(A_1) \otimes U_2^q(A_2) & \xrightarrow{F^{p-1,q}} & V^{p+q-1}(A_1 \otimes A_2) \\ \downarrow{\scriptstyle \delta \otimes i} & & \downarrow{\scriptstyle \Delta} \\ U_1^p(B) \otimes U_2^q(A_2) & \xrightarrow[F^{p,q}]{} & V^{p+q}(B \otimes A_2) \end{array}$$

Since $F^{p,q} = 0$ we have $\Delta F^{p-1,q} = 0$. However $V(Z(\Pi) \otimes A_1 \otimes A_2) = 0$ by assumption, so that Δ is an isomorphism. Thus $F^{p-1,q} = 0$. In exactly the same way arguing on the second variable, we prove that $F^{p,q-1} = 0$.

Next we consider the exact sequence

$$0 \longrightarrow A_1 \xrightarrow{\psi} \operatorname{Hom}(Z(\Pi),A_1) \longrightarrow B' \longrightarrow 0 \tag{2}$$

where $(\psi a)x = xa$ and $B' = \operatorname{Coker} \psi$. Again we obtain a splitting Z-homomorphism $\psi'\colon \operatorname{Hom}(Z(\Pi),A_1) \to A_1$ by setting $\psi' f = f1$. Consequently the sequence

$$0 \to A_1 \otimes A_2 \to \operatorname{Hom}(Z(\Pi),A_1) \otimes A_2 \to B' \otimes A_2 \to 0$$

is exact, and we obtain a commutative diagram

$$\begin{array}{ccc} U_1^p(B') \otimes U_2^q(A_2) & \xrightarrow{F^{p,q}} & V^{p+q}(B' \otimes A_2) \\ \downarrow{\scriptstyle \delta \otimes i} & & \downarrow{\scriptstyle \Delta} \\ U_1^{p+1}(A_1) \otimes U_2^q(A_2) & \xrightarrow[F^{p+1,q}]{} & V^{p+q+1}(A_1 \otimes A_2) \end{array}$$

Since $F^{p,q} = 0$ we have $F^{p+1,q}(\delta \otimes i) = 0$. Since $U_1(\text{Hom}\,(Z(\Pi),A_1) = 0$ by assumption, it follows that δ is an isomorphism. Thus $F^{p+1,q} = 0$. Similarly we prove $F^{p,q+1} = 0$. This completes the proof of the uniqueness theorem.

The uniqueness of the products asserted in 4.1 follows readily by taking $U_1 = U_2 = V = \hat{H}$.

For $a \in \hat{H}^p(\Pi,A)$, $b \in \hat{H}^q(\Pi,A')$ we shall denote the product $\xi^{p,q}(a \otimes b) \in \hat{H}^{p+q}(\Pi,A \otimes A')$ by the symbol ab. We shall regard the tensor product as a commutative and associative operation and thus identify $A \otimes A'$, $A \otimes (A' \otimes A'')$ with $A' \otimes A$, $(A \otimes A') \otimes A''$.

PROPOSITION 5.2. *For* $a \in \hat{H}^p(\Pi,A)$, $b \in \hat{H}^q(\Pi,A')$ *we have* $ab = (-1)^{pq}ba$; *more precisely, the elements* $ab \in \hat{H}^{p+q}(\Pi,A \otimes A')$ *and* $(-1)^{pq}ba \in \hat{H}^{p+q}(\Pi,A' \otimes A)$ *correspond to each other under the isomorphism induced by the natural isomorphism* $A \otimes A' \approx A' \otimes A$.

PROPOSITION 5.3. *For* $a \in \hat{H}^p(\Pi,A)$, $b \in \hat{H}^q(\Pi,A')$, $c \in \hat{H}^r(\Pi,A'')$ *we have* $a(bc) = (ab)c$.

To prove 5.2 it suffices to verify that $\xi^{p,q}(a \otimes b) = (-1)^{pq}ba$ verifies the axioms for a product. The proof of 5.3 follows from the uniqueness theorem by taking $U_1 = U_2 = U_3 = V = \hat{H}$ and $F(a \otimes b \otimes c) = a(bc)$, $G(a \otimes b \otimes c) = (ab)c$.

In the group

$$\hat{H}^0(\Pi,Z) = Z_r = Z/rZ \qquad r = (\Pi : 1)$$

we denote by 1 the element given by the coset $1 + rZ$ (i.e. the unit element of the ring Z_r).

PROPOSITION 5.4. *If* $a \in H^p(\Pi,A)$ *then* $1a = a = a1$ *provided we identify the modules* $Z \otimes A$, A *and* $A \otimes Z$.

The proof again follows from the uniqueness theorem by taking $U = V = \hat{H}$, $Fa = a$, $Ga = 1a$.

As usual, a Π-homomorphism $A \otimes A' \to B$ yields products

$$\hat{H}(\Pi,A) \otimes \hat{H}(\Pi,A') \to \hat{H}(\Pi,B)$$

by composition with the map $\hat{H}(\Pi,A \otimes A') \to \hat{H}(\Pi,B)$. In particular if A is a ring and Π operates on A in such a way that $x(a_1a_2) = (xa_1)(xa_2)$, then $\hat{H}(\Pi,A)$ becomes a ring. The unit element 1 of A is invariant, and its image in $H^0(\Pi,A) = A^\Pi/NA$ is a unit element for $\hat{H}(\Pi,A)$. If the multiplication in A is commutative then the multiplication in $\hat{H}(\Pi,A)$ is skew-commutative.

6. DUALITY

As in XI,8 we can use the Π-homomorphism

$$\varphi\colon \operatorname{Hom}(A,C) \otimes A \to C \qquad ({}_\Pi A, {}_\Pi C)$$

given by $f \otimes a \to fa$, to obtain the *modified product*

$$\hat{H}(\Pi, \operatorname{Hom}(A,C)) \otimes \hat{H}(\Pi,A) \to \hat{H}(\Pi,C) \tag{1}$$

where as usual Π operates on $\operatorname{Hom}(A,C)$ as $(xf)a = x(f(x^{-1}a))$. We shall still use the symbol ab to denote the image of $a \otimes b$ under (1).

PROPOSITION 6.1. *Let* $0 \longrightarrow A' \xrightarrow{i} A \xrightarrow{j} A'' \longrightarrow 0$ *be an exact sequence such that the sequence*

$$0 \longrightarrow \operatorname{Hom}(A'',C) \xrightarrow{j'} \operatorname{Hom}(A,C) \xrightarrow{i'} \operatorname{Hom}(A',C) \longrightarrow 0$$

is exact. Let $a \in \hat{H}^p(\Pi, \operatorname{Hom}(A',C))$, $b \in \hat{H}^q(\Pi,A'')$. *Then*

$$(\delta a) \,.\, b + (-1)^p a \,.\, \delta b = 0$$

where δ indicates the appropriate connecting homomorphisms.

PROOF. Let X be a complete projective resolution for Π. Let $g''\colon X_q \to A''$ be a cocycle in the class b. Then there is a cochain $g\colon X_q \to A$ with $jg = g''$ and a cocycle $g'\colon X_{q+1} \to A'$ with $ig' = dg$. The cocycle g' is then in the class δb. Similarly for a we have a cocycle $f'\colon X_p \to \operatorname{Hom}(A',C)$ in the class a, and a cochain $f\colon X_p \to \operatorname{Hom}(A,C)$ with $i'f = f'$ and a cocycle $f''\colon X_{p+1} \to \operatorname{Hom}(A'',C)$ with $j'f'' = df$. The cocycle f'' is in the class δa. Consequently (using the notation of § 4) we have

$$\begin{aligned} d(f \,.\, g) &= (df) \,.\, g + (-1)^p f \,.\, dg \\ &= j'f'' \,.\, g + (-1)^p f \,.\, ig' \\ &= f'' \,.\, jg + (-1)^p i'f \,.\, g' \\ &= f'' \,.\, g'' + (-1)^p f' \,.\, g' \end{aligned}$$

which implies the conclusion.

We introduce the mappings

$$\gamma_{p,q}\colon \hat{H}^p(\Pi, \operatorname{Hom}(A,C)) \to \operatorname{Hom}(\hat{H}^q(\Pi,A), \hat{H}^{p+q}(\Pi,C)) \tag{2}$$

by setting

$$(\gamma_{p,q} a)b = a \,.\, b.$$

PROPOSITION 6.2. *If for a fixed* Π*-module* C *and a pair of integers* p,q, *the mapping* $\gamma_{p,q}$ *is an isomorphism for all* Π*-modules* A, *then the same holds for all* $\gamma_{p',q'}$ *with* $p' + q' = p + q$.

PROOF. Consider the exact sequence

$$0 \to B \to Z(\Pi) \otimes A \to A \to 0$$

of § 5. Since this sequence Z-splits, the sequence

$$0 \to \operatorname{Hom}(A,C) \to \operatorname{Hom}(Z(\Pi) \otimes A,C) \to \operatorname{Hom}(B,C) \to 0$$

is exact. Thus by 6.1 we have the commutative diagram (with Π omitted)

$$\begin{array}{ccc} \hat{H}^p(\operatorname{Hom}(B,C)) & \xrightarrow{\gamma_{p,q}} & \operatorname{Hom}(\hat{H}^q(B),\hat{H}^{p+q}(C)) \\ \Big\downarrow{\scriptstyle (-1)^{p+1}\delta} & & \Big\downarrow{\scriptstyle \operatorname{Hom}(\delta,\hat{H}^{p+q}(C))} \\ \hat{H}^{p+1}(\operatorname{Hom}(A,C)) & \xrightarrow[\gamma_{p+1,q-1}]{} & \operatorname{Hom}(\hat{H}^{q-1}(A),\hat{H}^{p+q}(C)) \end{array}$$

Since $Z(\Pi) \otimes A$ is weakly projective, it follows from x,8.5 that $\operatorname{Hom}(Z(\Pi) \otimes A,C)$ is weakly injective. Consequently both connecting homomorphisms involved are isomorphisms. Since $\gamma_{p,q}$ is an isomorphism by assumption, it follows that $\gamma_{p+1,q-1}$ is an isomorphism. The proof that $\gamma_{p-1,q+1}$ is an isomorphism is similar and uses the exact sequence 5,(2).

PROPOSITION 6.3. *The mapping*

$$\gamma_{0,q}\colon \hat{H}^0(\Pi, \operatorname{Hom}(A,C)) \to \operatorname{Hom}(\hat{H}^q(\Pi,A),\hat{H}^q(\Pi,C))$$

composed with the natural epimorphism

$$\operatorname{Hom}_\Pi(A,C) \to \hat{H}^0(\Pi, \operatorname{Hom}(A,C))$$

yields a homomorphism

$$\operatorname{Hom}_\Pi(A,C) \to \operatorname{Hom}(\hat{H}^q(\Pi,A),\hat{H}^q(\Pi,C))$$

which to each $f \in \operatorname{Hom}_\Pi(A,C)$ *assigns the induced homomorphisms*

$$\hat{f}\colon \hat{H}^q(\Pi,A) \to \hat{H}^q(\Pi,C).$$

PROOF. Consider the map $g\colon Z \to \operatorname{Hom}(A,C)$ given by $g1 = f$, and let $h\colon A = Z \otimes A \to \operatorname{Hom}(A,C) \otimes A$ be induced by g. We obtain a commutative diagram

$$\begin{array}{ccccc} \hat{H}^0(\Pi,Z) \otimes \hat{H}^q(\Pi,A) & \longrightarrow & \hat{H}^q(\Pi, Z \otimes A) & \longrightarrow & \hat{H}^q(\Pi,A) \\ \Big\downarrow{\scriptstyle \hat{g}\otimes i} & & \Big\downarrow{\scriptstyle \hat{h}} & & \Big\downarrow{\scriptstyle \hat{f}} \\ \hat{H}^0(\Pi, \operatorname{Hom}(A,C)) \otimes \hat{H}^q(\Pi,A) & \longrightarrow & \hat{H}^q(\Pi, \operatorname{Hom}(A,C) \otimes A) & \xrightarrow{\hat{\varphi}} & \hat{H}^q(\Pi,C). \end{array}$$

If $a \in \hat{H}^0(\Pi, \mathrm{Hom}\,(A,C))$ is the element determined by f and $1 \in \hat{H}(\Pi,Z)$ is the unit element, then $\hat{g}1 = a$. Thus for each $b \in \hat{H}^q(\Pi,A)$ we have

$$(\gamma_{0,q}a)b = \hat{\varphi}(a\,.\,b) = \hat{\varphi}(\hat{g}1\,.\,b) = \hat{\varphi}\hat{h}(1\,.\,b) = \hat{f}(1\,.\,b) = \hat{f}b$$

since, by 5.3, $1b = b$.

THEOREM 6.4. (*Duality theorem*). *Let C be a group with trivial Π-operators and which is Z-injective (i.e. C is a divisible abelian group). Then for any Π-module A the homomorphism*

$$(4) \qquad \gamma_{p-1,-p}\colon \hat{H}^{p-1}(\Pi, \mathrm{Hom}\,(A,C)) \to \mathrm{Hom}\,(\hat{H}^{-p}(\Pi,A),\hat{H}^{-1}(\Pi,C))$$

given by $(\gamma a)b = a\,.\,b$, is an isomorphism.

We note that $\hat{H}^{-1}(\Pi,C) = {}_NC/IC = {}_rC$ is the subgroup of those elements $c \in C$ such that $rc = 0$ where $r = (\Pi : 1)$. Since $r\hat{H}(\Pi,A) = 0$, it follows that every homomorphism $\hat{H}(\Pi,A) \to C$ automatically is a homomorphism $\hat{H}(\Pi,A) \to {}_rC$. Thus (4) may be rewritten as follows

$$(4') \qquad \gamma_{p-1,-p}\colon \hat{H}^{p-1}(\Pi, \mathrm{Hom}\,(A,C)) \to \mathrm{Hom}\,(\hat{H}^{-p}(\Pi,A),C).$$

In view of 6.2, it suffices to show that $\gamma_{0,-1}$ is an isomorphism. Since $[\mathrm{Hom}\,(A,C)]^\Pi = \mathrm{Hom}_\Pi\,(A,C)$, we have

$$\gamma_{0,-1}\colon \mathrm{Hom}_\Pi\,(A,C)/N\,\mathrm{Hom}\,(A,C) \to \mathrm{Hom}\,({}_NA/I.A,C).$$

It follows from 6.3 that $\gamma_{0,-1}$ is obtained by restricting Π-homomorphisms $A \to C$ to the subgroup ${}_NA$. Consider any homomorphism $f\colon {}_NA \to C$ with $f(IA) = 0$. Since C is injective, f admits an extension $g\colon A \to C$. Since $g(IA) = 0$ we have $g \in \mathrm{Hom}_\Pi\,(A,C)$. Thus $\gamma_{0,-1}$ is an epimorphism. Next consider $g \in \mathrm{Hom}_\Pi\,(A,C)$ with $g({}_NA) = 0$. Since the sequence $0 \longrightarrow {}_NA \longrightarrow A \xrightarrow{N} A$ is exact and since C is Z-injective, it follows that

$$\mathrm{Hom}\,(A,C) \xrightarrow{N} \mathrm{Hom}\,(A,C) \longrightarrow \mathrm{Hom}\,({}_NA,C) \longrightarrow 0$$

is exact. Thus there exists $h \in \mathrm{Hom}\,(A,C)$ such that the composition $A \xrightarrow{N} A \xrightarrow{h} C$ is g. Then

$$(Ng)a = \sum xg(x^{-1}a) = \sum g(x^{-1}a) = g(Na) = ha$$

and $g \in N\,\mathrm{Hom}\,(A,C)$. Thus $\gamma_{0,-1}$ is a monomorphism. This concludes the proof of 6.4.

Taking $C = T = R/Z$ where R is the group of reals, and using the notation $D(A) = \mathrm{Hom}\,(A,T)$ introduced in VII,6, we obtain

COROLLARY 6.5. *The homomorphism*

$$(5) \qquad \gamma_{p-1,-p}\colon \hat{H}^{p-1}(\Pi, D(A)) \to D(\hat{H}^{-p}(\Pi,A))$$

is an isomorphism for all Π*-modules* A.

In particular, taking $A = Z$ we have $D(A) = T$, so that we obtain the isomorphism

$$(6) \qquad \gamma_{p-1,-p}\colon \hat{H}^{p-1}(\Pi,T) \approx D(\hat{H}^{-p}(\Pi,Z)).$$

Actually the group $D(\hat{H}^{-p}(\Pi,Z))$ should be replaced by $\mathrm{Hom}\,(\hat{H}^{-p}(\Pi,Z), \hat{H}^{-1}(\Pi,T))$ and $\hat{H}^{-1}(\Pi,T) = {}_rT$ is a cyclic group of order $r = (\Pi : 1)$.

THEOREM 6.6. (*Integral duality theorem*). *The mapping*

$$\gamma_{p,-p}\colon \hat{H}^{p}(\Pi,Z) \to \mathrm{Hom}\ (\hat{H}^{-p}(\Pi,Z),Z_r)$$

is an isomorphism. More exactly for every isomorphism $\varphi\colon \hat{H}^{-p}(\Pi,Z) \to Z_r = \hat{H}^0(\Pi,Z)$ *there is a unique* $a \in \hat{H}^p(\Pi,Z)$ *with*

$$\varphi b = ab, \qquad b \in \hat{H}^{-p}(\Pi,Z).$$

PROOF. Consider the exact sequence

$$0 \to Z \to R \to T \to 0$$

where R is the group of reals with trivial Π-operators. Let δ denote the connecting homomorphism $\hat{H}^p(\Pi,T) \to \hat{H}^{p+1}(\Pi,Z)$. Define the endomorphism $\rho\colon R \to R$ by setting $\rho t = r^{-1}t$ for $t \in R$, $r = (\Pi : 1)$. Then $N\rho = r\rho =$ identity. Thus, by 1.1, R is weakly projective and $\hat{H}(\Pi,R) = 0$. It follows that δ is an isomorphism. Since $(\delta a)b = \delta(ab)$ for $a \in \hat{H}^p(\Pi,T)$, $b \in \hat{H}^q(\Pi,Z)$, the conclusion follows directly from 6.5.

Using 6.6 we can supplement the list of values of $\hat{H}^q(\Pi,Z)$ given in (4) of § 2 by the following one

$$(7) \qquad \hat{H}^2(\Pi,Z) \approx \mathrm{Hom}\ (\Pi/[\Pi,\Pi],Z_r), \qquad r = (\Pi : 1).$$

7. EXAMPLES

Our first example is that of a cyclic group Π of order h with generator x. The ring $Z(\Pi)$ is then the quotient of the ring of polynomials $Z[x]$ by the ideal generated by the polynomial $x^h - 1$. In addition to the element $N = \sum_{0 \leq i < h} x^i$ we also consider the element $T = x - 1$. For every Π-module we thus obtain homomorphisms

$$N\colon A \to A, \qquad T\colon A \to A.$$

The kernel of T is A^Π while the image of T is $I.A$. These are independent of the choice of the generator x.

A complete resolution X for Π is obtained by setting

$$X_n = Z(\Pi), \qquad d_{2n} = N, \qquad d_{2n+1} = T,$$

the distinguished element of X_{-1} is the element $N = d_0 1$. The fact that the sequence

$$\cdots \longrightarrow Z(\Pi) \xrightarrow{T} Z(\Pi) \xrightarrow{N} Z(\Pi) \xrightarrow{T} Z(\Pi) \longrightarrow \cdots$$

is exact can be verified trivially. Also the following contracting homotopy s may be used for the proof:

$$sx^k = \begin{cases} 0 \ \text{ if } k = 0 \\ 1 + x + \cdots + x^{k-1} \ (k \geqq 1) \end{cases} \quad \text{in even degrees,}$$

$$sx^k = \begin{cases} 0 \ \text{ if } 0 \leqq k < h - 1 \\ 1 \ \text{ if } k = h - 1 \end{cases} \quad \text{in odd degrees.}$$

For any Π-module A the complex $\operatorname{Hom}_\Pi(X,A)$ is

$$\cdots \longleftarrow A \xleftarrow{N} A \xleftarrow{T} A \xleftarrow{N} A \xleftarrow{T} A \longleftarrow \cdots$$

with N appearing in odd and T in even dimensions. As a consequence we have

$$\hat{H}^{2n}(\Pi,A) = A^\Pi/NA, \qquad \hat{H}^{2n+1}(\Pi,A) = {}_NA/IA.$$

If A has trivial Π-operators then

$$\hat{H}^{2n}(\Pi,A) = A_h, \qquad \hat{H}^{2n+1}(\Pi,A) = {}_hA.$$

In particular

$$\hat{H}^{2n}(\Pi,Z) = Z_h, \qquad \hat{H}^{2n+1}(\Pi,Z) = 0.$$

To compute the products we must define a map $\Phi\colon X \to X \otimes X$, or rather a family of maps $\Phi_{p,q}\colon X_{p+q} \to X_p \otimes X_q$. These are obtained by setting

$$\Phi_{p,q}1 = 1 \otimes 1, \qquad p \text{ even}$$

$$\Phi_{p,q}1 = 1 \otimes x, \qquad p \text{ odd, } q \text{ even,}$$

$$\Phi_{p,q}1 = \sum_{0 \leqq m < n \leqq h-1} x^m \otimes x^n. \qquad p \text{ odd, } q \text{ odd.}$$

In verifying that these formulae satisfy the required identities we use the identity

$$(x \otimes x - 1 \otimes 1) \sum_{0 \leqq m < n \leqq h-1} x^m \otimes x^n = N \otimes 1 - 1 \otimes N$$

in the ring $\Lambda \otimes \Lambda$.

To exhibit the multiplication of cohomology classes, we consider an element of $\hat{H}^p(\Pi,A)$ represented by $a \in A$ with $a \in A^\Pi$ for p even and $a \in {}_NA$ for p odd. Similarly let $a' \in A'$ represent an element of $\hat{H}^q(\Pi,A')$. Then the product is an element of $\hat{H}^{p+q}(\Pi,A \otimes A')$ represented by

$$a \otimes a', \qquad p \text{ even or } q \text{ even,}$$

$$\sum_{0 \leqq m < n \leqq h} x^m a \otimes x^n a', \qquad p \text{ and } q \text{ odd.}$$

This last product may be simplified if A and A' have trivial Π-operators. Then for p and q odd we have $a \in {}_hA$, $a' \in {}_hA$ and the product is represented by the element $\dfrac{h(h-1)}{2} a \otimes a'$. Since the result lies in $\hat{H}^{p+q}(\Pi,A \otimes A') = (A \otimes A')_h$, it follows that the integer $\dfrac{h(h-1)}{2}$ may be reduced mod h. We thus obtain the product (for p and q odd):

$$a \,.\, a' = \frac{h}{2} a \otimes a' \qquad \text{if } h \text{ is even,}$$

$$a \,.\, a' = 0 \qquad \text{if } h \text{ is odd.}$$

If we wish to treat the cyclic group Π within the framework of the homology and cohomology theory of Ch. x, we must replace the complete resolution X by its positive part X_L. The homology and cohomology groups are as follows

$$H_0(\Pi,A) = A/I.A, \quad H_{2p+1}(\Pi,A) = A^\Pi/NA, \quad H_{2p+2}(\Pi,A) = {}_NA/IA,$$

$$H^0(\Pi,A) = A^\Pi, \qquad H^{2p+1}(\Pi,A) = {}_NA/IA, \qquad H^{2p+2}(\Pi,A) = A^\Pi/NA.$$

The products $\cup$ and $\cap$ may be computed using the map $\Phi_L\colon X_L \to X_L \otimes X_L$ induced by the map Φ above. In addition we have the products $\pitchfork$ and $\curlyvee$; to compute these we need a map $X_L \otimes_\Pi X_L \to X_L$. To define such a map we denote by y_p the unit element of $X_p = Z(\Pi)$. Now we convert X_L into a commutative graded $Z(\Pi)$-algebra with differentiation by setting

$$y_{2p} y_{2q} = \binom{p+q}{p} y_{2p+2q}$$

$$y_{2p+1} y_{2q} = y_{2q} y_{2p+1} = \binom{p+q}{p} y_{2p+2q+1}$$

$$y_{2p+1} y_{2q+1} = 0.$$

The description of this algebra may be simplified if we observe that $y_1 y_{2p} = y_{2p+1}$. We may now represent the algebra as a tensor product

$$Z(\Pi) \otimes E(y_1) \otimes P(y_2, y_4, \ldots, y_{2p}, \ldots)$$

where $E(y_1)$ is the exterior Z-algebra on the element y_1 of degree 1 and P is generated by the elements $y_2, y_4, \ldots$ with degrees indicated by the subscripts and with multiplication given by

$$y_{2p}y_{2q} = \binom{p+q}{p} y_{2p+2q}.$$

The differentiation in the algebra is given by $dx = 0$ for $x \in \Pi$, $dy_1 = T$, $dy_2 = Ny_1$, $dy_{2p+2} = Ny_1y_{2p}$.

Our next example is the group Π defined by two generators x and y with relations

$$x^t = y^2, \qquad xyx = y$$

where t is a non-negative integer. Iterating the second relation we find $x^tyx^t = y$ which implies that $x^{2t} = 1$. Any element $w \in \Pi$ has a unique canonical form

$$w = x^m y^\delta, \qquad 0 \leqq m < 2t, \qquad \delta = 0,1.$$

The group Π has order $4t$. The group Π may be regarded as a subgroup of the group of quaternions of absolute value 1 by setting

$$x \to e^{\pi i/t}, \qquad y \to j.$$

These groups are usually considered only when t is a power of 2 and are called the *generalized quaternion groups.*

A complete resolution X for the group Π described above is defined as follows, using abstract generators a_p, b_p, b'_p, c_p, c'_p, e_p:

$$\begin{aligned}
X_{4p} &= \Lambda a_p \quad \text{where} \quad \Lambda = Z(\Pi),\\
X_{4p+1} &= \Lambda b_p + \Lambda b'_p\\
X_{4p+2} &= \Lambda c_p + \Lambda c'_p\\
X_{4p+3} &= \Lambda e_p,\\
da_p &= Ne_{p-1}\\
db_p &= (x-1)a_p\\
db'_p &= (y-1)a_p\\
dc_p &= Lb_p - (y+1)b'_p, \qquad L = 1 + x + \cdots + x^{t-1}\\
dc'_p &= (xy+1)b_p + (x-1)b'_p\\
de_p &= (x-1)c_p - (xy-1)c'_p.
\end{aligned}$$

The selected element of X_{-1} is the element Ne_{-1}. The verification that $dd = 0$ is straightforward. The verification that the homology groups are trivial involves some computations which will be omitted.

The groups $\hat{H}(\Pi,A)$ with trivial operators of Π on A are as follows

$$\hat{H}^{4p}(\Pi,A) = A_{4t},$$

$$\hat{H}^{4p+1}(\Pi,A) = \begin{cases} {}_2A + {}_2A & t \text{ even}, \\ {}_4A & t \text{ odd}, \end{cases}$$

$$\hat{H}^{4p+2}(\Pi,A) = \begin{cases} A_2 + A_2 & t \text{ even}, \\ A_4 & t \text{ odd}, \end{cases}$$

$$\hat{H}^{4p+3}(\Pi,A) = {}_{4t}A.$$

A common feature of the cyclic groups and the generalized quaternion groups is the periodicity encountered in the complex X and the groups $\hat{H}(\Pi,A)$. A detailed study of the phenomenon of periodicity will be carried out in § 11.

8. RELATIONS WITH SUBGROUPS

Let π be a subgroup of a (finite) group Π. We shall use the letters A, A' etc. to denote Π-modules, which of course may also be regarded as π-modules.

Since $Z(\Pi)$ regarded as a left $Z(\pi)$-module is free on a finite base, it follows that every complete resolution X for Π also may be regarded as a complete resolution for π.

The inclusion

$$\mathrm{Hom}_\Pi (X,A) \subset \mathrm{Hom}_\pi (X,A)$$

induces a homomorphism (called *restriction*)

$$i(\pi,\Pi)\colon\ \hat{H}(\Pi,A) \to \hat{H}(\pi,A) \tag{1}$$

Next, consider two (left) Π-modules C and A. Define the homomorphism (called the *transfer*)

$$t\colon\ \mathrm{Hom}_\pi (C,A) \to \mathrm{Hom}_\Pi (C,A)$$

by setting for $f \in \mathrm{Hom}_\pi (C,A)$

$$(tf)c = \sum_i x_i f(x_i^{-1}c)$$

where $x_1\pi, \ldots, x_r\pi$, $r = (\Pi : \pi)$, are the left cosets of π in Π. If x_i is replaced by x_iy, for $y \in \pi$, then $x_iyf((x_iy)^{-1}c) = x_iyf(y^{-1}x_i^{-1}c) = x_if(x_i^{-1}c)$, so that the definition of tf is independent of the choice of the representatives $x_1, \ldots, x_r$. Further, for $x \in \Pi$, we have

$$(tf)(xc) = \sum x(x^{-1}x_i)f((x^{-1}x_i)^{-1}c) = x[(tf)c]$$

since $x^{-1}x_1, \ldots, x^{-1}x_r$ also is a system of representatives of left cosets of π in Π. Consequently tf is indeed a Π-homomorphism.

Replacing C by a complete resolution X of Π and passing to homology we obtain the *transfer homomorphisms*

$$t(\Pi,\pi)\colon\ \hat{H}(\pi,A) \to \hat{H}(\Pi,A). \tag{2}$$

In addition to the homomorphisms (1) and (2) we also have the isomorphisms

$$c_x\colon\ \hat{H}(\pi,A) \to \hat{H}(x\pi x^{-1},A) \tag{3}$$

defined using the homomorphism (studied in x,7)

$$c_x\colon\ \mathrm{Hom}_\pi(C,A) \to \mathrm{Hom}_{x\pi x^{-1}}(C,A)$$

given by $(c_x f)c = xf(x^{-1}c)$.

In degree zero the homomorphisms (1)–(3) yield

$$i(\pi,\Pi)\colon\ A^\Pi/N_\Pi A \to A^\pi/N_\pi A, \tag{1$_0$}$$

$$t(\Pi,\pi)\colon\ A^\pi/N_\pi A \to A^\Pi/N_\Pi A, \tag{2$_0$}$$

$$c_x\colon\ A^\pi/N_\pi A \to A^{x\pi x^{-1}}/N_{x\pi x^{-1}}A. \tag{3$_0$}$$

The map $(1)_0$ is induced by the inclusion $A^\Pi \subset A^\pi$; the map $(2)_0$ is induced by the map $A^\pi \to A^\Pi$ given by $a \to \sum x_i a$; the map $(3)_0$ is induced by the map $A^\pi \to A^{x\pi x^{-1}}$ given by $a \to xa$.

To verify these rules, we consider a complete resolution X for Π and consider the mapping $\varepsilon\colon\ X \longrightarrow Z$ given by $X_0 \xrightarrow{\varepsilon} Z$ of the factorization $X_0 \xrightarrow{\varepsilon} Z \longrightarrow X_{-1}$ of d_0. This induces a homomorphism $\mathrm{Hom}_\Pi(Z,A) \to \hat{H}^0(\Pi,A)$ which is easily seen to be the natural homomorphism $A^\Pi \to \hat{H}^0(\Pi,A) = A^\Pi/NA$. Thus the above formal rules follow trivially by replacing the complex X by Z.

The formal properties of the homomorphisms (1)–(3) will now be discussed. Clearly the homomorphisms are natural relative to maps $A \to A'$ and commute with connecting homomorphisms relative to exact sequences $0 \to A' \to A \to A'' \to 0$. Further we have

$$c_x c_y = c_{xy} \tag{4}$$

$$c_x = \text{identity if } x \in \pi \tag{5}$$

$$t(\Pi,\pi)i(\pi,\Pi)a = (\Pi : \pi)a, \qquad \text{for } a \in \hat{H}(\Pi,A). \tag{6}$$

If π' is a subgroup of π then we have the following rules

(7) $$i(\pi',\pi)i(\pi,\Pi) = i(\pi',\Pi)$$

(8) $$t(\Pi,\pi)t(\pi,\pi') = t(\Pi,\pi')$$

(9) $$c_x i(\pi',\pi) = i(x\pi'x^{-1},x\pi x^{-1})c_x$$

(10) $$c_x t(\pi,\pi') = t(x\pi x^{-1},x\pi' x^{-1})c_x.$$

All the above rules are straightforward consequences of the definitions.

To consider the rules connecting the homomorphisms (1)–(3) with the products, we consider elements $a \in \hat{H}(\Pi,A)$, $a' \in \hat{H}(\Pi,A')$, $b \in \hat{H}(\pi,A)$, $b' \in \hat{H}(\pi,A')$. Then

(11) $$i(\pi,\Pi)(a \,.\, a') = [i(\pi,\Pi)a] \,.\, [i(\pi,\Pi)a'],$$

(12) $$t(\Pi,\pi)(b \,.\, i(\pi,\Pi)a') = t(\Pi,\pi)b \,.\, a'$$

(13) $$t(\Pi,\pi)(i(\pi,\Pi)a \,.\, b') = a \,.\, t(\Pi,\pi)b',$$

(14) $$c_x(b \,.\, b') = c_x b \,.\, c_x b'.$$

We first use the rules given earlier for computing $i(\pi,\Pi)$, $t(\Pi,\pi)$ and c_x in degree zero, to verify that (11)–(14) hold if a, a', b, b' all have degree zero. Then we use the uniqueness theorem 5.1 to complete the proof. Taking rule (12) as an example, we introduce the functors

$$U_1(A) = \hat{H}(\pi,A), \qquad U_2(A) = \hat{H}(\Pi,A), \qquad V(A) = \hat{H}(\Pi,A)$$

and the maps

$$F,G\colon\ U_1(A) \otimes U_2(A') \to V(A \otimes A')$$

$$F(b \otimes a') = t(\Pi,\pi)(b \,.\, i(\pi,\Pi)a')$$

$$G(b \otimes a') = t(\Pi,\pi)b \,.\, a'.$$

We must verify that the maps F and G properly commute with the connecting homomorphisms; this is immediate, since F and G are compositions of maps which commute with connecting homomorphisms. Next we must show that U_1, U_2, V satisfy the conditions of 5.1. Thus we must show that $\hat{H}(\pi, \operatorname{Hom}(Z(\Pi),A)) = 0$. To prove this we observe that $Z(\Pi)$ is π-projective and therefore, by x,8.1, $\operatorname{Hom}(Z(\Pi),A)$ is weakly π-injective; consequently $\hat{H}(\pi, \operatorname{Hom}(Z(\Pi),A)) = 0$ by 2.2. We now can apply 5.1 to deduce that $F = G$.

9. DOUBLE COSETS

Let π and π' be subgroups of Π. We shall investigate the map

(1) $$i(\pi,\Pi)t(\Pi,\pi')\colon\ \hat{H}(\pi',A) \to \hat{H}(\pi,A)$$

for a Π-module A.

To this end we consider double cosets $\pi x\pi'$ with $x \in \Pi$. It can easily be seen that two such cosets are either equal or disjoint, so that we may represent Π as a disjoint union

$$\Pi = \cup_i \pi x_i \pi' \tag{2}$$

of such double cosets.

PROPOSITION 9.1. *Given a decomposition* (2) *of* Π *as a disjoint union of double cosets we have*

$$(\Pi : \pi') = \textstyle\sum_i (\pi : \pi \cap x_i \pi' x_i^{-1}) \tag{3}$$

$$i(\pi,\Pi)t(\Pi,\pi') = \textstyle\sum_i t(\pi,\pi \cap x_i\pi' x_i^{-1}) i(\pi \cap x_i\pi' x_i^{-1}, x_i\pi' x_i^{-1}) c_{x_i}. \tag{4}$$

PROOF. Let $\gamma_i = \pi \cap x_i\pi' x_i^{-1}$ and let

$$\pi = \cup_j y_{ji}\gamma_i$$

be a representation of π as a disjoint union of left γ_i cosets. Then

$$\pi x_i = \cup_j y_{ji}(\pi x_i \cap x_i \pi').$$

Multiplying by π' on the right we find

$$\pi x_i \pi' = \cup_j y_{ji}(\pi x_i \pi' \cap x_i \pi') = \cup_j y_{ji} x_i \pi'$$

and this union is still disjoint. Combining this with (2) we obtain a representation

$$\Pi = \bigcup_{i,j} y_{ji} x_i \pi'$$

of Π as a disjoint union of left cosets of π'. This implies (3).

Let $f \in \mathrm{Hom}_{\pi'}(X_n, A)$ where X is a complete resolution of Π. Then

$$i(\pi,\Pi)t(\Pi,\pi')f = \sum_{j,i} c_{y_{ji}} x_i f = \sum_i (\sum_j c_{y_{ji}} c_{x_i} f) = \sum_i t(\pi,\gamma_i) i(\gamma_i, x_i\pi' x_i) c_{x_i} f.$$

Passing to homology, we obtain (4).

COROLLARY 9.2. *If* π *is an invariant subgroup of* Π, *then for any* Π*-module* A *and any* $a \in \hat{H}(\pi,A)$,

$$i(\pi,\Pi)t(\Pi,\pi)a = \textstyle\sum xa, \qquad x \in \Pi/\pi.$$

An element $a \in \hat{H}(\pi,A)$ will be called *stable* if for each $x \in \Pi$ we have

$$i(\pi \cap x\pi x^{-1}, \pi)a = i(\pi \cap x\pi x^{-1}, x\pi x^{-1}) c_x a \tag{5}$$

or equivalently if

$$i(\pi \cap x\pi x^{-1}, \pi)a = c_x i(x^{-1}\pi x \cap \pi, \pi)a.$$

If π is an invariant subgroup, (5) reduces to $a = c_x a$. Thus in this case the stable elements are precisely those invariant under the operators of Π/π.

PROPOSITION 9.3. *If a is in the image of* $i(\pi,\Pi)$ *then a is stable.*

PROOF. Let $a = i(\pi,\Pi)b$ for some $b \in \hat{H}(\Pi,A)$. Then $c_x b = b$. Therefore

$$c_x a = c_x i(\pi,\Pi)b = i(x\pi x^{-1},\Pi)c_x b = i(x\pi x^{-1},\Pi)b$$

and thus

$$\begin{aligned} i(\pi \cap x\pi x^{-1}, x\pi x^{-1})c_x a &= i(\pi \cap x\pi x^{-1},\Pi)b \\ &= i(\pi \cap x\pi x^{-1},\pi)i(\pi,\Pi)b \\ &= i(\pi \cap x\pi x^{-1},\pi)a. \end{aligned}$$

PROPOSITION 9.4. *If* $a \in \hat{H}(\pi,A)$ *is stable then*

$$i(\pi,\Pi)t(\Pi,\pi)a = (\Pi : \pi)a.$$

PROOF. Applying formula (4) with $\pi' = \pi$ we have

$$\begin{aligned} i(\pi,\Pi)t(\Pi,\pi)a &= \sum_i t(\pi,\pi \cap x_i\pi x_i^{-1})i(\pi \cap x_i\pi x_i^{-1}, x_i\pi x_i^{-1})c_{x_i}a \\ &= \sum_i t(\pi,\pi \cap x_i\pi x_i^{-1})i(\pi \cap x_i\pi x_i^{-1},\pi)a \\ &= \sum_i (\pi : \pi \cap x_i\pi x_i^{-1})a. \end{aligned}$$

Thus formula (3) yields the desired result.

10. p-GROUPS AND SYLOW GROUPS

For each prime p we shall denote by $\hat{H}(\Pi,A,p)$ the p-primary component of $\hat{H}(\Pi,A)$. Clearly $\hat{H}(\Pi,A)$ is the direct sum of $\hat{H}(\Pi,A,p)$ for various primes p. Since the order of each element of $\hat{H}(\Pi,A)$ is a divisor of $(\Pi : 1)$ it follows that $\hat{H}(\Pi,A,p) = 0$ unless p is a divisor of $(\Pi : 1)$, and each element of $\hat{H}(\Pi,A,p)$ has an order which is a divisor of p^ν, where p^ν is the p-primary component of $(\Pi : 1)$. In particular, if Π is a p-group (i.e. $(\Pi : 1) = p^\nu$), then $\hat{H}(\Pi,A,p) = \hat{H}(\Pi,A)$.

The product of two elements $a \in \hat{H}(\Pi,A,p)$ and $b \in \hat{H}(\Pi,A',q)$ is zero if $p \neq q$ and is in $\hat{H}(\Pi,A \otimes A',p)$ if $p = q$.

Taking $A = Z$ we find that the ring $\hat{H}(\Pi,Z)$ is a direct product of the rings $\hat{H}(\Pi,Z,p)$ for p running through all the prime divisors of $(\Pi : 1)$. The unit element of $\hat{H}(\Pi,Z,p)$ will be denoted by 1_p.

THEOREM 10.1. *Let π be a p-Sylow subgroup of* Π *and let A be a* Π*-module. Then*

$$t(\Pi,\pi)\colon\ \hat{H}(\pi,A) \to \hat{H}(\Pi,A,p)$$

is an epimorphism and

$$i(\pi,\Pi)\colon\ \hat{H}(\Pi,A,p) \to \hat{H}(\pi,A)$$

is a monomorphism whose image consists of the stable elements of $\hat{H}(\pi,A)$. *Further we have a direct sum decomposition*

$$\hat{H}(\pi,A) = \operatorname{Im} i(\pi,\Pi) + \operatorname{Ker} t(\pi,\Pi).$$

If further π is an invariant subgroup of Π *then* Π/π *operates on* $\hat{H}(\pi,A)$ *and*

$$N\hat{H}(\pi,A) = [\hat{H}(\pi,A)]^{\Pi/\pi} = \operatorname{Im} i(\pi,\Pi) \approx \hat{H}(\Pi,A,p)$$

$$_N\hat{H}(\pi,A) = I(\Pi/\pi)\hat{H}(\pi,A) = \operatorname{Ker} t(\pi,\Pi)$$

$$[\hat{H}(\pi,A)]_{\Pi/\pi} = \operatorname{Coim} t(\pi,\Pi) \approx \hat{H}(\Pi,A,p).$$

PROOF. Let $(\pi : 1) = p^\nu$ and $(\Pi : \pi) = q$. Then p^ν and q are relatively prime so that there exists an integer l such that $ql \equiv 1 \bmod p^\nu$.

It follows from 9.3 that the elements of $\operatorname{Im} i(\pi,\Pi)$ are stable. Conversely assume that $a \in \hat{H}(\pi,A)$ is stable. Then, by 9.2.

$$li(\pi,\Pi)t(\Pi,\pi)a = l(\Pi : \pi)a = lqa = a.$$

Thus $a \in \operatorname{Im} i(\pi,\Pi)$. In view of (6) of § 8 we also have

$$lt(\Pi,\pi)i(\pi,\Pi)b = l(\Pi : \pi)b = lqb = b$$

for each $b \in \hat{H}(\Pi,A,p)$, and this yields all the conclusions of the first half of the theorem.

If π is an invariant subgroup of Π then Π/π operates on $\hat{H}(\pi,A)$ and the stable elements of $\hat{H}(\pi,A)$ are those invariant under the operators of Π/π. Thus $\operatorname{Im} i(\pi,\Pi) = [\hat{H}(\pi,A)]^{\Pi/\pi}$. Further, from 9.2, we have

$$i(\pi,\Pi)t(\Pi,\pi)a = Na$$

so that $\operatorname{Ker} t(\Pi,\pi) = {}_N\hat{H}(\pi,A)$.

Since $p^\nu\hat{H}(\pi,A) = 0$ and since Π/π has order relatively prime to p^ν it follows from 2.7 that $\hat{H}(\Pi/\pi,\hat{H}(\pi,A)) = 0$. In particular, ${}_NH(\pi,A) = I(\Pi/\pi)\,\hat{H}(\pi,A)$ and $[\hat{H}(\pi,A)]^{\Pi/\pi} = N\hat{H}(\pi,A)$. This concludes the proof.

11. PERIODICITY

We shall discuss here the finite groups Π for which the cohomology groups $\hat{H}^n(\Pi,A)$ show a periodicity with respect to n. This question is of interest for the problem of groups operating without fixed points on spheres (see XVI,9). The results of this section are due to Artin and Tate (unpublished).

An element $g \in \hat{H}^q(\Pi,Z)$ will be called a *maximal generator* if it is a generator and has order $(\Pi : 1)$.

PROPOSITION 11.1. *For each $g \in \hat{H}^q(\Pi,Z)$ the following properties are equivalent:*

(a) *g is a maximal generator;*

(b) *g has order* $(\Pi : 1)$;

(c) *there is an element $g^{-1} \in \hat{H}^{-q}(\Pi,Z)$ with $g^{-1}g = 1$;*

(d) *the map $a \to ag$ is an isomorphism*

$$\hat{H}^n(\Pi,A) \approx \hat{H}^{n+q}(\Pi,A) \text{ for all } n \text{ and } A.$$

PROOF. (a) $\Rightarrow$ (b) is obvious.

(b) $\Rightarrow$ (c). Assume g has order $(\Pi : 1)$. Since the order of any element of $\hat{H}^q(\Pi,Z)$ is a divisor of $(\Pi : 1)$, it easily follows that there exists a map φ: $\hat{H}^q(\Pi,Z) \to Z_r$, $r = (\Pi : 1)$, with $\varphi g = 1$. By 6.6 there is then an element $g^{-1} \in \hat{H}^{-q}(\Pi,Z)$ with $g^{-1}g = \varphi g = 1$.

(c) $\Rightarrow$ (d). Consider the maps

$$\hat{H}^n(\Pi,A) \xrightarrow{\alpha} \hat{H}^{n+q}(\Pi,A) \xrightarrow{\beta} \hat{H}^n(\Pi,A)$$

given by $\alpha a = ag$, $\beta a = ag^{-1}$. Then $\alpha\beta a = ag^{-1}g = a$ and $\beta\alpha a = agg^{-1} = (-1)^q ag^{-1}g = (-1)^q a$. Thus α and β are isomorphisms.

(d) $\Rightarrow$ (a). Consider the isomorphism $\hat{H}^0(\Pi,Z) \approx \hat{H}^q(\Pi,Z)$ given by $a \to ag$. Since $\hat{H}^0(\Pi,Z)$ is cyclic of order $(\Pi : 1)$ and generated by the element 1 it follows that $\hat{H}^q(\Pi,Z)$ also is cyclic of order $(\Pi : 1)$ and generated by the element g.

The uniqueness of the element g^{-1} with $g^{-1}g = 1$ follows from the following argument. For any $a \in \hat{H}^{-q}(\Pi,Z)$

$$a = ag^{-1}g = (-1)^q g^{-1}ag.$$

Thus $ag = 1$ implies $a = (-1)^q g^{-1}$. By the same reason $g^{-1} = (-1)^q g^{-1}$ so that $a = g^{-1}$. This justifies the notation g^{-1}.

PROPOSITION 11.2. *If $g \in H^q(\Pi,Z)$ is a maximal generator then so is $g^{-1} \in H^{-q}(\Pi,Z)$. If $h \in H^r(\Pi,Z)$ is another maximal generator, then $gh \in H^{q+r}(\Pi,Z)$ also is a maximal generator.*

PROOF. The first part follows from (c) above. The second one follows from (d) since the map $a \to agh$ is a composition of two isomorphisms.

An integer q will be called a *period* for the group Π if $\hat{H}^q(\Pi,Z)$ contains a maximal generator, i.e. if $\hat{H}^q(\Pi,Z)$ is cyclic of order $(\Pi : 1)$. It follows from 11.2 that the periods form a subgroup of Z. It can easily be seen that the periods are even if $\Pi \neq \{1\}$. Indeed assume that $g \in \hat{H}^q(\Pi,Z)$ is a maximal generator with q odd. Then $g = gg^{-1}g = -g^{-1}gg = -g$. Thus $2g = 0$ so that $(\Pi : 1) = 2$. However we know from § 7 that the group $\Pi = Z_2$ has only even periods.

PROPOSITION 11.3. *If Π has period q then so does every subgroup π. Further, if $g \in \hat{H}^q(\Pi,Z)$ is a maximal generator then so is $i(\pi,\Pi)g \in \hat{H}^q(\pi,Z)$.*

PROOF. We have $t(\Pi,\pi)i(\pi,\Pi)g = (\Pi : \pi)g$. Since $(\Pi : \pi)g$ has order precisely $(\pi : 1) = (\Pi : 1)/(\Pi : \pi)$, it follows that $i(\pi,\Pi)g$ has order at least $(\pi : 1)$. However no element of $\hat{H}^q(\pi,Z)$ has order exceeding $(\pi : 1)$. Thus $i(\pi,\Pi)g$ has order $(\pi : 1)$ and thus, by 11.1(b), it is a maximal generator.

PROPOSITION 11.4. *Let π be a p-Sylow subgroup of Π and let $g \in \hat{H}^q(\pi,Z)$ be a maximal generator. Let r be an integer such that*

$$k^r \equiv 1 \bmod (\pi : 1)$$

for all integers k prime to p. Then the element $g^r \in \hat{H}^{qr}(\pi,Z)$ is stable and $t(\Pi,\pi)g^r$ has order $(\pi : 1)$.

PROOF. Let $x\pi x^{-1}$ be a subgroup of Π conjugate to π. Since the mapping $a \to c_x a$ is an isomorphism it follows that $c_x g \in \hat{H}^q(x\pi x^{-1},Z)$ is a maximal generator. Consequently, by 11.3, the elements $g_1 = i(\pi \cap x\pi x^{-1},\pi)g$ and $g_2 = i(\pi \cap x\pi x^{-1},x\pi x^{-1})c_x g$ both are maximal generators of $\hat{H}^q(\pi \cap x\pi x^{-1},Z)$. There exists therefore an integer k prime to p such that $g_1 = kg_2$. This implies

$$g_1^r = k^r g_2^r = g_2^r.$$

However,

$$g_1^r = i(\pi \cap x\pi x^{-1},\pi)g^r$$

$$g_2^r = i(\pi \cap x\pi x^{-1},x\pi x^{-1})c_x g^r$$

which shows that g^r is stable.

Since g^r is stable it follows from 9.4 that

$$i(\pi,\Pi)t(\Pi,\pi)g^r = (\Pi : \pi)g^r.$$

Since g^r has order $(\pi : 1)$, and $(\Pi : \pi)$ is relatively prime to $(\pi : 1)$, it follows that $(\Pi : \pi)g^r$ also has order $(\pi : 1)$. Consequently $t(\Pi,\pi)g^r$

must have as order a multiple of $(\pi : 1)$. But $t(\Pi,\pi)g^r \in \hat{H}(\Pi,A,p)$ and every element of this last group has order at most $(\pi : 1)$. Thus $t(\Pi,\pi)g^r$ has order $(\pi : 1)$.

THEOREM 11.6. *For each finite group* Π *the following statements are equivalent:*

(a) Π *has a period* > 0;

(b) *every abelian subgroup of* Π *is cyclic*;

(c) *every p-subgroup of* Π *is either cyclic or is a generalized quaternion group*;

(d) *every Sylow subgroup of* Π *is either cyclic or is a generalized quaternion group.*

PROOF. (a) $\Rightarrow$ (b). A non-cyclic abelian group contains a subgroup of the form $Z_p + Z_p$ where p is a prime. In view of 11.3, it therefore suffices to show that $Z_p + Z_p$ has no period. Consider homomorphisms $Z_p \to Z_p + Z_p \to Z_p$ whose composition is the identity. This induces homomorphisms

$$H^q(Z_p,Z) \to H^q(Z_p + Z_p,Z) \to H^q(Z_p,Z)$$

whose composition is the identity. For each positive even integer q the group $H^q(Z_p,Z)$ is cyclic of order p. Consequently $H^q(Z_p + Z_p,Z)$ has a direct summand which is cyclic of order p. Thus $H^q(Z_p + Z_p,Z)$, for positive even integers q, is not cyclic of order p^2. Consequently $Z_p + Z_p$ does not have a period.

(b) $\Rightarrow$ (c). Let π be a p-subgroup of Π. Since the center of a p-group is non-trivial (see Zassenhaus, *The Theory of Groups*, New York, 1949, p. 110), π contains a central cyclic subgroup π' of order p. We claim that π' is the only subgroup of π of order p. Indeed if π'' is another such subgroup, then since $\pi' \cap \pi'' = \{1\}$ and since π' is in the center of π, it follows that π contains the direct sum $\pi' + \pi''$ which is a non-cyclic abelian group, contradicting (b). Thus π contains only one subgroup of order p. It is then known (Zassenhaus, *ibid.*, p. 118) that π is either cyclic or a generalized quaternion group.

(c) $\Rightarrow$ (d) is obvious.

(d) $\Rightarrow$ (a). We have seen in § 7 that a cyclic group has period 2 while the generalized quaternion groups have period 4.

Let $\pi_1, \dots, \pi_s$ be Sylow subgroups corresponding to the primes $p_1, \dots, p_s$ that occur in $(\Pi : 1)$. Assume that π_i has period q_i and maximal generator $g_i \in H^{q_i}(\pi_i,Z)$. By 11.4 there exists an integer u which is a common multiple of $q_1, \dots, q_s$ and such that the elements

$$t(\Pi,\pi_i)g_i^{u/q_i} \in H^u(\Pi,Z,p_i)$$

have order $(\pi_i : 1)$. It follows that the sum of these elements is an element of $H^u(\Pi,Z)$ of order $(\Pi : 1)$, i.e. a maximal generator. Thus Π has period u.

EXERCISES

1. Let Π be a group of order r. Show that a Π-module A, such that the multiplication by r is an isomorphism $r : A \approx A$, is weakly projective.

2. Show that if there exists an exact sequence $0 \to A_n \to \cdots \to A_0 \to A \to 0$ or $0 \to A \to A_0 \to \cdots \to A_n \to 0$ with $A_0, \ldots, A_n$ weakly projective, then $\hat{H}(\Pi,A) = 0$. In particular, $\hat{H}(\Pi,A) = 0$ whenever A has a finite projective or injective dimension. As an application show that if $\Pi \neq 1$ then the projective and injective dimensions of Z as a Π-module are infinite. Thus gl.dim $Z(\Pi) = \infty$.

3. Show that if A is finitely generated then $\hat{H}^q(\Pi,A)$ is finitely generated and hence finite.

4. Let Π be the cyclic group of order h with generator x and A a cyclic group of order k (written additively) with generator y. Assume that an integer l is given such that $l^h - 1 \equiv 0 \bmod k$. Then define the operators of Π on A as $xy = ly$. Show that A is weakly projective if and only if $(h,k) = 1$. Show that if $l^h - 1 = k$ then $\hat{H}(\Pi,A) = 0$. In particular for $h = 2$, $k = 8$, $l = 3$, we have $\hat{H}(\Pi,A) = 0$ without A being weakly projective.

5. Given a complete resolution X for Π, show that X^0, suitably relabelled, again is a complete resolution for Π. Using this result establish the isomorphism

$$\hat{H}^q(\Pi,A) \approx H_{-q-1}(A \otimes_\Pi X)$$

where A is regarded as a right Π-module by setting $ax = x^{-1}a$, $x \in \Pi$.

6. Show that the products

$$\cup : \ H^p(\Pi,A) \otimes H^q(\Pi,A') \to H^{p+q}(\Pi,A \otimes A')$$

$$\cap : \ H_{p+q}(\Pi,A \otimes A') \to \operatorname{Hom}(H^p(\Pi,A),H_q(\Pi,A'))$$

of Ch. XI, may be modified (for Π finite) so that H^0 and H_0 be replaced by $\hat{H}^0$ and $\hat{H}_0$. Show that after this modification we have

$$a \cup b = a \,.\, b$$

for $a \in \hat{H}^p(\Pi,A)$, $b \in \hat{H}^q(\Pi,A')$, $p \geqq 0$, $q \geqq 0$, and

$$a \cap b = (-1)^{\frac{p(p+1)}{2}}\, b \,.\, a$$

for $a \in \hat{H}_{-q}(\Pi,A)$, $b \in \hat{H}^p(\Pi,A')$, $q > p \geqq 0$.

7. Given a subgroup π of Π establish the isomorphism

$$\hat{H}(\pi,A) \approx \hat{H}(\Pi, \operatorname{Hom}_\pi (Z(\Pi),A))$$

for any Π-module A. Show that $\operatorname{Hom}_\pi (Z(\Pi),A)$ and $Z(\Pi) \otimes_\pi A$ are isomorphic. Show that the following diagram is commutative

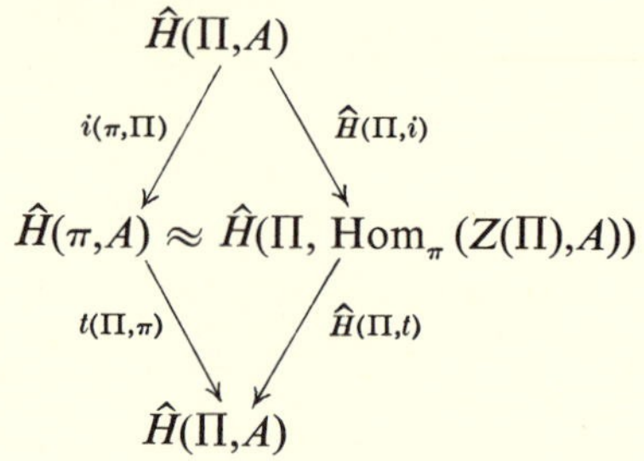

where

$$i\colon\ A = \operatorname{Hom}_\Pi (Z(\Pi),A) \to \operatorname{Hom}_\pi (Z(\Pi),A)$$

is the inclusion, while

$$t\colon\ \operatorname{Hom}_\pi (Z(\Pi),A) \to \operatorname{Hom}_\Pi (Z(\Pi),A) = A$$

is the transfer.

8. Let π be a subgroup of Π. In the situation $(A_\Pi, {}_\Pi C)$ define

$$t\colon\ A \otimes_\Pi C \to A \otimes_\pi C$$

by setting

$$t(a \otimes_\Pi c) = \sum a x_i \otimes_\pi x_i^{-1} c$$

where $x_1\pi, \ldots, x_r\pi$ are distinct cosets of π in Π with $r = (\Pi,\pi)$. Examine the formal properties of t and compare it with the natural epimorphism

$$j\colon\ A \otimes_\pi C \to A \otimes_\Pi C.$$

Replace C by a complete resolution X for Π and use Exer. 5 to show that t leads to the homomorphism $i(\pi,\Pi)\colon\ \hat{H}(\Pi,A) \to \hat{H}(\pi,A)$ while j leads to $t(\Pi,\pi)\colon\ \hat{H}(\pi,A) \to \hat{H}(\Pi,A)$.

9. Define the transfer homomorphisms

$$H^n(\pi,A) \to H^n(\Pi,A), \qquad H_n(\Pi,A) \to H_n(\pi,A)$$

where Π is any group and π is a subgroup of Π of finite index. Compare these homomorphisms with $t(\Pi,\pi)$ and $i(\pi,\Pi)$ for finite groups. Establish the analogue of Exer. 7.

10. Show that the map

$$i(\pi,\Pi)\colon\ H^{-2}(\Pi,Z) \to H^{-2}(\pi,Z)$$

which coincides with the transfer map

$$t(\pi,\Pi)\colon\ H_1(\Pi,Z) \to H_1(\pi,Z)$$

(see Exer. 9) coincides with the classical transfer map

$$t\colon\ \Pi/[\Pi,\Pi] \to \pi/[\pi,\pi]$$

as defined for instance in Zassenhaus (*Theory of Groups*, New York, 1949, p. 137).

11. Let $P = (p_1, \ldots, p_l)$ be a set of primes. We define $\hat{H}(\Pi,A,P)$ as the direct sum of $\hat{H}(\Pi,A,p_i)$, and define $(\Pi : 1)_P$ as the product of the p_i-primary components of $(\Pi,1)$ for $i = 1, \ldots, l$. An element $g \in \hat{H}^q(\Pi,Z,P)$ will be called a maximal P-generator if it generates $\hat{H}^q(\Pi,Z,P)$ and has order $(\Pi : 1)_P$. The integer q is then called a P-period for Π. Restate all the results of § 11 in this more general setting. In particular show that theorem 11.6 may be reformulated to assert the equivalence of the following four conditions:

(a) Π has a P-period > 0;

(b) every abelian subgroup of Π whose order is a divisor of $(\Pi : 1)_P$ is cyclic;

(c) every p-subgroup of Π with $p \in P$ is either cyclic or a generalized quaternion group;

(d) every p-Sylow subgroup of Π with $p \in P$ is either cyclic or is a generalized quaternion group.

11. Let p^ν be the order of the p-Sylow subgroup of Π. Show that the least integer r satisfying 11.4 is (using the Euler φ-function)

$$r = \varphi(p^\nu) = p^{\nu-1}(p-1) \qquad \text{if } p \neq 2$$

$$r = \varphi(p^\nu) = 2^{\nu-1} \qquad \text{if } p=2,\ \nu=1 \text{ or } 2$$

$$r = 1/2\varphi(p^\nu) = 2^{\nu-2} \qquad \text{if } p=2,\ \nu > 2.$$

As a consequence, show that if Π has a p-period, then $2\varphi(p^\nu)$ is a p-period. If $p = 2$ and π is cyclic of order $\geqq 8$ then Π has $\varphi(p^\nu)$ as a p-period.

If Π has a period, then $2\varphi(\Pi : 1)$ is a period for Π.

13. Show that if for some integers i and p the functors $\hat{H}^i(\Pi,A)$ and $\hat{H}^{i+q}(\Pi,A)$ are naturally equivalent (as functors of the Π-module A), then Π has period q.

14. Let Π have order r. Show that for each q there exists Π-modules C with $H^q(\Pi,C)$ cyclic of order r. [Hint: for $q = 0$ take $C = Z$, then use sequences of the type (1) and (2) of § 5.]

CHAPTER XIII

Lie Algebras

Introduction. In this chapter, Lie algebras are considered from a purely algebraical point of view, without reference to Lie groups and differential geometry. The "Jacobi identity" may be justified by the properties of the "bracket" operation $[x,y] = xy - yx$ in an associative algebra.

To each Lie algebra $\mathfrak{g}$ (over a commutative ring K) there corresponds a K-algebra $\mathfrak{g}^e$ (called the "enveloping algebra" of $\mathfrak{g}$), in such a way that the "representations" of $\mathfrak{g}$ in a K-module C are in a 1–1–correspondence with the $\mathfrak{g}^e$-module structures of C. Since $\mathfrak{g}^e$ has a natural augmentation $\varepsilon\colon \mathfrak{g}^e \to K$, it is a supplemented K-algebra. This at once leads to the homology and cohomology groups of $\mathfrak{g}$. To prove that these coincide with the ones hitherto considered(Chevalley-Eilenberg, *Trans. Am. Math. Soc.* 63 (1948), 85–124) we must assume that $\mathfrak{g}$ is K-free and apply the theorem of Poincaré-Witt (§ 3) which is an essential tool in the theory.

While the first two sections contain only definitions and results which are essentially trivial, because they do not use Jacobi's identity, this identity is essential for the theorem of Poincaré-Witt (§ 3). Once this theorem is established, the theory develops in a manner analogous to that for groups.

We do not touch upon the more advanced aspects of the homology theory of Lie algebras (Whitehead lemmas, Levi's theorem, semi-simple Lie algebras, etc.).

1. LIE ALGEBRAS AND THEIR ENVELOPING ALGEBRAS

We recall that a Lie algebra over a commutative ring K is a K-module $\mathfrak{g}$ together with a K-homomorphism $x \otimes y \to [x,y]$ of $\mathfrak{g} \otimes_K \mathfrak{g}$ into $\mathfrak{g}$ such that for $x, y, z \in \mathfrak{g}$:

(1) $$[x,x] = 0$$

(2) $$[x,[y,z]] + [y,[z,x]] + [z,[x,y]] = 0 \qquad \text{(Jacobi's identity).}$$

Condition (1) implies the condition

(1′) $$[x,y] + [y,x] = 0$$

and is equivalent with (1′) if in the ring K there is an element k with $2k = 1$.

A (left) $\mathfrak{g}$-representation of $\mathfrak{g}$ is a K-module A together with a K-homomorphism $x \otimes a \to xa$ of $\mathfrak{g} \otimes A$ into A such that

$$x(ya) - y(xa) = [x,y]a.$$

We now construct an associative K-algebra $\mathfrak{g}^e$ with the property that each (left) $\mathfrak{g}$-representation may be regarded as a (left) $\mathfrak{g}^e$-module and vice-versa. We shall call $\mathfrak{g}^e$ the *enveloping algebra* of $\mathfrak{g}$.

Let $T(\mathfrak{g})$ be the tensor algebra of the K-module $\mathfrak{g}$: this is the graded (associative) K-algebra such that $T_0(\mathfrak{g}) = K$ and $T_n(\mathfrak{g})$ is the n-fold tensor product (over K) of $\mathfrak{g}$ with itself. The product of elements $x_1 \otimes \cdots \otimes x_p$ and $y_1 \otimes \cdots \otimes y_q$ is $x_1 \otimes \cdots \otimes x_p \otimes y_1 \otimes \cdots y_q$. It is clear that a K-linear map $\mathfrak{g} \otimes_K A \to A$ admits a unique extension $T(\mathfrak{g}) \otimes_K A \to A$ satisfying $(x_1 \otimes \cdots \otimes x_n) \otimes a \to (x_1 \cdots (x_n a) \cdots)$. This converts A into a left $T(\mathfrak{g})$-module. Conversely any $T(\mathfrak{g})$-module A is obtained this way from a unique map $\mathfrak{g} \otimes A \to A$. In order that this map $\mathfrak{g} \otimes A \to A$ be a $\mathfrak{g}$-representation it is necessary and sufficient that the elements of $T(\mathfrak{g})$ of the form

$$x \otimes y - y \otimes x - [x,y] \qquad x,y \in \mathfrak{g} \tag{3}$$

annihilate A. Consequently, we are led to introduce the two-sided ideal $U(\mathfrak{g})$ of $T(\mathfrak{g})$ generated by the elements (3) and define the enveloping algebra of $\mathfrak{g}$ as $\mathfrak{g}^e = T(\mathfrak{g})/U(\mathfrak{g})$. Clearly left $\mathfrak{g}$-representations and left $\mathfrak{g}^e$-modules may be identified; we shall use the term *left $\mathfrak{g}$-module* to indicate either of the above.

We arrived at the enveloping algebra $\mathfrak{g}^e$ by the consideration of *left* representations $\mathfrak{g} \otimes A \to A$. A *right* representation $A \otimes \mathfrak{g} \to A$ with

$$(ax)y - (ay)x = a[x,y]$$

could equally well be used. Indeed, any K-homomorphism $A \otimes \mathfrak{g} \to A$ extends uniquely to a K-homomorphism $A \otimes T(\mathfrak{g}) \to A$ satisfying $a \otimes (x_1 \otimes \cdots \otimes x_n) = (\cdots (ax_1) \cdots x_n)$. This converts A into a right $T(\mathfrak{g})$-module. In order that $A \otimes \mathfrak{g} \to A$ be a right representation of A it is necessary and sufficient that the elements of the form (3) in $T(\mathfrak{g})$ annihilate A. We are thus led to the same enveloping algebra $\mathfrak{g}^e = T(\mathfrak{g})/U(\mathfrak{g})$. Thus right $\mathfrak{g}$-representations and right $\mathfrak{g}^e$-modules may be identified; we shall use the term *right $\mathfrak{g}$-module* to indicate either of the two.

The relation between $\mathfrak{g}$-representations and $\mathfrak{g}^e$-modules can be made more explicit by the use of the K-homomorphism

$$i\colon\ \mathfrak{g} \to \mathfrak{g}^e$$

defined by the fact that $\mathfrak{g} = T_1(\mathfrak{g})$. We then have

PROPOSITION 1.1. *Let f: $\mathfrak{g} \otimes A \to A$ be the map which defines A as a left $\mathfrak{g}$-representation. Then f admits a unique factorization $f = h(i \otimes A)$ where h: $\mathfrak{g}^e \otimes A \to A$ is a map defining A as a left $\mathfrak{g}^e$-module. Similarly for right representations and right modules.*

Since $T(\mathfrak{g})$ is a graded ring we have a natural augmentation ε: $T(\mathfrak{g}) \to T_0(\mathfrak{g}) = K$. Since ε is zero on $T_n(\mathfrak{g})$ for $n > 0$ it follows that the ideal $U(\mathfrak{g})$ is in the kernel of ε. Thus by passing to quotients we obtain the augmentation

$$\varepsilon\colon\ \mathfrak{g}^e \to K$$

which converts $\mathfrak{g}^e$ into a supplemented K-algebra. The augmentation ideal $I(\mathfrak{g})$ is generated by the image of i: $\mathfrak{g} \to \mathfrak{g}^e$.

As an example, consider the case of an *abelian* Lie algebra $\mathfrak{g}$ (i.e. $[x,y] = 0$ for $x,y \in \mathfrak{g}$). The enveloping algebra $\mathfrak{g}^e$ is then the quotient of $T(\mathfrak{g})$ by the two-sided ideal $U(\mathfrak{g})$ generated by the elements $x \otimes y - y \otimes x$; thus $\mathfrak{g}^e$ is the "symmetric algebra" of the K-module $\mathfrak{g}$. If $\mathfrak{g}$ is K-free with K-basis $\{x_\alpha\}$, then $\mathfrak{g}^e$ may be identified with the algebra $K[x_\alpha]$ of polynomials in the letters x_α.

A homomorphism f: $\mathfrak{g} \to \mathfrak{g}'$ of a Lie algebra $\mathfrak{g}$ into a Lie algebra $\mathfrak{g}'$ over the same ring K is a K-homomorphism satisfying $f([x,y]) = [fx,fy]$. Clearly f induces a map f^e: $\mathfrak{g}^e \to \mathfrak{g}'^e$ of supplemented algebras such that the diagram

$$\begin{array}{ccc} \mathfrak{g} & \xrightarrow{f} & \mathfrak{g}' \\ \downarrow{\scriptstyle i} & & \downarrow{\scriptstyle i'} \\ \mathfrak{g}^e & \xrightarrow[f^e]{} & \mathfrak{g}'^e \end{array}$$

is commutative.

Let $\mathfrak{g}$ and $\mathfrak{g}'$ be two Lie algebras over the same ring K. The direct sum $\mathfrak{g} + \mathfrak{g}'$ (also called "direct product") is defined as a Lie algebra by setting

$$[(x,x'),(y,y')] = ([x,y],[x',y']).$$

If we identify x with $(x,0)$ and x' with $(0,x')$ then $\mathfrak{g}$ and $\mathfrak{g}'$ become subalgebras of $\mathfrak{g} + \mathfrak{g}'$, and $[x,x'] = 0$ for $x \in \mathfrak{g}$, $x' \in \mathfrak{g}'$. The inclusion maps $\mathfrak{g} \to \mathfrak{g} + \mathfrak{g}'$, $\mathfrak{g}' \to \mathfrak{g} + \mathfrak{g}'$ induce homomorphisms

$$\mathfrak{g}^e \to (\mathfrak{g} + \mathfrak{g}')^e, \qquad \mathfrak{g}'^e \to (\mathfrak{g} + \mathfrak{g}')^e$$

which in turn define a homomorphism

$$\varphi\colon\ \mathfrak{g}^e \otimes \mathfrak{g}'^e \to (\mathfrak{g} + \mathfrak{g}')^e.$$

PROPOSITION 1.2. *The homomorphism φ is an isomorphism of supplemented algebras.*

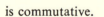

PROOF. The map $(x,x') \to x \otimes 1 + 1 \otimes x'$ of $\mathfrak{g} + \mathfrak{g}'$ into the tensor product of algebras $T(\mathfrak{g}) \otimes T(\mathfrak{g}')$ induces a homomorphism of K-algebras

$$\bar{\psi}\colon\ T(\mathfrak{g} + \mathfrak{g}') \to T(\mathfrak{g}) \otimes T(\mathfrak{g}').$$

After composing $\bar{\psi}$ with the natural map $T(\mathfrak{g}) \otimes T(\mathfrak{g}') \to \mathfrak{g}^e \otimes \mathfrak{g}'^e$ we find that $U(\mathfrak{g} + \mathfrak{g}')$ is mapped into zero. Thus we obtain a homomorphism

$$\psi\colon\ (\mathfrak{g} + \mathfrak{g}')^e \to \mathfrak{g}^e \otimes \mathfrak{g}'^e,$$

and it is trivial to verify that $\psi\varphi$ and $\varphi\psi$ are identity maps. Thus φ is an isomorphism.

The definition of a Lie subalgebra $\mathfrak{h}$ of a Lie algebra is obvious. We say that $\mathfrak{h}$ is an ideal if $[x,y] \in \mathfrak{h}$ for $x \in \mathfrak{g}, y \in \mathfrak{h}$. In view of the anti-commutativity of the bracket operation, there is no need to distinguish between left and right ideals. If $\mathfrak{h}$ is an ideal, then $\mathfrak{g}/\mathfrak{h}$ is again a Lie algebra with the bracket operation induced by that of $\mathfrak{g}$. Consider the composite map

$$\mathfrak{h} \xrightarrow{f} \mathfrak{g} \xrightarrow{i} \mathfrak{g}^e \tag{4}$$

where f is the inclusion, and let L denote the right ideal in $\mathfrak{g}^e$ generated by the image of if. Then L coincides with the left ideal generated by the image if, since in $\mathfrak{g}^e$ we have

$$if(x')i(x) = i(x)if(x') + if([x',x]) \qquad x' \in \mathfrak{h},\ x \in \mathfrak{g}$$

PROPOSITION 1.3. *Let $\mathfrak{h}$ be an ideal in $\mathfrak{g}$ and $\varphi\colon\ \mathfrak{g} \to \mathfrak{g}/\mathfrak{h}$ the natural homomorphism. Then $\varphi^e\colon\ \mathfrak{g}^e \to (\mathfrak{g}/\mathfrak{h})^e$ is an epimorphism and its kernel is the ideal L generated by the image of the composed map* (4).

PROOF. The fact that φ^e is an epimorphism is obvious. Clearly the image of if is in the kernel of φ^e. Thus φ^e induces a homomorphism $\bar{\varphi}\colon\ \mathfrak{g}^e/L \to (\mathfrak{g}/\mathfrak{h})^e$. We choose a function $u\colon\ \mathfrak{g}/\mathfrak{h} \to \mathfrak{g}$ (not a homomorphism) which followed by φ is the identity. It is easily seen that the composite map

$$\mathfrak{g}/\mathfrak{h} \xrightarrow{u} \mathfrak{g} \xrightarrow{i} \mathfrak{g}^e \longrightarrow \mathfrak{g}^e/L$$

is independent of the choice of u and is a K-homomorphism. There results a K-algebra homomorphism $T(\mathfrak{g}/\mathfrak{h}) \to \mathfrak{g}^e/L$ which maps $U(\mathfrak{g}/\mathfrak{h})$ into zero. We thus obtain a map $\psi\colon\ (\mathfrak{g}/\mathfrak{h})^e \to \mathfrak{g}^e/L$ for which both compositions $\bar{\varphi}\psi$ and $\psi\bar{\varphi}$ are identity maps. Thus $\bar{\varphi}$ is an isomorphism.

As in the case of groups we have an antipodism

$$\omega\colon\ \mathfrak{g}^e \approx (\mathfrak{g}^e)^*$$

defined by the map $x_1 \otimes \cdots \otimes x_p \to (-1)^p x_p^* \otimes \cdots \otimes x_1^*$ of $T(\mathfrak{g})$ into $T(\mathfrak{g})^*$. As in the case of groups this allows us to convert a right $\mathfrak{g}$-module A into a left one, by setting

$$xa = -ax.$$

2. HOMOLOGY AND COHOMOLOGY OF LIE ALGEBRAS

For each Lie algebra $\mathfrak{g}$ over K, the (associative) K-algebra $\mathfrak{g}^e$ is a supplemented K-algebra, and therefore, following x,1, we have homology and cohomology groups of $\mathfrak{g}^e$. We shall write

$$H_n(\mathfrak{g},A) = \operatorname{Tor}_n^{\mathfrak{g}^e}(A,K), \qquad H^n(\mathfrak{g},C) = \operatorname{Ext}_{\mathfrak{g}^e}^n(K,C)$$

for any right $\mathfrak{g}$-module A and any left $\mathfrak{g}$-module C. Thus the homology and cohomology groups of $\mathfrak{g}$ are defined as those of the supplemented algebra $\mathfrak{g}^e$.

If f: $\mathfrak{g} \to \mathfrak{h}$ is a homomorphism of Lie algebras, we have the induced homomorphism f^e: $\mathfrak{g}^e \to \mathfrak{h}^e$ which in turn induces homomorphisms

$$F^f\colon\ H_n(\mathfrak{g},A) \to H_n(\mathfrak{h},A), \qquad A_{\mathfrak{h}},$$

$$F_f\colon\ H^n(\mathfrak{h},C) \to H^n(\mathfrak{g},C), \qquad {}_{\mathfrak{h}}C.$$

The homology group $H_0(\mathfrak{g},A)$ is the K-module $A \otimes_{\mathfrak{g}^e} K \approx A/AI$ where $I = I(\mathfrak{g})$ is the augmentation ideal in $\mathfrak{g}^e$. Clearly $AI = A\mathfrak{g}$ and therefore

$$H_0(\mathfrak{g},A) = A/A\mathfrak{g}. \tag{1}$$

This K-module will also be denoted by $A_{\mathfrak{g}}$.

The cohomology group $H^0(\mathfrak{g},C)$ is the group $\operatorname{Hom}_{\mathfrak{g}^e}(K,C)$ which may be identified with the K-module of all *invariant* elements of C, i.e. all elements c such that $xc = 0$ for any $x \in \mathfrak{g}$. Denoting this module by $C^{\mathfrak{g}}$, we have

$$H^0(\mathfrak{g},C) = C^{\mathfrak{g}}. \tag{1a}$$

The group $H^1(\mathfrak{g},C)$ has been described in x,1 as the group of all crossed homomorphisms f: $\mathfrak{g}^e \to C$ modulo the subgroup of principal crossed homomorphisms. Composing f with the map i: $\mathfrak{g} \to \mathfrak{g}^e$ we obtain a K-homomorphism g: $\mathfrak{g} \to C$ such that

$$x(gy) - y(gx) = g([x,y]) \qquad x,y \in \mathfrak{g}$$

which we call a *crossed homomorphism* of $\mathfrak{g}$ into C. Clearly the crossed homomorphisms of $\mathfrak{g}$ and those of $\mathfrak{g}^e$ are in a 1–1-correspondence given by the relation $g = fi$. The principal crossed homomorphisms $\mathfrak{g} \to A$

are those of the form $gx = xc$ for some fixed $c \in C$. We thus obtain again that $H^1(\mathfrak{g},C)$ may be identified with the group of crossed homomorphisms $\mathfrak{g} \to C$ reduced modulo principal homomorphisms.

If A has trivial $\mathfrak{g}$-operators (i.e. $xa = 0$ for all $a \in A, x \in \mathfrak{g}$), then we find

$$H_0(\mathfrak{g},A) = A = H^0(\mathfrak{g},A) \tag{2}$$

$$H^1(\mathfrak{g},A) = \operatorname{Hom}(\mathfrak{g}/[\mathfrak{g},\mathfrak{g}],A) \tag{3}$$

where $[\mathfrak{g},\mathfrak{g}]$ is the image of $\mathfrak{g} \otimes \mathfrak{g} \to \mathfrak{g}$ under the map $x \otimes y \to [x,y]$.

We shall also interpret the group $H_1(\mathfrak{g},A)$ for A with trivial $\mathfrak{g}$-operators. We know from x,1,(4) that $H_1(\mathfrak{g},A) \approx A \otimes_K I/I^2$ where $I = I(\mathfrak{g})$ is the augmentation ideal. Since i maps $\mathfrak{g}$ into I and $[\mathfrak{g},\mathfrak{g}]$ into I^2 it defines a map φ: $\mathfrak{g}/[\mathfrak{g},\mathfrak{g}] \to I/I^2$. On the other hand the map $T(\mathfrak{g}) \to \mathfrak{g}$ which is the identity on $T_1(\mathfrak{g}) = \mathfrak{g}$ and is zero on $T_n(\mathfrak{g})$ for $n \neq 1$, maps $U(\mathfrak{g})$ into $[\mathfrak{g},\mathfrak{g}]$ thus defining a map $I \to \mathfrak{g}/[\mathfrak{g},\mathfrak{g}]$. Since this map is zero on I^2 we obtain a map ψ: $I/I^2 \to \mathfrak{g}/[\mathfrak{g},\mathfrak{g}]$. Both compositions $\varphi\psi$ and $\psi\varphi$ are identities and we obtain an isomorphism

$$I/I^2 \approx \mathfrak{g}/[\mathfrak{g},\mathfrak{g}]. \tag{4}$$

We thus have

$$H_1(\mathfrak{g},A) \approx A \otimes_K \mathfrak{g}/[\mathfrak{g},\mathfrak{g}] \tag{5}$$

if $\mathfrak{g}$ operates trivially on A.

3. THE POINCARÉ-WITT THEOREM

Throughout this section it will be assumed that the Lie algebra $\mathfrak{g}$ over K is K-free. A fixed K-base $\{x_\alpha\}$ will be chosen and it will be assumed that this K-base (or rather the set of indices) has been simply ordered.

We shall use the following notation: y_α will stand for the image of x_α under the map i: $\mathfrak{g} \to \mathfrak{g}^e$; if I is a finite sequence of indices $\alpha_1, \ldots, \alpha_p$ we shall write $y_I = y_{\alpha_1} \cdots y_{\alpha_p}$; we say that I is *increasing* if $\alpha_1 \leqq \cdots \leqq \alpha_p$; we define $y_I = 1$ if I is empty, and we regard the empty set as increasing; the integer p will be called the *length* of I.

THEOREM 3.1. *The elements y_I, corresponding to finite increasing sequences I, form a K-base of the enveloping algebra $\mathfrak{g}^e$.*

COROLLARY 3.2. *$\mathfrak{g}^e$ is K-free.*

Since by 3.1, the elements y_α are linearly independent in $\mathfrak{g}^e$ we obtain

COROLLARY 3.3. *The map i: $\mathfrak{g} \to \mathfrak{g}^e$ is a K-monomorphism.*

PROOF of 3.1. We first show that the elements y_I corresponding to finite increasing sequences generate $\mathfrak{g}^e$. We denote by $F_p(\mathfrak{g}^e)$ the image of the submodule $\sum_{i \leq p} T_i(\mathfrak{g})$ of $T(\mathfrak{g})$ under the natural mapping $T(\mathfrak{g}) \to \mathfrak{g}^e$. It suffices to show that the elements y_I corresponding to increasing sequences I of length $\leqq p$ generate $F_p(\mathfrak{g}^e)$. Clearly the elements y_I corresponding to all sequences I of length $\leqq p$ generate $F_p(\mathfrak{g}^e)$. The conclusion thus follows by recursion from the following lemma (in which the fact that $\mathfrak{g}$ is K-free is not needed):

LEMMA 3.4. *For each sequence* $a_1, \ldots, a_p \in \mathfrak{g}$ *and each permutation* π *of* $(1, \ldots, p)$ *we have*

$$i(a_1) \cdots i(a_p) - i(a_{\pi(1)}) \cdots i(a_{\pi(p)}) \in F_{p-1}(\mathfrak{g}^e).$$

As usual $i\colon \mathfrak{g} \rightarrow \mathfrak{g}^e$ is the natural map. It clearly suffices to consider the case when π interchanges two consecutive indices $j, j+1$. In this case the conclusion is evident from the relation

$$i(a_j)i(a_{j+1}) - i(a_{j+1})i(a_j) = i([a_j,a_{j+1}]).$$

We now come to the more difficult part of the proof which consists in showing that the elements y_I of 3.1 are K-linearly independent. We shall denote by P the polynomial algebra $K[z_\alpha]$ on letters $\{z_\alpha\}$ in a 1–1-correspondence with the base $\{x_\alpha\}$. For each finite sequence I of indices $\alpha_1, \ldots, \alpha_p$ we shall denote by z_I the element $z_{\alpha_1} \cdots z_{\alpha_p}$ of P.

LEMMA 3.5. *There exists a left representation of* $\mathfrak{g}$ *in* P *such that*

$$x_\alpha z_I = z_\alpha z_I \tag{1}$$

whenever $\alpha \leqq I$ *(i.e. whenever* $\alpha \leqq \beta$ *for all* $\beta \in I$*).*

Postponing the proof of the lemma, we can complete the proof of the theorem. The representation of $\mathfrak{g}$ in P induces a left $\mathfrak{g}^e$-module structure on P. If I is an increasing sequence of indices of length n it follows from (1) by recursion on n that $y_I \,.\, 1 = z_I$. Since the elements z_I are K-linearly independent in P, the same follows for the elements y_I of $\mathfrak{g}^e$.

PROOF of 3.5. In the graded algebra P, we denote as usual by P_p the K-module of homogeneous polynomials of degree p and set $Q_p = \sum_{i \leqq p} P_i$. Lemma 3.5 is an immediate consequence of the following inductive proposition:

(A_p). *For each integer* p *there is a unique homomorphism*

$$f\colon \mathfrak{g} \otimes Q_p \rightarrow P$$

such that

$$f(x_\alpha \otimes z_I) = z_\alpha z_I \qquad \alpha \leqq I,\ z_I \in Q_p \tag{1'}$$

$$f(x_\alpha \otimes z_I) \in Q_{q+1} \qquad z_I \in Q_q,\ q < p \tag{2}$$

$$f(x_\alpha \otimes f(x_\beta \otimes z_J)) = f(x_\beta \otimes f(x_\alpha \otimes z_J)) + f([x_\alpha,x_\beta] \otimes z_J),\ z_J \in Q_{p-1} \tag{3}$$

$$f(x_\alpha \otimes z_I) - z_\alpha z_I \in Q_q \qquad z_I \in Q_q,\ q \leqq p. \tag{4}$$

It is immediate that (2) is a consequence of (4); however we wrote (2) out explicitly in order to make it clear that the terms in (3) are well defined.

For $p = 0$, the definition $f(x_\alpha \otimes 1) = z_\alpha$ is forced by (1′) and trivially satisfies also (2)–(4).

Assume now that (A_{p-1}) is established for some $p > 0$. We shall show that the map f satisfying (A_{p-1}) admits a unique extension (also denoted by f) satisfying (A_p). We must define $f(x_\alpha \otimes z_I)$ for I of length p. If $\alpha \leqq I$, the definition is forced by (1′). If $\alpha \leqq I$ is false then I may be uniquely written as $I = (\beta, J)$ where $\alpha > \beta \leqq J$. Then $z_I = z_\beta z_J = f(x_\beta \otimes z_J)$ so that the left side of (3) is $f(x_\alpha \otimes z_I)$. In order to be able to use (3) as a definition we must verify that the right hand side of (3) is already defined. To this end we use (4) to write

$$f(x_\alpha \otimes z_J) = z_\alpha z_J + w, \qquad w \in Q_{p-1}.$$

Then the right hand side of (3) becomes

$$z_\beta z_\alpha z_J + f(x_\beta \otimes w) + f([x_\alpha, x_\beta] \otimes z_J).$$

This defines f in all cases, and (1′), (2) and (4) are clearly satisfied. As for (3) we only know that it holds if $\alpha > \beta \leqq J$. Because of the anti-symmetry of $[x_\alpha, x_\beta]$ it follows that (3) also holds if $\beta > \alpha \leqq J$. Since (3) trivially holds if $\alpha = \beta$, it follows that (3) is verified if either $\alpha \leqq J$ or $\beta \leqq J$. We shall show that this together with (1′) and (4) and together with the inductive assumption (A_{p-1}) implies (3) in all cases.

Indeed suppose that neither $\alpha \leqq J$ nor $\beta \leqq J$. Then J has positive length and $J = (\gamma, L)$ where $\gamma \leqq L$, $\gamma < \alpha$, $\gamma < \beta$. Using the abridged notation $f(x_\alpha \otimes z_I) = x_\alpha z_I$ we then have by the inductive assumption

$$\begin{aligned} x_\beta(z_J) = x_\beta(x_\gamma z_L) &= x_\gamma(x_\beta z_L) + [x_\beta, x_\gamma] z_L \\ &= x_\gamma(z_\beta z_L) + x_\gamma w + [x_\beta, x_\gamma] z_L \end{aligned}$$

where $w = x_\beta z_L - z_\beta z_L \in Q_{p-2}$. Applying x_α to both sides we have

$$x_\alpha(x_\beta z_J) = x_\alpha(x_\gamma(z_\beta z_L)) + x_\alpha(x_\gamma w) + x_\alpha([x_\beta, x_\gamma] z_L).$$

Since $\gamma \leqq (\beta, L)$, (3) may be applied to the term $x_\alpha(x_\gamma(z_\beta z_L))$; to the remaining two terms on the right we may apply (3) by the inductive assumption. Upon computation we obtain

$$(5) \quad x_\alpha(x_\beta z_J) = x_\gamma(x_\alpha(x_\beta z_L)) + [x_\alpha, x_\gamma](x_\beta z_L) + [x_\beta, x_\gamma](x_\alpha z_L) + [x_\alpha, [x_\beta, x_\gamma]] z_L.$$

Our assumptions on α and β were symmetric, so that (5) holds with α and β interchanged. Subtracting from (5) this yields

$$(6) \quad x_\alpha(x_\beta z_J) - x_\beta(x_\alpha z_J) = x_\gamma\{x_\alpha(x_\beta z_L) - x_\beta(x_\alpha z_L)\} + [x_\alpha, [x_\beta, x_\gamma]] z_L - [x_\beta, [x_\alpha, x_\gamma]] z_L.$$

Applying (3) we have

$$x\{x_{\gamma\alpha}(x_\beta z_L) - x_\beta(x_\alpha z_L)\} = x_\gamma([x_\alpha,x_\beta]z_L)$$
$$= [x_\alpha,x_\beta](x_\gamma z_L) + [x_\gamma,[x_\alpha,x_\beta]]z_L$$
$$= [x_\alpha,x_\beta]z_J + [x_\gamma,[x_\alpha,x_\beta]]z_L.$$

Substituting this in (6), we find that the three terms involving double brackets cancel by virtue of Jacobi's identity, and the final result is

$$x_\alpha(x_\beta z_J) - x_\beta(x_\alpha z_J) = [x_\alpha,x_\beta]z_J$$

as desired.

Theorem 3.1 was first proved by Poincaré (*Cambridge Philosophical Transactions* 18 (1899), 220–225, § III); a complete proof, based on the same principles, was given later by E. Witt (*Journ. für r.u.a. Math.* (*Crelle*) 177 (1937), 152–166; Hilfsatz, p. 153). The proof given here is modeled after Iwasawa.

4. SUBALGEBRAS AND IDEALS

If $\mathfrak{h}$ is a Lie subalgebra of a Lie algebra $\mathfrak{g}$ over K, then the inclusion map $\mathfrak{h} \to \mathfrak{g}$ induces a K-algebra homomorphism

(1) $$\varphi\colon \mathfrak{h}^e \to \mathfrak{g}^e$$

so that $\mathfrak{g}^e$ may be regarded either as a left or as a right $\mathfrak{h}^e$-module.

PROPOSITION 4.1. *If the K-modules $\mathfrak{h}$ and $\mathfrak{g}/\mathfrak{h}$ are both K-free, then φ is a monomorphism and $\mathfrak{g}^e$ regarded as a left or right $\mathfrak{h}^e$-module is $\mathfrak{h}^e$-free.*

PROOF. In the exact sequence $0 \to \mathfrak{h} \to \mathfrak{g} \to \mathfrak{g}/\mathfrak{h} \to 0$ of K-modules, the modules $\mathfrak{h}$ and $\mathfrak{g}/\mathfrak{h}$ are K-free. Therefore the sequence splits and $\mathfrak{g}$ also is K-free. Furthermore, we can find a K-base of $\mathfrak{g}$ composed of two disjoint sets $\{x_\alpha\}_{\alpha \in A}$, $\{y_\beta\}_{\beta \in B}$ such that $\{x_\alpha\}$ is a K-base for $\mathfrak{h}$. We simply order the union $A \cup B$ of the disjoint sets A and B so that each $\alpha \in A$ precedes each $\beta \in B$. If we identify each element of $\mathfrak{g}$ with its image in $\mathfrak{g}^e$ under the monomorphism $i\colon\ \mathfrak{g} \to \mathfrak{g}^e$, then it follows from 3.1 that the elements of the form

$$x_{\alpha_1}\cdots x_{\alpha_p}y_{\beta_1}\cdots y_{\beta_q} \qquad \alpha_1 \leqq \cdots \leqq \alpha_p \in A, \qquad \beta_1 \leqq \cdots \leqq \beta_q \in B$$

of $\mathfrak{g}^e$ form a K-base for $\mathfrak{g}^e$, while the elements $x_{\alpha_1}\cdots x_{\alpha_p}$ form a K-base for $\mathfrak{h}^e$. This implies that (1) is a monomorphism and that the elements $y_{\beta_1}\cdots y_{\beta_q}$ form a left $\mathfrak{h}^e$-base for $\mathfrak{g}^e$. The proof that these elements also form a right $\mathfrak{h}^e$-base is similar.

We may now apply x,7.2 and x,7.3. We obtain

PROPOSITION 4.2. *Under the hypotheses of* 4.1 *we have*

(2) $$H_n(\mathfrak{h},A) \approx H_n(\mathfrak{g},A \otimes_{\mathfrak{h}^e} \mathfrak{g}^e)$$

(2a) $$H^n(\mathfrak{h},C) \approx H^n(\mathfrak{g}, \operatorname{Hom}_{\mathfrak{h}^e}(\mathfrak{g}^e,C)),$$

for each right $\mathfrak{h}$*-module* A *and each left* $\mathfrak{h}$*-module* C.

PROPOSITION 4.3. *Under the hypotheses of* 4.1 *we have*

(3) $$H_n(\mathfrak{h},A) \approx \operatorname{Tor}_n^{\mathfrak{g}^e}(A,\mathfrak{g}^e \otimes_{\mathfrak{h}^e} K),$$

(3a) $$H^n(\mathfrak{h},C) \approx \operatorname{Ext}_{\mathfrak{g}^e}^n(\mathfrak{g}^e \otimes_{\mathfrak{h}^e} K,C),$$

for each right $\mathfrak{g}$*-module* A *and each left* $\mathfrak{g}$*-module* C.

The module $\mathfrak{g}^e \otimes_{\mathfrak{h}^e} K$ appearing in (3) and (3a) may also be written as $H_0(\mathfrak{h},\mathfrak{g}^e)$ which has been computed in § 2 to be $\mathfrak{g}^e/\mathfrak{g}^e\mathfrak{h}$. If $\mathfrak{h}$ is an *ideal* in $\mathfrak{g}$ then $\mathfrak{g}^e\mathfrak{h}$ coincides with the ideal L of 1.3. Thus if $\mathfrak{h}$ is an ideal in $\mathfrak{g}$ we have the isomorphism

$$\mathfrak{g}^e \otimes_{\mathfrak{h}^e} K \approx (\mathfrak{g}/\mathfrak{h})^e.$$

COROLLARY 4.4. *If* $\mathfrak{h}$ *is an ideal in* $\mathfrak{g}$ *and the hypotheses of* 4.1 *are satisfied then*

(4) $$H_n(\mathfrak{h},A) \approx \operatorname{Tor}_n^{\mathfrak{g}^e}(A,(\mathfrak{g}/\mathfrak{h})^e),$$

(4a) $$H^n(\mathfrak{h},C) \approx \operatorname{Ext}_{\mathfrak{g}^e}^n((\mathfrak{g}/\mathfrak{h})^e,C),$$

for each right $\mathfrak{g}$*-module* A *and each left* $\mathfrak{g}$*-module* C. *These isomorphisms may be used to define a right* $\mathfrak{g}/\mathfrak{h}$*-module structure on* $H_n(\mathfrak{h},A)$ *and a left* $\mathfrak{g}/\mathfrak{h}$*-module structure on* $H^n(\mathfrak{h},C)$.

In XVI,6 we shall establish closer relations between the homology (and cohomology) groups of $\mathfrak{g}$, $\mathfrak{h}$ and $\mathfrak{g}/\mathfrak{h}$.

5. THE DIAGONAL MAP AND ITS APPLICATIONS

For each Lie algebra $\mathfrak{g}$ over K, the diagonal map

$$D\colon \mathfrak{g}^e \to \mathfrak{g}^e \otimes \mathfrak{g}^e$$

is defined by the requirement

$$Dx = x \otimes 1 + 1 \otimes x, \qquad x \in \mathfrak{g}.$$

If we identify $\mathfrak{g}^e \otimes \mathfrak{g}^e$ with $(\mathfrak{g} + \mathfrak{g})^e$ as in 1.2, and consider the map $l\colon \mathfrak{g} \to \mathfrak{g} + \mathfrak{g}$ given by $lx = (x,x) = (x,0) + (0,x)$, then $D = l^e$. This diagonal map D is compatible with the augmentation (in the sense explained in XI,8) and is commutative and associative (in the sense defined in XI,4).

The diagonal map D may be combined with the antipodism ω: $\mathfrak{g}^e \approx \mathfrak{g}^{e*}$ defined in § 1, to obtain a map

$$E\colon\ \mathfrak{g}^e \to \mathfrak{g}^e \otimes \mathfrak{g}^{e*} = (\mathfrak{g}^e)^e$$

as the composition

$$\mathfrak{g}^e \xrightarrow{D} \mathfrak{g}^e \otimes \mathfrak{g}^e \xrightarrow{\mathfrak{g}^e \otimes \omega} \mathfrak{g}^e \otimes \mathfrak{g}^{e*}.$$

It follows that the map E satisfies

$$Ex = x \otimes 1 - 1 \otimes x^*, \qquad x \in \mathfrak{g},$$

and that this condition determines E uniquely.

We first verify that the map E satisfies condition (E.1) of x,6. To this end we denote by I and J the kernels of the respective augmentation maps

$$\varepsilon\colon\ \mathfrak{g}^e \to K, \qquad \rho\colon\ \mathfrak{g}^e \otimes \mathfrak{g}^{e*} \to \mathfrak{g}^e.$$

By IX,3.1, J is the left ideal generated by the elements $u \otimes 1 - 1 \otimes u^*$ for $u \in \mathfrak{g}^e$. In view of the relation

$$(uv) \otimes 1 - 1 \otimes (uv)^* = (u \otimes 1)(v \otimes 1 - 1 \otimes v^*) + (1 \otimes v^*)(u \otimes 1 - 1 \otimes u^*)$$

valid for $u,v \in \mathfrak{g}^e$, we find that J is the left ideal in $\mathfrak{g}^e \otimes \mathfrak{g}^{e*}$ generated by the elements

$$x \otimes 1 - 1 \otimes x^* = Ex \qquad x \in \mathfrak{g}.$$

Since the elements $x \in \mathfrak{g}$ generate the ideal I of $\mathfrak{g}^e$ it follows that J is the left ideal generated by the image of EI in $\mathfrak{g}^e \otimes \mathfrak{g}^{e*}$. This is precisely condition (E.1) of x,6.

We now introduce the assumption that the Lie algebra $\mathfrak{g}$ is K-free. Then, by 3.2, $\mathfrak{g}^e$ also is K-free. Consequently, the diagonal map D may be used to define $\cup$- and $\cap$-products as in XI,7. Further we find that the conditions (i)–(vi) of XI,8 are satisfied by the maps D and ω. Consequently the considerations of XI,8 and XI,9 (reduction theorems) are applicable to the homology and cohomology groups of a K-free Lie algebra $\mathfrak{g}$.

Next (still under the assumption that $\mathfrak{g}$ is K-free) we shall show that condition (E.2) of x,6 is satisfied, i.e. that $\mathfrak{g}^e \otimes \mathfrak{g}^{e*}$ regarded as a right $\mathfrak{g}^e$-module by means of the map E is $\mathfrak{g}^e$-projective. Since the map $\mathfrak{g}^e \otimes \omega$ is an isomorphism, it clearly suffices to show that $\mathfrak{g}^e \otimes \mathfrak{g}^e$ regarded as a right $\mathfrak{g}^e$-module using the map D, is $\mathfrak{g}^e$-free. To this end we identify $\mathfrak{g}^e \otimes \mathfrak{g}^e$ with $(\mathfrak{g} + \mathfrak{g})^e$ and notice that $D = l^e$, where l: $\mathfrak{g} \to \mathfrak{g} + \mathfrak{g}$ is the map of Lie algebras given by $lx = (x,x)$. Since l is a monomorphism and since Coker l is a K-module isomorphic with $\mathfrak{g}$ which is K-free, it follows from 4.1 that $(\mathfrak{g} + \mathfrak{g})^e$ is $\mathfrak{g}^e$-free.

Now that condition (E.1) and (E.2) of the "inverse process" have been verified, we may apply x,6.1. We obtain

THEOREM 5.1. *Let* $\mathfrak{g}$ *be a Lie algebra over* K *which is* K*-free, and let* A *be a two-sided* $\mathfrak{g}^e$*-module. Let* A_E *be the right* $\mathfrak{g}$*-module obtained from* A *by setting*

$$(a,x) \to ax - xa \qquad a \in A, x \in \mathfrak{g}$$

and let ${}_EA$ *be the left* $\mathfrak{g}$*-module obtained by*

$$(x,a) \to xa - ax \qquad a \in A, x \in \mathfrak{g}.$$

We then have isomorphisms

$$F^E\colon\ H_n(\mathfrak{g}^e,A) \approx H_n(\mathfrak{g},A_E)$$

$$F_E\colon\ H^n(\mathfrak{g},{}_EA) \approx H^n(\mathfrak{g}^e,A).$$

Furthermore if $\Lambda = \mathfrak{g}^e$ *and if* X *is a* Λ*-projective resolution of* K *(as a left* Λ*-module) then* $\Lambda^e \otimes_\Lambda X$ *is a* Λ^e*-projective resolution of* Λ *as a left* Λ^e*-module.*

In particular, let $\mathfrak{g}$ be the abelian Lie algebra with the letters $x_1, \ldots, x_n$ as a K-base. Then $\mathfrak{g}^e = K[x_1, \ldots, x_n] = \Lambda$, and we know from VIII,4 that $\Lambda \otimes E(x_1, \ldots, x_n)$ with a suitable differentiation operator is a $\mathfrak{g}^e$-projective resolution of K. It follows that $\Lambda^e \otimes E(x_1, \ldots, x_n)$ with a suitable differentiation operator is a Λ^e-projective resolution of Λ.

An application of x,6.2 gives

THEOREM 5.2. *If* $\mathfrak{g}$ *is a Lie algebra over* K *which is* K*-free then*

$$\dim \mathfrak{g}^e = \dim_{\mathfrak{g}^e} K.$$

If further the commutative ring K *is semi-simple, then*

$$\dim \mathfrak{g}^e = \text{gl.dim}\ \mathfrak{g}^e.$$

In view of the antipodism ω, there is no need to distinguish between $\text{l.dim}_{\mathfrak{g}^e} K$ and $\text{r.dim}_{\mathfrak{g}^e} K$ and between $\text{l.gl.dim}\ \mathfrak{g}^e$ and $\text{r.gl.dim}\ \mathfrak{g}^e$.

6. A RELATION IN THE STANDARD COMPLEX

For the purpose of the next section we shall establish here a relation valid in the normalized standard complex $N(\Lambda)$ of an arbitrary (associative) K-algebra Λ.

The notation $[x_1, \ldots, x_n]$ in the complex $N(\Lambda)$ introduced in IX,6 will be replaced here by $\{x_1, \ldots, x_n\}$ in order to avoid confusion with the brackets in the Lie algebras.

For each $y \in \Lambda$ we consider the Λ^e-endomorphisms $\sigma(y)$ and $\vartheta(y)$ of $N(\Lambda)$ defined by

$$\sigma(y)\{x_1, \ldots, x_n\} = \sum_{0 \leq i \leq n} (-1)^i \{x_1, \ldots, x_i, y, x_{i+1}, \ldots, x_n\} \tag{1}$$

$$\vartheta(y)\{x_1, \ldots, x_n\} = y\{x_1, \ldots, x_n\} - \{x_1, \ldots, x_n\}y - \sum_{1 \leq i \leq n} \{x_1, \ldots, x_{i-1}, [y,x_i], x_{i+1}, \ldots, x_n\} \tag{2}$$

where $[y,x] = yx - xy$.

PROPOSITION 6.1. *For each $y \in \Lambda$ we have the identity*

$$d\sigma(y) + \sigma(y)d - \vartheta(y) = 0, \tag{3}$$

where d is the differentiation operator of $N(\Lambda)$.

PROOF. Let $A(y)$ denote the left hand side of (3). We must show that for all $n \geqq 0$,

$$A(y)\{x_1, \ldots, x_n\} = 0. \tag{4}$$

This is immediate if $n = 0$. We now assume, by induction, that (4) holds for $n - 1$ $(n > 0)$. In the complex $N(\Lambda)$ we have the contracting homotopy s defined in IX,6 and satisfying the identity

$$ds + sd = \text{identity}$$

when applied to any element of degree > 0. Thus for $n > 0$, relation (4) is equivalent to the pair of relations

$$sA(y)\{x_1, \ldots, x_n\} = 0, \tag{5}$$

$$sdA(y)\{x_1, \ldots, x_n\} = 0. \tag{6}$$

We recall that in the normalized complex we have $s(y\{x_1, \ldots, x_n\}y') = \{y, x_1, \ldots, x_n\}y'$ and that the right hand side is zero if $y = 1$. This rule easily implies

$$sd\sigma(y)\{x_1, \ldots, x_n\} = s(y\{x_1, \ldots, x_n\} - x_1\sigma(y)\{x_2, \ldots, x_n\})$$

$$s\sigma(y)d\{x_1, \ldots, x_n\} = s(x_1\sigma(y)\{x_2, \ldots, x_n\})$$

$$-s\vartheta(y)\{x_1, \ldots, x_n\} = -s(y\{x_1, \ldots, x_n\}).$$

Adding these relations yields (5).

To prove (6) we first compute the element

$$z = dA(y)\{x_1, \ldots, x_n\} = d\sigma(y)d\{x_1, \ldots, x_n\} - d\vartheta(y)\{x_1, \ldots, x_n\}.$$

An application of the inductive assumption yields

$$z = \vartheta(y)d\{x_1, \ldots, x_n\} - d\vartheta(y)\{x_1, \ldots, x_n\}.$$

We must show that $z \equiv 0$ mod the kernel of s. Calculating modulo this kernel we find that $d\vartheta(y)\{x_1, \ldots, x_n\}$ gives

$$yd\{x_1, \ldots, x_n\} - x_1\{x_2, \ldots, x_n\}y - [y,x_1]\{x_2, \ldots, x_n\} - \sum_{2 \leq i \leq n} x_1\{x_2, \ldots, x_{i-1},[y,x_i],x_{i+1}, \ldots, x_n\}$$

while $\vartheta(y)d\{x_1, \ldots, x_n\}$ gives

$$x_1 y\{x_2, \ldots, x_n\} + y(d\{x_1, \ldots, x_n\} - x_1\{x_2, \ldots, x_n\})$$
$$-x_1\{x_2, \ldots, x_n\}y - \sum_{2 \leq i \leq n} x_1\{x_2, \ldots, x_{i-1},[y,x_i],x_{i+1}, \ldots, x_n\}.$$

The two results coincide and this concludes the proof.

Suppose now that Λ is a supplemented K-algebra with augmentation $\varepsilon\colon \Lambda \to K$. In the normalized standard complex $N(\Lambda,\varepsilon) = N(\Lambda) \otimes_\Lambda K$ we have endomorphisms induced by $\sigma(y)$ and $\vartheta(y)$. These will still be denoted by $\sigma(y)$ and $\vartheta(y)$. These operators are left Λ-endomorphisms of $N(\Lambda,\varepsilon)$ and we still have the relation (3). The explicit definition of $\sigma(y)$ is still given by formula (1), while the definition of $\vartheta(y)$ gets replaced by

$$(2') \qquad \vartheta(y)\{x_1, \ldots, x_n\} = y\{x_1, \ldots, x_n\} - \{x_1, \ldots, x_n\}(\varepsilon y) - \sum_{1 \leq i \leq n} \{x_1, \ldots, x_{i-1},[y,x_i],x_{i+1}, \ldots, x_n\}.$$

7. THE COMPLEX $V(\mathfrak{g})$

Throughout this section $\mathfrak{g}$ will denote a Lie algebra over K which is K-free.

We denote by $E(\mathfrak{g})$ the exterior algebra of the K-module $\mathfrak{g}$. The tensor product (over K)

$$V(\mathfrak{g}) = \mathfrak{g}^e \otimes E(\mathfrak{g})$$

is a left $\mathfrak{g}^e$-module and is $\mathfrak{g}^e$-free since $E(\mathfrak{g})$ is K-free. Using the grading of $E(\mathfrak{g})$ we define a grading in $V(\mathfrak{g})$ as

$$V_n(\mathfrak{g}) = \mathfrak{g}^e \otimes E_n(\mathfrak{g}).$$

Since $E_0(\mathfrak{g}) = K$ it follows that $V_0(\mathfrak{g}) = \mathfrak{g}^e$, and the augmentation $\varepsilon\colon \mathfrak{g}^e \to K$ defines an augmentation $\varepsilon\colon V(\mathfrak{g}) \to K$ which is zero on $V_n(\mathfrak{g})$, $n > 0$.

For $u \in \mathfrak{g}^e$, $x_1, \ldots, x_n \in \mathfrak{g}$, the element $u \otimes (x_1 \cdots x_n) \in \mathfrak{g}^e \otimes E(\mathfrak{g}) = V(\mathfrak{g})$ will be written as $u\langle x_1, \ldots, x_n\rangle$. If $u = 1$ we shall simply write $\langle x_1, \ldots, x_n\rangle$. Consequently the symbol $\langle\,\rangle$ will denote the element $1 \otimes 1$ of $\mathfrak{g}^e \otimes E(\mathfrak{g})$.

We now consider the normalized standard complex $N(\mathfrak{g}^e,\varepsilon)$ of the supplemented algebra $\mathfrak{g}^e$. Let

$$f\colon\ V(\mathfrak{g}) \to N(\mathfrak{g}^e,\varepsilon)$$

be the $\mathfrak{g}^e$-homomorphism defined by the requirement

$$f\langle x_1,\ldots,x_n\rangle = \sum_{\pi}(-1)^{\tau(\pi)}\{x_{\pi(1)},\ldots,x_{\pi(n)}\},$$

where the summation extends over all permutations π of $(1,\ldots,n)$ and $\tau(\pi)$ is the signature of π. To verify that f is well defined we only need to observe that $f\langle x_1,\ldots,x_n\rangle = 0$ if $x_i = x_j$ for some $0 \leqq i < j \leqq n$. In particular, the definition yields $f\langle\ \rangle = \{\ \}$.

If we choose a simply ordered K-base for $\mathfrak{g}$, we obtain in the usual fashion a K-base for $E(\mathfrak{g})$ which in turn induces a $\mathfrak{g}^e$-base for $V(\mathfrak{g})$. It follows then by inspection that f maps this $\mathfrak{g}^e$-base of $V(\mathfrak{g})$ into elements of $N(\mathfrak{g}^e,\varepsilon)$ which are $\mathfrak{g}^e$-linearly independent. Consequently f is a monomorphism. In the sequel we shall identify $V(\mathfrak{g})$ with a $\mathfrak{g}^e$-submodule of $N(\mathfrak{g}^e,\varepsilon)$ and regard f as an inclusion map.

THEOREM 7.1. *The submodule $V(\mathfrak{g})$ of $N(\mathfrak{g}^e,\varepsilon)$ is a subcomplex. The differentiation in $V(\mathfrak{g})$ is given by the formula*

$$(1)\qquad d\langle x_1,\ldots,x_n\rangle = \sum_{1\leqq i\leqq n}(-1)^{i+1}x_i\langle x_1,\ldots,\hat{x}_i,\ldots,x_n\rangle$$
$$+ \sum_{1\leqq i<j\leqq n}(-1)^{i+j}\langle[x_i,x_j],x_1,\ldots,\hat{x}_i,\ldots,\hat{x}_j,\ldots,x_n\rangle.$$

With the augmentation $\varepsilon\colon\ V(\mathfrak{g}) \to K$, the complex $V(\mathfrak{g})$ is a $\mathfrak{g}^e$-free resolution of K as a left $\mathfrak{g}^e$-module.

PROOF. Once formula (1) is proved, it will follow that $V(\mathfrak{g})$ is a subcomplex of $N(\mathfrak{g}^e,\varepsilon)$. For $n = 0$ formula (1) needs $d\langle\ \rangle = 0$ which is obviously correct. We now proceed by induction and assume that (1) holds for n.

We shall use the endomorphisms $\sigma(x)$ and $\vartheta(x)$ of the complex $N(\mathfrak{g}^e,\varepsilon)$ as defined by formulas (1) and (2′) of § 6. For $y,x_1,\ldots,x_n \in \mathfrak{g}$, we obtain

$$(2)\qquad \sigma(y)\langle x_1,\ldots,x_n\rangle = \langle y,x_1,\ldots,x_n\rangle$$

$$(3)\quad \vartheta(y)\langle x_1,\ldots,x_n\rangle = y\langle x_1,\ldots,x_n\rangle - \sum_{1\leqq i\leqq n}\langle x_1,\ldots,[y,x_i],\ldots,x_n\rangle.$$

The formula $d\sigma(y) + \sigma(y)d = \vartheta(y)$ (established in 6.1) together with (2) yields

$$d\langle y,x_1,\ldots,x_n\rangle = d\sigma(y)\langle x_1,\ldots,x_n\rangle = \vartheta(y)\langle x_1,\ldots,x_n\rangle - \sigma(y)d\langle x_1,\ldots,x_n\rangle.$$

Using (2) and (3) and the inductive assumption, this implies

$$\begin{aligned} d\langle y,x_1,\ldots,x_n\rangle = y\langle x_1,\ldots,x_n\rangle &+ \sum_{1\leq i\leq n}(-1)^i\langle[y,x_i],x_1,\ldots,\hat{x}_i,\ldots,x_n\rangle \\ &+ \sum_{1\leq i\leq n}(-1)^i x_i\langle y,x_1,\ldots,\hat{x}_i,\ldots,x_n\rangle \\ &- \sum_{1\leq i<j\leq n}(-1)^{i+j}\langle y,[x_i,x_j],x_1,\ldots,\hat{x}_i,\ldots,\hat{x}_j,\ldots,x_n\rangle. \end{aligned}$$

This is precisely the desired formula for $d\langle y,x_1,\ldots,x_n\rangle$. Thus (1) is proved.

We have already exhibited a $\mathfrak{g}^e$-base for $V(\mathfrak{g})$, which is thus $\mathfrak{g}^e$-free.

The kernel of the augmentation $V_0(\mathfrak{g})\to K$ is the K-module generated by the elements of the form $x_1\cdots x_p\langle\ \rangle$, with $x_i\in\mathfrak{g}$, $p>0$. Since $x_1\cdots x_p\langle\ \rangle = d(x_1\cdots x_{p-1}\langle x_p\rangle)$, it follows that $V_1(\mathfrak{g})\to V_0(\mathfrak{g})\to K\to 0$ is exact. Thus to conclude the proof of the theorem it suffices to show that $H_q(V(\mathfrak{g}))=0$ for $q>0$. The following proof is due to J. L. Koszul.

We choose a simply ordered K-base $\{x_\alpha\}$ for $\mathfrak{g}$. The elements $\langle x_{\alpha_1},\ldots,x_{\alpha_n}\rangle$ with $\alpha_1<\cdots<\alpha_n$, $n\geqq 0$ form a K-base for $E(\mathfrak{g})$. The elements $x_{\beta_1}\cdots x_{\beta_m}$ with $\beta_1\leqq\cdots\leqq\beta_m$, $m\geqq 0$, form by 3.1, a K-base for $\mathfrak{g}^e$. Consequently we obtain a K-base of $V(\mathfrak{g})=\mathfrak{g}^e\otimes E(\mathfrak{g})$:

$$(4)\quad x_{\beta_1}\cdots x_{\beta_m}\langle x_{\alpha_1},\ldots,x_{\alpha_n}\rangle,\qquad \begin{aligned} &\alpha_1<\cdots<\alpha_n, && n\geqq 0\\ &\beta_1\leqq\cdots\leqq\beta_m, && m\geqq 0. \end{aligned}$$

We introduce the submodule $F_pV(\mathfrak{g})$ generated by the elements (4) with $m+n\leqq p$. In the quotient module $W_p=F_pV(\mathfrak{g})/F_{p-1}V(\mathfrak{g})$ we then have the K-base represented by the elements (4) with $m+n=p$. Furthermore, it follows from 3.4 that the class represented in W_p by an element (4) is independent of the order in which the elements $x_{\beta_1},\ldots,x_{\beta_m}$ are written. The formula (1) for the differentiation d in $V(\mathfrak{g})$ implies

$$(5)\quad \begin{aligned} &d(x_{\beta_1}\cdots x_{\beta_m}\langle x_{\alpha_1},\ldots,x_{\alpha_n}\rangle)\\ &\quad=\sum_{1\leq i\leq n}(-1)^{i+1}x_{\beta_1}\cdots x_{\beta_m}x_{\alpha_i}\langle x_{\alpha_1},\ldots,\hat{x}_{\alpha_i},\ldots,x_{\alpha_m}\rangle \end{aligned}$$

modulo $F_{m+n-1}V(\mathfrak{g})$. This implies that the modules $F_pV(\mathfrak{g})$ are subcomplexes and that the differentiation induced in W_p is given by the formula (5).

It is now clear that the complex $W=\sum_p W_p$ is the complex

$$K[x_\alpha]\otimes E(x_\alpha)$$

with the differentiation given by (5). This complex is isomorphic to the projective resolution of K as a left $K[x_\alpha]$-module constructed in VIII,4. It follows that $H_q(W)=0$ for $q>0$, and therefore that $H_q(W_p)=0$ for $q>0$.

Now consider the exact sequence

$$H_q(F_{p-1}V(\mathfrak{g})) \to H_q(F_pV(\mathfrak{g})) \to H_q(W_p), \qquad q > 0.$$

This implies that $H_q(F_{p-1}V(\mathfrak{g})) \to H_q(F_pV(\mathfrak{g}))$ is an epimorphism. Since $F_{-1}V(\mathfrak{g}) = 0$ we obtain $H_q(F_pV(\mathfrak{g})) = 0$ for $q > 0$ and all p. Since $V(\mathfrak{g}) = \cup_p F_pV(\mathfrak{g})$ it follows that $H_q(V(\mathfrak{g})) = 0$ for $q > 0$. This concludes the proof of the theorem.

8. APPLICATIONS OF THE COMPLEX $V(\mathfrak{g})$

We first show how the homology and cohomology groups of $\mathfrak{g}$ may be computed using the complex $V(\mathfrak{g})$.

If A is a right $\mathfrak{g}$-module, then the homology groups $H_q(\mathfrak{g},A)$ are the homology groups of the complex

$$A \otimes_{\mathfrak{g}^e} V(\mathfrak{g}) = A \otimes_{\mathfrak{g}^e} \mathfrak{g}^e \otimes E(\mathfrak{g}) = A \otimes E(\mathfrak{g}).$$

The differentiation operator in this complex is

$$d(a \otimes \langle x_1, \dots, x_n\rangle) = \sum_{1 \leqq i \leqq n} (-1)^{i+1}(ax_i) \otimes \langle x_1, \dots, \hat{x}_i, \dots, x_n\rangle$$
$$+ \sum_{1 \leqq i < j \leqq n} (-1)^{i+j} a \otimes \langle [x_i,x_j], x_1, \dots, \hat{x}_i, \dots, \hat{x}_j, \dots, x_n\rangle.$$

If C is a left $\mathfrak{g}$-module, the cohomology groups $H^q(\mathfrak{g},C)$ are the homology groups of the complex

$$\operatorname{Hom}_{\mathfrak{g}^e}(V(\mathfrak{g}),C) = \operatorname{Hom}_{\mathfrak{g}^e}(\mathfrak{g}^e \otimes E(\mathfrak{g}),C) = \operatorname{Hom}(E(\mathfrak{g}),C).$$

In this last complex, a q-cochain $f\colon E_q(\mathfrak{g}) \to C$ is simply a K-linear alternating function $f(x_1, \dots, x_q)$ of q variables in $\mathfrak{g}$, with values in C. The coboundary δf of such a cochain is the $q + 1$-cochain given by the formula

$$(\delta f)(x_1, \dots, x_{q+1}) = \sum_{1 \leqq i \leqq q+1} (-1)^{i+1} x_i f(x_1, \dots, \hat{x}_i, \dots, x_{q+1})$$
$$+ \sum_{1 \leqq i < j \leqq q+1} (-1)^{i+j} f([x_i,x_j], x_1, \dots, \hat{x}_i, \dots, \hat{x}_j, \dots, x_{q+1}).$$

This description of the cohomology groups $H^q(\mathfrak{g},C)$ shows directly that these coincide with the cohomology groups of $\mathfrak{g}$ considered hitherto (C. Chevalley and S. Eilenberg, *Trans. Am. Math. Soc.* 63 (1948), 85–124).

We recall that the complex $V(\mathfrak{g})$ is a subcomplex of the normalized standard complex $N(\mathfrak{g}^e,\varepsilon)$. In this connection the following proposition will be useful.

PROPOSITION 8.1. *Every cochain* $f \in \operatorname{Hom}_{\mathfrak{g}^e}(V(\mathfrak{g}),C)$ *admits an extension* $f' \in \operatorname{Hom}_{\mathfrak{g}^e}(N(\mathfrak{g}^e,\varepsilon),C)$. *If f is a cocycle then f' may be chosen to be a cocycle.*

PROOF. The first fact follows from the observation that $V(\mathfrak{g})$ as a $\mathfrak{g}^e$-module is a direct summand of $N(\mathfrak{g}^e,\varepsilon)$. This is clear from the bases exhibited in § 7. Now assume that $\delta f = 0$. Since the cohomology groups obtained using $V(\mathfrak{g})$ and $N(\mathfrak{g}^e,\varepsilon)$ are isomorphic under the inclusion map, there exists a cocycle $g' \in \mathrm{Hom}_{\mathfrak{g}^e}(N(\mathfrak{g}^e,\varepsilon),C)$ whose restriction g to $V(\mathfrak{g})$ is cohomologous to f; then $f - g = \delta h$. Let h' be an extension of the cochain h. It follows that $f' = g' + \delta h'$ is an extension of f and $\delta f' = 0$ as desired.

The next application of the complex $V(\mathfrak{g})$ has to do with dimension.

THEOREM 8.2. *If $\mathfrak{g}$ has a K-base composed of n elements, then*

$$\dim \mathfrak{g}^e = \dim_{\mathfrak{g}^e} K = n.$$

If further the commutative ring K is semi-simple then

$$\mathrm{gl.dim}\ \mathfrak{g}^e = n.$$

PROOF. In view of 5.2, we only need to prove $\dim_{\mathfrak{g}^e} K = n$. Since $E_q(\mathfrak{g}) = 0$ for $q > n$, it follows that the complex $V(\mathfrak{g})$ is n-dimensional and thus $\dim_{\mathfrak{g}^e} K \leqq n$. Now consider the group $E_n(\mathfrak{g})$, with $\mathfrak{g}$ operating on the left by

$$y.\langle x_1, \ldots, x_n\rangle = \sum_{1 \leqq i \leqq n} \langle x_1, \ldots, [y,x_i], \ldots, x_n\rangle.$$

Let f be a $(n-1)$-cochain with values in $E_n(\mathfrak{g})$; an easy computation (cf. Exer. 12) shows that $\delta f = 0$; thus $H^n(\mathfrak{g},E_n(\mathfrak{g}))$ is isomorphic to the K-module of n-cochains $E_n(\mathfrak{g}) \to E_n(\mathfrak{g})$, which is obviously isomorphic to K. Hence $\dim_{\mathfrak{g}^e} K \geqq n$.

Next we pass to the question of computing the products using the complexes $V(\mathfrak{g})$. We begin with the external products for two Lie algebras $\mathfrak{g}$ and $\mathfrak{h}$ over K, both of which are K-free. As agreed upon in § 1, we shall systematically identify $(\mathfrak{g} + \mathfrak{h})^e$ with $\mathfrak{g}^e \otimes \mathfrak{h}^e$. As we have seen in XI,5, to compute the products $\perp$ and $\top$ we need a map

$$f\colon\ V(\mathfrak{g}) \otimes V(\mathfrak{h}) \to V(\mathfrak{g} + \mathfrak{h})$$

while for the products $\vee$ and $\wedge$ we need a map

$$g\colon\ V(\mathfrak{g} + \mathfrak{h}) \to V(\mathfrak{g}) \otimes V(\mathfrak{h}).$$

The answer to both of these problems is quite trivial here since the identification $(\mathfrak{g} + \mathfrak{h})^e = \mathfrak{g}^e \otimes \mathfrak{h}^e$ and the natural isomorphism $E(\mathfrak{g} + \mathfrak{h}) \approx E(\mathfrak{g}) \otimes E(\mathfrak{h})$ imply a natural isomorphism

$$V(\mathfrak{g} + \mathfrak{h}) \approx V(\mathfrak{g}) \otimes V(\mathfrak{h}) \tag{1}$$

compatible with the $(\mathfrak{g} + \mathfrak{h})^e$-operators and the differentiations.

For the internal products $\smile$ and $\frown$ we assume that $\mathfrak{g}$ is an abelian Lie algebra. Then $\mathfrak{g}^e$ is a commutative algebra. As we have seen in XI,5, to compute the products $\smile$ and $\frown$ we need a map

$$V(\mathfrak{g}) \otimes_{\mathfrak{g}^e} V(\mathfrak{g}) \to V(\mathfrak{g}).$$

To obtain such a map it suffices to regard $V(\mathfrak{g}) = \mathfrak{g}^e \otimes E(\mathfrak{g})$ as a $\mathfrak{g}^e$-algebra, and verify that this map is compatible with the differentiation (cf. Exer. 15).

We finally consider the products $\cup$ and $\cap$ defined using the diagonal map D: $\mathfrak{g}^e \to \mathfrak{g}^e \otimes \mathfrak{g}^e = (\mathfrak{g} + \mathfrak{g})^e$. According to XI,5, we need a map

$$V(\mathfrak{g}) \to V(\mathfrak{g}) \otimes V(\mathfrak{g}).$$

This is given by the maps $\mathfrak{g}^e \to \mathfrak{g}^e \otimes \mathfrak{g}^e$ and $E(\mathfrak{g}) \to E(\mathfrak{g}) \otimes E(\mathfrak{g})$ both defined by $x \to x \otimes 1 + 1 \otimes x$, $x \in \mathfrak{g}$. If we carry out the explicit computation and apply this map to find the cup product of cochains we obtain the classical formula for the multiplication of alternating multilinear forms. Explicitly, consider cochains $f \in \operatorname{Hom}(E_p(\mathfrak{g}),C)$, $f' \in \operatorname{Hom}(E_q(\mathfrak{g}),C')$, where C and C' are left $\mathfrak{g}$-modules. If $C \otimes C'$ is regarded as a left $\mathfrak{g}$-module by means of the map D, we find that the cochain $f \cup f' \in \operatorname{Hom}(E_{p+q}(\mathfrak{g}),C \otimes C')$ is given by

$$(f \cup f')(x_1, \ldots, x_{p+q}) = \Sigma \pm f(x_{i_1}, \ldots, x_{i_p}) \otimes f'(x_{j_1}, \ldots, x_{j_q}),$$

the sum being extended over all partitions of the sequence $(1, \ldots, p+q)$ into two increasing sequences $(i_1, \ldots, i_p)$ and $(j_1, \ldots, j_q)$. The sign is the signature of the permutation $(i_1, \ldots, i_p, j_1, \ldots, j_q)$.

EXERCISES

1. Given an associative K-algebra Λ, define

$$[x,y] = xy - yx \qquad x,y \in \Lambda$$

and prove that this assigns to Λ the structure of a Lie algebra, denoted by $\mathfrak{l}(\Lambda)$. Show that for any Lie algebra $\mathfrak{g}$ (over K), the map i is a Lie algebra homomorphism

$$i\colon \mathfrak{g} \to \mathfrak{l}(\mathfrak{g}^e).$$

2. Given a Lie algebra $\mathfrak{g}$ and an associative algebra Λ (both over the same ring K), show that any Lie algebra homomorphism f: $\mathfrak{g} \to \mathfrak{l}(\Lambda)$ admits a unique factorization

$$\mathfrak{g} \xrightarrow{i} \mathfrak{g}^e \xrightarrow{h} \Lambda$$

where h is a K-algebra homomorphism. Show further that this property of the pair $(\mathfrak{g}^e,i)$ characterizes this pair uniquely up to an isomorphism.

Show that in order that there exists an associative K-algebra Λ and a Lie algebra monomorphism $f\colon \mathfrak{g} \to \mathfrak{l}(\Lambda)$, it is necessary and sufficient that $i\colon \mathfrak{g} \to \mathfrak{g}^e$ be a monomorphism.

3. For a given Lie algebra $\mathfrak{g}$, let $\bar{\mathfrak{g}}$ denote the image of the homomorphism $i\colon \mathfrak{g} \to \mathfrak{g}^e$; we regard $\bar{\mathfrak{g}}$ as a Lie algebra. Show that the inclusion map $\bar{\mathfrak{g}} \to \mathfrak{g}^e$ satisfies the criterion of Exer. 2 and thus we may identify $\mathfrak{g}^e$ with $(\bar{\mathfrak{g}})^e$.

4. Given a Lie algebra $\mathfrak{g}$ over K, consider the associative K-algebra $\Lambda = \operatorname{Hom}_K(\mathfrak{g},\mathfrak{g})$ and the map

$$\rho\colon \mathfrak{g} \to \mathfrak{l}(\Lambda)$$

given by

$$(\rho x)y = [x,y].$$

Show that ρ is a homomorphism of Lie algebras and that $\rho = 0$ if and only if $\mathfrak{g}$ is a commutative (i.e. $[\mathfrak{g},\mathfrak{g}] = 0$). As an application show that if $\mathfrak{g} \neq 0$ then the natural map $i\colon \mathfrak{g} \to \mathfrak{g}^e$ is not zero.

5. Let M be a K-module. Consider the graded K-module $A(M) = \sum_{k \geqq 1} A^k(M)$, where

$$A^1(M) = M, \qquad A^k(M) = \sum_{0<i<k} A^i(M) \otimes_K A^{k-i}(M) \qquad \text{for } k > 1.$$

Define the mapping $A(M) \otimes_K A(M) \to A(M)$ by the inclusion maps $A^k(M) \otimes_K A^h(M) \to A^{k+h}(M)$. We call $A(M)$ the *free non-associative K-algebra* (without unit element) *over M*. In $A(M)$ consider the two-sided ideal $J(M)$ generated by the elements

$$xx \text{ and } x(yz) + y(zx) + z(xy), \qquad x,y,z \in A(M).$$

Show that the quotient $L(M) = A(M)/J(M)$ is a (graded) Lie algebra; we call $L(M)$ the *free Lie algebra over M*. Show that the map $j\colon M \to L(M)$, defined by composition $M = A^1(M) \to A(M) \to L(M)$, is a monomorphism. Show that every K-homomorphism $f\colon M \to \mathfrak{g}$ into a Lie algebra $\mathfrak{g}$ admits a unique factorization $M \xrightarrow{j} L(M) \longrightarrow \mathfrak{g}$, where φ is a homomorphism of Lie algebras over K.

6. Let M be a K-module, and k a K-homomorphism of M into a Lie K-algebra $\mathfrak{l}$. Suppose that each K-homomorphism $f\colon M \to \mathfrak{g}$ into a Lie K-algebra $\mathfrak{g}$ admits a unique factorization

$$M \xrightarrow{k} \mathfrak{l} \xrightarrow{\psi} \mathfrak{g},$$

where ψ is a homomorphism of Lie K-algebras. Prove that there exists a unique isomorphism $\alpha\colon \mathfrak{l} \approx L(M)$ such that $\alpha k = j$. This gives an axiomatic description of the pair $(L(M),j)$.

7. Consider the tensor algebra $T(M)$ of the K-module M. Show that the natural injection $M \to T(M)$ admits a unique factorization $M \xrightarrow{j} L(M) \xrightarrow{i} T(M)$, where i is a homomorphism of the Lie algebra $L(M)$ into the Lie algebra $\mathfrak{l}(T(M))$. This mapping i is compatible with with the gradings in $L(M)$ and $T(M)$. Show that $T(M)$ may be identified with the enveloping algebra $L(M)^e$ of $L(M)$. If $\bar{L}(M)$ denotes the image of i, show that $\bar{L}(M)$ is the Lie subalgebra of $\mathfrak{l}(T(M))$ generated by the elements of degree 1 in $T(M)$, i.e. by M.

8. Prove the following theorem: if M is a K-free module, then $L(M)$ is K-free and $i: L(M) \to T(M)$ is a monomorphism; thus the Lie subalgebra $\bar{L}(M)$ of $\mathfrak{l}(T(M))$, generated by M, is K-free and isomorphic to $L(M)$.

[Hint: if $L(M)$ is K-free, then, by 3.3 and Exer. 7, i is a monomorphism. Hence the theorem is proved when K is a field. For any commutative ring K, and any K-free module M, there exists a free abelian group A such that $M = A \otimes K$; show that $L(M) = L(A) \otimes K$. This reduces the proof to showing that $L(A)$ is Z-free when A is Z-free; it will be sufficient to prove that $i: L(A) \to T(A)$ is a monomorphism. Let A_I be the subgroup of A generated by any finite subset I of the base of A; then $T(A_I) \to T(A)$ is a monomorphism, which reduces the proof to the case of a finitely generated free abelian group. Let now A be an abelian group with a finite base; for proving that $i: L(A) \to T(A)$ is a monomorphism, observe that, for each prime p, $L(A) \otimes Z_p \to T(A) \otimes Z_p$ is a monomorphism of degree zero, since the theorem is proved for a field; then apply VII, Exer. 12 to each graded component $L_k(A) \otimes Z_p \to T_k(A) \otimes Z_p$.]

9. Show that any representation satisfying 3.5 automatically satisfies condition (4) and therefore is unique.

10. Show that if $\mathfrak{g}$ is K-free and $\mathfrak{g}^e$ is commutative then $\mathfrak{g}$ is an abelian Lie algebra.

11. Given a map $K \to L$ (of commutative rings) examine the effects of this change of ground ring upon the homology and cohomology groups of a Lie algebra.

12. Let $\mathfrak{g}$ be a Lie algebra with a K-base $x_1, \ldots, x_n$. Define the *constants of structure* c_{ijk} by the relations

$$[x_i, x_j] = \sum_k c_{ijk} x_k.$$

Express the axioms of the Lie algebra in terms of c_{ijk}. Prove that in the complex $K \otimes_{\mathfrak{g}^e} V(\mathfrak{g})$ we have

$$d\langle x_1, \ldots, x_n \rangle = \sum_{\substack{1 \leq i \leq n \\ 1 \leq j \leq n}} (-1)^i c_{ijj} \langle x_1, \ldots, \hat{x}_i, \ldots, x_n \rangle.$$

13. Under the conditions of Exer. 12, $\mathfrak{g}$ is said to be *unimodular* if for any $y \in \mathfrak{g}$, the relation

$$\sum_{1 \leqq j \leqq n} x_1 \cdots x_{j-1}[y,x_j]x_{j+1} \cdots x_n = 0$$

holds in $E(\mathfrak{g})$. Show that this is equivalent with

$$d\langle x_1, \ldots, x_n\rangle = 0$$

in the complex $K \otimes_{\mathfrak{g}^e} V(\mathfrak{g})$.

14. (Alternative description of the complex $V(\mathfrak{g})$). Let $\Lambda = (K,d)$ be the ring of dual numbers over K and consider the K-module $\Lambda \otimes_K \mathfrak{g}$ with endomorphism d. Let $T(\Lambda \otimes \mathfrak{g})$ be the tensor algebra over K of the K-module $\Lambda \otimes \mathfrak{g}$. The map i: $x \to 1 \otimes x$ will be used to identify $\mathfrak{g}$ with a submodule of $\Lambda \otimes \mathfrak{g}$ and thus also of $T(\Lambda \otimes \mathfrak{g})$. In $T(\Lambda \otimes \mathfrak{g})$ introduce a grading written with lower indices in which the elements $x \in \mathfrak{g}$ have degree 1 and the elements dx ($x \in \mathfrak{g}$) have degree 0. The endomorphism d of $\Lambda \otimes \mathfrak{g}$ may now be extended uniquely to an antiderivation d of $T(\Lambda \otimes \mathfrak{g})$, i.e. a K-endomorphism satisfying

$$d(uv) = (du)v + (-1)^p u(dv)$$

for u of degree p in $T(\Lambda \otimes \mathfrak{g})$. This operator d satisfies $dd = 0$ and is of degree -1 (with respect to the lower indices).

Let L be the two-sided ideal in $T(\Lambda \otimes \mathfrak{g})$ generated by the elements

$$xx \tag{1}$$

$$(dx)y - y(dx) - [x,y] \tag{2}$$

$$(dx)(dy) - (dy)(dx) - d[x,y] \tag{3}$$

for $x,y \in \mathfrak{g}$.

Prove that L is a homogeneous ideal and is stable under d. Consider the K-algebra

$$W(\mathfrak{g}) = T(\Lambda \otimes \mathfrak{g})/L$$

which is a left $\mathfrak{g}^e$-complex over K.

Use the maps

$$i\colon \mathfrak{g} \to \Lambda \otimes \mathfrak{g}, \qquad j = di\colon \mathfrak{g} \to \Lambda \otimes \mathfrak{g}$$

to obtain maps

$$i'\colon T(\mathfrak{g}) \to T(\Lambda \otimes \mathfrak{g}), \qquad j'\colon T(\mathfrak{g}) \to T(\Lambda \otimes \mathfrak{g})$$

$$i^*\colon E(\mathfrak{g}) \to W(\mathfrak{g}), \qquad j^*\colon \mathfrak{g}^e \to W(\mathfrak{g})$$

$$\varphi = j^* \otimes i^*\colon \mathfrak{g}^e \otimes E(\mathfrak{g}) \to W(\mathfrak{g}).$$

Prove that φ is an isomorphism of graded K-modules and is an isomorphism of the complexes $V(\mathfrak{g})$ and $W(\mathfrak{g})$.

[Hint to the last part: denote by M the ideal of $T(\Lambda \otimes \mathfrak{g})$ generated by the elements (2). Prove that $j' \otimes i'\colon T(\mathfrak{g}) \otimes T(\mathfrak{g}) \to T(\Lambda \otimes \mathfrak{g})$ induces an isomorphism $T(\mathfrak{g}) \otimes T(\mathfrak{g}) \approx T(\Lambda \otimes \mathfrak{g})/M$.]

15. Let $\mathfrak{g}$ be a Lie algebra with a K-base; then $W(\mathfrak{g})$ (Exer. 14) is a graded differential algebra and $\mathfrak{g}^e$ a subalgebra of degree 0, thus $W(\mathfrak{g})$ is a two-sided $\mathfrak{g}^e$-module. The multiplication of $W(\mathfrak{g})$ defines a map

$$W(\mathfrak{g}) \otimes_{\mathfrak{g}^e} W(\mathfrak{g}) \to W(\mathfrak{g}) \tag{1}$$

which is compatible with the structures of two-sided $\mathfrak{g}^e$-modules. Let A be a left $\mathfrak{g}^e$-module; (1) defines

$$W(\mathfrak{g}) \otimes_{\mathfrak{g}^e} A \to \operatorname{Hom}_{\mathfrak{g}^e}(W(\mathfrak{g}), W(\mathfrak{g}) \otimes_{\mathfrak{g}^e} A), \tag{2}$$

where $\operatorname{Hom}_{\mathfrak{g}^e}$ is related to the left $\mathfrak{g}^e$-module structures. Let n be the number of the elements of the K-base of $\mathfrak{g}$; (2) induces

$$W_{n-k}(\mathfrak{g}) \otimes_{\mathfrak{g}^e} A \to \operatorname{Hom}_{\mathfrak{g}^e}(W_k(\mathfrak{g}), W_n(\mathfrak{g}) \otimes_{\mathfrak{g}^e} A) \tag{3}$$

for any integer k; this is a map φ_k of the module of $(n - k)$-chains (with coefficients in A) into the module of k-cochains (with coefficients in $W_n(\mathfrak{g}) \otimes_{\mathfrak{g}^e} A \approx E_n(\mathfrak{g}) \otimes_K A$). Show that the collection of maps φ_k commute (up to the sign) with the boundary and coboundary operators, and that each φ_k is an isomorphism. Compute explicitly the left operations of $\mathfrak{g}$ on $E_n(\mathfrak{g}) \otimes_K A$, and establish the natural isomorphisms

$$H_{n-k}(\mathfrak{g}, A) \approx H^k(\mathfrak{g}, E_n(\mathfrak{g}) \otimes_K A).$$

For $k = n$ and $A = K$ (with trivial operators) we find again $H^n(\mathfrak{g}, E_n(\mathfrak{g})) = K$ (cf. 8.2).

CHAPTER XIV

Extensions

Introduction. In general an extension over A is given by an epimorphism $f\colon X \to A$. This concept may be considered for various kinds of algebraic structures:

(1) X and A are Λ-modules, and f is an epimorphism of Λ-modules.

(2) Γ and Λ are K-algebras, and $f\colon \Gamma \to \Lambda$ is a K-algebra epimorphism.

(3) W and Π are groups, and $f\colon W \to \Pi$ is an epimorphism of groups.

(4) $\mathfrak{h}$ and $\mathfrak{g}$ are Lie algebras, and $f\colon \mathfrak{h} \to \mathfrak{g}$ is an epimorphism of Lie algebras.

In the case (1), the kernel C of f is a Λ-module. The knowledge of A and C does not yet determine the extension even up to "equivalence." Indeed, the set of equivalence classes of extensions is in a 1–1-correspondence with the group $\operatorname{Ext}^1_\Lambda(A,C)$; this was the origin of the notation "Ext".

Cases (2), (3), (4) are more complicated, and will be studied here only under restrictive conditions which permit the introduction of a suitable structure into the kernel C of f. In the case (2), C is assumed to be a two-sided Λ-module; in the case (3), C is assumed to be a Π-module (C is then an abelian subgroup of W); in the case (4), C is assumed to be a $\mathfrak{g}$-module (C is then an abelian ideal in $\mathfrak{h}$). In each of these cases, the set of all equivalence classes is in a-1-1-correspondence with a 2-dimensional cohomology group with coefficients in C. These are: the Hochschild cohomology group $H^2(\Lambda,C)$ in the case (2), the group $H^2(\Pi,C)$ in the case (3), and the group $H^2(\mathfrak{g},C)$ in the case (4).

The four problems of extensions listed above are inter-related and some of these relations are studied in detail.

1. EXTENSIONS OF MODULES

Let A and C be (left) Λ-modules. An *extension* over A with kernel C is an exact sequence

$$(E) \qquad 0 \longrightarrow C \xrightarrow{\psi} X \xrightarrow{\varphi} A \longrightarrow 0$$

where X is a Λ-module and φ and ψ are Λ-homomorphisms. The extension (E) is said to be *equivalent* with an extension

$$(E') \qquad 0 \to C \to X' \to A \to 0$$

if there is a Λ-homomorphism k: $X \to X'$ such that the diagram

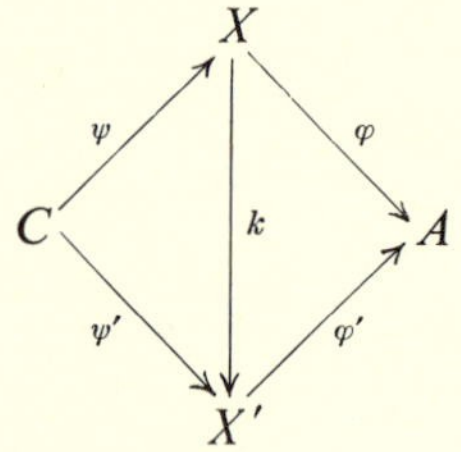

is commutative. It is clear that such a k is an isomorphism. We denote by $E(A,C)$ the set of all equivalence classes of extensions of A and C. All split exact sequences are in the same class, called the "split class" of $E(A,C)$.

Following Baer (*Math. Zeit.* 38 (1934), 375–416) we define a multiplication in the set $E(A,C)$: given extensions (E) and (E'), we define their product as an extension $0 \to C \to Y \to A \to 0$ as follows. In the direct sum $X + X'$ consider the submodule B consisting of pairs (x,x') with $\varphi x = \varphi' x'$, and the submodule D of pairs of the form $(-\psi c, \psi' c)$ for $c \in C$. Then $D \subset B$; we define $Y = B/D$. The maps Ψ and Φ are defined by

$$\Psi c = \text{class of } (\psi c, 0) = \text{class of } (0, \psi' c),$$

$$\Phi\ (\text{class of } (x,x')) = \varphi x = \varphi' x',$$

where by "class" we mean congruence class of B mod D.

The verification that the sequence $0 \longrightarrow C \xrightarrow{\Psi} Y \xrightarrow{\Phi} A \longrightarrow 0$ is exact is immediate. It is also clear that this multiplication defines a multiplication in the set $E(A,C)$.

An extension

$$(E) \qquad 0 \longrightarrow C \xrightarrow{\psi} X \xrightarrow{\varphi} A \longrightarrow 0$$

defines a connecting homomorphism Θ_E: $\operatorname{Hom}(C,C) \to \operatorname{Ext}^1(A,C)$ which maps the identity element $j \in \operatorname{Hom}(C,C)$ into the "characteristic class" of the extension (E) (cf. XI,9). Equivalent extensions have the same characteristic class.

THEOREM 1.1. *Given two Λ-modules A and C the mapping $(E) \to \Theta_E j$ establishes a 1–1-correspondence Θ between $E(A,C)$ and $\operatorname{Ext}^1_\Lambda(A,C)$. Baer multiplication in $E(A,C)$ is carried into the addition in $\operatorname{Ext}^1_\Lambda(A,C)$; the split class of $E(A,C)$ is carried into the zero element of $\operatorname{Ext}^1_\Lambda(A,C)$.*

COROLLARY 1.2. *The set $E(A,C)$ with the Baer multiplication is an abelian group with the split class as zero element.*

(This assertion could be proved directly although the proof is somewhat laborious.)

Proof of 1.1. We choose once and for all an exact sequence

$$0 \longrightarrow M \xrightarrow{\beta} P \xrightarrow{\alpha} A \longrightarrow 0$$

with P projective. For each extension (E) we can find homomorphisms γ and τ such that the diagram

$$\begin{array}{ccccccccc} 0 & \longrightarrow & M & \xrightarrow{\beta} & P & \xrightarrow{\alpha} & A & \longrightarrow & 0 \\ & & \downarrow{\scriptstyle \gamma} & & \downarrow{\scriptstyle \tau} & & \downarrow{\scriptstyle i} & & \\ 0 & \longrightarrow & C & \xrightarrow{\psi} & X & \xrightarrow{\varphi} & A & \longrightarrow & 0 \end{array} \tag{1}$$

is commutative (i denotes the identity map of A).

Diagram (1) gives rise to a commutative diagram

$$\begin{array}{ccccccc} & & \operatorname{Hom}_\Lambda(C,C) & & & & \\ & & \downarrow{\scriptstyle \operatorname{Hom}(\gamma,C)} & \searrow{\scriptstyle \Theta_E} & & & \\ \operatorname{Hom}_\Lambda(P,C) & \longrightarrow & \operatorname{Hom}_\Lambda(M,C) & \xrightarrow[\vartheta]{} & \operatorname{Ext}^1_\Lambda(A,C) & \longrightarrow & 0 \end{array} \tag{2}$$

in which the row is exact and where ϑ is the connecting homomorphism induced by the top row of (1). Since $\operatorname{Hom}(\gamma,C)$ maps j into γ, it follows that $\vartheta\gamma$ is the characteristic class $\Theta_E j$ of the extension (E).

Now consider the direct sum $C + P$ and define an exact sequence

$$0 \longrightarrow M \xrightarrow{\mu} C + P \xrightarrow{\pi} X \longrightarrow 0 \tag{3}$$

by setting

$$\mu m = (-\gamma m, \beta m), \qquad \pi(c,p) = \psi c + \tau p. \tag{4}$$

Using the exact sequence (3) to identify X with $\operatorname{Coker} \mu$ we obtain

$$\begin{cases} \psi c = \text{class of } (c,0), \\ \tau p = \text{class of } (0,p), \\ \varphi\,(\text{class of } (c,p)) = \alpha p, \end{cases} \tag{5}$$

where "class" means "congruence class mod the image of μ."

Now assume that a homomorphism $\gamma \in \operatorname{Hom}(M,C)$ is given. We may then define μ by formula (4), take $X = \operatorname{Coker} \mu$, and define ψ, τ, φ by (5). The resulting sequence

$$0 \longrightarrow C \xrightarrow{\psi} X \xrightarrow{\varphi} A \longrightarrow 0 \tag{E_γ}$$

is then exact, and diagram (1) is valid with the prescribed map γ. The characteristic class of (E_γ) is therefore $\vartheta\gamma$. Since $\vartheta\colon \operatorname{Hom}_\Lambda(M,C) \to \operatorname{Ext}^1_\Lambda(A,C)$ is an epimorphism, it follows that the correspondence Θ maps $E(A,C)$ *onto* $\operatorname{Ext}^1_\Lambda(A,C)$.

In order to prove that the correspondence Θ is 1–1 we must show that $\vartheta\gamma_1 = \vartheta\gamma_2$ implies that (E_{γ_1}) and (E_{γ_2}) are equivalent. From the exactness of the row in (2) it follows that $\vartheta(\gamma_1 - \gamma_2) = 0$ is equivalent with $\gamma_1 - \gamma_2 = \omega\beta$ for some $\omega \in \operatorname{Hom}_\Lambda(P,C)$. Using ω we define an automorphism $\Omega\colon C + P \to C + P$ by setting

$$\Omega(c,p) = (c + \omega p,p).$$

It follows readily that $\mu_2 = \Omega\mu_1$; thus Ω induces an isomorphism $\Omega'\colon X_1 \to X_2$ (where $X_i = \operatorname{Coker} \mu_i$, $i = 1,2$), such that $\Omega'\psi_1 = \psi_2$, $\varphi_2\Omega' = \varphi_1$. Therefore (E_{γ_1}) and (E_{γ_2}) are equivalent.

If the extension (E) splits, then by v,4.5 the connecting homomorphism Θ_E is zero, and therefore $\Theta_E j = 0$. An alternative proof is obtained by taking $\gamma = 0$ in the construction of (E_γ). Then $\mu m = (0,\beta m)$ so that $\operatorname{Coker} \mu = C + A$.

It remains to be proved that Θ carries the Baer multiplication in $E(A,C)$ into the addition in $\operatorname{Ext}^1_\Lambda(A,C)$. To this end we consider the Baer product (E) of two extensions (E_{γ_1}) and (E_{γ_2}); define $\gamma\colon M \to C$ and $\tau\colon P \to X$ by

$$\gamma = \gamma_1 + \gamma_2, \qquad \tau p = \text{class of } (\tau_1 p,\tau_2 p).$$

Then $\varphi\tau = \alpha$ and

$$\begin{aligned}\psi\gamma m = \psi\gamma_1 m + \psi\gamma_2 m &= \text{class of } (\psi_1\gamma_1 m,0) + \text{class of } (0,\psi_2\gamma_2 m)\\ &= \text{class of } (\tau_1\beta_1 m,\tau_2\beta_2 m) = \tau\beta m.\end{aligned}$$

Thus we have a commutative diagram like (1), which proves that the extension (E) is defined by $\gamma = \gamma_1 + \gamma_2$.

This concludes the proof of theorem 1.1.

REMARK. Instead of using the connecting homomorphism $\Theta_E\colon \operatorname{Hom}_\Lambda(C,C) \to \operatorname{Ext}^1_\Lambda(A,C)$ we could use the connecting homomorphism

$$\Theta'_E\colon \operatorname{Hom}_\Lambda(A,A) \to \operatorname{Ext}^1_\Lambda(A,C)$$

induced by (E). If $i \in \operatorname{Hom}_\Lambda(A,A)$ denotes the identity element, it can be shown (see Exer. 1) that $\Theta'_E i + \Theta_E j = 0$. There exists a proof of 1.1, dual to the one given above, and adjusted to the connecting homomorphism Θ'_E. We choose once and for all an exact sequence

$$0 \longrightarrow C \xrightarrow{\delta} Q \xrightarrow{\eta} N \longrightarrow 0$$

with Q injective. Diagram (1) is then replaced by

$$(1') \qquad \begin{array}{ccccccccc} 0 & \longrightarrow & C & \xrightarrow{\psi} & X & \xrightarrow{\varphi} & A & \longrightarrow & 0 \\ & & \downarrow{\scriptstyle j} & & \downarrow{\scriptstyle \zeta} & & \downarrow{\scriptstyle \varepsilon} & & \\ 0 & \longrightarrow & C & \xrightarrow{\delta} & Q & \xrightarrow{\eta} & N & \longrightarrow & 0 \end{array}$$

As before, the map ε: $A \to N$ determines essentially the rest of the diagram. Indeed if we define ν: $A + Q \to N$ by $\nu(a,q) = -\varepsilon a + \eta q$ and identify X with Ker ν, then

$$(5') \qquad \begin{cases} \psi c = (0,\delta c), \\ \zeta(a,q) = q, \\ \varphi(a,q) = a. \end{cases}$$

2. EXTENSIONS OF ASSOCIATIVE ALGEBRAS

Let K be a commutative ring. An epimorphism of K-algebras

$$f\colon\ \Gamma \to \Lambda,$$

will be called an *extension* over Λ. The extension is called *inessential* if there exists a K-algebra homomorphism u: $\Lambda \to \Gamma$ with $fu =$ identity.

The kernel C of f is a two-sided ideal in Γ and therefore is also a two-sided Γ-module. In particular, the multiplication in Γ induces a multiplication in C. If this multiplication is zero, i.e. $c_1c_2 = 0$ for all $c_1,c_2 \in C$, then the structure of C as a two-sided Γ-module induces on C the structure of a two-sided Λ-module:

$$(1) \qquad \lambda c = \gamma c, \qquad c\lambda = c\gamma \qquad\qquad c \in C, \gamma \in \Gamma, \lambda = f\gamma \in \Lambda.$$

Conversely if C carries the structure of a two-sided Λ-module satisfying (1) then the multiplication in C induced by that of Γ is zero.

Definition. Let Λ be a K-algebra and C a two-sided Λ-module. An *extension over* Λ *with kernel* C is an exact sequence

$$(F) \qquad C \xrightarrow{g} \Gamma \xrightarrow{f} \Lambda$$

where Γ is a K-algebra, f is a K-algebra epimorphism, g is a monomorphism of K-modules and

$$(2) \qquad g(\lambda c) = \gamma(gc), \qquad g(c\lambda) = (gc)\gamma, \qquad c \in C, \gamma \in \Gamma, \lambda = f\gamma \in \Lambda.$$

These last conditions are simply a translation of (1).

Two extensions (F) and (F') over A with kernel C are *equivalent* if there exists a K-algebra homomorphism k: $\Gamma \to \Gamma'$ such that the diagram

(3)

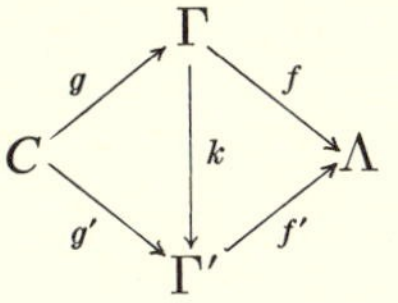

is commutative. The map k is then necessarily an isomorphism.

The set of all equivalence classes of extensions over Λ with kernel C will be denoted by $F(\Lambda,C)$.

From now on we shall assume that Λ is K-projective. Under this assumption the exact sequence $0 \to C \to \Gamma \to \Lambda \to 0$ regarded as a sequence of K-modules, splits. Therefore without loss of generality we may assume that Γ as a K-module coincides with the direct sum

$$\Gamma = C + \Lambda$$

and that

$$gc = (c,0), \qquad f(c,\lambda) = \lambda.$$

The multiplication in Γ has then necessarily the form

$$\begin{cases} (c_1,0)(c_2,0) = 0 \\ (c_1,0)(0,\lambda_2) = (c_1\lambda_2,0) \\ (0,\lambda_1)(c_2,0) = (\lambda_1 c_2,0) \\ (0,\lambda_1)(0,\lambda_2) = (a(\lambda_1,\lambda_2),\, \lambda_1\lambda_2) \qquad a(\lambda_1,\lambda_2) \in C. \end{cases}$$

The first of these relations expresses the fact that $(gc_1)(gc_2) = 0$, the second and third conditions are translations of (2), while the last condition expresses the fact that f is multiplicative. The function a is a K-homomorphism $a\colon \Lambda \otimes \Lambda \to C$ and will be regarded as a 2-cochain in the standard complex $S(\Lambda)$ with coefficients in the two-sided Λ-module C. The multiplication table above may be summarized in the single formula

$$(4) \qquad (c_1,\lambda_1)(c_2,\lambda_2) = (c_1\lambda_2 + \lambda_1 c_2 + a(\lambda_1,\lambda_2),\, \lambda_1\lambda_2).$$

Conversely, consider an arbitrary 2-cochain a, define a multiplication in $\Gamma = C + \Lambda$ by (4) and examine the associativity of this multiplication. Upon calculation we obtain

$$\begin{aligned} (5) \qquad (c_1,\lambda_1)((c_2,\lambda_2)(c_3,\lambda_3)) &- ((c_1,\lambda_1)(c_2,\lambda_2))(c_3,\lambda_3) \\ &= (\lambda_1 a(\lambda_2,\lambda_3) + a(\lambda_1,\lambda_2\lambda_3) - a(\lambda_1\lambda_2,\lambda_3) - a(\lambda_1,\lambda_2)\lambda_3,\, 0) \\ &= (\delta a(\lambda_1,\lambda_2,\lambda_3),0). \end{aligned}$$

This shows that the multiplication (4) is associative if and only if a is a 2-cocycle. If this is the case then

$$\lambda a(1,1) = a(\lambda,1), \qquad a(1,1)\lambda = a(1,\lambda)$$

which implies

$$(-a(1,1),1)(c,\lambda) = (c,\lambda) = (c,\lambda)(-a(1,1),1).$$

It follows that the element $(-a(1,1),1)$ is the unit element of Γ. Then, defining $g\colon C \to \Gamma$ and $f\colon \Gamma \to C$ by $gc = (c,0)$, $f(c,\lambda) = \lambda$, it is immediately verified that (2) holds.

Summarizing, we find that (4) defines a mapping of the group $Z^2(S(\Lambda),C)$ *onto* the set $F(\Lambda,C)$. There remains to find when two cocycles a and a' yield equivalent extensions (F) and (F'). The commutativity of the diagram (3) is equivalent with

$$k(c,\lambda) = (c + b(\lambda),\lambda), \qquad b(\lambda) \in C.$$

Since

$$[k(c_1,\lambda_1)][k(c_2,\lambda_2)] = (c_1\lambda_2 + \lambda_1 c_2 + \lambda_1 b(\lambda_2) + b(\lambda_1)\lambda_2 + a'(\lambda_1,\lambda_2), \lambda_1\lambda_2)$$

$$k[(c_1,\lambda_1)(c_2,\lambda_2)] = (c_1\lambda_2 + \lambda_1 c_2 + a(\lambda_1,\lambda_2) + b(\lambda_1\lambda_2), \lambda_1\lambda_2)$$

we find that k is multiplicative if and only if

$$\lambda_1 b(\lambda_2) + b(\lambda_1)\lambda_2 + a'(\lambda_1,\lambda_2) = a(\lambda_1,\lambda_2) + b(\lambda_1\lambda_2)$$

i.e. if

$$a - a' = \delta b.$$

This condition implies $a(1,1) - a'(1,1) = b(1)$, so that k maps the unit element of Γ into that of Γ'.

Finally let us examine the conditions under which the extension given by a cocycle a is inessential. A K-homomorphism u: $\Lambda \to \Gamma$ such that $fu =$ identity, must have the form

$$u\lambda = (b(\lambda),\lambda) \qquad b(\lambda) \in C.$$

Then

$$(u\lambda_1)(u\lambda_2) - u(\lambda_1\lambda_2) = (b(\lambda_1)\lambda_2 + \lambda_1 b(\lambda_2) + a(\lambda_1,\lambda_2 - b(\lambda_1\lambda_2), \lambda_1\lambda_2),$$

so that u is multiplicative if and only if $a = -\delta b$. If this is the case then $a(1,1) = -b(1)$ so that $u1 = 1$.

We thus obtain

THEOREM 2.1. *The set $F(\Lambda,C)$ of all equivalence classes of extensions over Λ with kernel C is in a* 1–1-*correspondence with the group $H^2(\Lambda,C)$. This correspondence ω: $H^2(\Lambda,C) \to F(\Lambda,C)$ is obtained by assigning to each cocycle $a \in Z^2(S(\Lambda),C)$, the extension given by the multiplication* (4). *The inessential extensions form a single class of $F(\Lambda,C)$ and correspond to the zero element of $H^2(\Lambda,C)$.*

This exposition follows Hochschild (*Ann. of Math.* 46 (1945), 58–67).

3. EXTENSIONS OF SUPPLEMENTED ALGEBRAS

Let Λ be a supplemented K-algebra with ε: $\Lambda \to K$ as augmentation. We shall always assume that Λ is K-projective. If f: $\Gamma \to \Lambda$ is an extension over Λ, then Γ may be regarded as a supplemented K-algebra with εf: $\Gamma \to K$ as augmentation.

Let C be a left Λ-module; we denote by C_ε the two-sided Λ-module obtained from C by defining the right operators by the formula

$$c\lambda = c(\varepsilon\lambda) = (\varepsilon\lambda)c \qquad c \in C, \lambda \in \Lambda. \tag{1}$$

We may then consider the set $F(\Lambda,C_\varepsilon)$ as defined in § 2. We are thus led to consider exact sequences

$$C \xrightarrow{g} \Gamma \xrightarrow{f} \Lambda \tag{F}$$

where Γ is a K-algebra, f is a K-algebra epimorphism, g is a monomorphism of K-modules, and

$$g(\lambda c) = \gamma(gc), \qquad (gc)\gamma = g(c(\varepsilon\lambda)) \qquad c \in C, \gamma \in \Gamma, \lambda = f\gamma \in \Lambda. \tag{2}$$

Such a sequence will be called an *extension over* Λ *with kernel* C. The set of all equivalence classes $F(\Lambda,C_\varepsilon)$ of such extensions will be denoted simply by $F(\Lambda,C)$. The discussion carried out in § 2 now applies without change. In particular, the basic formula (4) may be written as

$$(c_1,\lambda_1)(c_2,\lambda_2) = (c_1(\varepsilon\lambda_2) + \lambda_1 c_2 + a(\lambda_1,\lambda_2), \lambda_1\lambda_2). \tag{3}$$

In the calculation (5) of § 2 the term $a(\lambda_1,\lambda_2)\lambda_3$ is therefore to be replaced by $a(\lambda_1,\lambda_2)(\varepsilon\lambda_3)$. It follows that a is to be regarded as a cocycle in the complex $S(\Lambda,\varepsilon) = S(\Lambda) \otimes_\Lambda K$.

There is one further improvement that can be introduced. Since as a K-module, Λ is the direct sum $K + I(\Lambda)$, it follows that $I(\Lambda)$ is K-projective. Therefore for each extension $C \xrightarrow{g} \Gamma \xrightarrow{f} \Lambda$ the K-homomorphism u: $\Lambda \to \Gamma$ which shows that the sequence $0 \to C \to \Gamma \to \Lambda \to 0$ splits (over K) may be chosen so that $u1 = 1$. Thus the identification of Γ with $C + \Lambda$ may be chosen so that the unit element of Γ corresponds to the element $(0,1)$. This implies that the cocycle a satisfies $a(\lambda,1) = 0 = a(1,\lambda)$. We thus find that a is a cocycle in the *normalized* standard complex $N(\Lambda,\varepsilon)$.

We shall now relate the extensions over supplemented algebras with the extension theory for Λ-modules of § 1. Consider an extension

$$C \xrightarrow{g} \Gamma \xrightarrow{f} \Lambda \tag{F}$$

over Λ with kernel C. Let X denote the set of all $x \in \Gamma$ with $fx \in I(\Lambda)$. Then $g(C) \subset X$ and there results a commutative diagram

$$\begin{array}{ccccccccc}
0 & \longrightarrow & C & \xrightarrow{\psi} & X & \xrightarrow{\varphi} & I(\Lambda) & \longrightarrow & 0 \\
 & & \downarrow j & & \downarrow k & & \downarrow i & & \\
0 & \longrightarrow & C & \xrightarrow[g]{} & \Gamma & \xrightarrow[f]{} & \Lambda & \longrightarrow & 0
\end{array} \tag{4}$$

where k and i are inclusion maps, j is the identity, and ψ and φ are induced by g and f. The top row clearly is exact. Since X is an ideal in Γ we may regard X as a left Γ-module. For each $c \in C$ we have

$$(gc)x = g(c(fx)) = g(c(\varepsilon fx)) = 0.$$

This implies that X may be regarded as a left Λ-module, and that ψ and φ are Λ-homomorphisms. It follows that the top row of the diagram is an extension over $I(\Lambda)$ with kernel C. We have thus obtained a mapping

$$\eta\colon F(\Lambda,C) \to E(I(\Lambda),C). \tag{5}$$

THEOREM 3.1. *If Λ is a K-projective supplemented K-algebra and C is a left Λ-module, then the following diagram is anticommutative*

$$\begin{array}{ccc} F(\Lambda,C) & \xrightarrow{\omega^{-1}} & H^2(\Lambda,C) = \operatorname{Ext}^2_\Lambda(K,C) \\ \eta\downarrow & & \uparrow\vartheta \\ E(I(\Lambda),C) & \xrightarrow[\Theta]{} & \operatorname{Ext}^1_\Lambda(I(\Lambda),C) \end{array}$$

where Θ and ω are the correspondences of 1.1 *and* 2.1, *and ϑ is the connecting homomorphism corresponding to the exact sequence*

$$0 \to I(\Lambda) \to \Lambda \to K \to 0.$$

Since ϑ is an isomorphism, and Θ and ω are 1–1-correspondences, we obtain

COROLLARY 3.2. *The correspondence* (5) *is a* 1–1-*correspondence.*

PROOF of 3.1. To simplify the notation, let N denote the complex $N(\Lambda,\varepsilon)$. The extension (F) will be assumed given in the form $\Gamma = C + \Lambda$, with the multiplication described by a cocycle $a \in \operatorname{Hom}_\Lambda(N_2,C)$. Then X consists of all elements $(c,\lambda - \varepsilon\lambda)$. Let M denote the image of $d_2\colon N_2 \to N_1$. The maps $d_2\colon N_2 \to N_1$ and $d_1\colon N_1 \to N_0 = \Lambda$ then admit factorizations

$$N_2 \xrightarrow{d_2'} M \xrightarrow{i_2} N_1, \qquad N_1 \xrightarrow{d_1'} I(\Lambda) \xrightarrow{i} \Lambda.$$

Consider the commutative diagram

$$\begin{array}{ccccccccc} 0 & \longrightarrow & M & \xrightarrow{i_2} & N_1 & \xrightarrow{d_1'} & I(\Lambda) & \longrightarrow & 0 \\ & & \downarrow w & & \downarrow v & & \updownarrow & & \\ 0 & \longrightarrow & C & \xrightarrow[\psi]{} & X & \xrightarrow[\varphi]{} & I(\Lambda) & \longrightarrow & 0 \end{array} \tag{6}$$

with

$$v[\lambda] = (0,\lambda - \varepsilon\lambda)$$

and with w defined by v. The lower row is the exact sequence $(E) = \eta(F)$. The element $\Theta(E) \in \operatorname{Ext}^1_\Lambda(I(\Lambda),C)$ defined in § 1, is the image of the identity element $j \in \operatorname{Hom}_\Lambda(C,C)$ under the connecting homomorphism corresponding to the lower row of (6). It then follows from the commutativity in (6) that $\Theta(E)$ is the image of $w \in \operatorname{Hom}_\Lambda(M,C)$ under the connecting homomorphism corresponding to the upper row of (6). It follows that the composition $\vartheta\Theta(E)$ is the image of w under the iterated connecting homomorphism

$$\delta\colon \operatorname{Hom}_\Lambda(M,C) \to \operatorname{Ext}^2_\Lambda(K,C)$$

corresponding to the exact sequence

$$0 \to M \to N_1 \to \Lambda \to K \to 0.$$

We are now in the situation described in v,7.1. The element $\delta w = \vartheta\Theta(E)$ is thus the negative of the cohomology class given by the 2-cocycle

$$N_2 \xrightarrow{d_2'} M \xrightarrow{w} C.$$

To complete the proof it suffices to show that $wd_2' = a$. Since ψ is a monomorphism it suffices to show that $\psi wd_2' = \psi a$. Since $\psi wd_2' = vi_2d_2' = vd$, it suffices to show that $vd = \psi a$. We have

$$\begin{aligned} vd\{\lambda_1,\lambda_2\} &= v[\lambda_1\{\lambda_2\} - \{\lambda_1\lambda_2\} + \{\lambda_1\}\varepsilon\lambda_2] \\ &= (0,\lambda_1)(0,\lambda_2 - \varepsilon\lambda_2) - (0,\lambda_1\lambda_2 - \varepsilon\lambda_1\varepsilon\lambda_2) + (0,\lambda_1 - \varepsilon\lambda_1)(0,\varepsilon\lambda_2) \\ &= (a(\lambda_1,\lambda_2),\lambda_1\lambda_2 - \lambda_1\varepsilon\lambda_2) - (0,\lambda_1\lambda_2 - \varepsilon\lambda_1\varepsilon\lambda_2) \\ &\quad + (0,\lambda_1\varepsilon\lambda_2 - \varepsilon\lambda_1\varepsilon\lambda_2) \\ &= (a(\lambda_1,\lambda_2),0) = \psi a(\lambda_1,\lambda_2). \end{aligned}$$

This concludes the proof.

PROPOSITION 3.3. *The composite map*

$$E(I(\Lambda),C) \xrightarrow{\Theta} \operatorname{Ext}^1_\Lambda(I(\Lambda),C) \xrightarrow{\vartheta} \operatorname{Ext}^2_\Lambda(K,C) = H^2(\Lambda,C)$$

may be described as follows. For each extension

$$(E) \qquad 0 \to C \to X \to I(\Lambda) \to 0$$

the element $\vartheta\Theta(E)$ is the characteristic element $\delta_S j$ (see XI,9) *of the exact sequence*

$$(S) \qquad 0 \to C \to X \to \Lambda \to K \to 0$$

obtained by joining (E) with the exact sequence

$$(L) \qquad 0 \to I(\Lambda) \to \Lambda \to K \to 0.$$

PROOF. We recall that $j \in \mathrm{Hom}_\Lambda(C,C)$ is the identity map $C \to C$ and that δ_S is the iterated connecting homomorphism. Thus

$$\delta_S j = \delta_L \delta_E j = \delta_L \Theta(E) = \vartheta\Theta(E),$$

by the definition of the map Θ (§ 1).

4. EXTENSIONS OF GROUPS

Consider two groups W and Π (not necessarily commutative, and written multiplicatively). An epimorphism

$$f\colon\ W \to \Pi$$

will be called an *extension* over Π. The extension is called *inessential* if there exists a group homomorphism $u\colon\ \Pi \to W$ with $fu =$ identity.

The kernel of f (i.e. the set of all $w \in W$ with $fw = 1$) is an invariant subgroup C of W. The mapping $(w,c) \to wcw^{-1}$ of $W \times C$ into C defines operators of W on C. If C is abelian, then C (written additively) is a left W-module, and since the elements of $C \subset W$ operate trivially, we find that C is a left Π-module satisfying

$$(1) \qquad xc = wcw^{-1} \qquad c \in C,\ w \in W,\ x = fw \in \Pi.$$

Conversely, if Π operates on C so that (1) holds, C is necessarily abelian.

DEFINITION. Let Π be a (multiplicative) group and C a left Π-module. An *extension over* Π *with kernel* C is a sequence

$$(\Sigma) \qquad C \xrightarrow{g} W \xrightarrow{f} \Pi$$

where W is a (multiplicative) group, f is a group epimorphism, g is a monomorphism of the additive structure of C into the multiplicative structure of W, the image of g is the kernel of f, and

$$(2) \qquad g(xc) = w(gc)w^{-1} \qquad c \in C,\ w \in W,\ x = fw \in \Pi.$$

Two extensions (Σ) and (Σ') over Π with kernel C are *equivalent* if there exists a group homomorphism $k\colon\ W \to W'$ such that the diagram

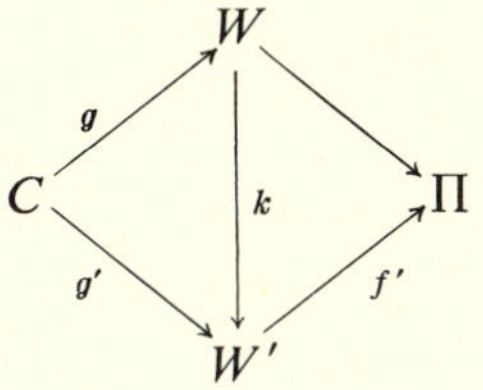

is commutative; k is then necessarily an isomorphism.

The set of all equivalence classes of extensions over Π with kernel C will be denoted by $\Sigma(\Pi, C)$.

With the group Π and the left Π-module C given, consider the supplemented algebra $Z(\Pi)$ (with the unit augmentation). We shall consider the set $F(Z(\Pi), C)$ of equivalence classes of extensions over $Z(\Pi)$ with kernel C. Such an extension is an exact sequence

$$(F) \qquad C \xrightarrow{\bar{g}} \Gamma \xrightarrow{\bar{f}} Z(\Pi)$$

where Γ is a Z-algebra, $\bar{f}$ is an epimorphism of algebras, $\bar{g}$ is a monomorphism of abelian groups, and

$$(3) \qquad \gamma(\bar{g}c) = \bar{g}(\lambda c), \quad (\bar{g}c)\gamma = \bar{g}(c(\varepsilon\lambda)), \qquad c \in C, \gamma \in \Gamma, \lambda = \bar{f}\gamma \in Z(\Pi).$$

PROPOSITION 4.1. *Let (F) be an extension as above. The set W of elements* $w \in \Gamma$ *with* $\bar{f}w \in \Pi$ *is a group under the multiplication defined by that of the ring* Γ. *If f denotes the map* $W \to \Pi$ *induced by* $\bar{f}$ *and* $g\colon C \to W$ *is given by* $gc = \bar{g}c + 1$, *then the sequence*

$$(\Sigma) \qquad C \xrightarrow{g} W \xrightarrow{f} \Pi$$

is an extension over Π *with kernel C.*

PROOF. Clearly W is closed under multiplication, is associative and has a unit element, namely the element $1 \in \Gamma$. To show that $w \in W$ has an inverse, choose $v \in W$ such that $f(w)f(v) = 1 = f(v)f(w)$. Since the elements $1 - wv$ and $1 - vw$ yield zero in $Z(\Pi)$, there exist elements $c_1, c_2 \in C$ with $\bar{g}c_i = 1 - wv$, $\bar{g}c_2 = 1 - vw$. The second of the relations (3) then yields $(1 - wv)w = 1 - wv$, $(1 - vw)w = 1 - vw$. This implies

$$w(1 - v(w - 1)) = (1 - wv)w + wv = 1 - wv + wv = 1$$

$$(1 - v(w - 1))w = (1 - vw)w + vw = 1 - vw + vw = 1$$

which shows that w has an inverse. Thus W is a group.

It is clear that $f\colon W \to \Pi$ is a group epimorphism whose kernel is the image of g. Since

$$\begin{aligned}(gc_1)(gc_2) = (\bar{g}c_1 + 1)(\bar{g}c_2 + 1) &= \bar{g}c_1 + \bar{g}c_2 + 1 \\ &= \bar{g}(c_1 + c_2) + 1 = g(c_1 + c_2)\end{aligned}$$

it follows that g is a homomorphism of the additive structure of C into the multiplicative structure of W. Finally for $c \in C$, $w \in W$, $x = fw$ we have, using (3)

$$w(gc)w^{-1} = w(\bar{g}c)w^{-1} + 1 = \bar{g}(xc)w^{-1} + 1 = \bar{g}(xc) + 1 = g(xc).$$

This proves (2) and concludes the proof.

Proposition 4.1 assigns to each extension (F) an extension (Σ). Clearly if (F) and (F') are equivalent, then so are the corresponding extensions (Σ) and (Σ'). There results a mapping

$$\varphi\colon\ F(Z(\Pi),C) \to \Sigma(\Pi,C).$$

THEOREM 4.2. *The mapping φ establishes a* 1–1-*correspondence between the set $F(Z(\Pi),C)$ and the set $\Sigma(\Pi,C)$. The inessential extensions (Σ) form one equivalence class, corresponding to the inessential class in $F(Z(\Pi),C)$.*

PROOF. We first show that if two extensions (F) and (F') yield two equivalent extensions $\varphi(F)$ and $\varphi(F')$, then (F) and (F') are equivalent. To do this, we shall give a complete description (up to an equivalence) of any extension (F) using the extension $(\Sigma) = \varphi(F)$.

We have the commutative diagram

$$\begin{array}{ccccc} C & \xrightarrow{g} & W & \xrightarrow{f} & \Pi \\ \downarrow & & \downarrow{\scriptstyle j} & & \downarrow{\scriptstyle i_\Pi} \\ C & \xrightarrow{\bar g} & \Gamma & \xrightarrow{\bar f} & Z(\Pi) \end{array}$$

with $jw = w - 1$, $i_\Pi x = x - 1$. The map j may be factored as follows

$$W \xrightarrow{i_W} Z(W) \xrightarrow{k} \Gamma$$

where $i_W(w) = w - 1$ and k is a homomorphism of Z-algebras defined by $k(w) = w$. We obtain a commutative diagram

$$\begin{array}{ccccc} C & \xrightarrow{g} & W & \xrightarrow{f} & \Pi \\ \updownarrow & & \downarrow{\scriptstyle i_W} & & \downarrow{\scriptstyle i_\Pi} \\ C & \xrightarrow{i_W g} & Z(W) & \xrightarrow{f^*} & Z(\Pi) \\ \updownarrow & & \downarrow{\scriptstyle k} & & \updownarrow \\ C & \xrightarrow{\bar g} & \Gamma & \xrightarrow{\bar f} & Z(\Pi) \end{array}$$

Let $\bar C = i_W gC$ and let $\bar C \,.\, Z(W)$ (resp. $\bar C \,.\, I(W)$) denote the set of all linear combinations of elements $\bar c w$ (resp. $\bar c(w - 1)$) for $\bar c \in \bar C$, $w \in W$. Then $\bar C \,.\, Z(W)$ is precisely the kernel of f^*, while $\bar C \,.\, I(W)$ is, by virtue of the second relation (3), in the kernel of k. There results a commutative diagram

$$(4)\qquad \begin{array}{ccccccccc} 0 & \longrightarrow & C & \xrightarrow{l} & Z(W)/\bar C \,.\, I(W) & \xrightarrow{f'} & Z(\Pi) & \longrightarrow & 0 \\ & & \updownarrow & & \downarrow{\scriptstyle k'} & & \updownarrow & & \\ 0 & \longrightarrow & C & \xrightarrow{\bar g} & \Gamma & \xrightarrow{\bar f} & Z(\Pi) & \longrightarrow & 0 \end{array}$$

We can now show that l is additive. Indeed we have

$$\begin{aligned} i_W g(c_1+c_2) &= (gc_1)(gc_2) - 1 = (gc_1 - 1) + (gc_2 - 1) + (gc_1 - 1)(gc_2 - 1) \\ &= i_W gc_1 + i_W gc_2 + (i_W gc_1)(gc_2 - 1). \end{aligned}$$

Since the last term is in $\bar{C} \,.\, I(W)$, we obtain that $l(c_1+c_2)=lc_1+lc_2$. The kernel of f' is $\bar{C} \,.\, Z(W)/\bar{C} \,.\, I(W)$ and this is precisely the image of l. Finally l is a monomorphism since $\bar{g} = k'l$ is one. Thus the top row is an exact sequence of Z-modules. Since the lower row is exact by hypothesis, k' is an isomorphism.

This shows that $\bar{C} \,.\, I(W)$ is the kernel of k which was a homomorphism of Z-algebras. Thus $\bar{C} \,.\, I(W)$ is a two-sided ideal of $Z(W)$ (this could be seen directly). Consequently k' is an isomorphism of Z-algebras.

We now see that the top row of (4) is an extension (F_0) described entirely in terms of the extension (Σ) and equivalent with (F). This proves our assertion.

We now show that (F) is inessential if and only if $(\Sigma) = \varphi(F)$ is inessential. Suppose (F) is inessential and let $\bar{u} : Z(\Pi) \to \Gamma$ be a homomorphism of Z-algebras such that $\bar{f}\bar{u} =$ identity. The induced group homomorphism u: $\Pi \to W$ then satisfies $fu =$ identity. Conversely given a group homomorphism u: $\Pi \to W$ with $fu =$ identity, the homomorphism ku^*: $Z(\Pi) \to \Gamma$ shows that (F) is inessential.

There remains the proof that φ maps $F(Z(\Pi),C)$ *onto* $\Sigma(\Pi,C)$. Given an extension (Σ) over Π with kernel C, choose a function u: $\Pi \to W$ such that $fu =$ identity and $u(1) = 1$. Then each element of W may be written uniquely as a product $(gc)(ux)$ with $c \in C$, $x \in \Pi$; we shall denote this element by (c,x). Then the unit element of W is $(0,1)$ and

$$gc = (c,1), \qquad f(c,x) = x.$$

Let us find the product of two elements (c_1,x_1) and (c_2,x_2). Using (2) we have

$$\begin{aligned} (c_1,x_1)(c_2,x_2) &= (gc_1)(ux_1)(gc_2)(ux_2) \\ &= (gc_1)(ux_1)(gc_2)(ux_1)^{-1}(ux_1)(ux_2) \\ &= (gc_1)g(x_1c_2)(ux_1)(ux_2). \end{aligned}$$

To calculate $(ux_1)(ux_2)$ we observe that this element has the same image in Π as $u(x_1x_2)$. Thus there is a unique element $a(x_1,x_2) \in C$ such that

$$(ux_1)(ux_2) = ga(x_1,x_2)u(x_1x_2).$$

Since g maps the additive structure of C into the multiplicative structure of W, we have the final result

$$(5) \qquad (c_1,x_1)(c_2,x_2) = (c_1 + x_1c_2 + a(x_1,x_2), x_1x_2)$$

Because of the choice $u(1) = 1$, we have $a(x_1,1) = 0 = a(1,x_2)$. We shall treat a as a 2-cochain on the normalized complex $N(\Pi)$ with coefficients in C.

Let us express the fact that the multiplication given by formula (5) is associative. We have

$$(c_1x_1)((c_2,x_2)(c_3,x_3)) = (c_1 + x_1c_2 + x_1x_2c_3 + x_1a(x_2,x_3) + a(x_1,x_2x_3), x_1x_2x_3)$$

$$((c_1,x_1)(c_2,x_2))(c_3,x_3) = (c_1 + x_1c_2 + x_1x_2c_3 + a(x_1,x_2) + a(x_1x_2,x_3), x_1x_2x_3).$$

In order that the two results coincide it is necessary and sufficient that

$$x_1a(x_2,x_3) - a(x_1x_2,x_3) + a(x_1,x_2x_3) - a(x_1,x_2) = 0$$

i.e. that $\delta a(x_1,x_2,x_3) = 0$. We thus find that a is a 2-cocycle of the complex $N(\Pi) = N(Z(\Pi),\varepsilon)$.

We now use this cocycle a to construct an extension

$$(F) \qquad C \xrightarrow{\bar{g}} \Gamma \xrightarrow{\bar{f}} Z(\Pi)$$

with $\Gamma = C + Z(\Pi)$, $\bar{g}c = (c,0)$, $\bar{f}(c,\lambda) = \lambda$, and with multiplication given by formula (3) of § 3 as

$$(c_1,\lambda_1)(c_2,\lambda_2) = (c_1(\varepsilon\lambda_2) + \lambda_1c_2 + a(\lambda_1,\lambda_2), \lambda_1\lambda_2).$$

It is clear that if we apply φ to this extension (F) we find exactly the extension (Σ) with which we started, with multiplication given by (5). This concludes the proof of 4.2.

REMARK. Our results so far may be summarized in the following diagram

$$\begin{array}{ccccc} & & H^2(\Pi,C) & & \\ & \swarrow^{\omega} & & \nwarrow^{\vartheta} & \\ \Sigma(\Pi,C) \xleftarrow{\varphi} F(Z(\Pi),C) & & & & \operatorname{Ext}^1_{Z(\Pi)}(I(\Pi),C) \\ & \searrow_{\eta} & & \nearrow_{\Theta} & \\ & & E(I(\Pi),C) & & \end{array}$$

in which all the maps are 1-1 correspondences and the square is commutative (in the sense that the composition of any four consecutive maps is the identity). The preceding proof shows that the map $\varphi\omega$: $H^2(\Pi,C) \to \Sigma(\Pi,C)$ is obtained by assigning to each cocycle $a \in Z^2(N(\Pi),C)$ the extension (Σ) given by formula (5). This is the familiar method of describing group extensions by means of "factor sets." The composition $\vartheta\Theta$ was described in 3.3. For a discussion of the composition

$$\varphi\eta^{-1}: E(I(\Pi),C) \to \Sigma(\Pi,C),$$

see Exer. 3.

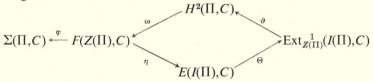

5. EXTENSIONS OF LIE ALGEBRAS

Let $\mathfrak{h}$ and $\mathfrak{g}$ be Lie algebras over a commutative ring K. An epimorphism of Lie algebras

$$f\colon \mathfrak{h} \to \mathfrak{g}$$

will be called an *extension* over $\mathfrak{g}$. The extension is called *inessential* if there exists a Lie algebra homomorphism $u\colon \mathfrak{g} \to \mathfrak{h}$ with $fu =$ identity.

The kernel C of f is an ideal in $\mathfrak{h}$. The mapping $y \otimes c \to [y,c]$ of $\mathfrak{h} \otimes C$ into C defines C as a left $\mathfrak{h}$-module, because of Jacobi's identity

$$[y_1,[y_2,c]] - [y_2,[y_1,c]] = [[y_1,y_2],c].$$

If the Lie algebra C is abelian, then the structure of C as a left $\mathfrak{h}$-module induces on C the structure of a left $\mathfrak{g}$-module:

$$xc = [y,c], \qquad c \in C,\ y \in \mathfrak{h},\ x = fy \in \mathfrak{g}. \tag{1}$$

Conversely, if C carries the structure of a left $\mathfrak{g}$-module satisfying (1), then C is an abelian Lie algebra.

DEFINITION: Let $\mathfrak{g}$ be a K-Lie algebra and C a left $\mathfrak{g}$-module. An *extension over* $\mathfrak{g}$ *with kernel* C is an exact sequence

$$C \xrightarrow{g} \mathfrak{h} \xrightarrow{f} \mathfrak{g} \tag{Σ}$$

where $\mathfrak{h}$ is a K-Lie algebra, f is a Lie algebra epimorphism, g is a monomorphism of K-modules, and

$$g(xc) = [y,gc] \qquad \text{for } c \in C,\ y \in \mathfrak{h},\ x = fy \in \mathfrak{g}. \tag{2}$$

Two extensions (Σ), (Σ') over $\mathfrak{g}$ with kernel C are *equivalent* if there exists a K-Lie algebra homomorphism $k\colon \mathfrak{h} \to \mathfrak{h}'$ such that the diagram

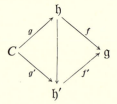

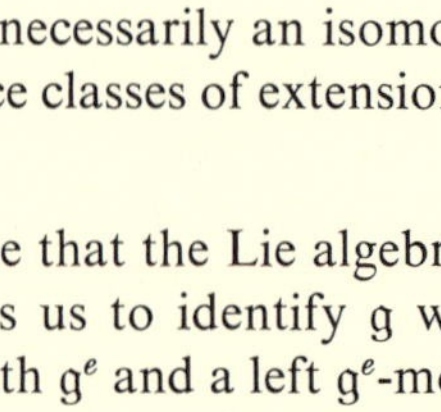

is commutative; k is then necessarily an isomorphism.

The set of all equivalence classes of extensions over $\mathfrak{g}$ with kernel C will be denoted by $\Sigma(\mathfrak{g},C)$.

We shall always suppose that the Lie algebra $\mathfrak{g}$ is K-free. As we have seen in XIII,3.3 this permits us to identify $\mathfrak{g}$ with a K-submodule of the enveloping algebra $\mathfrak{g}^e$. With $\mathfrak{g}^e$ and a left $\mathfrak{g}^e$-module C given, consider the set $F(\mathfrak{g}^e,C)$ of equivalence classes of extensions over $\mathfrak{g}^e$ with kernel C.

Such an extension is an exact sequence

$$(F) \qquad C \xrightarrow{\bar{g}} \Gamma \xrightarrow{\bar{f}} \mathfrak{g}^e$$

where Γ is a K-algebra, $\bar{f}$ is an epimorphism of K-algebras, $\bar{g}$ is a monomorphism of K-modules, and

$$(3) \qquad \gamma(\bar{g}c) = \bar{g}(\lambda c), \quad (\bar{g}c)\gamma = \bar{g}(c(\varepsilon\lambda)) \qquad c \in C, \gamma \in \Gamma, \lambda = \bar{f}\gamma \in \mathfrak{g}^e.$$

PROPOSITION 5.1. *Let* (F) *be an extension as above. The set* $\mathfrak{h}$ *of elements* $y \in \Gamma$ *with* $\bar{f}y \in \mathfrak{g}$ *is a Lie algebra over* K *with the bracket operation* $[y_1,y_2] = y_1y_2 - y_2y_1$. *If* f *denotes the map* $f\colon \mathfrak{h} \to \mathfrak{g}$ *induced by* $\bar{f}$, *and* $g\colon C \to \mathfrak{h}$ *is defined by* $\bar{g}$, *then the sequence*

$$(\Sigma) \qquad C \xrightarrow{g} \mathfrak{h} \xrightarrow{f} \mathfrak{g}$$

is an extension over $\mathfrak{g}$ *with kernel* C.

PROOF. If $\bar{f}y_1 = x_1$, $\bar{f}y_2 = x_2$ for $x_1,x_2 \in \mathfrak{g}$, then $\bar{f}(y_1y_2 - y_2y_1) = x_1x_2 - x_2x_1$; this last term is equal to $[x_1,x_2]$ because of the relations in the enveloping algebra $\mathfrak{g}^e$. Thus $y_1y_2 - y_2y_1 \in \mathfrak{h}$, so that $\mathfrak{h}$ is closed under the bracket operation. Thus $\mathfrak{h}$ is a K-Lie algebra. Since $\bar{f}\bar{g} = 0$, it follows that g maps C into $\mathfrak{h}$ and its image is the kernel of f. Relations (3) with $\gamma = y$, $\lambda = x$ give $y(\bar{g}c) = \bar{g}(xc)$ and $(\bar{g}c)y = 0$, which implies $[y,gc] = y(\bar{g}c) - (\bar{g}c)y = g(xc)$. This proves (2) and concludes the proof.

Proposition 5.1 assigns to each extension (F) an extension (Σ). Clearly if (F) and (F') are equivalent, then so are the corresponding extensions (Σ) and (Σ'). There results a mapping

$$\varphi\colon F(\mathfrak{g}^e,C) \to \Sigma(\mathfrak{g},C)$$

defined whenever $\mathfrak{g}$ is K-free.

THEOREM 5.2. *The mapping* φ *establishes a* 1–1-*correspondence between the set* $F(\mathfrak{g}^e,C)$ *and the set* $\Sigma(\mathfrak{g},C)$. *The inessential extensions* (Σ) *form one equivalence class, corresponding to the inessential class in* $F(\mathfrak{g}^e,C)$.

PROOF. We first show that if two extensions (F) and (F') yield two equivalent extensions $\varphi(F)$ and $\varphi(F')$, then (F) and (F') are equivalent. To do this, we shall give a complete description (up to an equivalence) of any extension (F) using the extension $(\Sigma) = \varphi(F)$.

We have the commutative diagram

$$\begin{array}{ccccc} C & \xrightarrow{g} & \mathfrak{h} & \xrightarrow{f} & \mathfrak{g} \\ \updownarrow & & \downarrow j & & \downarrow i_{\mathfrak{g}} \\ C & \xrightarrow{\bar{g}} & \Gamma & \xrightarrow{\bar{f}} & \mathfrak{g}^e \end{array}$$

where j is given by inclusion and $i_{\mathfrak{g}}$ is the natural injection of $\mathfrak{g}$ into $\mathfrak{g}^e$. The map j may be factored as follows

$$\mathfrak{h} \xrightarrow{a} T(\mathfrak{h}) \xrightarrow{b} \Gamma$$

where a is the natural injection of $\mathfrak{h}$ into its tensor algebra and b is a K-algebra homomorphism. Since b is zero on all elements of the form $y_1 \otimes y_2 - y_2 \otimes y_1 - [y_1,y_2]$, we find a factorization

$$\mathfrak{h} \xrightarrow{i_{\mathfrak{h}}} \mathfrak{h}^e \xrightarrow{k} \Gamma$$

of j, where $i_{\mathfrak{h}}$ is the natural injection and k is a K-algebra homomorphism. There results a commutative diagram

$$\begin{array}{ccccc}
C & \xrightarrow{g} & \mathfrak{h} & \xrightarrow{f} & \mathfrak{g} \\
\updownarrow & & \downarrow{\scriptstyle i_{\mathfrak{h}}} & & \downarrow \\
C & \xrightarrow{i_{\mathfrak{h}}g} & \mathfrak{h}^e & \xrightarrow{f^e} & \mathfrak{g}^e \\
\updownarrow & & \downarrow{\scriptstyle k} & & \updownarrow \\
C & \xrightarrow{\bar{g}} & \Gamma & \xrightarrow{\bar{f}} & \mathfrak{g}^e
\end{array}$$

Let $\bar{C} = i_{\mathfrak{h}}gC$. The kernel of f^e is then, by XIII,1.3, the ideal $\bar{C} \,.\, \mathfrak{h}^e$. We shall also consider the K-module $\bar{C} \,.\, I(\mathfrak{h})$ where $I(\mathfrak{h})$ is the augmentation ideal of $\mathfrak{h}^e$. In virtue of the second relation (3), $\bar{C} \,.\, I(\mathfrak{h})$ is in the kernel of k. There results a commutative diagram

$$\begin{array}{ccccccccc}
0 & \longrightarrow & C & \xrightarrow{l} & \mathfrak{h}^e/\bar{C} \,.\, I(\mathfrak{h}) & \xrightarrow{f'} & \mathfrak{g}^e & \longrightarrow & 0 \\
 & & \updownarrow & & \downarrow{\scriptstyle k'} & & \updownarrow & & \\
0 & \longrightarrow & C & \xrightarrow{\bar{g}} & \Gamma & \xrightarrow{\bar{f}} & \mathfrak{g}^e & \longrightarrow & 0
\end{array} \tag{4}$$

The kernel of f' is $\bar{C} \,.\, \mathfrak{h}^e/\bar{C} \,.\, I(\mathfrak{h})$ which is precisely the image of l. Finally l is a monomorphism since $\bar{g} = k'\,l$ is one. Thus the top row is an exact sequence of K-modules. Since the lower row also is exact, this implies that k' is an isomorphism.

This shows that $\bar{C} \,.\, I(\mathfrak{h})$ is the kernel of k which was a homomorphism of K-algebras. Thus $\bar{C} \,.\, I(\mathfrak{h})$ is a two-sided ideal of $\mathfrak{h}^e$ (this could be seen directly). Consequently k' is an isomorphism of K-algebras.

We now see that the top row of (4) is an extension (F_0) described entirely in terms of the extension (Σ) and equivalent with (F). This proves our assertion.

We now show that (F) is inessential if and only if $(\Sigma) = \varphi(F)$ is inessential. Suppose (F) is inessential and let $\bar{u} : \mathfrak{g}^e \to \Gamma$ be a homomorphism of

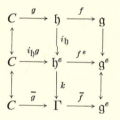

K-algebras such that $\bar{f}\bar{u} =$ identity. The induced Lie algebra homomorphism u: $\mathfrak{g} \to \mathfrak{h}$ then satisfies $fu =$ identity. Conversely given a Lie algebra homomorphism u: $\mathfrak{g} \to \mathfrak{h}$ with $fu =$ identity, the K-algebra homomorphism ku^e: $\mathfrak{g}^e \to \Gamma$ shows that (F) is inessential.

There remains the proof that φ maps $F(\mathfrak{g}^e,C)$ *onto* $\Sigma(\mathfrak{g},C)$. Consider an extension (Σ) over $\mathfrak{g}$ with kernel C. Since $\mathfrak{g}$ is K-free we may assume that, as a K-module, $\mathfrak{h}$ is the direct sum $C + \mathfrak{g}$ with

$$gc = (c,0), \qquad f(c,x) = x.$$

The bracket in $\mathfrak{h}$ then necessarily has the form

$$[(c_1,0),(c_2,0)] = 0$$

$$[(c_1,0),(0,x_2)] = (-x_2c_1,0)$$

$$[(0,x_1),(c_2,0)] = (x_1c_2,0)$$

$$[(0,x_1),(0,x_2)] = (a(x_1,x_2),[x_1,x_2]), \qquad a(x_1,x_2) \in C.$$

The first relation expresses the fact that C is an abelian Lie algebra, the second and the third conditions follow from (2), while the last one expresses the fact that f is a Lie algebra homomorphism. Combining these relations we obtain

$$(5) \qquad [(c_1,x_1),(c_2,x_2)] = (-x_2c_1 + x_1c_2 + a(x_1,x_2),[x_1,x_2]).$$

Let us now express the conditions that the bracket operation in $C + \mathfrak{g}$ given by (5) satisfies the axioms of a Lie algebra. The K-bilinearity of the bracket implies the K-bilinearity of $a(x_1,x_2)$. The condition $[(c,x),(c,x)] = 0$ is equivalent with $a(x,x) = 0$. Thus $a(x_1,x_2)$ is an alternating bilinear function, and we may regard $a(x_1,x_2)$ as a 2-cochain of the complex $V(\mathfrak{g})$ of XIII,7. Let us now express the Jacobi identity. We have

$$[(c_1,x_1),[(c_2x_2),(c_3,x_3)]]$$
$$=(-[x_2,x_3]c_1 - x_1x_3c_2 + x_1x_2c_3 + x_1a(x_2,x_3) + a(x_1,[x_2x_3]),[x_1,[x_2,x_3]]).$$

Permuting cyclically and adding we find that the sum is zero if and only if

$$x_1a(x_2,x_3) + x_2a(x_3,x_1) + x_3a(x_1,x_2) + a(x_1,[x_2,x_3])$$
$$+ a(x_2,[x_3,x_1]) + a(x_3,[x_1,x_2]) = 0.$$

Since a is alternating, this is equivalent with $\delta a(x_1,x_2,x_3) = 0$. Thus a is a 2-cocycle.

By XIII,8.1, there exists a cocycle $\bar{a}$ of the normalized standard complex $N(\mathfrak{g}^e,\varepsilon)$ which induces a on the subcomplex $V(\mathfrak{g})$. With this cocycle $\bar{a}$ let us construct the extension

$$(F) \qquad C \xrightarrow{\bar{g}} \Gamma \xrightarrow{\bar{f}} \mathfrak{g}^e$$

with $\Gamma = C + \mathfrak{g}^e$, $\bar{g}c = (c,0)$, $\bar{f}(c,x) = x$, and with multiplication given by the formula (3) of § 3 as

$$(c_1,\lambda_1)(c_2,\lambda_2) = (c_1(\varepsilon\lambda_2) + \lambda_1 c_2 + \bar{a}(\lambda_1,\lambda_2), \lambda_1\lambda_2).$$

In particular, for $x_1,x_2 \in \mathfrak{g}$ we obtain

$$(c_1,x_1)(c_2,x_2) = (x_1c_2 + \bar{a}(x_1,x_2), x_1x_2).$$

If we compute the bracket of (c_1,x_1) and (c_2,x_2) and recall that $a(x_1,x_2) = \bar{a}(x_1,x_2) - \bar{a}(x_2,x_1)$, we find precisely formula (5). This shows that if we apply φ to the extension (F) we find exactly the extension (Σ) with which we started. This concludes the proof of 5.2.

REMARK. The remark at the end of § 4 may be repeated here with Π and $N(\Pi)$ replaced by $\mathfrak{g}$ and $V(\mathfrak{g})$.

EXERCISES

1. Consider an extension of Λ-modules

$$(E) \qquad 0 \to C \to X \to A \to 0$$

and the connecting homomorphisms

$$\Theta_E\colon \operatorname{Hom}(C,C) \to \operatorname{Ext}^1(A,C)$$

$$\Theta'_E\colon \operatorname{Hom}(A,A) \to \operatorname{Ext}^1(A,C)$$

Show that $\Theta'_E i + \Theta_E j = 0$ where $i \in \operatorname{Hom}(A,A)$ and $j \in \operatorname{Hom}(C,C)$ are identity maps. [Hint: use Exer. VI,18.]

2. Given a K-algebra Λ and a two-sided Λ-module C, introduce a Baer multiplication in the set $F(\Lambda,C)$ of § 2. Show that the correspondence established in 2.1 carries the Baer multiplication in $F(\Lambda,C)$ into the addition in $H^2(\Lambda,C)$. Carry a similar discussion for the sets $\Sigma(\Pi,C)$ and $\Sigma(\mathfrak{g},C)$ of §§ 4 and 5.

3. Let Π be a group and C a left Π-module. Consider a group extension

$$(\Sigma) \qquad C \xrightarrow{g} W \xrightarrow{f} \Pi$$

and an extension of left $Z(\Pi)$-modules

$$(E) \qquad 0 \longrightarrow C \xrightarrow{\beta} X \xrightarrow{\alpha} I(\Pi) \longrightarrow 0.$$

We shall say that (Σ) and (E) are *related* if there is a map $k\colon W \to X$ satisfying

$$k(w_1 w_2) = kw_1 + (fw_1)(kw_2)$$

and such that the following diagram is commutative

$$\begin{array}{ccccc} C & \xrightarrow{g} & W & \xrightarrow{f} & \Pi \\ \updownarrow & & \downarrow{\scriptstyle k} & & \downarrow{\scriptstyle i_\Pi} \\ C & \xrightarrow{\beta} & X & \xrightarrow{\alpha} & I(\Pi) \end{array}$$

where $i_\Pi(x) = x - 1$ $(x \in \Pi)$. Show that (Σ) and (E) are related if and only if (Σ) and (E) correspond to the same class of $F(Z(\Pi),C)$ under the correspondences

$$\eta\colon\ F(Z(\Pi),C) \to E(I(\Pi),C)$$

$$\varphi\colon\ F(Z(\Pi),C) \to \Sigma(\Pi,C)$$

established in §§ 3 and 4, i.e. if and only if $(\Sigma) = \varphi\eta^{-1}(E)$.

4. Let $\mathfrak{g}$ be a K-Lie algebra which is K-free, and C a left $\mathfrak{g}$-module. Consider a Lie algebra extension

$$(\Sigma) \qquad C \xrightarrow{g} \mathfrak{h} \xrightarrow{f} \mathfrak{g}$$

and an extension of left $\mathfrak{g}^e$-modules

$$(E) \qquad 0 \longrightarrow C \xrightarrow{\beta} X \xrightarrow{\alpha} I(\mathfrak{g}) \longrightarrow 0.$$

We shall say that (Σ) and (E) are *related* if there is a map $k\colon \mathfrak{h} \to X$ satisfying

$$k([y_1,y_2]) = (fy_1)(ky_2) - (fy_2)(ky_1)$$

and such that the following diagram is commutative:

$$\begin{array}{ccccc} C & \xrightarrow{g} & \mathfrak{h} & \xrightarrow{f} & \mathfrak{g} \\ \updownarrow & & \downarrow{\scriptstyle k} & & \downarrow{\scriptstyle i_\mathfrak{g}} \\ C & \xrightarrow{\beta} & X & \xrightarrow{\alpha} & I(\mathfrak{g}) \end{array}$$

where $i_\mathfrak{g}$ denotes the natural map. Show that (Σ) and (E) are related if and only if (Σ) and (E) correspond to the same class of $F(\mathfrak{g}^e,C)$ under the correspondences

$$\eta\colon\ F(\mathfrak{g}^e,C) \to E(I(\mathfrak{g}),C)$$

$$\varphi\colon\ F(\mathfrak{g}^e,C) \to \Sigma(\mathfrak{g},C)$$

established in §§ 3 and 5, i.e. if and only if $(\Sigma) = \varphi\eta^{-1}(E)$.

5. Let

$$(E) \qquad 0 \longrightarrow C \xrightarrow{\psi} X \xrightarrow{\varphi} A \longrightarrow 0$$

be an exact sequence of (left) Λ-modules, where Λ is a K-algebra. Assume that A is K-projective. Then $\Lambda \otimes_K A$ is a projective Λ-module; let N be the kernel of $p\colon\ \Lambda \otimes A \to A$ defined by setting

$$p(\lambda \otimes a) = \lambda a.$$

Choose a K-homomorphism $u\colon\ A \to X$ such that $\varphi u =$ identity, and consider the commutative diagram

$$\begin{array}{ccccccccc}
0 & \longrightarrow & N & \longrightarrow & \Lambda \otimes A & \xrightarrow{p} & A & \longrightarrow & 0 \\
 & & \downarrow{\scriptstyle v} & & \downarrow{\scriptstyle f} & & \downarrow{\scriptstyle i} & & \\
0 & \longrightarrow & C & \xrightarrow{\psi} & X & \xrightarrow{\varphi} & A & \longrightarrow & 0
\end{array}$$

with exact rows, f being defined by setting

$$f(\lambda \otimes a) = \lambda(ua);$$

v is induced by f, and i denotes the identity map of A.

Show that the image of $v \in \operatorname{Hom}_\Lambda(N,C)$ under the connecting homomorphism $\operatorname{Hom}_\Lambda(N,C) \to \operatorname{Ext}^1_\Lambda(A,C)$ is the characteristic class of the extension (E).

[Hint: identify X with $A + C$ by using the map

$$x \to (\varphi x, \psi^{-1}(x - u\varphi x)).]$$

6. Consider an extension over Λ with kernel C:

$$(F) \qquad C \xrightarrow{g} \Gamma \xrightarrow{f} \Lambda$$

in the sense of § 2. Let i_Γ and i_Λ be the maps

$$i_\Gamma\colon\ \Gamma \to J(\Gamma), \qquad i_\Lambda\colon\ \Lambda \to J(\Lambda)$$

defined by $i_\Gamma(\gamma) = \gamma \otimes 1 - 1 \otimes \gamma^*$, $i_\Lambda(\lambda) = \lambda \otimes 1 - 1 \otimes \lambda^*$. Let I be the sub-K-module of $J(\Gamma)$ generated by the elements $(gc)u$ and $u(gc)$ $(c \in C,\ u \in J(\Gamma))$; show that I is a two-sided Γ-module, using (2) of § 2. Let $X = J(\Gamma)/I$ be the quotient which is now a two-sided Λ-module; and let k be the natural epimorphism $J(\Gamma) \to X$. Then the map $\bar{f}\colon\ J(\Gamma) \to J(\Lambda)$ induces $\alpha\colon\ X \to J(\Lambda)$. Define $j = ki_\Gamma\colon\ \Gamma \to X$ and $\beta = jg\colon\ C \to X$. Show that α and β are homomorphisms of two-sided Λ-modules, and that the bottom row of the commutative diagram

$$\begin{array}{lccccccccc}
(F) & & & C & \xrightarrow{g} & \Gamma & \xrightarrow{f} & \Lambda & & \\
 & & & \updownarrow & & \downarrow{\scriptstyle j} & & \downarrow{\scriptstyle i_\Lambda} & & \\
(E) & 0 & \longrightarrow & C & \xrightarrow{\beta} & X & \xrightarrow{\alpha} & J(\Lambda) & \longrightarrow & 0
\end{array}$$

is exact. Thus we obtain a mapping

$$\eta\colon\ F(\Lambda,C) \to E(J(\Lambda),C)$$

which to each extension (F) over Λ with kernel C associates an extension (E) of Λ^e-modules.

Assume now that Λ is K-projective, and prove that the following diagram is anticommutative

$$\begin{array}{ccc} F(\Lambda,C) & \xrightarrow{\omega^{-1}} & H^2(\Lambda,C) = \operatorname{Ext}^2_{\Lambda^e}(\Lambda,C) \\ \downarrow{\scriptstyle\eta} & & \uparrow{\scriptstyle\vartheta} \\ E(J(\Lambda),C) & \xrightarrow{\Theta} & \operatorname{Ext}^1_{\Lambda^e}(J(\Lambda),C), \end{array}$$

where Θ and ω are the correspondences of 1.1 and 2.1, and ϑ is the connecting homomorphism corresponding to the exact sequence

$$0 \longrightarrow J(\Lambda) \longrightarrow \Lambda^e \xrightarrow{\rho} \Lambda \longrightarrow 0.$$

[Hint: use the standard complex and proceed as in the proof of 3.1.]

7. Let Λ be a K-projective K-algebra, and let C be a two-sided Λ-module. Consider an extension over Λ with kernel C

$$(F) \qquad C \xrightarrow{g} \Gamma \xrightarrow{f} \Lambda$$

and an extension of two-sided Λ-modules

$$(E) \qquad 0 \longrightarrow C \xrightarrow{\beta} X \xrightarrow{\alpha} J(\Lambda) \longrightarrow 0.$$

We shall say that (F) and (E) are *related* if there is a map $k\colon\ \Gamma \to X$ satisfying

$$k(\gamma_1\gamma_2) = (f\gamma_1)(k\gamma_2) + (k\gamma_1)(f\gamma_2)$$

and such that the following diagram is commutative:

$$\begin{array}{ccccc} C & \xrightarrow{g} & \Gamma & \xrightarrow{f} & \Lambda \\ \updownarrow & & \downarrow{\scriptstyle k} & & \downarrow{\scriptstyle i_\Lambda} \\ C & \xrightarrow{\beta} & X & \xrightarrow{\alpha} & J(\Lambda) \end{array}$$

where $i_\Lambda(\lambda) = \lambda \otimes 1 - 1 \otimes \lambda^*$. Show that (F) and (E) are related if and only if their classes in $F(\Lambda,C)$ and $E(J(\Lambda),C)$ correspond to each other under the correspondence η of Exer. 6.

8. Let $\mathfrak{g}$ be a Lie K-algebra and C a (left) $\mathfrak{g}$-module. In the direct sum $C + \mathfrak{g}$ we introduce a Lie algebra structure by setting

$$[(c_1,x_1),(c_2,x_2)] = (x_1c_2 - x_2c_1, [x_1,x_2]).$$

There results an extension

$$(\Sigma) \qquad C \xrightarrow{g} C + \mathfrak{g} \xrightarrow{h} \mathfrak{g}$$

with the maps g and h defined in the obvious way. Show that this extension is the inessential extension in $\Sigma(\mathfrak{g},C)$. Show that each Lie algebra homomorphism u: $\mathfrak{g} \to C + \mathfrak{g}$ satisfying $hu =$ identity, is of the form $ux = (\varphi x,x)$ where φ: $\mathfrak{g} \to C$ is a crossed homomorphism. Note that the above is valid without the assumption that $\mathfrak{g}$ is K-free.

9. Apply the above exercise to the case when $\mathfrak{g} = L(M)$ is the free Lie algebra of a K-module M (XIII, Exer. 5). Show that for any (left) $L(M)$-module C, every K-homomorphism $M \to C$ admits a unique factorization $M \xrightarrow{j} L(M) \xrightarrow{\varphi} C$, where φ is a crossed homomorphism. [Hint: Consider the Lie algebra homomorphism $L \to C + L(M)$ which when composed with $C + L(M) \to L(M)$ gives the identity.]

10. Formulate exercises analogous to Exer. 8 and 9 for groups. Here M will be an arbitrary set, and $L(M)$ will be replaced by the free group generated by the elements of M.

11. Let f: $\Gamma \to \Lambda$ be an epimorphism of K-algebras, Λ being a supplemented algebra with ε: $\Lambda \to K$. For a left Λ-module A, we shall say that a K-homomorphism

$$\alpha\colon\ \Gamma \to A$$

is a *crossed homomorphism with respect* to f (or simply an f-crossed homomorphism) if

$$\alpha(\gamma_1\gamma_2) = (\alpha\gamma_1)(\varepsilon f\gamma_2) + (f\gamma_1)(\alpha\gamma_2).$$

We shall say that Γ is *projective with respect to* f if, for any epimorphism of left Λ-modules

$$g\colon\ A \to A'',$$

any f-crossed homomorphism α'': $\Gamma \to A''$ may be factored $\alpha'' = g\alpha$, where α: $\Gamma \to A$ is an f-crossed homomorphism.

Let now C be a left Λ-module. Assuming that an extension of algebras

$$(F) \qquad C \to \Gamma \to \Lambda$$

corresponds to an extension of left Λ-modules

$$(E) \qquad 0 \to C \to X \to I(\Lambda) \to 0$$

under the correspondence η of 3.1, show that Γ is projective with respect to f, if and only if X is a projective Λ-module.

12. Let f: $W \to \Pi$ be a homomorphism of groups. We shall say that W is *f-projective* if, given any Π-epimorphism g: $A \to A''$ of left Π-modules, and any crossed homomorphism (rel. to f) α'': $W \to A''$, there exists a crossed homomorphism (rel. to f) α: $W \to A$ with $g\alpha = \alpha''$.

Let now C be a left Π-module. Assuming that a group extension

$$(\Sigma) \qquad C \longrightarrow W \xrightarrow{f} \Pi$$

and an extension of left Π-modules

$$(E) \qquad 0 \to C \to X \to I(\Pi) \to 0$$

are "related" (cf. Exer. 3), show that W is f-projective if and only if X is a projective $Z(\Pi)$-module.

13. Let $f\colon F \to \Pi$ be an epimorphism of groups, F being a free group, and $R = \operatorname{Ker} f$. Let $[R,R]$ denote the commutator subgroup of R; then $[R,R]$ is an invariant subgroup of F. Show that, in the induced exact sequence

$$R/[R,R] \longrightarrow F/[R,R] \xrightarrow{f'} \Pi,$$

the group $F/[R,R]$ is f'-projective (in the sense of Exer. 12).

14. Formulate the analogue of Exer. 12 for Lie algebras. Consider a Lie algebra epimorphism $f\colon L(M) \to \mathfrak{g}$ with kernel $\mathfrak{h}$. Show that $[\mathfrak{h},\mathfrak{h}]$ is an ideal in $\mathfrak{h}$ and in $L(M)$, and that in the extension

$$(\Sigma) \qquad \mathfrak{h}/[\mathfrak{h},\mathfrak{h}] \longrightarrow L(M)/[\mathfrak{h},\mathfrak{h}] \xrightarrow{f'} \mathfrak{g}$$

with f' induced by f, the Lie algebra $L(M)/[\mathfrak{h},\mathfrak{h}]$ is f'-projective. [Hint: use Exer. 9.]

15. Consider a group extension

$$(\Sigma) \qquad C \longrightarrow W \xrightarrow{f} \Pi$$

and let $l = \omega^{-1}\varphi^{-1}(\Sigma) \in H^2(\Pi,C)$ be the cohomology class determined by (Σ). Consider the homomorphisms

$$(1) \qquad H^p(\Pi, \operatorname{Hom}(C,D)) \to H^{p+2}(\Pi,D), \qquad {}_{\Pi}D$$

$$(1a) \qquad H_{p+2}(\Pi,D') \to H_p(\Pi,D' \otimes D), \qquad D'_{\Pi}$$

given by the products $h \to h \cup l$ and $h' \to h' \cap l$. The Π-operators on $\operatorname{Hom}(C,D)$ and $D' \otimes C$ are given by

$$(xf)c = xf(x^{-1}c), \qquad f \in \operatorname{Hom}(C,D),$$

$$x(d' \otimes c) = d'x^{-1} \otimes xc.$$

The cup- and cap-products are those of XI,8, (8) and (8a).

Prove that if W is f-projective, then (1) and (1a) are isomorphisms for $p > 0$, while for $p = 0$ we have exact sequences

$$\operatorname{Hom}_{\Pi}(W,D) \to \operatorname{Hom}_{\Pi}(C,D) \to H^2(\Pi,D) \to 0$$

$$0 \to H_2(\Pi,D') \to D' \otimes_{\Pi} C \to D' \otimes_{\Pi} W$$

where $\mathrm{Hom}_\Pi(W,D)$ denotes the module of crossed homomorphisms $W \to D$ with respect to f, and $D' \otimes_\Pi W$ is the group generated by pairs $d' \otimes \omega$ with relations $(d_1' + d_2') \otimes \omega = d_1' \otimes \omega + d_2' \otimes \omega$. [Hint: pass to the sequence (E): $0 \to C \to X \to I(\Pi) \to 0$ related to (Σ), consider the sequence $0 \to C \to X \to Z(\Pi) \to Z \to 0$ in which X is Π-projective, and apply XI,9.5.]

Apply the above result to the sequences given in Exer. 13.

16. Formulate exercises analogous to Exer. 13 and 15 for Lie algebras.

17. Let (Σ): $C \longrightarrow W \xrightarrow{f} \Pi$ be a group extension with W being f-projective and with Π finite of order r. Let $l \in H^2(\Pi,C)$ be the corresponding cohomology class. Using Exer. 15 and XII, Exer. 14, prove that l has order r.

18. Consider a diagram of Λ-modules and Λ-homomorphisms

$$\begin{array}{ccccccccc} 0 & \longrightarrow & A' & \longrightarrow & A & \longrightarrow & A'' & \longrightarrow & 0 \\ & & \downarrow{\scriptstyle \varphi} & & & & \downarrow{\scriptstyle \psi} & & \\ 0 & \longrightarrow & C' & \longrightarrow & C & \longrightarrow & C'' & \longrightarrow & 0 \end{array}$$

with exact rows. It is asked whether a homomorphism f: $A \to C$ can be found so as to give commutativity in this diagram. A necessary condition is the commutativity of

$$\begin{array}{ccc} \mathrm{Ext}^n(C',D) & \xrightarrow{\delta_C} & \mathrm{Ext}^{n+1}(C'',D) \\ \downarrow{\scriptstyle \varphi^*} & & \downarrow{\scriptstyle \psi^*} \\ \mathrm{Ext}^n(A',D) & \xrightarrow{\delta_A} & \mathrm{Ext}^{n+1}(A'',D) \end{array}$$

for any Λ-module D and any integer n. In particular, taking $D = C'$, $n = 0$, we have a necessary condition

$$\delta_A \varphi^* j = \psi^* \delta_C j,$$

with $j \in \mathrm{Hom}\,(C',C')$ denoting the identity map.

Is this last condition sufficient?

Give the dual procedure.

19. Let A be an abelian group with a finite torsion group tA. Show that tA is a direct summand of A. [Hint: use VII, 6.2.]

CHAPTER XV

Spectral Sequences

Introduction. This chapter is devoted to a purely algebraical study of spectral sequences, which arise whenever a complex is given with a filtration (i.e. a sequence of subcomplexes ordered by inclusion). In particular, every double complex gives rise to two spectral sequences (§ 6). The applications will be presented in Chapters XVI and XVII. In all these considerations the multiplicative structure is left aside (see exercises).

Spectral sequences arose in connection with topological investigations concerned with fiber bundles (Leray, *C.R. Acad. Sci. Paris*, 222 (1945), 1419–1422). The main applications are still in the domain of algebraic topology.

The notion of a spectral sequence was first algebraicized by Koszul (*C.R. Acad. Sci. Paris*, 225 (1947), 217–219); our exposition involves some modifications. The theory could equally well be presented using the "exact couples" of Massey (*Ann. of Math.* 56 (1952), 363–396).

1. FILTRATIONS AND SPECTRAL SEQUENCES

A *filtration* F of a module A is a family of submodules $\{F^pA\}$, p running through all integers, subject to the conditions

$$\cdots \supset F^pA \supset F^{p+1}A \supset \cdots \tag{1}$$

$$\cup\, F^pA = A. \tag{2}$$

It is convenient to set $F^{\infty}A = 0$ and $F^{-\infty}A = A$. We also sometimes lower the index by setting $F_pA = F^{-p}A$.

With each module A with a filtration F we associate a graded module $E_0(A)$ defined by

$$E_0^p(A) = F^pA/F^{p+1}A.$$

Suppose that A is a module with differentiation d and a filtration F *compatible* with d, i.e. such that

$$d(F^pA) \subset F^pA.$$

The inclusion $F^pA \subset A$ induces a homomorphism

$$H(F^pA) \to H(A)$$

whose image we denote by $F^p H(A)$. In this way we obtain a filtration, also denoted by F, of $H(A)$.

We are particularly interested in the associated graded module $E_0(H(A))$. It is the direct sum of the modules

$$E_0^p(H(A)) = F^p H(A)/F^{p+1} H(A).$$

In what follows, we shall frequently encounter commutative diagrams

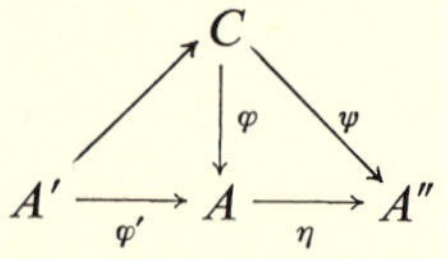

in which the row is exact.

LEMMA 1.1.　*The map η defines an isomorphism*

$$\operatorname{Im} \varphi / \operatorname{Im} \varphi' \simeq \operatorname{Im} \psi.$$

Indeed we have $\operatorname{Im} \varphi / \operatorname{Im} \varphi' = \operatorname{Im} \varphi / \operatorname{Ker} \eta$ which is mapped by η isomorphically onto $\operatorname{Im} (\eta\varphi) = \operatorname{Im} \psi$.

If we apply this lemma to the diagram

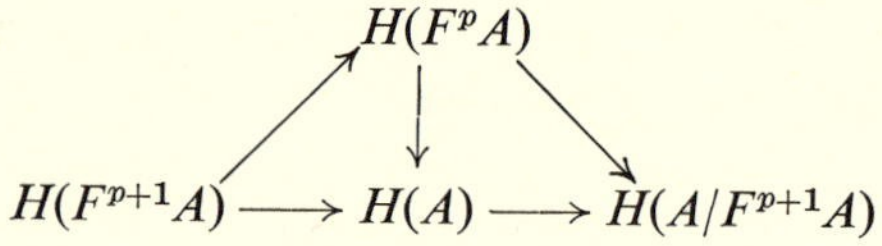

$$
\begin{array}{ccc}
 & H(F^p A) & \\
\nearrow & \downarrow & \searrow \\
H(F^{p+1} A) \longrightarrow & H(A) \longrightarrow & H(A/F^{p+1} A)
\end{array}
$$

we obtain an isomorphism

(3)　　　　　　$E_0^p(H(A)) \approx \operatorname{Im} (H(F^p A) \to H(A/F^{p+1} A)).$

We define

$$Z_\infty^p(A) = \operatorname{Im} (H(F^p A) \to H(F^p A/F^{p+1} A)),$$

$$B_\infty^p(A) = \operatorname{Im} (H(A/F^p A) \to H(F^p A/F^{p+1} A)),$$

$$E_\infty^p(A) = Z_\infty^p(A)/B_\infty^p(A).$$

Applying 1.1 to the diagram

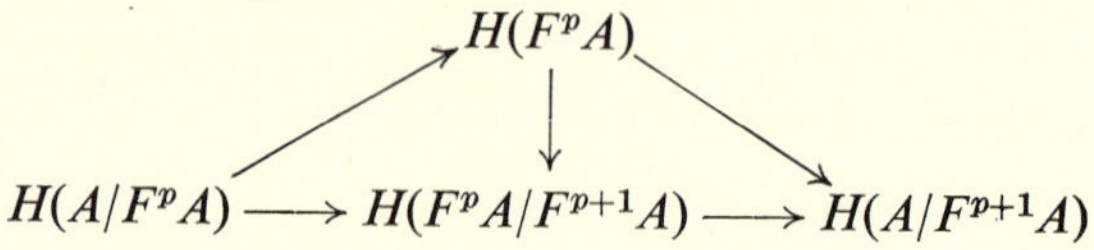

we obtain an isomorphism

(4)　　　　　　$E_\infty^p(A) \approx \operatorname{Im} (H(F^p A) \to H(A/F^{p+1} A))$

Combining this with the isomorphism (3) we obtain

$$E_0^p(H(A)) \approx E_\infty^p(A). \tag{5}$$

We denote by $E_\infty(A)$ the graded module $\sum_p E_\infty^p(A)$; it is isomorphic with the graded module $E_0(H(A))$ associated with $H(A)$.

We shall now introduce modules $Z_r^p(A)$, $B_r^p(A)$ and $E_r^p(A)$ which, in a sense that will be specified later, approximate the modules $Z_\infty^p(A)$, $B_\infty^p(A)$ and $E_\infty^p(A)$. We define for each $r \geqq 1$

$$Z_r^p(A) = \operatorname{Im}\,(H(F^pA/F^{p+r}A) \to H(F^pA/F^{p+1}A))$$

$$B_r^p(A) = \operatorname{Im}\,(H(F^{p-r+1}A/F^pA) \to H(F^pA/F^{p+1}A))$$

$$E_r^p(A) = Z_r^p(A)/B_r^p(A).$$

Setting $r = \infty$ and using the conventions $F^\infty A = 0$, $F^{-\infty}A = A$ we find the previous definitions. We have the inclusions

$$\cdots \subset B_r^p \subset B_{r+1}^p \subset \cdots \subset B_\infty^p \subset Z_\infty^p \subset \cdots \subset Z_{r+1}^p \subset Z_r^p \subset \cdots$$

Further since A/F^pA is a direct limit of $F^{p-r+1}A/F^pA$ and since the functor H commutes with direct limits, we have

$$B_\infty^p = \cup_r B_r^p.$$

In general, it is not true that Z_∞^p is the intersection $\cap Z_r^p$.

Applying 1.1 to the diagrams (in which we write F^p for F^pA)

$$\begin{array}{ccccc} & \nearrow & H(F^p/F^{p+r}) & \searrow & \\ & & \downarrow & & \\ H(F^p/F^{p+r+1}) & \longrightarrow & H(F^p/F^{p+1}) & \longrightarrow & H(F^{p+1}/F^{p+r+1}) \end{array}$$

$$\begin{array}{ccccc} & \nearrow & H(F^p/F^{p+r}) & \searrow & \\ & & \downarrow & & \\ H(F^{p+1}/F^{p+r}) & \longrightarrow & H(F^{p+r}/F^{p+r+1}) & \longrightarrow & H(F^{p+1}/F^{p+r+1}) \end{array}$$

we obtain isomorphisms

$$Z_r^p/Z_{r+1}^p \approx \operatorname{Im}\,(H(F^p/F^{p+r}) \to H(F^{p+1}/F^{p+r+1})) \approx B_{r+1}^{p+r}/B_r^{p+r}$$

which yield an isomorphism

$$\delta_r^p\colon\ Z_r^p/Z_{r+1}^p \approx B_{r+1}^{p+r}/B_r^{p+r}. \tag{6}$$

We define the homomorphism

$$d_r^p\colon\ E_r^p(A) \to E_r^{p+r}(A)$$

as the composition

$$E_r^p = Z_r^p/B_r^p \longrightarrow Z_r^p/Z_{r+1}^p \xrightarrow{\delta_r^p} B_{r+1}^{p+r}/B_r^{p+r} \longrightarrow Z_r^{p+r}/B_r^{p+r} = E_r^{p+r}.$$

It follows that

$$\text{(7)} \qquad \operatorname{Ker} d_r^p = Z_{r+1}^p/B_r^p, \qquad \operatorname{Im} d_r^p = B_{r+1}^{p+r}/B_r^{p+r}.$$

Thus in the diagram

$$E_r^{p-r}(A) \xrightarrow{d_r^{p-r}} E_r^p(A) \xrightarrow{d_r^p} E_r^{p+r}(A)$$

we have

$$\operatorname{Im} d_r^{p-r} = B_{r+1}^p/B_r^p \subset Z_{r+1}^p/B_r^p = \operatorname{Ker} d_r^p$$

which yields the natural isomorphism

$$\operatorname{Ker} d_r^p/\operatorname{Im} d_r^{p-r} \approx Z_{r+1}^p/B_{r+1}^p = E_{r+1}^p(A).$$

Therefore if we introduce the graded modules

$$E_r(A) = \textstyle\sum_p E_r^p(A)$$

and the endomorphisms d_r of $E_r(A)$ defined by d_r^p we obtain:

THEOREM 1.2. *For each $r \geqq 1$, the endomorphism d_r of $E_r(A)$ is a differentiation of degree r. The graded homology module $H(E_r(A))$ relative to the differentiation d_r, is naturally isomorphic with $E_{r+1}(A)$.*

For $r = 1$ we have $B_1^p = 0$ and $E_1^p = Z_1^p = H(F^p/F^{p+1})$. Since $E_0^p = F^p/F^{p+1}$ we find that if we denote by d_0 the differentiation induced in $E_0 = \sum E_0^p$ by the differentiation d of A, then $H(E_0) = E_1$. Thus the theorem remains valid also for $r = 0$.

An alternative description of d_r^p may be obtained as follows. Applying 1.1 to the diagram

$$\begin{array}{ccccc} & & H(F^p/F^{p+r}) & & \\ & \nearrow & \downarrow & \searrow & \\ H(F^{p-r+1}/F^p) & \longrightarrow & H(F^p/F^{p+1}) & \longrightarrow & H(F^{p-r+1}/F^{p+1}) \end{array}$$

we obtain an isomorphism (for $1 \leqq r \leqq \infty$)

$$\text{(8)} \qquad E_r^p(A) \approx \operatorname{Im}\,(H(F^p/F^{p+r}) \to H(F^{p-r+1}/F^{p+1})).$$

From the commutative diagram

$$\begin{array}{ccc} H(F^p/F^{p+r}) & \xrightarrow{\varphi_r^p} & H(F^{p-r+1}/F^{p+1}) \\ \downarrow & & \downarrow \\ H(F^{p+r}/F^{p+2r}) & \xrightarrow[\varphi_r^{p+r}]{} & H(F^{p+1}/F^{p+r+1}) \end{array}$$

we obtain a homomorphism

$$\operatorname{Im} \varphi_r^p \to \operatorname{Im} \varphi_r^{p+r}$$

which when combined with the isomorphisms (8) yields d_r^p. If $r = 1$ then φ_1^p is the identity and so is the isomorphism (8). Thus we see that d_1^p is simply the connecting homomorphism $H(F^p/F^{p+1}) \to H(F^{p+1}/F^{p+2})$ from the exact sequence $0 \to F^{p+1}/F^{p+2} \to F^p/F^{p+2} \to F^p/F^{p+1} \to 0$.

The sequence of graded modules $E_2(A), E_3(A), \ldots$, with the differentiations $d_2, d_3, \ldots$ and the isomorphisms $H(E_r(A)) \approx E_{r+1}(A)$ $(r \geqq 2)$ is called the *spectral sequence* of the module with differentiation A corresponding to the filtration F. The reason for not including $E_1(A)$ into the spectral sequence will appear later.

2. CONVERGENCE

We shall now investigate the problem of the sense in which the spectral sequence $E_2, E_3, \ldots$ approximates the module $E_\infty(A) \approx E_0(H(A))$.

The filtration F is said to be *weakly convergent* if

(1) $$Z_\infty^p(A) = \bigcap_r Z_r^p(A).$$

We shall now show how in the case of a convergent filtration, the spectral sequence "determines" the module E_∞, which is its "limit." Consider any term E_k of the spectral sequence. In E_k^p we have the following relations:

$$E_\infty^p \approx (Z_\infty^p/B_k^p)/(B_\infty^p/B_k^p)$$

$$Z_\infty^p/B_k^p = \bigcap_{r \geqq k} (Z_r^p/B_k^p)$$

$$B_\infty^p/B_k^p = \bigcup_{r \geqq k} (B_r^p/B_k^p).$$

Thus in order to show that the sequence $E_k, E_{k+1}, \ldots$ determines E_∞ it suffices to show how the modules

$$Z_r^p/B_k^p, \qquad B_r^p/B_k^p \qquad r \geqq k$$

can be reconstructed from the spectral sequence. If $r = k$, these terms reduce to E_k^p and 0. For $r > k$ we first observe that Z_r^p/B_k^p is in the kernel of the operator d_k. Further, the natural homomorphism ψ, mapping $Z(E_k^p)$ onto E_{k+1}^p and which has B_{k+1}^p/B_k^p as kernel, satisfies

$$Z_r^p/B_k^p = \psi^{-1}[Z_r^p/B_{k+1}^p]$$

$$B_r^p/B_k^p = \psi^{-1}[B_r^p/B_{k+1}^p].$$

This yields a recursive description of the desired modules.

We shall now derive two characterizations of weakly convergent filtrations. Applying 1.1 to the diagram

$$\begin{array}{ccccc} & & H(F^p/F^{p+r}) & & \\ & \nearrow & \downarrow & \searrow & \\ H(F^p) & \to & H(F^p/F^{p+1}) & \to & H(F^{p+1}) \end{array}$$

we obtain an isomorphism

$$Z_r^p(A)/Z_\infty^p(A) \approx \operatorname{Im}\,(H(F^p/F^{p+r}) \to H(F^{p+1})) \tag{2}$$

where the latter homomorphism is obtained by composition $H(F^p/F^{p+r}) \to H(F^{p+r}) \to H(F^{p+1})$. Since (1) is equivalent with $\bigcap_r Z_r^p/Z_\infty^p = 0$ we obtain:

PROPOSITION 2.1. *In order that the filtration F be weakly convergent it is necessary and sufficient that for each p the intersection of the images of the homomorphisms*

$$H(F^pA/F^{p+r}A) \to H(F^{p+1}A) \qquad r \geqq 1$$

be zero.

We define

$$R^p = \bigcap_r \operatorname{Im}\,(H(F^{p+r}A) \to H(F^pA)) \qquad r > 1$$

$$R^{-\infty} = \bigcap_p F^pH(A) = \bigcap_p \operatorname{Im}\,(H(F^pA) \to H(A)).$$

The homomorphisms $H(F^{p+1}A) \to H(F^pA)$ and $H(F^pA) \to H(A)$ induce homomorphisms

$$R^{p+1} \to R^p, \qquad R^p \to R^{-\infty}.$$

PROPOSITION 2.2. *The filtration F is weakly convergent if and only if each $R^{p+1} \to R^p$ is a monomorphism.*

PROOF. Consider the diagram

$$\begin{array}{ccccc} & & H(F^{p+1}) & & \\ & \nearrow & \uparrow & \searrow & \\ H(F^p/F^{p+r}) & \to & H(F^{p+r}) & \to & H(F^p) \end{array}$$

and let $x \in R^{p+1}$. Then x is in the image of $H(F^{p+r})$, and it follows that x is in the image of $H(F^p/F^{p+r})$ if and only if the image of x in R^p is zero. We thus obtain the relation

$$\bigcap_{r\geqq 1} \operatorname{Im}\,(H(F^p/F^{p+r}) \to H(F^{p+1})) = \operatorname{Ker}\,(R^{p+1} \to R^p).$$

Thus 2.2 follows from 2.1.

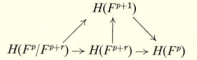

We shall say that the filtration F of A is *convergent* if it is weakly convergent and

$$\cap_p F^p H(A) = 0$$

(i.e. $R^{-\infty} = 0$). If we consider the homomorphism

$$u\colon\ H(A) \to \varprojlim H(A)/F^p H(A)$$

defined by $u_p\colon\ H(A) \to H(A)/F^p H(A)$, we find that $R^{-\infty} = \operatorname{Ker} u$. Thus for a convergent filtration, u is a monomorphism. We shall say that the filtration F of A is *strongly convergent*, if it is weakly convergent and if u is an *isomorphism*. Clearly a strongly convergent filtration is convergent.

In addition we can also consider the homomorphism

$$v\colon\ \varprojlim H(A)/F^p H(A) \to \varprojlim H(A/F^p A)$$

induced by

$$v_p\colon\ H(A)/F^p H(A) \to H(A/F^p A).$$

Since each v_p is a monomorphism, it follows that v is a monomorphism. It can be proved that if $R^p = 0$ for all p, then v is an isomorphism.

3. MAPS AND HOMOTOPIES

Let $f\colon\ A \to A'$ be a map of modules with differentiation, and let filtrations of A, A' compatible with the differentiations be given. We say that f is compatible with the filtrations if

$$f(F^p A) \subset F^p A'.$$

Such a map clearly induces homomorphisms

$$f^*\colon\ H(A) \to H(A')$$

$$f_r^*\colon\ E_r(A) \to E_r(A')$$

$$f_\infty^*\colon\ E_\infty(A) \to E_\infty(A')$$

and $d_r f_r^* = f_r^* d_r$ where d_r denotes the differentiation both in $E_r(A)$ and $E_r(A')$.

If f and g are two such maps, we define a homotopy $s\colon\ f \simeq g$ of order $\leqq k$ as a homomorphism $s\colon\ A \to A'$ satisfying

$$ds + sd = g - f, \qquad s(F^p A) \subset F^{p-k} A'.$$

Proposition 3.1. *If $s\colon\ f \simeq g$ is a homotopy of order $\leqq k$, then $f^* = g^*$, $f_\infty^* = g_\infty^*$ and $f_r^* = g_r^*$ for $r > k$.*

Proof. The fact that $f^* = g^*$ is trivial and well known. To show that $f_r^* = g_r^*$ for $k < r \leqq \infty$ we utilize the natural isomorphism (8) of § 1.

We consider the commutative diagram

$$\begin{array}{ccc} H(F^pA/F^{p+r}A) & \xrightarrow{\alpha} & H(F^{p-r+1}A/F^{p+1}A) \\ \downarrow & & \downarrow \beta \\ H(F^pA'/F^{p+r}A') & \xrightarrow[\alpha']{} & H(F^{p-r+1}A'/F^{p+1}A') \end{array}$$

where the vertical maps are induced by the map $g - f\colon A \to A'$. In view of (8) of § 1 it suffices to show that $\beta\alpha = 0$ if $k < r \leqq \infty$. Let then $x \in F^pA$ be such that $dx \in F^{p+r}A$. Since

$$gx - fx = sdx + dsx$$

and $sdx \in F^{p+r-k}A'$ and $sx \in F^{p-k}A'$ we have

$$sdx \in F^{p+1}A', \qquad sx \in F^{p-r+1}A'.$$

This expresses the fact that $gx - fx$ yields the zero element of $H(F^{p-r+1}A'/F^{p+1}A')$.

THEOREM 3.2. *Let $f\colon A \to A'$ be a map of modules with differentiations and with filtrations compatible with f. If for a certain index k, $f_k^*\colon E_k(A) \to E_k(A')$ is an isomorphism then the same holds for every finite index $r \geqq k$. If the filtrations are weakly convergent, then $f_\infty^*\colon E_\infty(A) \to E_\infty(A')$ also is an isomorphism. Finally, if the filtrations are strongly convergent then $f^*\colon H(A) \to H(A')$ also is an isomorphism.*

PROOF. Since f_k^* is an isomorphism and commutes with the differentiation operators d_k in $E_k(A)$ and $E_k(A')$ it follows that f_{k+1}^* also is an isomorphism. Thus f_r^* is an isomorphism for all finite $r \geqq k$. The weak convergence conditions of F and F' imply then that f_∞^* also is an isomorphism.

Since the homomorphisms

$$F^{p-1}H(A)/F^pH(A) \to F^{p-1}H(A')/F^pH(A')$$

are isomorphisms for all p, it follows by recursion that the homomorphisms

$$F^{p-r}H(A)/F^pH(A) \to F^{p-r}H(A')/F^pH(A')$$

are isomorphisms for all p and all $r \geqq 1$. It follows that

$$H(A)/F^pH(A) \to H(A')/F^pH(A')$$

is an isomorphism. Therefore in the commutative diagram

$$\begin{array}{ccc} H(A) & \xrightarrow{u} & \varprojlim H(A)/F^pH(A) \\ \downarrow f^* & & \downarrow g \\ H(A') & \xrightarrow[u']{} & \varprojlim H(A')/F^pH(A') \end{array}$$

g is an isomorphism. If the filtrations are strongly convergent, then u and u' are isomorphisms and thus f^* also is an isomorphism.

For the last argument it suffices to assume that the filtration of A is strongly convergent while the filtration of A' is convergent. Indeed u is then an isomorphism while u' is a monomorphism. Since g is an isomorphism, it follows that gu is an epimorphism so that u' is an epimorphism. Thus u' is an isomorphism and the filtration of A' also is strongly convergent.

4. THE GRADED CASE

Suppose that A is a complex (i.e., A is graded and the differentiation d of A has degree 1). We then require that each module F^pA of the filtration of A be *homogeneous*, i.e., that F^pA be the direct sum of the submodules $A^{p+q} \cap F^pA$. We introduce the notations

$$F^{p,q}A = A^{p+q} \cap F^pA = F^pA^{p+q}$$

$$E_0^{p,q}(A) = F^{p,q}A/F^{p+1,q-1}A.$$

The module $E_0^p(A)$ may be identified with the direct sum $\sum_q E_0^{p,q}(A)$, so that the module $E_0(A)$ is doubly graded

$$E_0(A) = \sum_{p,q} E_0^{p,q}(A).$$

Similarly the module $E_0(H(A))$ is bigraded by the modules

$$E_0^{p,q}(H(A)) = F^pH^{p+q}(A)/F^{p+1}H^{p+q}(A).$$

As in § 1 we define for $1 \leqq r \leqq \infty$

$$Z_r^{p,q}(A) = \operatorname{Im}(H^{p+q}(F^pA/F^{p+r}A) \to H^{p+q}(F^pA/F^{p+1}A)),$$

$$B_r^{p,q}(A) = \operatorname{Im}(H^{p+q-1}(F^{p-r+1}A/F^pA) \to H^{p+q}(F^pA/F^{p+1}A)),$$

$$E_r^{p,q}(A) = Z_r^{p,q}(A)/B_r^{p,q}(A).$$

Each of the modules $E_r^p(A)$ may be identified with the direct sum $\sum_q E_r^{p,q}$, so that $E_r(A)$ is doubly graded. The isomorphism $E_\infty^{p,q}(A) \approx E_0^{p,q}(H(A))$ still holds. The differentiation operator $d_r\colon E_r \to E_r$ is composed of homomorphisms

$$d_r^{p,q}\colon E_r^{p,q} \to E_r^{p+r,q-r+1}$$

i.e. d_r has bidegree $(r, 1-r)$. In all these bigraded modules the first degree, p, is called the *degree of the filtration*, the second degree, q, is called the *complementary degree*; $p+q$ is the *total degree*.

Sometimes to avoid negative numbers we find it convenient to lower the indices using the rule

$$E^r_{p,q} = E_r^{-p,-q}.$$

The differentiation then becomes

$$d^r_{p,q}\colon E^r_{p,q} \to E^r_{p-r,q+r-1}$$

and has bidegree $(-r, r-1)$.

The filtration F of the complex A will be called *regular* if for each n there exists an integer $u(n)$ such that

(1) $$H^n(F^p(A)) = 0 \qquad \text{for } p > u(n).$$

We shall show that (1) implies

(2) $$Z_r^{p,q}(A) = Z_\infty^{p,q}(A) \qquad \text{for } r > u(p+q+1) - p.$$

Indeed, from (2) of § 2 (in graded form) we have

$$Z_r^{p,q}(A)/Z_\infty^{p,q}(A) \approx \operatorname{Im}\,(H^{p+q}(F^p/F^{p+r}) \to H^{p+q+1}(F^{p+1})).$$

This last homomorphism admits a factorization

$$H^{p+q}(F^p/F^{p+r}) \to H^{p+q+1}(F^{p+r}) \to H^{p+q+1}(F^{p+1})$$

and the term in the middle is zero if $p + r > u(p+q+1)$. This proves (2).

In § 2 we introduced the notion of a strongly convergent filtration by requiring that the homomorphism

$$u\colon H(A) \to \varprojlim H(A)/F^pH(A)$$

be an isomorphism. In the graded case we may consider the homomorphisms

$$u^n\colon H^n(A) \to \varprojlim H^n(A)/F^pH^n(A).$$

If u is an isomorphism so is each of u^n; the converse is false. But for our purpose it suffices that each u^n be an isomorphism; therefore, in the graded case, we shift to this weaker definition of strong convergence. The last part of Theorem 3.2 remains then valid in the graded case, provided the map has degree zero.

PROPOSITION 4.1. *A regular filtration of a complex A is strongly convergent.*

PROOF. (2) implies that the filtration is weakly convergent. We must now verify that for each n

$$u^n\colon H^n(A) \to \varprojlim H^n(A)/F^pH^n(A)$$

is an isomorphism. Since $F^pH^n(A)$ is the image of $H^n(F^pA) \to H^n(A)$, it follows from (1) that $F^pH^n(A) = 0$ for $p > u(n)$. Thus u^n is an isomorphism.

For a regular filtration we can give a better interpretation of the way in which the spectral sequence $\{E_r(A)\}$ "tends" to $E_\infty(A)$ as "limit." Indeed, for $r > u(p+q+1) - p$ we have the relations

$$B_r^{p,q} \subset B_{r+1}^{p,q} \subset \cdots \subset B_\infty^{p,q} \subset Z_\infty^{p,q} = \cdots = Z_{r+1}^{p,q} = Z_r^{p,q}$$

and $B_\infty^{p,q} = \bigcup_r B_r^{p,q}$. There results a direct sequence of groups and epimorphisms

$$E_r^{p,q} \to E_{r+1}^{p,q} \to \cdots \tag{3}$$

with $E_\infty^{p,q}$ as direct limit. Each homomorphism in (3) is given by the spectral sequence, since

$$E_{r+1}^{p,q} = E_r^{p,q}/\operatorname{Im} d_r^{p-r,q+r-1}.$$

5. INDUCED HOMOMORPHISMS AND EXACT SEQUENCES

Let A be a complex with a filtration F. We shall derive certain homomorphisms and exact sequences involving the terms of the spectral sequence of F, the modules $E_\infty^{p,q}(A)$ and $H^n(A)$. We shall abbreviate the notation and write $E_r^{p,q}$, H^n, etc. instead of $E_r^{p,q}(A)$, $H^n(A)$, etc.

PROPOSITION 5.1. *$E_r^{p,q} = 0$ implies $E_s^{p,q} = 0$ for all $s > r$ ($s \leqq \infty$).* Indeed since $B_r^{p,q} = Z_r^{p,q}$ it follows that the modules

$$B_r^{p,q} \subset B_s^{p,q} \subset Z_s^{p,q} \subset Z_r^{p,q}$$

are all equal.

PROPOSITION 5.2. *Let $r < s \leqq \infty$; if $E_r^{u,v} = 0$ for $u + v = p + q - 1$, $p - s < u \leqq p - r$, then $B_r^{p,q} = B_s^{p,q}$. There results a monomorphism $E_s^{p,q} \to E_r^{p,q}$.*

PROOF. Take an integer t such that $r \leqq t < s$. By (7) of § 1 we have

$$B_{t+1}^{p,q}/B_t^{p,q} = \operatorname{Im} d_t^{p-t,q+t-1}.$$

Since $E_r^{p-t,q+t-1} = 0$, thus by 5.1, $E_t^{p-t,q+t-1} = 0$, so that $d_t^{p-t,q+t-1} = 0$ and $B_{t+1}^{p,q} = B_t^{p,q}$. If $s = \infty$, the same conclusion follows from $B_\infty^{p,q} = \bigcup_t B_t^{p,q}$.

PROPOSITION 5.2a. *Let $r < s \leqq \infty$. If $E_r^{u,v} = 0$ for $u + v = p + q + 1$, $p + r \leqq u < p + s$, and if moreover s is finite or the filtration is weakly convergent, then $Z_r^{p,q} = Z_s^{p,q}$. There results an epimorphism $E_r^{p,q} \to E_s^{p,q}$.*

The proof is dual to the preceding one, when s is finite. If s is infinite and the filtration is weakly convergent, we have $Z_\infty^{p,q} = \bigcap_t Z_t^{p,q}$ and this concludes the proof.

By combining 5.2 and 5.2a, one obtains conditions for $E_r^{p,q} \approx E_s^{p,q}$ ($r < s \leqq \infty$).

PROPOSITION 5.3. *If* $E_\infty^{u,n-u} = 0$ *for* $u < p$, *then* $F^pH^n = H^n$. *There results an epimorphism* $H^n \to E_\infty^{p,n-p}$.

PROOF. The conditions imply $F^uH^n = F^{u+1}H^n$ *for* $u < p$. *Since* $H^n = \bigcup_u F^uH^n$, we have $H^n = F^pH^n$. Since $E_\infty^{p,n-p} \approx F^pH^n/F^{p+1}H^n$, the conclusion follows.

PROPOSITION 5.3a. *If the filtration is convergent and* $E_\infty^{u,n-u} = 0$ *for* $u > p$, *then* $F^{p+1}H^n = 0$. *There results a monomorphism* $E_\infty^{p,n-p} \to H^n$.

PROOF. The conditions imply $F^uH^n = F^{u+1}H^n$ for $u > p$. Since the filtration is convergent we have $\bigcap_u F^uH^n = 0$, so that $F^{p+1}H^n = 0$. Since $E_\infty^{p,n-p} \approx F^pH^n/F^{p+1}H^n$, the conclusion follows.

COROLLARY 5.4. *If the filtration is convergent and* $E_\infty^{u,n-u} = 0$ *for* $u \neq p$, *then* $H^n \approx E_\infty^{p,n-p}$.

More generally we have

PROPOSITION 5.5. *If the filtration is convergent and, for some integers* n, p *and* k $(k > 0)$, *we have* $E_\infty^{u,n-u} = 0$ *for* $u \neq p, p + k$, *then there is an exact sequence*

$$0 \to E_\infty^{p+k,n-p-k} \to H^n \to E_\infty^{p,n-p} \to 0.$$

PROOF. The homomorphism $E_\infty^{p+k,n-p-k} \to H^n$ is defined by 5.3a and and has $F^{p+k}H^n$ as image. The homomorphism $H^n \to E_\infty^{p,n-p}$ is defined by 5.3 and has $F^{p+1}H^n$ as kernel. Since $0 = E_\infty^{u,n-u} \approx F^uH^n/F^{u+1}H^n$ for $p < u < p + k$, it follows that $F^{p+1}H^n = F^{p+k}H^n$.

PROPOSITION 5.6. (a) *If* $E_r^{u,v} = 0$ *for* $u + v = n - 1$, $u \leqq p - r$, *and for* $u + v = n$, $u < p$, *there is a homomorphism*

$$H^n \to E_r^{p,n-p}. \tag{1}$$

(b) *If the filtration is convergent and* $E_r^{u,v} = 0$ *for* $u + v = n + 1$, $u \geqq p + r$, *and for* $u + v = n$, $u > p$, *there is a homomorphism*

$$E_r^{p,n-p} \to H^n. \tag{2}$$

(c) *If* (a) *and* (b) *both hold, then* (1) *and* (2) *are reciprocal isomorphisms.*

PROOF. (a) follows from 5.1, 5.3 and 5.2. Similarly (b) follows from 5.2a, 5.1 and 5.3a. Finally, (c) follows from 5.1, 5.2 and 5.2a, and 5.4.

PROPOSITION 5.7. *If the filtration is convergent, and if* $E_r^{u,v} = 0$ *for*

$$u + v = n, \qquad u \neq p \text{ and } \neq p - k \ (k > 0, \textit{given})$$

$$u + v = n + 1, \ u \geqq p + r$$

$$u + v = n - 1, \ u \leqq p - k - r,$$

then the following sequence is exact

$$E_r^{p,n-p} \to H^n \to E_r^{p-k,n-p+k}. \tag{I}$$

PROOF. By 5.5, we have an exact sequence

$$0 \to E_{\infty}^{p,n-p} \to H^n \to E_{\infty}^{p-k,n-p+k} \to 0.$$

Moreover, 5.2a yields an epimorphism $E_r^{p,n-p} \to E_{\infty}^{p,n-p}$, and 5.2 yields a monomorphism $E_{\infty}^{p-k,n-p+k} \to E_r^{p-k,n-p+k}$. This concludes the proof.

We shall now give a new series of propositions. First we define a generalization of the homomorphism

$$d_r^{p,q}\colon\ E_r^{p,q} \to E_r^{p+r,q-r+1}.$$

DEFINITION: Given a finite integer $s > r$, suppose that $E_r^{u,v} = 0$ for

$$u + v = p + q + 1, \qquad p + r \leqq u < p + s,$$

and for

$$u + v = p + q \quad , \qquad p < u \leqq p + s - r.$$

We define a homomorphism $d_{r,s}^{p,q}\colon\ E_r^{p,q} \to E_r^{p+s,q-s+1}$ as the composition $E_r^{p,q} \xrightarrow{\alpha} E_s^{p,q} \xrightarrow{\beta} E_s^{p+s,q-s+1} \xrightarrow{\gamma} E_r^{p+s,q-s+1}$, where β is $d_s^{p,q}$, and γ and α are defined because of 5.2 and 5.2a.

LEMMA 5.8. *If* $E_r^{p-s,q+s-1} = 0$ $(r \leqq s < \infty)$, *we have the exact sequence*

$$(3) \qquad 0 \longrightarrow E_{s+1}^{p,q} \longrightarrow E_s^{p,q} \xrightarrow{d_s^{p,q}} E_s^{p+s,q-s+1}.$$

PROOF. By 5.1, $E_s^{p-s,q+s-1} = 0$, thus $d_s^{p-s,q+s-1} = 0$, and consequently $B_{s+1}^{p,q} = B_s^{p,q}$. Thus $E_{s+1}^{p,q} = Z_{s+1}^{p,q}/B_s^{p,q}$, and this last module is the kernel of $d_s^{p,q}$, by (7) of § 1.

PROPOSITION 5.9. *If the filtration is weakly convergent, and if, for some integer* $s \geqq r$, $E_r^{u,v} = 0$ *for*

$$u + v = p + q - 1, \qquad u \leqq p - r,$$

and for

$$u + v = p + q, \qquad p \neq u \leqq p + s - r,$$

and for

$$u + v = p + q + 1, \qquad p + r \leqq u \neq p + s,$$

then we have the exact sequence

$$\text{(II)} \qquad H^{p+q} \longrightarrow E_r^{p,q} \xrightarrow{d_{r,s}^{p,q}} E_r^{p+s,q-s+1}.$$

PROOF. Since $E_r^{p-s,q+s-1} = 0$, we can apply lemma 5.8. Thus we have the exact sequence (3). But $E_{s+1}^{p,q} \approx E_{\infty}^{p,q}$ and $E_s^{p,q} \approx E_r^{p,q}$ by 5.2 and 5.2a. Moreover, by 5.2, we have a monomorphism $E_s^{p+s,q-s+1} \to E_r^{p+s,q-s+1}$. Thus (3) yields an exact sequence

$$0 \longrightarrow E_{\infty}^{p,q} \longrightarrow E_r^{p,q} \xrightarrow{d_{r,s}^{p,q}} E_r^{p+s,q-s+1}.$$

Finally, by 5.3 we have an epimorphism $H^{p+q} \to E_{\infty}^{p,q}$. This yields the exact sequence (II).

PROPOSITION 5.9a. *If the filtration is convergent, and if, for some integer* $s \geqq r$, $E_r^{u,v} = 0$ *for*

$$u + v = p + q + 1, \qquad u \geqq p + r$$

and for

$$u + v = p + q, \qquad p + r - s \leqq u \neq p,$$

and for

$$u + v = p + q - 1, \qquad p - s \neq u \leqq p - r,$$

then we have the exact sequence

$$\text{(III)} \qquad E_r^{p-s,q+s-1} \xrightarrow{d_{r,s}^{p-s,q+s-1}} E_r^{p,q} \longrightarrow H^{p+q}.$$

The proof is dual to the preceding one.

It is now possible to combine the cases in which exact sequence such as (I), (II) or (III) hold. For example:

THEOREM 5.10. *Assuming that* $r \geqq 1$, *let* p *and* p' *be two integers such that* $p - p' \geqq r$. *If the filtration is convergent and* $E_r^{u,v} = 0$ *for* $u \neq p,p'$, *then we have an exact sequence*

$$\cdots \to E_r^{p,n-p} \to H^n \to E_r^{p',n-p'} \to E_r^{p,n+1-p} \to H^{n+1} \to E_r^{p',n+1-p'} \to \cdots$$

THEOREM 5.11. *Assuming that* $r \geqq 2$, *let* q *and* q' *be two integers such that* $q' - q \geqq r - 1$. *If the filtration is convergent and* $E_r^{u,v} = 0$ *for* $v \neq q,q'$, *then we have an exact sequence*

$$\cdots \to E_r^{n-q,q} \to H^n \to E_r^{n-q',q'} \to E_r^{n+1-q,q} \to H^{n+1} \to E_r^{n+1-q',q'} \to \cdots$$

THEOREM 5.12. *Assume that the filtration is convergent and that* $E_2^{p,q} = 0$ *if* $p < 0$ *or* $q < 0$. *Assume further that* $E_2^{p,q} = 0$ *for* $0 < q < n$ $(n > 0)$. *We then have isomorphisms*

$$E_2^{i,0} \approx H^i \qquad i < n$$

and an exact sequence

$$0 \to E_2^{n,0} \to H^n \to E_2^{0,n} \to E_2^{n+1,0} \to H^{n+1}.$$

PROOF. The isomorphisms follow from 5.6(c). The exact sequence follows from 5.7, 5.9 and 5.9a.

The dual result is

THEOREM 5.12a. *Assume that the filtration is convergent and that* $E_{p,q}^2 = 0$ *if* $p < 0$ *or* $q < 0$. *Assume further that* $E_{p,q}^2 = 0$ *for* $0 < q < n$ $(n > 0)$. *We then have isomorphisms*

$$E_{i,0}^2 \approx H_i \qquad i < n$$

and an exact sequence

$$H_{n+1} \to E_{n+1,0}^2 \to E_{0,n}^2 \to H_n \to E_{n,0}^2 \to 0.$$

To conclude, we list a number of special cases needed in the sequel.

Case A. $H(F^pA/F^{p+1}A) = 0$ for $p < 0$. Then $E_r^{p,q} = 0$ for $p < 0$ $(r \leqq \infty)$, and 5.6(a) yields the homomorphisms

$$H^n \to E_2^{0,n}.$$

Case B. $H^{p+q}(F^pA) = 0$ for $q < 0$. The filtration is regular (and thus convergent), $E_r^{p,q} = 0$ for $q < 0$, $(r \leqq \infty)$, and 5.6(b) yields the homomorphisms

$$E_2^{n,0} \to H^n.$$

Case C. $H(F^pA/F^{p+1}A) = 0$ for $p < 0$ and $H^{p+q}(F^pA) = 0$ for $q < 0$. The filtration is regular, $E_r^{p,q} = 0$ for $p < 0$ and for $q < 0$ $(r \leqq \infty)$, we have the homomorphisms

$$E_2^{n,0} \to H^n \to E_2^{0,n}$$

and the exact sequence

$$0 \to E_2^{1,0} \to H^1 \to E_2^{0,1} \to E_2^{2,0} \to H^2.$$

There are three dual cases.

Case A′. $H(F^pA) = 0$ for $p > 0$. The filtration is regular, $E^r_{p,q} = 0$ for $p < 0$ $(r \leqq \infty)$ and we have homomorphisms

$$E^2_{0,n} \to H_n.$$

Case B′. $H^{p+q}(F^pA/F^{p+1}A) = 0$ for $q > 0$. Then $E^r_{p,q} = 0$ for $q < 0$ $(r \leqq \infty)$ and we have homomorphisms

$$H_n \to E^2_{n,0}$$

Case C′. $H(F^pA) = 0$ for $p > 0$ and $H^{p+q}(F^pA/F^{p+1}A) = 0$ for $q > 0$. The filtration is regular, $E^r_{p,q} = 0$ for $p < 0$ and for $q < 0$ $(r \leqq \infty)$, we have the homomorphisms

$$E^2_{0,n} \to H_n \to E^2_{n,0}$$

and the exact sequence

$$H_2 \to E^2_{2,0} \to E^2_{0,1} \to H_1 \to E^2_{1,0} \to 0.$$

We shall consider two more cases:

Case D^k. $H^{p+q}(F^pA) = 0$ for $q < k$ and $H^{p+q}(F^pA/F^{p+1}A) = 0$ for $q > k+1$. The filtration is then regular and 5.11 yields the exact sequence

$$\cdots \to E_2^{n-k,k} \to H^n \to E_2^{n-k-1,k+1} \to E_2^{n-k+1,k} \to H^{n+1} \to E_2^{n-k,k+1} \to \cdots$$

Case E^k. $H(F^pA/F^{p+1}A) = 0$ for $p < k$ and $H(F^pA) = 0$ for $p > k+1$. The filtration is then regular. By 5.2 and 5.2a, we have $E_2 = E_\infty$, and 5.5 yields an exact sequence

$$0 \to E_2^{k+1,n-k-1} \to H^n \to E_2^{k,n-k} \to 0.$$

The homomorphisms obtained in cases A, B, A', B' will be called *edge homomorphisms.* The exact sequences obtained in cases C and C' will be called the *exact sequences of terms of low degree.*

So far we have dealt with spectral sequences connected with a specific complex with a filtration. In the applications that will be given in the next two chapters, the situation will be somewhat different. We shall encounter situations in which the complexes and the filtrations will be constructed with a large degree of arbitrariness. It will however turn out that the homology modules of these complexes, the filtrations of these homology modules, and the spectral sequences involved will be "independent" of the choices involved in the construction of the complexes.

Because of this it will be necessary to develop a notation and terminology which will allow us to handle spectral sequences without explicit reference to the complex and its filtration from which the spectral sequence results.

Let $\sum B^{p,q}$ be a doubly graded module and $\sum D^n$ a graded module. We shall use the notation

$$B^{p,q} \underset{p}{\Rightarrow} D^n$$

to say that there exists a complex A with a regular filtration F such that $H^n(A)$ is isomorphic with D^n for all n, and such that the terms $E_2^{p,q}(A)$ of the spectral sequence of the filtration F are isomorphic to $B^{p,q}$. We indicate the degree of the filtration under the arrow, because when the terms $B^{p,q}$ have an explicit (and sometimes complicated) form it may be impossible to tell which of the two integers involved is the degree of the filtration and which is the complementary degree.

6. APPLICATION TO DOUBLE COMPLEXES

Let $A = \sum A^{p,q}$ be a double complex with differentiations d_1 and d_2, as defined in IV,4. With this double complex, there is associated a (single) complex with total differentiation d and homology modules $H^n(A)$.

We introduce the bigraded module $H_{\mathrm{I}}(A)$ which is the homology module with respect to the differentiation d_1. We regard $H_{\mathrm{I}}(A)$ as a double complex with differentiation $d_1 = 0$ and with d_2 induced by the differentiation d_2 in A. Similarly we define the double complex $H_{\mathrm{II}}(A)$ which is the homology of A with respect to d_2. We may thus also consider the modules $H_{\mathrm{II}}H_{\mathrm{I}}(A)$ and $H_{\mathrm{I}}H_{\mathrm{II}}(A)$, bigraded by $H_{\mathrm{II}}^{p,q}H_I(A)$ and $H_{\mathrm{I}}^{p,q}H_{\mathrm{II}}(A)$ respectively.

We introduce two filtrations F_{I} (the *first filtration*) and F_{II} (the *second filtration*) as follows

$$F_{\mathrm{I}}^{p}A = \sum_{r \geqq p} \sum_{q} A^{r,q}, \qquad F_{\mathrm{II}}^{q}A = \sum_{s \geqq q} \sum_{p} A^{p,s}.$$

These filtrations regarded as filtrations of the (single) complex associated with A are compatible with the total differentiation operator d. There result two filtrations of $H(A)$ and two spectral sequences which we call the *first* and the *second spectral sequence* of the double complex A. The term $E_r^{p,q}$ of the first spectral sequence will be denoted by $\mathrm{I}_r^{p,q}$. The term $E_r^{p,q}$ of the second spectral sequence will be denoted by $\mathrm{II}_r^{q,p}$; the switch of indices is justified by what follows. The module $H(A)$ with its graduation and two filtrations, and the two spectral sequences will be called the *invariants* of the double complex A.

The module $F_{\mathrm{I}}^pA/F_{\mathrm{I}}^{p+1}A$ may be identified with $\sum_q A^{p,q}$. Thus the module I_0 associated with A by the filtration F_{I} may be identified with A itself. The differentiation operator in I_0 is then easily seen to be given by the homomorphism $d_2^{p,q}$. Consequently, the homology module of I_0 i.e. the module I_1 may be identified with $H_{\mathrm{II}}(A)$.

The differentiation operator $d_{\mathrm{I},1}$ of I_1 (i.e. of the term $E_1(A)$ for the first filtration) is the connecting homomorphism for the homology modules of the exact sequence

$$0 \to F_{\mathrm{I}}^{p+1}/F_{\mathrm{I}}^{p+2} \to F_{\mathrm{I}}^{p}/F_{\mathrm{I}}^{p+2} \to F_{\mathrm{I}}^{p}/F_{\mathrm{I}}^{p+1} \to 0.$$

Let $x \in H_{\mathrm{II}}^{p,q}(A)$ be an element of bidegree (p,q) of $H_{\mathrm{II}}(A)$, and let $a \in A^{p,q}$ be an element representing x. Then $d_2x = 0$ so that $dx = d_1x$. It follows that the connecting homomorphism $H(F_{\mathrm{I}}^p/F_{\mathrm{I}}^{p+1}) \to H(F_{\mathrm{I}}^{p+1}/F_{\mathrm{I}}^{p+2})$ is induced by d_1. Thus if I_1 and $H_{\mathrm{II}}(A)$ are identified it follows that the differentiation $d_{\mathrm{I},1}$ of I_1 coincides with the differentiation d_1 of $H_{\mathrm{II}}(A)$. There results the identification

$$\mathrm{I}_2(A) = H_{\mathrm{I}}H_{\mathrm{II}}(A). \tag{1}$$

This identification is compatible with the double gradings of the modules involved.

To compute the initial term E_2 of the spectral sequence for the second filtration we resort to the following trick. We introduce the transposed double complex ${}^tA$ defined by

$${}^tA^{p,q} = A^{q,p}, \qquad {}^td_1^{p,q} = d_2^{q,p}, \qquad {}^td_2^{p,q} = d_1^{q,p}.$$

The double complexes A and ${}^tA$ have the same associated single complex A. Further, the second filtration of A is the first filtration of ${}^tA$. Thus the required term E_2 is $H_{\mathrm{I}}H_{\mathrm{II}}({}^tA)$. This latter term coincides with the transposed of the bigraded module $H_{\mathrm{II}}H_{\mathrm{I}}(A)$. Thus if we denote by II_2 the transposed of the term E_2 for the filtration F_{II} of A we obtain

$$\mathrm{II}_2(A) = H_{\mathrm{II}}H_{\mathrm{I}}(A). \tag{2}$$

This is the reason why earlier in defining $\mathrm{II}_r^{p,q}$ we transposed the degree of the filtration and supplementary degree.

A map $f\colon A \to A'$ of double complexes is always compatible with the filtrations F_{I} and F_{II}. It therefore induces maps of the respective spectral sequences.

PROPOSITION 6.1. *Two homotopic maps of double complexes induce the same maps of the invariants.*

PROOF. Let $f,g\colon A \to A'$ be the maps in question and let $(s_1,s_2)\colon f \simeq g$ be a homotopy as defined in IV,4. There results a total homotopy s in the associated (single) complexes. This homotopy is of order $\leqq 1$ with respect to the filtrations F_{I} and F_{II}. Thus 3.1 implies that f and g induce the same homomorphisms $\mathrm{I}_r^{p,q}(A) \to \mathrm{I}_r^{p,q}(A')$ and $\mathrm{II}_r^{p,q}(A) \to \mathrm{II}_r^{p,q}(A')$ for $r \geqq 2$.

Incidentally, the above fact is the main reason for defining the spectral sequence as beginning with the term E_2, rather than with the term E_1 (or even with the term E_0 composed of $F^pA/F^{p+1}A$).

We shall derive various relations between the modules $\mathrm{I}_2^{p,q}$, $\mathrm{II}_2^{p,q}$ and $H^n(A)$ of a double complex A under the assumption that some terms $A^{p,q}$ are zero. All the results follow from those of § 5.

Case 1. $A^{p,q} = 0$ if $q < 0$. The filtration F_{I} is regular. We are in Case B for the filtration F_{I} and in Case A for the filtration F_{II}. There result edge homomorphisms

$$\mathrm{I}_2^{n,0} \to H^n \to \mathrm{II}_2^{n,0}$$

Case 2. $A^{p,q} = 0$ if $p < 0$ or $q < 0$ (i.e. the double complex A is positive). Both filtrations are regular and we are in Case C. We obtain edge homomorphisms

$$\mathrm{I}_2^{n,0} \to H^n \to \mathrm{II}_2^{n,0}$$

$$\mathrm{II}_2^{0,n} \to H^n \to \mathrm{I}_2^{0,n}$$

and exact sequences for terms of low degree

$$0 \to \mathrm{I}_2^{1,0} \to H^1 \to \mathrm{I}_2^{0,1} \to \mathrm{I}_2^{2,0} \to H^2$$

$$0 \to \mathrm{II}_2^{0,1} \to H^1 \to \mathrm{II}_2^{1,0} \to \mathrm{II}_2^{0,2} \to H^2$$

Case 3. $A^{p,q} = 0$ for $q \neq 0,1$. Both filtrations are regular, F_{I} is in **Case D^0** and F_{II} is in Case E^0. We thus obtain exact sequences

$$\cdots \to \mathrm{I}_2^{n,0} \to H^n \to \mathrm{I}_2^{n-1,1} \to \mathrm{I}_2^{n+1,0} \to H^{n+1} \to \mathrm{I}_2^{n,1} \to \cdots$$

$$0 \to \mathrm{II}_2^{n-1,1} \to H^n \to \mathrm{II}_2^{n,0} \to 0.$$

Case 4. $A^{p,q} = 0$ for $p \neq 0,1$ or $q \neq 0,1$. The sequences above collapse to the exact sequences

$$0 \to \mathrm{I}_2^{1,0} \to H^1 \to \mathrm{I}_2^{0,1} \to 0,$$

$$0 \to \mathrm{II}_2^{0,1} \to H^1 \to \mathrm{II}_2^{1,0} \to 0,$$

and the isomorphisms

$$\mathrm{I}_2^{1,1} \approx H^2 \approx \mathrm{II}_2^{1,1}.$$

The above cases were geared towards applications to cohomology and right derived functors. For applications to homology and left derived functors, we need four dual cases. These can be most conveniently stated using the principle of lowered indices:

$$A_{p,q} = A^{-p,-q}, \quad \mathrm{I}^r_{p,q} = \mathrm{I}_r^{-p,-q}, \quad \mathrm{II}^r_{p,q} = \mathrm{II}_r^{-p,-q}, \quad H_n = H^{-n}.$$

We then consider

Case 1′: $A_{p,q} = 0$ if $q < 0$

Case 2′: $A_{p,q} = 0$ if $p < 0$ or $q < 0$.

Case 3′: $A_{p,q} = 0$ if $q \neq 0,1$.

Case 4′: $A_{p,q} = 0$ if $p \neq 0,1$ or $q \neq 0,1$.

The conclusions in these cases are obtained from the cases 1–4 by lowering indices and reversing all arrows. The only difference is that in Case 1′ it is the filtration F_{II} that is regular.

REMARK. If the first differentiation operator d_1 in A is zero then we have $H_{\mathrm{I}}(A) = A$ and

$$H(A) = H_{\mathrm{II}}(A) = H_{\mathrm{I}}H_{\mathrm{II}}(A) = H_{\mathrm{II}}H_{\mathrm{I}}(A).$$

The module $H(A)$ is then bigraded and coincides with I_2 and II_2. All the differentiation operators in the spectral sequences are zero. Further, in Case 1, the composite map $\mathrm{I}_2^{n,0} \to \mathrm{II}_2^{n,0}$ is the identity map.

7. A GENERALIZATION

We shall indicate here a more general setting in which the theory of spectral sequences may be developed. This generalization is particularly interesting for geometrical applications (see below).

We shall assume that for each pair of integers (p,q) such that $-\infty \leqq p \leqq q \leqq \infty$ a module $H(p,q)$ is given. We shall write $H(p)$ instead of $H(p,\infty)$ and we shall write H instead of $H(-\infty,\infty)$.

Given two pairs (p,q), (p',q') such that $p \leqq p'$, $q \leqq q'$ (notation: $(p,q) \leqq (p',q')$) we shall assume that a homomorphism

$$H(p',q') \to H(p,q) \tag{1}$$

is defined.

Given any triple (p,q,r) such that $-\infty \leqq p \leqq q \leqq r \leqq \infty$ we shall assume a *connecting homomorphism*

$$\delta\colon\ H(p,q) \to H(q,r) \tag{2}$$

is defined.

The above three primitive concepts are subjected to the following axioms

(SP.1) $H(p,q) \to H(p,q)$ is the identity.

(SP.2) If $(p,q) \leqq (p',q') \leqq (p'',q'')$ then the diagram

$$\begin{array}{ccc} H(p'',q'') & \longrightarrow & H(p,q) \\ & \searrow \quad \nearrow & \\ & H(p',q') & \end{array}$$

is commutative.

(SP.3) If $(p,q,r) \leqq (p',q',r')$ then the diagram

$$\begin{array}{ccc} H(p',q') & \longrightarrow & H(q',r') \\ \downarrow & & \downarrow \\ H(p,q) & \longrightarrow & H(q,r) \end{array}$$

is commutative.

(SP.4) For each triple (p,q,r) the sequence

$$\cdots \longrightarrow H(q,r) \longrightarrow H(p,r) \longrightarrow H(p,q) \xrightarrow{\delta} H(q,r) \longrightarrow \cdots$$

is exact.

(SP.5) For a fixed q the direct system of modules

$$H(q,q) \to H(q-1,q) \to \cdots \to H(p,q) \to H(p-1,q) \to \cdots$$

has $H(-\infty,q)$ as direct limit (under the mappings $H(p,q) \to H(-\infty,q)$.

From (SP.1) and (SP.4) we deduce that $H(p,p)=0$. From (SP.3) we deduce that (2) admits a factorization

$$H(p,q) \xrightarrow{\delta} H(q) \longrightarrow H(q,r). \tag{2'}$$

This indicates that we could postulate (2) only with $r = \infty$ and define (2) in general using (2′). Axioms (SP.3) and (SP.4) may then be weakened by replacing r by ∞. It can be shown that the weaker system of axioms implies the stronger one.

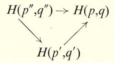

Usually the modules $H(p,q)$ will be graded. It is then assumed that (1) is a map of degree zero while (2) has degree 1.

Example 1. Let A be a module with differentiation d and filtration F. Define $H(p,q) = H(F^p/F^q)$ where $F^{-\infty} = A$, $F^\infty = 0$. For $(p,q) \leqq (p',q')$ we have a natural map $F^{p'}/F^{q'} — F^p/F^q$ which induces (1). For each triple (p,q,r) we have an exact sequence

$$0 \to F^q/F^r \to F^p/F^r \to F^p/F^q \to 0$$

which induces (2). Axioms (SP.1)—(SP.5) are readily verified. This is the case studied earlier in the chapter.

Example 2. Let X be a topological space and $\{X^p\}$ a family of subspaces defined for all integers p such that $X^p \subset X^{p+1}$. We set $X^{-\infty} = \Phi$, $X^\infty = X$ and define

$$H(p,q) = \textstyle\sum_n H^n(X^q,X^p)$$

where $H^n(X^q,X^p)$ is the n-th *cohomology* of the pair (X^q,X^p) with respect to some fixed cohomology theory. The maps (1) and (2) are then the induced homomorphisms and the coboundary operations of the cohomology theory. Axioms (SP.1)–(SP.4) are consequences of usual properties of cohomology groups. Axiom (SP.5) is not valid in general, but depends on the spaces involved and the cohomology theory that is being used.

Example 3. In the situation of example 2 set

$$H(p,q) = \textstyle\sum_n H_n(X^{-p},X^{-q})$$

using the relative *homology* groups of the pair (X^{-p},X^{-q}).

We now return to the abstract situation governed by the axioms (SP.1)–(SP.5) and define

$$F^pH = \operatorname{Im}(H(p) \to H)$$

$$Z_r^p = \operatorname{Im}(H(p,p+r) \to H(p,p+1))$$

$$B_r^p = \operatorname{Im}(H(p-r+1,p) \to H(p,p+1))$$

$$E_r^p = Z_r^p/B_r^p$$

where $1 \leqq r \leqq \infty$, $-\infty < p < \infty$. It follows readily that F^pH is a filtration of H. All the results of § 1, 2, 4, 5 now carry over without any change. The questions studied in § 3 require some care.

We consider two systems $\{H(p,q)\}$ and $\{H'(p,q)\}$ with homomorphisms (1) and (2). A map f of the first system into the second is a family of maps (p,q): $H(p,q) \to H'(p,q)$ which properly commute with (1) and (2).

Clearly f induces maps $f_r^*\colon E_r \to E'_r$ for $1 \leqq r \leqq \infty$. If g is another map of $\{H(p,q)\}$ into $\{H'(p,q)\}$ then we say that f and g are *k-equivalent*, $k \geqq 0$, (notation $f \underset{k}{\sim} g$) if the composition

$$H(p,q) \xrightarrow{\gamma} H'(p,q) \longrightarrow H'(p-k,q-k)$$

is zero for all (p,q), where $\gamma = g(p,q) - f(p,q)$. As an analogue of 3.1 we prove that $f \underset{k}{\sim} g$ implies $f_r^* = g_r^*$ for $r > k$. As in the proof of 3.1 we must show that the composition

$$H(p,p+r) \xrightarrow{\gamma} H'(p,p+r) \longrightarrow H'(p-r+1,p+1)$$

is zero. This, however, is immediate, since the second map can be factored as follows

$$H'(p,p+r) \to H'(p-k,p+r-k) \to H'(p-r+1,p+1).$$

This argument, incidentally, gives a new proof for 3.1. The remaining results of § 3 carry over without change.

To conclude we observe that the exact sequences

$$\cdots \to H(p+1) \to H(p) \to H(p,p+1) \to H(p+1) \to \cdots$$

taken for all finite integers p may be recorded in a single diagram

$$\begin{array}{ccc} C & \longrightarrow & C \\ & \nwarrow \quad \swarrow & \\ & E_1 & \end{array} \tag{3}$$

where $C = \sum_p H(p)$, $E_1 = \sum_p H(p,p+1)$. Diagram (3) is an *exact couple* in the sense of Massey (*Ann. of Math.* 56 (1952), 363–396, 1952). This again provides an alternative exposition of spectral sequences.

EXERCISES

1. Let Λ and Λ' be two K-algebras, and $\Gamma = \Lambda \otimes_K \Lambda'$. Let A (resp. A', A'') be a Λ-complex (resp. Λ'-complex, resp. Γ-complex). Suppose we are given a map of Γ-complexes

$$\varphi\colon A \otimes_K A' \to A''$$

such that φ maps $F^p(A) \otimes F^{p'}(A')$ into $F^{p+p'}(A'')$.

Show that the homomorphism $\alpha\colon H(A) \otimes H(A') \to H(A'')$ induces maps

$$\bar{\varphi}^{p,p'}\colon E_0^p(H(A)) \otimes E_0^{p'}(H(A')) \to E_0^{p+p'}(H(A''))$$

compatible with the gradings.

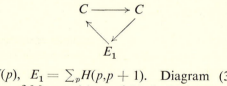

2. Using isomorphism (4) of § 1, give another definition of the map $\bar{\varphi}^{p,p'}$ of Exer. 1, by defining maps

$$Z^p_\infty(A) \otimes Z^{p'}_\infty(A') \to Z^{p+p'}_\infty(A'')$$

which induce maps of

$$B^p_\infty(A) \otimes Z^{p'}_\infty(A') \qquad \text{and} \qquad Z^p_\infty(A) \otimes B^{p'}_\infty(A')$$

into $B^{p+p'}_\infty(A'')$.

Show that the maps

$$\bar{\varphi}^{p,p'}\colon\ E^p_\infty(A) \otimes E^{p'}_\infty(A') \to E^{p+p'}_\infty(A'')$$

may be obtained by "passing to the limit" from maps

$$\varphi^{p,p'}_r\colon\ E^p_r(A) \otimes E^{p'}_r(A') \to E^{p+p'}_r(A'')$$

obtained by defining maps

$$Z^p_r(A) \otimes Z^{p'}_r(A') \to Z^{p+p'}_r(A'').$$

Define now on $E_r(A) \otimes E_r(A')$ a differentiation d_r by setting, for $a \in E^p_r(A)$, $a' \in E^{p'}_r(A')$,

$$d_r(a \otimes a') = (d_r a) \otimes a' + (-1)^p a \otimes (d_r a');$$

show that the maps $\varphi^{p,p'}_r$ are compatible with the differentiations of $E_r(A) \otimes E_r(A')$ and $E_r(A'')$. Passing to homology, the maps

$$H^p(E_r(A)) \otimes H^{p'}(E_r(A')) \to H^{p+p'}(E_r(A''))$$

are precisely the maps $\varphi^{p,p'}_{r+1}$ (by using the natural isomorphisms $H^p(E_r(A)) \approx E^p_{r+1}(A)$, $H^{p'}(E_r(A')) \approx E^{p'}_{r+1}(A')$, $H^{p+p'}(E_r(A'')) \approx E^{p+p'}_{r+1}(A'')$),

3. Let A be a graded K-algebra with a differentiation satisfying

$$d(aa') = (da)a' + (-1)^p a(da') \qquad p = \deg. a.$$

Then, taking $\Lambda = \Lambda' = K$, $A' = A'' = A$, apply Exer. 1 and 2 to the present situation.

4. Let $\{A^{p,q}\}$, $\{A'^{p',q'}\}$, $\{A''^{s,t}\}$ be three double complexes. Consider the double complex

$$C^{s,t} = \Sigma A^{p,q} \otimes A'^{p',q'}, \qquad p + p' = s, \qquad q + q' = t$$

with differentiation operators δ_1 and δ_2 defined as in IV,4. Assume we are given a map of double complexes

$$C^{s,t} \to A''^{s,t}.$$

Consider the first filtration on $\{A^{p,q}\}$, $\{A'^{p',q'}\}$ and $\{A''^{s,t}\}$; show that Exer. 1 and 2 may be applied to corresponding spectral sequences. The same result applies for the second filtration.

5. Let A be a double complex with differentiation operators d_1 and d_2. Denote $B_{\rm I} = \operatorname{Im} d_1$, $Z_{\rm I} = \operatorname{Ker} d_1$, and show that these are double complexes in which the first differentiation operator is zero. Similarly introduce $B_{\rm II}(A)$ and $Z_{\rm II}(A)$. Consider the doubly graded module

$$C(A) = Z_{\rm I}(A) \cap Z_{\rm II}(A) = Z(Z_{\rm II}(A)) = Z(Z_{\rm I}(A)).$$

Clearly $C^{p,q}(A) = A^{p,q} \cap Z(A)$. Define the maps in the diagram

$$\begin{array}{ccc} & C(A) = & Z_{\rm I}(Z_{\rm II}(A)) \\ & \downarrow f & \downarrow g \\ H(B_{\rm I}(A)) \xrightarrow{k} & H(A) & H_{\rm I}(H_{\rm II}(A)) \end{array}$$

where k and f are defined by inclusions $B_{\rm I}(A) \subset A$, $C(A) \subset A$, while g is defined by the natural homomorphisms $Z_{\rm II} \to H_{\rm II}$ and $Z_{\rm I} \to H_{\rm I}$. Show that g is a map of doubly graded groups, while k and f are compatible with both and the first and the second filtration. We denote by

$$f_{\rm I}: C(A) \to I_\infty(A) = \sum_p F_{\rm I}^p H(A)/F_{\rm I}^{p+1}H(A)$$

the induced map of the associated graded groups. Similarly we may introduce $f_{\rm II}$, $k_{\rm I}$ and $k_{\rm II}$.

6. With the notations as above, establish the equivalence of the following two propositions:

(*a*) If $a,b \in A$ and $d_1a = d_2b$, $d_2a = 0$, then there is $c \in A$ such that $d_1d_2c = d_1a$.

(*a'*) g is an epimorphism.

Assuming that the above conditions hold, show that:

(i) $f(F_{\rm I}^pC) = F_{\rm I}^pH(A)$, $\quad f(F_{\rm II}^qC) = F_{\rm II}^qH(A)$.

(ii) f, $f_{\rm I}$ and $f_{\rm II}$ are epimorphisms.

(iii) the first filtration of A is weakly convergent and satisfies $Z_\infty(A) = Z_2(A)$.

All the differentiation operators in the first spectral sequence of A are then zero and one obtains an isomorphism $\varphi: I_\infty(A) \approx H_{\rm I}(H_{\rm II}(A))$. This isomorphism satisfies $\varphi f_{\rm I} = g_{\rm I} = g$. [Hint to (i): if $a \in \sum_{0 \leqq i \leqq r} A^{p+i,q-i}$ is a d-cycle, then a is d-homologous to some element of $\sum_{0 \leqq i \leqq r} C^{p+i,q-i}$; proof by induction on r.]

7. With the notations of Exer. 5 show that an element $u \in C(A)$ is in the kernel of g if and only if there exist elements $b,c \in A$ such that $u = d_1b + d_2c$, $d_2b = 0$. Establish the equivalence of the following conditions:

(b) If $a \in A$ is such that $d_2 d_1 a = 0$, then there exist elements $b, c \in A$ satisfying $d_1 a = d_1 b + d_2 c$, $d_2 b = 0$.

(b') $g(Z(B_{\mathrm{I}}(A))) = 0$.

Assuming that the above conditions hold, prove:

(iv) $\operatorname{Ker} g = Z(B_{\mathrm{I}}(A)) + \operatorname{Ker} f$ [Hint: prove $\operatorname{Ker} f \subset \operatorname{Ker} g$ and $\operatorname{Ker} g \subset Z(B_{\mathrm{I}}(A)) + \operatorname{Ker} f$].

(v) $k(H^{p,q}(B_{\mathrm{I}}(A))) \subset k(H^{p+1,q-1}(B_{\mathrm{I}}(A)))$.

(vi) $k_{\mathrm{I}} = 0$.

(vii) $\operatorname{Im} k \subset \bigcap_p F_{\mathrm{I}}^p H(A)$.

(viii) If the filtration F_{I} of A is convergent, then $k = 0$.

8. Assume that in the double complex A conditions (a) and (b) of Exer. 6 and 7 are both satisfied. Establish the exact sequence

$$H(B_{\mathrm{I}}(A)) \xrightarrow{k} H(A) \xrightarrow{l} H_{\mathrm{I}}(H_{\mathrm{II}}(A)) \longrightarrow 0$$

where $l = gf^{-1}$, and show that the maps are compatible with both filtrations. In particular, $k_{\mathrm{I}} = 0$ and $l_{\mathrm{I}} = \varphi\colon \mathrm{I}_{\infty}(A) \approx H_{\mathrm{I}}(H_{\mathrm{II}}(A))$.

Show that if $k = 0$ then both filtrations of $H(A)$ are those obtained from the double grading of $H_{\mathrm{I}}H_{\mathrm{II}}(A)$ using the map l^{-1}.

Establish the equivalence of the following conditions:

(c) $k = 0$;

(d) The filtration F_{I} of A is convergent.

9. Suppose that the double complex A is the direct limit of double complexes A_α such that for each index α:

(1) A_α satisfies condition (b) of Exer. 7;

(2) the filtration F_{I} of A_α is convergent.

Then condition (b) holds in the complex A and the map $k\colon H(B_{\mathrm{I}}(A)) \to H(A)$ is zero. If moreover A satisfies condition (a) of Exer. 6, then l is an isomorphism $H(A) \approx H_{\mathrm{I}}(H_{\mathrm{II}}(A))$, which is compatible with both filtrations.

CHAPTER XVI

Applications of Spectral Sequences

Introduction. In v,8, we have considered functors of several variables and have studied the relations between the derived functors and the partial derived functors. Our results there were rather incomplete, because a complete treatment of the problem requires the use of spectral sequences (see § 1).

These spectral sequences arise each time we have an associativity relation of the type

$$\text{Hom}(A, \text{Hom}(B,C)) \approx \text{Hom}(A \otimes B,C).$$

There result two spectral sequences with essentially the same "limit," and with terms E_2 given by

$$\text{Ext}^p(A, \text{Ext}^q(B,C)), \qquad \text{Ext}^q(\text{Tor}_p(A,B),C).$$

This method provides a large number of spectral sequences. Among others, spectral sequences are obtained for the homology of a group, an invariant subgroup and the quotient group, as well as for the homology of a Lie algebra, an ideal and the quotient algebra.

§ 9 is a modest attempt to show how these spectral sequences can be applied to various problems in topology involving groups of operators.

1. PARTIAL DERIVED FUNCTORS

Let $T(A,C)$ be a functor covariant in A and contravariant in C. In addition to the right derived functors R^nT we shall consider the partial right derived functors $R^n_{(1)}T$ and $R^n_{(2)}T$ obtained by regarding one of the variables as active and the other one as passive. According to v,8 we then obtain natural maps

$$(1) \qquad R^n_{(1)}T \to R^nT, \qquad R^n_{(2)}T \to R^nT.$$

Let X be an injective resolution of A and Y a projective resolution of C. As described in iv,5, $T(X, Y)$ is then a double complex. An argument similar to the one given in v,3 shows that the invariants of this double complex are independent of the choice of the resolutions and are functors of A and C. The homology module of $T(X, Y)$ clearly is $RT(A,C)$.

Using the method of xv,6 for computing the initial terms of the spectral sequences we find

$$H_{\mathrm{II}}(T(X,Y)) = R_{(2)}T(X,C)$$

so that

$$\mathrm{I}_2 = H_{\mathrm{I}}H_{\mathrm{II}}(T(X,Y)) = H(R_{(2)}T(X,C)) = R_{(1)}R_{(2)}T(A,C).$$

With the double grading indicated we have

$$\mathrm{I}_2^{p,q} = R^p_{(1)}R^q_{(2)}T \tag{2}$$

and similarly

$$\mathrm{II}_2^{p,q} = R^q_{(2)}R^p_{(1)}T. \tag{3}$$

Since the double complex $T(X,Y)$ is positive, both filtrations are regular and we thus have

$$R^p_{(1)}R^q_{(2)}T \underset{p}{\Rightarrow} R^nT, \tag{4}$$

$$R^q_{(2)}R^p_{(1)}T \underset{q}{\Rightarrow} R^nT. \tag{5}$$

The above spectral sequences give rise to edge homomorphisms and exact sequences for terms of low degree. It will be convenient to assume that T is left exact so that $R^0T = R^0_{(1)}T = R^0_{(2)}T = T$. Then the edge homomorphisms are

$$R^n_{(1)}T \to R^nT, \qquad R^n_{(2)}T \to R^nT \tag{6}$$

$$R^nT \to R^0_{(1)}R^n_{(2)}T, \qquad R^nT \to R^0_{(2)}R^n_{(1)}T \tag{7}$$

while the exact sequences take the form

$$0 \to R^1_{(1)}T \to R^1T \to R^0_{(1)}R^1_{(2)}T \to R^2_{(1)}T \to R^2T \tag{8}$$

$$0 \to R^1_{(2)}T \to R^1T \to R^0_{(2)}R^1_{(1)}T \to R^2_{(2)}T \to R^2T. \tag{9}$$

We shall now show that the homomorphisms (1) and (6) coincide. To this end we consider the (single) complex $B = T(X,C)$ and define the filtration F of B by $F^pB = \sum B^n$ for $n \geq p$. The augmentation maps define a map $T(X,C) \to T(X,Y)$ which maps the filtration F into the filtration F_{I} of the complex $T(X,Y)$. There results a commutative diagram

$$\begin{array}{ccc} E_2^{n,0}(B) & \to & H^n(B) \\ \downarrow & & \downarrow \\ R^n_{(1)}T(A,C) & \to & R^nT(A,C). \end{array}$$

For the filtration F of B we find $E_0(B) = B$ and $E_1^{n,0}(B) = E_2^{n,0}(B) = H^n(B) = R^n_{(1)}T(A,C)$. The upper horizontal map is thus the identity and so is the left vertical map. It follows that the lower horizontal map and the right vertical map coincide. These are precisely the two maps $R^n_{(1)}T \to R^nT$ given in (6) and (1).

We can now give a short alternative proof of the fact established in v,8 that for a right balanced functor T, (1) are isomorphisms. Indeed, T being right balanced we have $R^q_{(2)}T(A,C) = 0$ for $q > 0$ and A injective. Therefore $R^p_{(1)}R^q_{(2)}T = 0$ for $q > 0$. Thus the first spectral sequence collapses and yields $R^n_{(1)}T \approx R^nT$.

The discussion generalizes easily to the case when A and C each represent a set of variables, some of which may be covariant and some contravariant.

Similar results may be obtained for left derived functors. The sequences (4) and (5) become

(4a) $$L_p^{(1)}L_q^{(2)}T \underset{p}{\Rightarrow} L^nT,$$

(5a) $$L_q^{(2)}L_p^{(1)}T \underset{q}{\Rightarrow} L_nT.$$

2. FUNCTORS OF COMPLEXES

Let $T(A,C)$ be a right balanced functor covariant in A and contravariant in C. We shall consider here the case when A is a complex and C is a module. Given a projective resolution Y of C, we obtain a double complex $T(A,Y)$. The invariants of this double complex are independent of the choice of Y and are functors of A and C.

We introduce the notation $\mathscr{R}^nT(A,C)$ for the homology module $H^n(T(A,Y))$. Next, we have $H_{\mathrm{II}}(T(A,Y)) = RT(A,C)$ so that

(1) $$\mathrm{I}_2^{p,q} = H^p(R^qT(A,C)).$$

Since $H_{\mathrm{I}}^{p,q}(T(A,Y)) = H^p(T(A,Y_q))$, and since the functor $T(A,Y_q)$ is exact for Y_q projective, it follows from IV,7.2 that we may identify $H^p(T(A,Y_q))$ with $T(H^p(A),Y_q)$. Thus $H_{\mathrm{I}}^{p,q}(T(A,Y)) = T(H^p(A),Y_q)$. Consequently applying H_{II} we obtain

(2) $$\mathrm{II}_2^{p,q} = R^qT(H^p(A),C)$$

The double complex $T(A,Y)$ falls into the case 1 of xv,6 so that we have the edge homomorphisms $\mathrm{I}_2^{n,0} \to H^n \to \mathrm{II}_2^{n,0}$ which in this case become

(3) $$H^n(T(A,C)) \to \mathscr{R}^nT(A,C) \to T(H^n(A),C).$$

PROPOSITION 2.1. *The composition of the homomorphisms* (3) *coincides with the homomorphism*

$$\alpha'\colon\ H^n(T(A,C)) \to T(H^n(A),C)$$

of IV,6.1a.

PROOF. Let $\bar{\alpha}$ denote the composition of the homomorphisms (3). Clearly $\bar{\alpha}$ is natural relative to maps $A \to A'$ of complexes. In view of IV,6.1a it therefore suffices to show that $\bar{\alpha}$ is the identity if A has zero differentiation. In this case the double complex $T(A,Y)$ has the first differentiation zero. Thus, by the final remark of xv,6, $\bar{\alpha}$ is the identity.

In all the applications that we shall encounter in this chapter, the complex A will be positive. In this case the double complex $T(A, Y)$ also will be positive and both filtrations will be regular. Thus we obtain

$$H^p(R^qT(A,C)) \underset{p}{\Rightarrow} \mathscr{R}^nT(A,C) \tag{4}$$

$$R^qT(H^p(A),C) \underset{q}{\Rightarrow} \mathscr{R}^nT(A,C). \tag{5}$$

A similar discussion can be carried out when A is a module and C is a (positive) complex. Using an injective resolution X of A we obtain a double complex $T(X,C)$ in which we regard the degree of C as the first degree and the degree of X as the second. This yields modules $\mathscr{R}^nT(A,C) = H^n(T(X,C))$ and spectral sequences

$$R^pT(A,H^q(C)) \underset{p}{\Rightarrow} \mathscr{R}^nT(A,C) \tag{4'}$$

$$H^q(R^pT(A,C)) \underset{q}{\Rightarrow} \mathscr{R}^nT(A,C). \tag{5'}$$

Both these cases are special cases of the more general situation considered in chapter XVII in which both A and C will be allowed to be complexes.

Quite analogous results are obtained for left balanced functors. The details are omitted.

3. COMPOSITE FUNCTORS

As a preparation for the later sections we shall consider here a functor

$$T(A,C) = U(A,V(C)) = U'(V'(A),C) \tag{1}$$

represented in two different ways as a composite functor.

We first consider the case when U and U' are both right balanced, contravariant in the first variable, covariant in the second variable, V is covariant and left exact while V' is covariant and right exact. Then T is left exact, contravariant in the first variable, and covariant in the second. We wish to compute $R^p_{(1)}R^q_{(2)}T(A,C)$ and $R^q_{(2)}R^p_{(1)}T(A,C)$.

If in (1) we replace A by a projective resolution X and C by an injective resolution Y, (1) yields a double complex. We have

$$\begin{aligned} H^{p,q}_{\text{II}}(T(X,Y)) = H^{p,q}_{\text{II}}(U(X,V(Y)) = H^q(U(X_p,V(Y)) &\approx U(X_p,H^qV(Y)) \\ &= U(X_p,R^qV(C)), \end{aligned}$$

where the isomorphism is given by the map α' of IV,6.1a, applied to the functor $U(X_p,C)$ which is exact since X_p is projective and U is right balanced. Consequently we find

$$H^{p,q}_{\text{I}}H_{\text{II}}(T(X,Y)) \approx R^pU(A,R^qV(C)).$$

Thus we have a spectral sequence

$$R^pU(A,R^qV(C)) \underset{p}{\Rightarrow} R^nT(A,C). \tag{2}$$

In the same way we have a spectral sequence

$$R^qU'(L_pV'(A),C) \underset{q}{\Rightarrow} R^nT(A,C). \tag{3}$$

The spectral sequences (2) and (3) yield edge homomorphisms

$$R^nU(A,V(C)) \to R^nT(A,C), \tag{4}$$

$$R^nU'(V'(A),C) \to R^nT(A,C), \tag{5}$$

$$R^nT(A,C) \to U(A,R^nV(C)), \tag{6}$$

$$R^nT(A,C) \to U'(L_nV'(A),C). \tag{7}$$

We shall show how these homomorphisms can be computed. We have

$$R^nU(A,V(C)) = H^n(U(X,V(C)) = H^n(T(X,C)) = R^n_{(1)}T(A,C),$$

and it follows from § 1 that (4) coincides with the natural homomorphism $R^n_{(1)}T \to R^nT$. Similarly (5) may be identified with the natural homomorphism $R^n_{(2)}T \to R^nT$.

Next we consider the composition of (4) and (7)

$$R^nU(A,V(C)) \to U'(L_nV'(A),C). \tag{8}$$

We have

$$R^nU(A,V(C)) = H^n(U(X,V(C)) = H^n(U'(V'(X),C))$$

$$U'(L_nV'(A),C) = U'(H_n(V'(X)),C)$$

and it follows from 2.1 that (8) coincides with the homomorphism α' of IV,6.1a. The composition of (5) and (6) can be computed similarly.

We also note here the exact sequences for terms of low degree that result from (2) and (3)

$$\begin{aligned} 0 \to R^1U(A,V(C)) \to R^1T(A,C) \to U(A,R^1V(C)) \\ \to R^2U(A,V(C)) \to R^2T(A,C) \end{aligned} \tag{9}$$

$$\begin{aligned} 0 \to R^1U'(V'(A),C) \to R^1T(A,C) \to U'(L_1V'(A),C) \\ \to R^2U(V'(A),C) \to R^2T(A,C) \end{aligned} \tag{10}$$

We shall also have occasion to consider the case when in (1) all the functors are covariant, U and U' are left balanced and V and V' are right exact. We then take projective resolutions X and Y of the variables A and C. The invariants of $T(X,Y)$ yield spectral sequences

(2a) $$L_pU(A,L_qV(C)) \underset{p}{\Rightarrow} L_nT(X,Y),$$

(3a) $$L_qU'(L_pV'(A),C) \underset{q}{\Rightarrow} L_nT(X,Y).$$

These yield edge homomorphisms

(4a) $$L_nT(A,C) \to L_nU(A,V(C)),$$

(5a) $$L_nT(A,C) \to L_nU'(V'(A),C),$$

(6a) $$U(A,L_nV(C)) \to L_nT(A,C),$$

(7a) $$U'(L_nV'(A),C) \to L_nT(A,C).$$

The rules for computing these homomorphisms are similar to the previous case. The exact sequences for terms of low degree are

(9a) $$L_2T(A,C) \to L_2U(A,V(C)) \to U(A,L_1V(C)) \to L_1T(A,C) \to L_1U(A,V(C)) \to 0$$

(10a) $$L_2T(A,C) \to L_2U'(V'(A),C) \to U'(L_1V'(A),C) \to L_1T(A,C) \to L_1U'(V'(A),C) \to 0.$$

4. ASSOCIATIVITY FORMULAE

We shall use the term "associativity formulae" for the type of isomorphisms established in II,5 and IX,2.

We begin with the situation described by the symbol $(A_{\Lambda\text{-}\Gamma},{}_\Lambda B_\Sigma,C_{\Gamma\text{-}\Sigma})$ where Λ, Γ and Σ are K-algebras. The identification of IX,2.2 yields two expressions for the functor

(1) $$T(A,C) = \operatorname{Hom}_{\Lambda\otimes\Gamma}(A, \operatorname{Hom}_\Sigma(B,C)) = \operatorname{Hom}_{\Gamma\otimes\Sigma}(A\otimes_\Lambda B,C).$$

We are thus exactly in the situation described in § 3 with

$$V(C) = \operatorname{Hom}_\Sigma(B,C), \qquad V'(A) = A\otimes_\Lambda B.$$

$$R^qV(C) = \operatorname{Ext}^q_\Sigma(B,C), \qquad L_pV'(A) = \operatorname{Tor}^\Lambda_p(A,B).$$

Now assume that Γ is K-projective, and let X be a $\Lambda\otimes\Gamma$-projective resolution of A, and Y a $\Gamma\otimes\Sigma$-injective resolution of C. It then follows from IX,2.4 and IX,2.4a that X is also a Λ-projective resolution of A and Y is a Σ-injective resolution of C. Consequently the spectral sequences (2) and (3) of § 3 become (if Γ is K-projective):

(2) $$\operatorname{Ext}^p_{\Lambda\otimes\Gamma}(A, \operatorname{Ext}^q_\Sigma(B,C)) \underset{p}{\Rightarrow} R^nT(A,C)$$

(3) $$\operatorname{Ext}^q_{\Gamma\otimes\Sigma}(\operatorname{Tor}^\Lambda_p(A,B),C) \underset{q}{\Rightarrow} R^nT(A,C).$$

Assuming that

$$\operatorname{Tor}_n^\Lambda(A,B) = 0 = \operatorname{Ext}_\Sigma^n(B,C) \qquad \text{for } n > 0,$$

both spectral sequences (2) and (3) collapse, and we obtain

$$\operatorname{Ext}_{\Lambda\otimes\Gamma}(A, \operatorname{Hom}_\Sigma(B,C)) \approx \operatorname{Ext}_{\Gamma\otimes\Sigma}(A \otimes_\Lambda B,C). \tag{4}$$

This is a generalization of IX,2.8a.

If we replace A by Λ and Γ by Λ^*, then the spectral sequence (3) collapses and (2) becomes

$$H^p(\Lambda, \operatorname{Ext}_\Sigma^q(B,C)) \underset{p}{\Rightarrow} \operatorname{Ext}_{\Lambda^*\otimes\Sigma}^n(B,C) \tag{5}$$

in the situation $({}_\Lambda B_{\Sigma}, {}_\Lambda C_\Sigma)$ and under the assumption that Λ is K-projective. This generalizes IX,4.3.

We now replace (Λ,Γ,Σ) by (K,Γ^e,Σ^e) and (A,B,C) by (Γ,Σ,C) with C a two-sided $\Gamma \otimes \Sigma$-module. Assuming that Γ is K-projective, it follows that Γ^e also is K-projective so that the spectral sequences (2) and (3) apply. They become

$$H^p(\Gamma,H^q(\Sigma,C)) \underset{p}{\Rightarrow} R^nT(\Gamma,C)$$

$$\operatorname{Ext}^q_{(\Gamma\otimes\Sigma)^e}(\operatorname{Tor}_p^K(\Gamma,\Sigma),C) \underset{q}{\Rightarrow} R^nT(\Gamma,C).$$

Since Γ is K-projective, the second sequence collapses to the isomorphism

$$R^nT(\Gamma,C) \approx \operatorname{Ext}^n_{(\Gamma\otimes\Sigma)^e}(\Gamma \otimes \Sigma,C) = H^n(\Gamma \otimes \Sigma,C).$$

Thus we obtain the spectral sequence

$$H^p(\Gamma,H^q(\Sigma,C)) \underset{p}{\Rightarrow} H^n(\Gamma \otimes \Sigma,C) \tag{6}$$

under the assumption that Γ is K-projective.

In view of X,2.1, the spectral sequence (6) applies also to the case when Γ and Σ are supplemented K-algebras provided both Γ and Σ are K-projective. However a stronger result may be obtained by directly substituting $\Lambda = A = B = K$ in (2) and (3). Then (3) collapses and (2) becomes

$$\operatorname{Ext}_\Gamma^p(K, \operatorname{Ext}_\Sigma^q(K,C)) \underset{p}{\Rightarrow} \operatorname{Ext}^n_{\Gamma\otimes\Sigma}(K,C), \qquad {}_{\Gamma-\Sigma}C \tag{7}$$

under the assumption that Γ is K-projective.

There is an analogous discussion for homology. We consider the situation $(A_{\Lambda\text{-}\Gamma}, {}_\Lambda B_\Sigma, {}_{\Gamma\text{-}\Sigma}C)$ and, using the identification of IX,2.1, define

$$T(A,C) = A \otimes_{\Lambda\otimes\Gamma}(B \otimes_\Sigma C) = (A \otimes_\Lambda B) \otimes_{\Gamma\otimes\Sigma} C. \tag{1a}$$

Then, under the assumption that Γ is K-projective, we obtain the spectral sequences

$$\operatorname{Tor}_p^{\Lambda\otimes\Gamma}(A, \operatorname{Tor}_q^{\Sigma}(B,C)) \underset{p}{\Rightarrow} L_nT(A,C) \tag{2a}$$

$$\operatorname{Tor}_q^{\Gamma\otimes\Sigma}(\operatorname{Tor}_p^{\Lambda}(A,B),C) \underset{q}{\Rightarrow} L_nT(A,C). \tag{3a}$$

If $\operatorname{Tor}_n^{\Lambda}(A,B) = 0 = \operatorname{Tor}_n^{\Sigma}(B,C)$ for $n > 0$, then both (2a) and (3a) collapse and we obtain

$$\operatorname{Tor}^{\Lambda\otimes\Gamma}(A,B \otimes_\Sigma C) \approx \operatorname{Tor}^{\Gamma\otimes\Sigma}(A \otimes_\Lambda B,C) \tag{4a}$$

which is a generalization of IX,2.8.

If we replace C by Σ and Γ by Σ^*, then (2a) collapses and (3a) becomes

$$H_q(\Sigma, \operatorname{Tor}_p^{\Lambda}(A,B)) \underset{q}{\Rightarrow} \operatorname{Tor}_n^{\Lambda\otimes\Sigma^*}(A,B) \tag{5a}$$

in the situation $({}_\Sigma A_\Lambda, {}_\Lambda B_\Sigma)$ and under the assumption that Σ is K-projective.

We now replace (Λ,Γ,Σ) by (Λ^e,Γ^e,K) and (A,B,C) by (A,Λ,Γ) with A a two-sided $\Lambda \otimes \Gamma$-module. Assuming that Γ is K-projective, the sequence (2a) collapses and (3a) becomes

$$H_q(\Gamma,H_p(\Lambda,A)) \underset{q}{\Rightarrow} H_n(\Lambda \otimes \Gamma,A). \tag{6a}$$

For supplemented algebras we obtain similarly

$$\operatorname{Tor}_q^{\Gamma}(\operatorname{Tor}_p^{\Lambda}(A,K),K) \underset{q}{\Rightarrow} \operatorname{Tor}_n^{\Lambda\otimes\Gamma}(A,K) \qquad A_{\Lambda\text{-}\Gamma} \tag{7a}$$

under the assumption that Γ is K-projective.

5. APPLICATIONS TO THE CHANGE OF RINGS

We apply the results of § 4 to obtain more detailed results for the "change of rings" as discussed in II,6 and VI,4. We assume a ring homomorphism

$$\varphi\colon \Lambda \to \Gamma$$

and adopt the various notations introduced in II,6. We shall treat Λ and Γ as Z-algebras and apply the results of § 4.

Case 1. $(A_\Lambda, {}_\Lambda\Gamma_\Gamma, {}_\Gamma C)$. Then (1a) of § 4 with (Λ,Γ,Σ) replaced by (Λ,Z,Γ) reduces to

$$T(A,C) = A \otimes_\Lambda C = A_{(\varphi)} \otimes_\Gamma C. \tag{1$_1$}$$

The spectral sequence (2a) of § 4 collapses to the isomorphisms $\operatorname{Tor}_n^{\Lambda}(A,C) \approx L_nT(A,C)$. Thus the spectral sequence (3a) of § 4 yields

$$\operatorname{Tor}_q^{\Gamma}(\operatorname{Tor}_p^{\Lambda}(A,\Gamma),C) \underset{q}{\Rightarrow} \operatorname{Tor}_n^{\Lambda}(A,C). \tag{2$_1$}$$

The edge homomorphisms are

$$(3)_1 \qquad \mathrm{Tor}_n^\Lambda (A,C) \to \mathrm{Tor}_n^\Gamma (A_{(\varphi)},C)$$

$$(4)_1 \qquad \mathrm{Tor}_n^\Lambda (A,\Gamma) \otimes_\Gamma C \to \mathrm{Tor}_n^\Lambda (A,C).$$

The homomorphism $(3)_1$ coincides with the homomorphism $f_{1,n}$ of VI,4. If $\mathrm{Tor}_p^\Lambda (A,\Gamma) = 0$ for $p > 0$, then $(2)_1$ collapses and $(3)_1$ is an isomorphism. We thus obtain a new proof of VI,4.1.1.

Case 2. $(A_\Gamma, {}_\Gamma\Gamma_\Lambda, {}_\Lambda C)$. Then (1a) of § 4, with (Λ,Γ,Σ) replaced by (Γ,Z,Λ) reduces to

$$(1)_2 \qquad T(A,C) = A \otimes_\Gamma ({}_{(\varphi)}C) = A \otimes_\Lambda C.$$

The spectral sequence (3a) of § 4 collapses to the isomorphism $\mathrm{Tor}_n^\Lambda (A,C) \approx L_nT(A,C)$. Thus (2a) of § 4 yields

$$(2)_2 \qquad \mathrm{Tor}_p^\Gamma (A, \mathrm{Tor}_q^\Lambda (\Gamma,C)) \underset{p}{\Rightarrow} \mathrm{Tor}_n^\Lambda (A,C).$$

The edge homomorphisms are

$$(3)_2 \qquad \mathrm{Tor}_n^\Lambda (A,C) \to \mathrm{Tor}_n^\Gamma (A,{}_{(\varphi)}C)$$

$$(4)_2 \qquad A \otimes_\Gamma \mathrm{Tor}_n^\Lambda (\Gamma,C) \to \mathrm{Tor}_n^\Lambda (A,C).$$

The homomorphism $(3)_2$ coincides with the homomorphism $f_{2,n}$ of VI,4. If $\mathrm{Tor}_q^\Lambda (\Gamma,C) = 0$ for $q > 0$, then $(2)_2$ collapses and $(3)_2$ becomes an isomorphism. We thus obtain a new proof of VI,4.1.2.

Case 3. $({}_\Lambda A, {}_\Gamma\Gamma_\Lambda, {}_\Gamma C)$. Then (1) of § 4 with (Λ,Γ,Σ) replaced by (Λ^*,Z,Γ^*) reduces to

$$(1)_3 \qquad T(A,C) = \mathrm{Hom}_\Lambda (A,C) = \mathrm{Hom}_\Gamma (A_{(\varphi)},C).$$

The spectral sequence (2) of § 4 collapses to the isomorphisms $\mathrm{Ext}_\Lambda^n (A,C) \approx R^nT(A,C)$, so that the spectral sequence (3) yields

$$(2)_3 \qquad \mathrm{Ext}_\Gamma^q (\mathrm{Tor}_p^\Lambda (\Gamma,A),C) \underset{q}{\Rightarrow} \mathrm{Ext}_\Lambda^n (A,C).$$

The edge homomorphisms are

$$(3)_3 \qquad \mathrm{Ext}_\Gamma^n ({}_{(\varphi)}A,C) \to \mathrm{Ext}_\Lambda^n (A,C)$$

$$(4)_3 \qquad \mathrm{Ext}_\Lambda^n (A,C) \to \mathrm{Hom}_\Gamma (\mathrm{Tor}_n^\Lambda (\Gamma,A),C).$$

The homomorphism $(3)_3$ coincides with the homomorphism $f_{3,n}$ of VI,4. If $\mathrm{Tor}_p^\Lambda (\Gamma,A) = 0$ for $p > 0$ then $(2)_3$ collapses and $(3)_3$ becomes an isomorphism. We thus obtain a new proof of VI,4.1.3.

Case 4. $({}_\Gamma A, {}_\Lambda\Gamma_\Gamma, {}_\Lambda C)$. Then (1) of § 4 with (Λ,Γ,Σ) replaced by (Γ^*,Z,Λ^*) reduces to

$$(1)_4 \qquad T(A,C) = \mathrm{Hom}_\Gamma (A,{}^{(\varphi)}C) = \mathrm{Hom}_\Lambda (A,C).$$

The spectral sequence (2) of § 4 collapses to the isomorphisms $\mathrm{Ext}^n_\Lambda(A,C) \approx R^nT(A,C)$, so that the spectral sequence (2) yields

$$(2)_4 \qquad \mathrm{Ext}^p_\Gamma(A, \mathrm{Ext}^q_\Lambda(\Gamma,C)) \underset{p}{\Rightarrow} \mathrm{Ext}^n_\Lambda(A,C).$$

The edge homomorphisms are

$$(3)_4 \qquad \mathrm{Ext}^n_\Gamma(A, {}^{(\varphi)}C) \to \mathrm{Ext}^n_\Lambda(A,C)$$

$$(4)_4 \qquad \mathrm{Ext}^n_\Lambda(A,C) \to \mathrm{Hom}_\Gamma(A, \mathrm{Ext}^n_\Lambda(\Gamma,C)).$$

The homomorphism $(3)_4$ coincides with the homomorphism $f_{4,n}$ of VI,4. If $\mathrm{Ext}^q_\Lambda(\Gamma,C) = 0$ for $q > 0$, then $(2)_4$ collapses and $(3)_4$ is an isomorphism. We thus obtain a new proof of VI,4.1.4.

6. NORMAL SUBALGEBRAS

Let Λ and Γ be supplemented K-algebras and consider a K-algebra homomorphism

$$\varphi\colon \Lambda \to \Gamma$$

compatible with the augmentations. In Γ we may consider the left ideal $\Gamma . I(\Lambda)$ generated by the image of $I(\Lambda)$ under φ. We shall say that the map φ is (right) *normal* if the left ideal $\Gamma . I(\Lambda)$ is also a right ideal. If φ is normal then $\Gamma . I(\Lambda)$ is a two-sided ideal contained in $I(\Gamma)$. Therefore $\Gamma/\Gamma . I(\Lambda)$ is again a supplemented K-algebra which will be denoted by $\Gamma//\varphi$. From the exact sequence $0 \to I(\Lambda) \to \Lambda \to K \to 0$ we deduce the exact sequence

$$\Gamma \otimes_\Lambda I(\Lambda) \to \Gamma \to \Gamma \otimes_\Lambda K \to 0$$

where Γ is regarded as a right Λ-module. Since $\Gamma . I(\Lambda) = \mathrm{Im}(\Gamma \otimes_\Lambda I(\Lambda) \to \Gamma)$ it follows that $\Gamma//\varphi$ may alternatively be defined by

$$\Gamma//\varphi = \Gamma \otimes_\Lambda K.$$

THEOREM 6.1. *If the map $\varphi\colon \Lambda \to \Gamma$ is (right) normal and if Γ, regarded as a right Λ-module, is Λ-projective, then setting $\Omega = \Gamma//\varphi$ we have the spectral sequences*

$$(1) \qquad \mathrm{Ext}^p_\Omega(A, \mathrm{Ext}^q_\Lambda(K,C)) \underset{p}{\Rightarrow} \mathrm{Ext}^n_\Gamma(A,C), \qquad ({}_\Omega A, {}_\Gamma C),$$

$$(1a) \qquad \mathrm{Tor}^\Omega_p(\mathrm{Tor}^\Lambda_q(A,K),C) \underset{p}{\Rightarrow} \mathrm{Tor}^\Gamma_n(A,C), \qquad (A_\Gamma, C_\Omega).$$

The operators of $\Gamma//\varphi$ on $\mathrm{Ext}^q_\Lambda(K,C)$ and $\mathrm{Tor}^\Lambda_q(A,K)$ will be defined below.

PROOF. With $\Omega = \Gamma//\varphi$ we consider the situations

$$({}_\Omega A, {}_\Gamma\Omega_\Omega, {}_\Gamma C), \qquad (A_\Gamma, {}_\Gamma\Omega_\Omega, {}_\Omega C).$$

These are cases 4 and 1 of the change of rings corresponding to the natural map $\psi\colon \Gamma \to \Omega$. Thus the spectral sequences $(2)_4$ and $(2)_1$ yield

$$\text{(2)} \qquad \operatorname{Ext}^p_\Omega(A, \operatorname{Ext}^q_\Gamma(\Omega,C)) \underset{p}{\Rightarrow} \operatorname{Ext}^n_\Gamma(A,C),$$

$$\text{(2a)} \qquad \operatorname{Tor}^\Omega_p(\operatorname{Tor}^\Gamma_q(A,\Omega),C) \underset{p}{\Rightarrow} \operatorname{Tor}^\Gamma_n(A,C).$$

Since Γ is right Λ-projective, VI,4.1.3 and VI,4.1.2 imply isomorphisms:

$$\operatorname{Ext}^q_\Gamma(\Omega,C) = \operatorname{Ext}^q_\Gamma(\Gamma \otimes_\Lambda K,C) \approx \operatorname{Ext}^q_\Lambda(K,C),$$

$$\operatorname{Tor}^\Gamma_q(A,\Omega) = \operatorname{Tor}^\Gamma_q(A,\Gamma \otimes_\Lambda K) \approx \operatorname{Tor}^\Lambda_q(A,K).$$

This introduces left Ω-operators in $\operatorname{Ext}^q_\Lambda(K,C)$ and right Ω-operators in $\operatorname{Tor}^\Lambda_q(A,K)$. Carrying out the appropriate replacements in (2) and (2a) we obtain (1) and (1a).

In most interesting applications Λ will be a subalgebra of Γ and φ will be the inclusion map. In this case we say that "Λ is normal" instead of "φ is normal" and write $\Gamma//\Lambda$ instead of $\Gamma//\varphi$.

Consider an invariant subgroup π of a group Π. Then $Z(\pi)$ is a subring of $Z(\Pi)$. For $x \in \Pi$, $y \in \pi$ we have

$$x(y-1) = (xyx^{-1}-1)x, \qquad (y-1)x = x(x^{-1}yx-1).$$

This shows that $Z(\Pi) \,.\, I(\pi) = I(\pi) \,.\, Z(\Pi)$, and thus $Z(\pi)$ is a normal subalgebra of $Z(\Pi)$. It is further clear that $Z(\Pi)//Z(\pi) = Z(\Pi) \otimes_\pi Z = Z(\Pi/\pi)$. Since $Z(\Pi)$ is π-projective it follows that 6.1 applies and we obtain spectral sequences

$$\text{(3)} \qquad \operatorname{Ext}^p_{\Pi/\pi}(A,H^q(\pi,C)) \underset{p}{\Rightarrow} \operatorname{Ext}^n_\Pi(A,C), \qquad ({}_{\Pi/\pi}A, {}_\Pi C),$$

$$\text{(3a)} \qquad \operatorname{Tor}^{\Pi/\pi}_p(H_q(\pi,A),C) \underset{p}{\Rightarrow} \operatorname{Tor}^\Pi_n(A,C), \qquad (A_\Pi, {}_{\Pi/\pi}C).$$

Taking $A = Z$ in (3) and $C = Z$ in (3a) we obtain

$$\text{(4)} \qquad H^p(\Pi/\pi,H^q(\pi,C)) \underset{p}{\Rightarrow} H^n(\Pi,C), \qquad {}_\Pi C,$$

$$\text{(4a)} \qquad H_p(\Pi/\pi,H_q(\pi,A)) \underset{p}{\Rightarrow} H_n(\Pi,A), \qquad A_\Pi.$$

The sequence (4) is that of Hochschild-Serre (*Trans. Am. Math. Soc.* 74 (1953), 110–134).

As a second example consider an ideal $\mathfrak{h}$ of a Lie algebra $\mathfrak{g}$ over K, and assume that $\mathfrak{h}$ and $\mathfrak{g}/\mathfrak{h}$ are K-free. Take

$$\Lambda = \mathfrak{h}^e, \qquad \Gamma = \mathfrak{g}^e, \qquad \Omega = (\mathfrak{g}/\mathfrak{h})^e.$$

Then we have the natural maps

$$\varphi\colon \Lambda \to \Gamma, \qquad \psi\colon \Gamma \to \Omega.$$

It follows from XIII,4.1 that Γ regarded as a right Λ-module is free. In particular φ is a monomorphism which may be regarded as an inclusion. In XIII,1.3 we have studied the kernel of the epimorphism ψ and have proved that its kernel is the left ideal $\Gamma.\, I(\Lambda)$ (which coincides with $I(\Lambda).\Gamma$). This proves that the subalgebra Λ of Γ is (both left and right) normal and that $\Omega = \Gamma//\Lambda$. It follows that 6.1 applies; we obtain spectral sequences

(5) $$\operatorname{Ext}^p_\Omega (A, H^q(\mathfrak{h},C)) \underset{p}{\Rightarrow} \operatorname{Ext}^n_\Gamma (A,C) \qquad ({}_{\mathfrak{g}/\mathfrak{h}}A, {}_{\mathfrak{g}}C)$$

(5a) $$\operatorname{Tor}^\Omega_p (H_q(\mathfrak{h},A),C) \underset{p}{\Rightarrow} \operatorname{Tor}^\Gamma_n (A,C) \qquad (A_{\mathfrak{g}}, {}_{\mathfrak{g}/\mathfrak{h}}C).$$

Taking $A = K$ in (5) and $C = K$ in (5a) we obtain

(6) $$H^p(\mathfrak{g}/\mathfrak{h}, H^q(\mathfrak{h},C)) \underset{p}{\Rightarrow} H^n(\mathfrak{g},C), \qquad {}_{\mathfrak{g}}C$$

(6a) $$H_p(\mathfrak{g}/\mathfrak{h}, H_q(\mathfrak{h},A)) \underset{p}{\Rightarrow} H_n(\mathfrak{g},A), \qquad A_{\mathfrak{g}}.$$

The sequence (6) is that of Hochschild-Serre (*Ann. of Math.* 57 (1953), 591–603).

7. ASSOCIATIVITY FORMULAE USING DIAGONAL MAPS

We shall consider a supplemented K-algebra Λ together with a diagonal map $D\colon \Lambda \to \Lambda \otimes \Lambda$ and an antipodism $\omega\colon \Lambda \to \Lambda^*$ satisfying conditions (i)–(vi) of XI,8. In the situation $({}_\Lambda A, {}_\Lambda B, {}_\Lambda C)$ we may use the identification of XI,8.1 to obtain two representations for the functor

(1) $$T(A,C) = \operatorname{Hom}_\Lambda (A, \operatorname{Hom} (B,C)) = \operatorname{Hom}_\Lambda (A \otimes B,C).$$

Thus applying the procedure of § 3 we obtain spectral sequences

(2) $$\operatorname{Ext}^p_\Lambda (A, \operatorname{Ext}^q (B,C)) \underset{p}{\Rightarrow} R^nT(A,C),$$

(3) $$\operatorname{Ext}^q_\Lambda (\operatorname{Tor}_p (A,B),C) \underset{q}{\Rightarrow} R^nT(A,C).$$

Here $\operatorname{Ext}^q (B,C)$ and $\operatorname{Tor}_p (A,B)$ are regarded first as left $\Lambda \otimes \Lambda^*$- and $\Lambda \otimes \Lambda$-modules and then converted into left Λ-modules using the maps $E\colon \Lambda \to \Lambda \otimes \Lambda^*$ and $D\colon \Lambda \to \Lambda \otimes \Lambda$. We recall that Ext^q and Tor_p stand for Ext^q_K, Tor^K_p.

Assuming that $\operatorname{Tor}_p (A,B) = 0$ for $p > 0$ the spectral sequence (3) collapses and (2) becomes

(4) $$\operatorname{Ext}^p_\Lambda (A, \operatorname{Ext}^q (B,C)) \underset{p}{\Rightarrow} \operatorname{Ext}^n_\Lambda (A \otimes B,C).$$

In particular, for $A = K$ we obtain

(5) $$\operatorname{Ext}^p_\Lambda (K, \operatorname{Ext}^q (B,C)) \underset{p}{\Rightarrow} \operatorname{Ext}^n_\Lambda (B,C).$$

If further $\operatorname{Ext}^q(B,C) = 0$ for $q > 0$ then (5) collapses to the isomorphism

$$\operatorname{Ext}^n_\Lambda(K, \operatorname{Hom}(B,C)) \approx \operatorname{Ext}^n_\Lambda(B,C). \tag{6}$$

A similar discussion applies to homology. In the situation $(A_\Lambda, {}_\Lambda B, {}_\Lambda C)$ we use XI,8.1a to define

$$T(A,C) = A \otimes_\Lambda (B \otimes C) = (A \otimes B) \otimes_\Lambda C. \tag{1a}$$

This yields spectral sequences

$$\operatorname{Tor}^\Lambda_p(A, \operatorname{Tor}_q(B,C)) \underset{p}{\Rightarrow} L_n T(A,C), \tag{2a}$$

$$\operatorname{Tor}^\Lambda_q(\operatorname{Tor}_p(A,B),C) \underset{q}{\Rightarrow} L_n T(A,C). \tag{3a}$$

If $\operatorname{Tor}_q(B,C) = 0$ for $q > 0$, then (2a) collapses and (3a) becomes

$$\operatorname{Tor}^\Lambda_q(\operatorname{Tor}_p(A,B),C) \underset{q}{\Rightarrow} \operatorname{Tor}^\Lambda_n(A,B \otimes C). \tag{4a}$$

In particular, for $C = K$ we obtain

$$\operatorname{Tor}^\Lambda_q(\operatorname{Tor}_p(A,B),K) \underset{q}{\Rightarrow} \operatorname{Tor}^\Lambda_n(A,B). \tag{5a}$$

If further $\operatorname{Tor}_p(A,B) = 0$ for $p > 0$, then

$$\operatorname{Tor}^\Lambda_n(A \otimes B,K) \approx \operatorname{Tor}^\Lambda_n(A,B). \tag{6a}$$

The considerations of this section are applicable to groups and to Lie algebras; we only need to replace Λ by $Z(\Pi)$ or by $\mathfrak{g}^e$.

8. COMPLEXES OVER ALGEBRAS

Let Λ be a K-projective supplemented K-algebra. We shall consider a positive complex C composed of left Λ-modules and in which the differentiation is a Λ-homomorphism. To the functor

$$T(A,C) = \operatorname{Hom}_\Lambda(A,C)$$

where A is a left Λ-module, we apply the considerations of § 2. The spectral sequences (4′) and (5′) of § 2 then give

$$\operatorname{Ext}^p_\Lambda(A,H^q(C)) \underset{p}{\Rightarrow} \mathscr{R}^n T(A,C)$$

$$H^q(\operatorname{Ext}^p_\Lambda(A,C)) \underset{q}{\Rightarrow} \mathscr{R}^n T(A,C).$$

If we take $A = K$ and denote $\mathscr{R}^n T(K,C)$ by $\mathscr{H}^n(\Lambda,C)$ we obtain

$$H^p(\Lambda,H^q(C)) \underset{p}{\Rightarrow} \mathscr{H}^n(\Lambda,C) \tag{1}$$

$$H^q(H^p(\Lambda,C)) \underset{q}{\Rightarrow} \mathscr{H}^n(\Lambda,C). \tag{2}$$

We recall that in virtue of the definition of $\mathscr{R}^nT$ we have

$$\mathscr{H}^n(\Lambda,C) = H^n(\operatorname{Hom}_\Lambda(X,C))$$

where X is a Λ-projective resolution of K and $\operatorname{Hom}_\Lambda(X,C)$ is regarded as a double complex.

PROPOSITION 8.1. *If $H^q(C) = 0$ for $q \neq 0$, then we have the spectral sequence*

$$H^q(H^p(\Lambda,C), \underset{q}{\Rightarrow} H^n(\Lambda,H^0(C))). \tag{3}$$

Indeed, in this case the spectral sequence (1) collapses to the isomorphisms $\mathscr{H}^n(\Lambda,C) \approx H^n(\Lambda,H^0(C))$.

PROPOSITION 8.2. *If C is weakly injective, then we have the spectral sequence*

$$H^p(\Lambda,H^q(C)) \underset{p}{\Rightarrow} H^n(\operatorname{Hom}_\Lambda(K,C)). \tag{4}$$

Indeed, in this case $H^p(\Lambda,C) = 0$ for $p > 0$ so that the spectral sequence (2) collapses to isomorphisms $\mathscr{H}^n(\Lambda,C) \approx H^n(H^0(\Lambda,C)) = H^n(\operatorname{Hom}_\Lambda(K,C))$.

PROPOSITION 8.3. *If the conditions of* 8.1 *and* 8.2 *are simultaneously verified then*

$$H^n(\Lambda,H^0(C)) \approx H^n(\operatorname{Hom}_\Lambda(K,C)). \tag{5}$$

This result may be interpreted as follows. The complex C may be called a weakly injective resolution of $H^0(C)$. The isomorphism (5) then generalizes the usual rule for computing $H^n(\Lambda,H^0(C))$ using an injective resolution of $H^0(C)$.

Similar considerations apply to homology. We denote by A a negative right Λ-complex and consider the functor

$$T(A,C) = A \otimes_\Lambda C$$

where C is a left Λ-module. In particular, taking $C = K$, we obtain spectral sequences

$$H_p(H_q(\Lambda,A)) \underset{p}{\Rightarrow} \mathscr{H}_n(\Lambda,A) \tag{1a}$$

$$H_q(\Lambda,H_p(A)) \underset{q}{\Rightarrow} \mathscr{H}_n(\Lambda,A) \tag{2a}$$

where

$$\mathscr{H}_n(\Lambda,A) = H_n(A \otimes_\Lambda X)$$

for any Λ-projective resolution X of K.

PROPOSITION 8.1a. *If $H_p(A) = 0$ for $p > 0$, then we have the spectral sequence*

(3a) $$H_p(H_q(\Lambda,A)) \underset{p}{\Rightarrow} H_n(\Lambda,H_0(A)).$$

PROPOSITION 8.2a. *If A is weakly projective, then we have the spectral sequence*

(4a) $$H_q(\Lambda,H_p(A)) \underset{q}{\Rightarrow} H_n(A \otimes_\Lambda Z).$$

PROPOSITION 8.3a. *If the conditions of* 8.1a *and* 8.2a *are simultaneously verified then*

(5a) $$H_n(\Lambda,H_0(A)) \approx H_n(A \otimes_\Lambda Z).$$

We shall apply the above results to the case $\Lambda = Z(\Pi)$ where Π is a group with unit augmentation. We consider a negative Π-complex X. Thus (using lower indices) $X_n = 0$ for $n < 0$.

THEOREM 8.4. *If X is weakly projective then for any left* Π*-module C we have the spectral sequence*

(6) $$H^p(\Pi,H^q(\mathrm{Hom}\,(X,C))) \underset{p}{\Rightarrow} H^n(\mathrm{Hom}_\Pi\,(X,C))$$

while for any right Π*-module A we have*

(6a) $$H_p(\Pi,H_q(A \otimes X)) \underset{p}{\Rightarrow} H_n(A \otimes_\Pi X).$$

We recall that $s \in \Pi$ operates on $\mathrm{Hom}\,(X,C)$ and $A \otimes X$ as follows

$$(sf)x = s[f(s^{-1}x)], \qquad (a \otimes x)s = as \otimes s^{-1}x.$$

For the proof, we first observe that, by x,8.5, $\mathrm{Hom}\,(X,C)$ is weakly injective and $A \otimes X$ is weakly projective. Thus 8.2 and 8.2a yield

$$H^p(\Pi,H^q(\mathrm{Hom}\,(X,C)) \underset{p}{\Rightarrow} H^n(\mathrm{Hom}_\Pi\,(Z, \mathrm{Hom}\,(X,C))$$

$$H_p(\Pi,H_q(A \otimes X)) \underset{p}{\Rightarrow} H_n((A \otimes X) \otimes_\Pi Z).$$

Since by associativity relations

$$\mathrm{Hom}_\Pi\,(K, \mathrm{Hom}\,(X,C)) \approx \mathrm{Hom}_\Pi\,(X,C)$$

$$(A \otimes X) \otimes_\Pi K \approx A \otimes_\Pi X,$$

the spectral sequences (6) and (6a) follow.

The spectral sequence (6) was first indicated by H. Cartan and J. Leray (*Colloque Top. Alg.*, Paris, 1947, pp. 83–85) for the case when Π is a finite group, and by H. Cartan (*C.R. Acad. Sci. Paris* 226 (1948), 303–305) for the general case.

In the next section we shall give a number of topological applications of theorem 8.4.

9. TOPOLOGICAL APPLICATIONS

Let $\mathscr{X}$ be a topological space on which a group Π operates on the left. We shall assume that the operators of Π are *proper*, i.e. that for each point $x \in \mathscr{X}$ there is a neighborhood U such that

$$U \cap sU = \emptyset \qquad \text{for } s \in \Pi, s \neq 1.$$

In particular, for $s \neq 1$, the transformation $s\colon \mathscr{X} \to \mathscr{X}$ admits no fixed points.

We shall denote by $\mathscr{X}_\Pi$, the space obtained from $\mathscr{X}$ by identifying each point x with its images sx, $s \in \Pi$. If we assume that $\mathscr{X}$ is arcwise connected, then $\mathscr{X}$ is a regular covering of $\mathscr{X}_\Pi$ and the fundamental group of $\mathscr{X}$ may be identified with an invariant subgroup of the fundamental group of $\mathscr{X}_\Pi$. The factor group is then the group Π.

Let X denote the total singular complex of the space $\mathscr{X}$. Clearly X has left Π-operators and, since the transformations of Π on $\mathscr{X}$ have no fixed points, X is Π-free.

Given a right Π-module A and a left Π-module C we may consider the homology and cohomology groups

$$H_n(\mathscr{X};A) = H_n(A \otimes X), \qquad H^n(\mathscr{X};C) = H^n(\operatorname{Hom}(X,C)).$$

The operators of Π on A and C are not used in this definition. However, they are used (together with the operators of Π on X) to convert $H_n(\mathscr{X};A)$ into a right Π-module and $H^n(\mathscr{X};C)$ into a left Π-module.

The modules A and C may also be regarded as *local coefficient systems* on the space $\mathscr{X}_\Pi$ and can be used in defining the homology and cohomology groups

$$H_n(\mathscr{X}_\Pi;A), \qquad H^n(\mathscr{X}_\Pi;C)$$

with local coefficients. It is well known that we have natural isomorphisms

$$H^n(\mathscr{X}_\Pi;C) \approx H^n(\operatorname{Hom}_\Pi(X,C))$$

$$H_n(\mathscr{X}_\Pi;A) \approx H_n(A \otimes_\Pi X).$$

Since X is Π-free, we may apply theorem 8.4. We thus obtain spectral sequences

$$H^p(\Pi,H^q(\mathscr{X};C)) \underset{p}{\Rightarrow} H^n(\mathscr{X}_\Pi;C), \tag{1}$$

$$H_p(\Pi,H_q(\mathscr{X};A)) \underset{p}{\Rightarrow} H_n(\mathscr{X}_\Pi;A). \tag{1a}$$

We shall now examine various applications of these spectral sequences.

Application 1. Assume that $\mathscr{X}$ is pathwise connected and that for some integer n

$$H^q(\mathscr{X};C) = 0 \qquad \text{for } 0 < q < n. \tag{2}$$

Then $H^0(\mathscr{X};C) = C$. By xv,5.12, the spectral sequence (1) gives isomorphisms

$$H^q(\mathscr{X}_\Pi;C) \approx H^q(\Pi,C), \qquad q < n, \tag{3}$$

and an exact sequence

$$0 \to H^n(\Pi,C) \to H^n(\mathscr{X}_\Pi;C) \to [H^n(\mathscr{X};C)]^\Pi \to H^{n+1}(\Pi,C) \to H^{n+1}(\mathscr{X}_\Pi;C). \tag{4}$$

For analogous homology results we assume that $\mathscr{X}$ is pathwise connected and that

$$H_q(\mathscr{X};A) = 0 \qquad \text{for } 0 < q < n. \tag{2a}$$

Then $H_0(\mathscr{X};A) = A$ and using xv,5.12a we obtain isomorphisms

$$H_q(\mathscr{X}_\Pi;A) \approx H_q(\Pi,A) \qquad q < n \tag{3a}$$

and the exact sequence

$$H_{n+1}(\mathscr{X}_\Pi;A) \to H_{n+1}(\Pi,A) \to [H_n(\mathscr{X};A)]_\Pi \to H_n(\mathscr{X}_\Pi;A) \to H_n(\Pi,A) \to 0. \tag{4a}$$

These results include various results of Eckmann (*Comment. Math. Helv.* 18 (1945), 232–282), Eilenberg-MacLane (*Proc. Nat. Acad. Sci. U.S.A.* 29 (1943), 155–158; *Trans. Am. Math. Soc.* 65 (1949), 49–99; *Ann. of Math.* 51 (1950), 514–533) and Hopf (*Comment. Math. Helv.* 17 (1944), 39–79). Not included are the results of Eilenberg-MacLane dealing with the invariant $\boldsymbol{k}^{n+1}$ (see Exer. 9). The knowledge of this invariant yields a complete determination of $H^n(\mathscr{X}_\Pi;C)$ and $H_n(\mathscr{X}_\Pi;A)$ rather than the partial information contained in the exact sequences (4) and (4a).

Application 2. Assume that Π is cyclic infinite with generator s. Then a Π-projective resolution of Z is given by

$$0 \longrightarrow Z(\Pi) \xrightarrow{d} Z(\Pi) \longrightarrow Z \longrightarrow 0$$

where d is multiplication by $s - 1$. It follows that

$$H^0(\Pi,C) = C^\Pi, \quad H^1(\Pi,C) = C_\Pi, \quad H^p(\Pi,C) = 0 \qquad \text{for } p > 1$$

$$H_0(\Pi,A) = A_\Pi, \quad H_1(\Pi,A) = A^\Pi, \quad H_p(\Pi,A) = 0 \qquad \text{for } p > 1.$$

Therefore in the spectral sequences (1) and (1a) non-zero terms occur only for $p = 0$, 1 and case E of XV,5 applies. We thus obtain exact sequences

$$0 \to [H^{n-1}(\mathscr{X};A)]_\Pi \to H^n(\mathscr{X}_\Pi;A) \to [H^n(\mathscr{X};A)]^\Pi \to 0$$

$$0 \to [H_n(\mathscr{X};A)]_\Pi \to H_n(\mathscr{X}_\Pi,A) \to [H_{n-1}(\mathscr{X};A)]^\Pi \to 0$$

For a direct proof see Serre (*Ann. of Math.* 54 (1951), 503).

Application 3. Assume that $\mathscr{X}$ is an n-dimensional manifold which is acyclic, i.e.

$$H_0(\mathscr{X};Z) = Z, \qquad H_q(\mathscr{X};Z) = 0 \qquad \text{for } q > 0.$$

The Euclidean n-space is an example of such a manifold. It follows from the Künneth relations (VI,3.1 and VI,3.1a) that

$$H^0(\mathscr{X};C) = C, \qquad H^q(\mathscr{X};C) = 0 \qquad \text{for } q > 0$$

$$H_0(\mathscr{X};A) = A, \qquad H_q(\mathscr{X};A) = 0 \qquad \text{for } q > 0$$

for any coefficient group. This can also be deduced from the fact that the singular complex X of $\mathscr{X}$ is a Z-projective resolution of Z.

The spectral sequences (1) and (1a) collapse, therefore, to the isomorphisms

$$H^p(\mathscr{X}_\Pi;C) \approx H^p(\Pi;C),$$

$$H_p(\mathscr{X}_\Pi;A) \approx H_p(\Pi;A).$$

Since $\mathscr{X}_\Pi$ is an n-dimensional manifold, we have $H^p(\mathscr{X}_\Pi;C) = 0$ for $p > n$. Therefore

$$H^p(\Pi,C) = 0 \qquad \text{for } p > n$$

for all coefficient modules C. This means that (cf. X,6.2)

$$\dim_{Z(\Pi)} Z = \dim Z(\Pi) \leqq n.$$

If further $\mathscr{X}_\Pi$ is compact, then $H_n(\Pi,Z_2) \approx H_n(\mathscr{X}_\Pi,Z_2) \neq 0$, and thus

$$\dim_{Z(\Pi)} Z = \dim Z(\Pi) = n.$$

This imposes severe limitations upon the groups Π that can operate properly on $\mathscr{X}$. In particular, all finite groups (except for $\Pi = \{1\}$) are excluded (see XII, Exer. 2).

Application 4. Assume that $\mathscr{X} = S^n$ is the n-sphere. Since $\mathscr{X}$ is compact and Π operates properly, it follows that Π must be finite. If n is even, then because of well known fixed point theorems every element $s \in \Pi$, $s \neq 1$ must reverse the orientation of S^n. Therefore either

$\Pi = \{1\}$ or $\Pi = Z_2$. Eliminating this not too interesting case, we may assume that *n is odd*. Then each element $s \in \Pi$ preserves the orientation of S^n. Further

$$H^q(S^n;C) = C \qquad \text{for } q = 0,n$$

$$H_q(S^n;A) = A \qquad \text{for } q = 0,n$$

and the remaining groups are zero. In the spectral sequence (1) and (1a) non-zero terms are obtained only for $q = 0,n$. Thus XV,5.11 implies the exact sequences

$$\cdots \to H^p(\Pi,C) \to H^p(S^n_\Pi;C) \to H^{p-n}(\Pi,C) \to H^{p+1}(\Pi,C) \to \cdots$$

$$\cdots \to H_{p+1}(\Pi,A) \to H_{p-n}(\Pi,A) \to H_p(S^n_\Pi;A) \to H_p(\Pi,A) \to \cdots$$

Since S^n_Π is an n-dimensional manifold, we have

$$H^p(S^n_\Pi;C) = 0 = H_p(S^n_\Pi;A) \qquad \text{for } p > n.$$

The exact sequences above thus yield the isomorphisms

$$H^p(S^n_\Pi;C) \approx H^p(\Pi;C), \qquad 0 \leqq p < n$$

$$H_p(S^n_\Pi;A) \approx H_p(\Pi;A), \qquad 0 \leqq p < n$$

$$H^i(\Pi,C) \approx H^{i+n+1}(\Pi,C), \qquad i > 0$$

$$H_i(\Pi,A) \approx H_{i+n+1}(\Pi,A) \qquad i > 0.$$

We now consider the complete derived sequence $\hat{H}^q(\Pi,C)$ of XII,2. We have the natural isomorphism $\hat{H}^1(\Pi,C) \approx \hat{H}^{n+2}(\Pi,C)$. Since $\hat{H}^0$ and $\hat{H}^{n+1}$ are left satellites of $\hat{H}^1$ and $\hat{H}^{n+2}$ there results an isomorphism $\hat{H}^{n+1}(\Pi,C) \approx \hat{H}^0(\Pi,C)$. Therefore the finite group Π has period $n + 1$, in the sense of XII,11. We have seen in XII,11 that this imposes severe limitations on Π. It is an open question whether every finite group Π with period $n + 1$ (n odd) can operate properly on S^n.

10. THE ALMOST ZERO THEORY

Let Π be a group and X a negative (left) Π-complex (i.e. $X_n = 0$ for $n < 0$) and A an abelian group. An n-dimensional cochain f: $X_n \to A$ is called Π-*finite* if for every $x \in X_n$ we have $f(sx) = 0$ for all but a finite number of elements $s \in \Pi$. The Π-finite cochains form a subcomplex $\overline{\text{Hom}}\,(X,A)$ of Hom (X,A).

With each Π-finite cochain f: $X_n \to A$ we associate the cochain $f' \in \text{Hom}_\Pi\,(X_n,Z(\Pi) \otimes A)$ given by

$$f'x = \sum_{s \in \Pi} s \otimes f(s^{-1}x).$$

Here $Z(\Pi) \otimes A$ is regarded as a left Π-module using the left Π-operators of $Z(\Pi)$. Conversely any f' has the form $f'x = \sum_{s \in \Pi} s \otimes g(s,x)$ and we may define f by setting $fx = g(1,x)$. It is now clear that we obtain an isomorphism

$$(1) \qquad \overline{\mathrm{Hom}}\,(X,A) \approx \mathrm{Hom}_{\Pi}\,(X,Z(\Pi) \otimes A).$$

If X is a projective Π-resolution of Z, then we define

$$(2) \qquad \bar{H}^n(\Pi,A) = H^n(\overline{\mathrm{Hom}}\,(X,A)).$$

These are the cohomology groups of Π in the "almost zero theory"; they were considered by Eckmann (*Proc. Nat. Aca. Sci. U.S.A.* 33 (1947), 275–281, 372–376; 39 (1953), 35–42). Combining this with (1) we obtain

$$\bar{H}^n(\Pi,A) \approx H^n(\Pi,Z(\Pi) \otimes A).$$

This reduces the "almost zero theory" to the usual cohomology theory of groups.

We now drop the assumption that X was a Π-projective resolution of Z and assume only that X is weakly projective. Then 8.4 may be applied to yield the spectral sequence

$$(4) \qquad H^p(\Pi,H^q(\mathrm{Hom}\,(X,Z(\Pi) \otimes A))) \underset{p}{\Rightarrow} H^n(\overline{\mathrm{Hom}}\,(X,A)).$$

Assume now that, for each n, the Π-module X_n is Π-free on a *finite* base $\{\sigma_{n,\alpha}\}$. The elements $s\sigma_{n,\alpha}$, for $s \in \Pi$, form a Z-base for X_n. A cochain f: $X_n \to A$ is Π-finite if and only if $f(s\sigma_{n,\alpha}) = 0$ except for a finite number of pairs (s,α). Thus the Π-finite cochains coincide with the finite cochains on X relative to the system of cells $s\sigma_{n,\alpha}$. Thus the spectral sequence (4) becomes

$$(4') \qquad H^p(\Pi,H^q(\mathrm{Hom}\,(X,Z(\Pi) \otimes A))) \underset{p}{\Rightarrow} \mathfrak{H}^n(X,A)$$

where $\mathfrak{H}^n(X,A)$ is the cohomology group of X based on *finite cochains*.

These considerations may be applied in the following topological situation. Let $\mathscr{X}$ be a topological space with Π as a group of left operators. Assume further that a cellular decomposition of $\mathscr{X}$ is given which is invariant under the operations of Π and such that no cell is transformed onto itself except by the element 1 of Π. We finally assume that $\mathscr{X}_{\Pi}$ is compact. If we denote by X the Π-complex of the chains of the cellular decomposition of $\mathscr{X}$, we find that each X_n is Π-free on a finite base. Therefore $\mathfrak{H}^n(X,A)$ is the cohomology group based on the finite cochains of the cell complex. This group, denoted by $\mathfrak{H}^n(\mathscr{X},A)$, is known to be independent of the choice of the cell decomposition, and is known as the "cohomology group of $\mathscr{X}$ with compact supports." The spectral sequence (4′) may now be rewritten as

$$(4'') \qquad H^p(\Pi, H^q(\mathscr{X}, Z(\Pi) \otimes A)) \underset{p}{\Rightarrow} \mathfrak{H}^n(\mathscr{X}, A).$$

EXERCISES

1. In the situation $({}_{\Lambda\text{-}\Sigma}A, B_{\Gamma\text{-}\Sigma}, {}_{\Lambda}C_{\Gamma})$ where Λ, Γ and Σ are K-algebras, define the isomorphism

$$\operatorname{Hom}_{\Lambda\otimes\Sigma}(A, \operatorname{Hom}_{\Gamma}(B,C)) \approx \operatorname{Hom}_{\Gamma\otimes\Sigma}(B, \operatorname{Hom}_{\Lambda}(A,C))$$

and derive appropriate spectral sequences.

2. Show that the homomorphism ρ of VI,5 is an edge homomorphism in one of the spectral sequences of § 4. Use this to generalize VI,5.1.

3. Show that the maps $\cup j$ and $\cap j$ of XI,9 are edge homomorphisms in two of the spectral sequences of § 7. Use this to generalize XI,9.2.

4. Prove VI,3.5, VI,3.5a and VI, Exer. 14 using the spectral sequences of § 4 and Exer. 1.

5. Let $\varphi\colon \Lambda \to \Gamma$ be a ring homomorphism and let A be a left Γ-module. Show that

$$\text{l.w.dim}_\Lambda A \leqq \text{l.w.dim}_\Lambda \Gamma + \text{l.w.dim}_\Gamma A,$$
$$\text{l.inj.dim}_\Lambda A \leqq \text{r.w.dim}_\Lambda \Gamma + \text{l.inj.dim}_\Gamma A,$$
$$\text{l.dim}_\Lambda A \leqq \text{l.dim}_\Lambda \Gamma + \text{l.dim}_\Gamma A.$$

6. Let Λ be a K-projective supplemented algebra. Let C be a positive complex, composed of left Λ-modules and in which the differentiation is a Λ-homomorphism. Assuming that

$$H^q(C) = 0 \qquad \text{for } 0 < q < n,$$

define an exact sequence

$$0 \longrightarrow H^n(\Lambda, H^0(C)) \longrightarrow \mathscr{H}^n(\Lambda, C) \longrightarrow H^0(\Lambda, H^n(C)) \xrightarrow{\varphi} H^{n+1}(\Lambda, H^0(C)) \longrightarrow \mathscr{H}^{n+1}(\Lambda, C).$$

[Hint: use the spectral sequence (1) of § 8.]

Give a dual statement.

7. Let $\mathscr{X}$ and $\mathscr{X}'$ be two n-dimensional spheres (n odd), with the cyclic group Π of order p operating on $\mathscr{X}$ and on $\mathscr{X}'$ (p odd prime). Let f be a continuous mapping $\mathscr{X} \to \mathscr{X}'$ compatible with the operations of Π. Then, using Exer. 6, define a commutative diagram

$$(\Delta) \qquad \begin{array}{ccc} H^n(\mathscr{X}'; Z_p) & \xrightarrow{\varphi'} & H^{n+1}(\Pi, H^0(\mathscr{X}'; Z_p)) \\ \downarrow{\scriptstyle f^n} & & \downarrow{\scriptstyle f^0} \\ H^n(\mathscr{X}; Z_p) & \xrightarrow{\varphi} & H^{n+1}(\Pi, H^0(\mathscr{X}; Z_p)) \end{array}$$

Assume now that Π operates properly on $\mathscr{X}$; then, using $H^{n+1}(\mathscr{X}_\Pi;Z_p) = 0$ and Exer. 6 (or the exact sequence (4) of § 9), prove that φ is an isomorphism. Using the diagram (Δ), show that the homomorphism f^n does not depend on the choice of the mapping f. In other words, the "degrees" of any two mappings $\mathscr{X} \to \mathscr{X}'$ (compatible with Π) are congruent mod. p.

8. Let $\mathscr{X}$ be a connected topological space, with a group Π operating on $\mathscr{X}$; assume that

$$H_q(\mathscr{X}) = 0 \qquad \text{for } 0 < q < n.$$

X denoting the singular complex of $\mathscr{X}$, show that

$$H^q(\operatorname{Hom}_Z(X,H_n(\mathscr{X}))) = 0 \qquad 0 < q < n$$

$$H^n(\operatorname{Hom}_Z(X,H_n(\mathscr{X}))) = \operatorname{Hom}_Z(H_n(\mathscr{X}),H_n(\mathscr{X})).$$

Applying Exer. 6, define a homomorphism

$$\varphi\colon \operatorname{Hom}_\Pi(H_n(\mathscr{X}),H_n(\mathscr{X})) \to H^{n+1}(\Pi,H_n(\mathscr{X})).$$

Let i denote the identity map: $H_n(\mathscr{X}) \to H_n(\mathscr{X})$. Then $\varphi(i)$ is the Eilenberg-MacLane invariant

$$\boldsymbol{k}^{n+1} \in H^{n+1}(\Pi,H_n(\mathscr{X})).$$

Let $\mathscr{X}'$ be another space satisfying the same conditions, with the same group Π operating on $\mathscr{X}'$. Let f be a continuous mapping $\mathscr{X} \to \mathscr{X}'$ compatible with the operations of Π. Show that the corresponding homomorphism

$$H^{n+1}(\Pi,H_n(\mathscr{X})) \to H^{n+1}(\Pi,H_n(\mathscr{X}'))$$

maps the invariant $\boldsymbol{k}^{n+1}$ of $\mathscr{X}$ into the invariant $\boldsymbol{k}'^{n+1}$ of $\mathscr{X}'$.

Using the invariant $\boldsymbol{k}^{n+1}$ give a new proof of the final result of Exer. 7.

CHAPTER XVII

Hyperhomology

Introduction. In Chapter v a resolution of a module A was defined to be a complex with suitable properties. If A itself is a complex the resolution must be defined as a double complex satisfying rather strong conditions (§ 1). Given a functor T of one variable, a complex A and a resolution X of A, it turns out that the invariants of the double complex $T(X)$ are independent of the choice of X and yield the "hyperhomology invariants" of $T(A)$. There result two spectral sequences with essentially the same "limit" and with terms E_2 given by

$$H^p(R^qT(A)) \qquad \text{and} \qquad (R^qT)(H^p(A)).$$

Similar results hold for functors of any number of variables.

The spectral sequences obtained may be regarded as a general solution of the problem partially solved earlier by the Künneth relations (IV,8 and VI,3).

1. RESOLUTIONS OF COMPLEXES

In the sequel we shall have to consider modules, complexes and double complexes all in the same context. The following conventions will simplify matters. Given a double complex A and an integer p, we denote by $A^{p,*}$ the complex B defined by $B^q = A^{p,q}$ and the second differentiation operator d_2 of A. Similarly $A^{*,q}$ is the complex C defined by $C^p = A^{p,q}$ and the first differentiation operator d_1 of A. We shall refer to $A^{p,*}$ and $A^{*,q}$ as the p-th *row* and q-th *column* of A, respectively. The differentiation operators of A yield maps

$$A^{p,*} \to A^{p+1,*}, \qquad A^{*,q} \to A^{*,q+1}.$$

For each double complex A we defined in XV,6 the double complexes $H_{\text{I}}(A)$ and $H_{\text{II}}(A)$. The double complex $H_{\text{I}}(A)$ is obtained by taking homology modules with respect to the first differentiation operator in A. Thus in $H_{\text{I}}(A)$ the first differentiation operator is zero and the second one is induced by d_2. In $H_{\text{II}}(A)$ it is the other way around. Clearly

$$H_{\text{II}}^{p,*}(A) = H(A^{p,*}), \qquad H_{\text{I}}^{*,q}(A) = H(A^{*,q})$$

In quite the same way we define the double complexes

$$B_{\mathrm{I}}(A), \qquad B'_{\mathrm{I}}(A), \qquad Z_{\mathrm{I}}(A), \qquad Z'_{\mathrm{I}}(A)$$

and similar double complexes with I replaced by II.

As in v,1, a module A will be regarded as a complex with $A^0 = A$, $A^n = 0$ for $n \neq 0$. A complex A will as a rule be regarded as a double complex such that $A^{p,0} = A^p$, $A^{p,q} = 0$ for $q \neq 0$. Thus $A^{*,0} = A$ and the given complex appears as a 0-th column in the double complex.

Let A be a complex. A *left double complex X over A* consists of a double complex X such that $X^{p,q} = 0$ for $q > 0$, and of an augmentation map ε: $X \to A$. The augmentation actually is given by the map ε: $X^{*,0} \to A$ such that the composition $X^{*,-1} \to X^{*,0} \to A$ is zero.

Let f: $A \to A'$ be a map of complexes and let X, X' be left double complexes over A,A' with augmentations ε, ε'. A map F: $X \to X'$ such that $\varepsilon' F = f\varepsilon$ is called a *map over f*.

Let X be a left double complex over the complex A. There result the following left complexes:

$(1)_p$ $\qquad X^{p,*}$ over A^p

$(2)_p$ $\qquad Z_{\mathrm{I}}^{p,*}(X)$ over $Z^p(A)$

$(3)_p$ $\qquad Z_{\mathrm{I}}'^{p,*}(X)$ over $Z'^p(A)$

$(4)_p$ $\qquad B_{\mathrm{I}}^{p,*}(X)$ over $B^p(A)$

$(5)_p$ $\qquad B_{\mathrm{I}}'^{p,*}(X)$ over $B'^p(A)$

$(6)_p$ $\qquad H_{\mathrm{I}}^{p,*}(X)$ over $H^p(A)$

We shall say that X is a *projective resolution* of the *complex A* if for all p, (1)–(6) are projective resolutions.

PROPOSITION 1.1. *If for all p,* (4) *and* (6) *are projective resolutions, then X is a projective resolution of A.*

PROOF. Since $(4)_{p+1}$ and $(5)_p$ are naturally isomorphic, it follows that $(5)_p$ is a projective resolution. The sequences $0 \to (4)_p \to (2)_p \to (6)_p \to 0$, $0 \to (6)_p \to (3)_p \to (5)_p \to 0$ being exact, it follows from v,2.1 that $(2)_p$ and $(3)_p$ are projective resolutions. For the same reason the exactness of the sequence $0 \to (4)_p \to (1)_p \to (3)_p \to 0$ implies that $(1)_p$ is a projective resolution.

PROPOSITION 1.2. *Each complex has a projective resolution. If X and Y are projective resolutions of complexes A and C, and f: $A \to C$ is a map, then there is a map F: $X \to Y$ over f. If F,G: $X \to Y$ are maps over homotopic maps f,g: $A \to C$ then F and G are homotopic* (*in the sense of* IV,4).

PROOF. Let A be a complex. For each p select projective resolutions $X^{p,B}$ and $X^{p,H}$ of $B^p(A)$ and $H^p(A)$. By v,2.2 we may find for each p an exact sequence

$$0 \to X^{p,B} \to X^{p,Z} \to X^{p,H} \to 0$$

where $X^{p,Z}$ is a projective resolution of $Z^p(A)$. Applying v,2.2 again we obtain exact sequences

$$0 \to X^{p,Z} \to X^{p,A} \to X^{p+1,B} \to 0$$

where $X^{p,A}$ is a projective resolution of A^p. We define X to be the doubly graded module with $X^{p,A}$ as the p-th row. The first differentiation operator d_1 is defined by composition

$$X^{p,A} \to X^{p+1,B} \to X^{p+1,Z} \to X^{p+1,A}.$$

The second differentiation d_2 is defined for each row $X^{p,A}$ as the differentiation in $X^{p,A}$ with the sign $(-1)^p$. Then $d_1d_2 + d_2d_1 = 0$ and X is indeed a double complex. The augmentation $X \to A$ is defined by the augmentations $X^{p,A} \to A^p$. For this double complex X, the complexes $B_{\mathrm{I}}^{p,*}(A)$ and $H_{\mathrm{I}}^{p,*}(A)$ are isomorphic with $X^{p,B}$ and $X^{p,H}$. Thus it follows from 1.1 that X is a projective resolution of A.

Let X and Y be projective resolutions of the complexes A and C and let $f\colon A \to C$ be a map. Consider the maps

$$f^{p,B}\colon B^p(A) \to B^p(C), \quad f^{p,Z}\colon Z^p(A) \to Z^p(C), \quad f^{p,H}\colon H^p(A) \to H^p(C)$$

induced by f. By v,1.2 there exist maps

$$F^{p,B}\colon B_{\mathrm{I}}^{p,*}(X) \to B_{\mathrm{I}}^{p,*}(Y), \qquad F^{p,H}\colon H_{\mathrm{I}}^{p,*}(X) \to H_{\mathrm{I}}^{p,*}(Y)$$

over $f^{p,B}$ and $f^{p,H}$. By v,2.3, there exist maps

$$F^{p,Z}\colon Z_{\mathrm{I}}^{p,*}(X) \to Z_{\mathrm{I}}^{p,*}(Y)$$

over $f^{p,Z}$ such that the diagrams

$$\begin{array}{ccccccccc} 0 \to & B_{\mathrm{I}}^{p,*}(X) & \to & Z_{\mathrm{I}}^{p,*}(X) & \to & H_{\mathrm{I}}^{p,*}(X) & \to 0 \\ & \downarrow & & \downarrow & & \downarrow & \\ 0 \to & B_{\mathrm{I}}^{p,*}(Y) & \to & Z_{\mathrm{I}}^{p,*}(Y) & \to & H_{\mathrm{I}}^{p,*}(Y) & \to 0 \end{array}$$

are commutative. Applying v,2.3 again we find maps

$$F^{p,*}\colon X^{p,*} \to Y^{p,*}$$

over f^p such that the diagrams

$$\begin{array}{ccccccccc} 0 \to & Z_{\mathrm{I}}^{p,*}(X) & \to & X^{p,*} & \to & B_{\mathrm{I}}^{p+1,*}(X) & \to 0 \\ & \downarrow & & \downarrow & & \downarrow & \\ 0 \to & Z_{\mathrm{I}}^{p,*}(Y) & \to & Y^{p,*} & \to & B_{\mathrm{I}}^{p+1,*}(Y) & \to 0 \end{array}$$

are commutative. It follows that the diagrams

$$\begin{array}{ccc} X^{p,*} & \xrightarrow{d_1} & X^{p+1,*} \\ \downarrow{\scriptstyle F^{p,*}} & & \downarrow{\scriptstyle F^{p+1,*}} \\ Y^{p,*} & \xrightarrow[d_1]{} & Y^{p+1,*} \end{array}$$

are commutative. Thus the maps $F^{p,*}$ yield a map F: $X \to Y$ over f as desired.

Finally let F,G: $X \to Y$ be maps over f,g: $A \to C$ and let s: $f \simeq g$ be a homotopy. By v,1.2 there exist maps $S^{p,*}$: $X^{p,*} \to Y^{p+1,*}$ over s^p: $A^p \to C^{p+1}$. The maps $S^{p,q}$ yield a homomorphism S: $X \to Y$ of bidegree (1,0), which commutes with the augmentation and anticommutes with the second differentiation. Setting

$$J = F + d_1 S + S d_1$$

we find that J: $X \to Y$ is a map over g and that $(S,0)$ is a homotopy $F \simeq J$. It thus remains to be shown that the maps G and J over the same map g are homotopic.

In each of the rows $X^{p,*}$, $Y^{p,*}$, $B_{\mathrm{I}}^{p,*}(X)$ etc., we consider the differentiation operator given by $(-1)^p d_2$. By v,1.2 we may choose homotopies,

$$T^{p,B}\colon J^{p,B} \simeq G^{p,B}, \qquad T^{p,H}\colon J^{p,H} \simeq G^{p,H}.$$

Then v,2.3 yields a homotopy

$$T^{p,Z}\colon J^{p,Z} \simeq G^{p,Z}$$

which properly commutes with the above two. Applying v,2.3 we obtain homotopies

$$T^{p,*}\colon J^{p,*} \simeq G^{p,*}$$

which commute with the above. The maps $(-1)^p T^{p,*}$ yield a homomorphism T: $X \to Y$ of bidegree (0,1) and such that

$$d_1 T + T d_1 = 0, \qquad d_2 T + T d_2 = G - J.$$

Thus $(0,T)$ is the desired homotopy $J \simeq G$. This concludes the proof of 1.2.

A *right double complex over a complex A* is a double complex X such that $X^{p,q} = 0$ for $q < 0$ and an augmentation map ε: $A \to X$. The definition of an *injective resolution* of A and the formulation and proof of the analogue of 1.2 are left to the reader.

PROPOSITION 1.3. *Let A be a complex such that $A^n = 0$ for some set N of integers. Then the projective resolution X of A and the injective resolution Y of A may be chosen so that $X^{n,q} = 0$ and $Y^{n,q} = 0$ for all indices q and all integers $n \in N$.*

This is a direct corollary of the construction of X given in the proof of 1.2.

PROPOSITION 1.4. *Let Λ be a ring such that for some integer n the functor* Ext^{n+1} *is zero. Then for any Λ-complex A the projective resolution X and the injective resolution Y may be chosen so that $X^{p,q} = 0$* and $Y^{p,q} = 0$ for $|q| > n$.

PROOF. The condition $\text{Ext}^{n+1} = 0$ implies, by VI,2.1 and VI,2.1a, that all Λ-modules have projective (injective) dimension $\leqq n$. Thus all projective and injective resolutions of Λ-modules may be chosen of dimension $\leqq n$. Thus the conclusion of 1.4 again follows from the construction given in the proof of 1.2.

2. THE INVARIANTS

Consider the (additive) functor $T(A,C)$ covariant in A, contravariant in C, where A is a Λ_1-module, C is a Λ_2-module and $T(A,C)$ is a Λ-module.

Let A be a Λ_1-complex and X an injective resolution of A, and let C be a Λ_2-complex with a projective resolution Y. Then $T(X,Y)$ is a quadruple complex. We pass from this quadruple complex to a double complex by grouping (see IV,4) the first and the third index and the second and the fourth index. Thus

$$T^{p,q}(X,Y) = \sum T(X^{p_1,q_1}, Y_{p_2,q_2}), \qquad p_1 + p_2 = p, \qquad q_1 + q_2 = q.$$

The differentiation operators δ_1 and δ_2 on $T(X,Y)$ are defined on $T(X^{p_1,q_1}, Y_{p_2,q_2})$ as

$$\delta_1 = T(d_1, Y_{p_2,q_2}) + T(X^{p_1,q_1}, d_1)$$

$$\delta_2 = T(d_2, Y_{p_2,q_2}) + T(X^{p_1,q_1}, d_2).$$

We shall show in a moment, that the invariants of the double complex $T(X,Y)$, consisting of the graded module $\Sigma H^n(T(X,Y))$, its two filtrations, and the two spectral sequences belonging to these two filtrations, are independent of the choice of the resolutions of A and C. We shall refer to these as the "cohomology invariants of the functor T and the complexes A and C" or, by abuse of notations, as the cohomology invariants of $T(A,C)$. The module $H^n(T(X,Y))$ will be written as $\mathscr{R}^n T(A,C)$ and will be called the n-th *hypercohomology* module of $T(A,C)$.

Consider another pair A', C' of complexes and their resolutions X', Y'. Given maps f,f_1: $A \to A'$ and g,g_1: $C' \to C$, we can find maps F,F_1: $X \to X'$, G,G_1: $Y' \to Y$ over f,f_1, g,g_1 respectively. These induce maps

$J = T(F,G)$, $J_1 = T(F_1,G_1)$ of the double complex $T(X,Y)$ into $T(X',Y')$. Thus J and J_1 induce maps of the invariants of these double complexes. Suppose now that we have homotopies $f \simeq f_1$, $g \simeq g_1$. Then by 1.2 we also have homotopies $F \simeq F_1$, $G \simeq G_1$. As was shown in IV,4, this yields a homotopy $J \simeq J_1$ and therefore by XV,6.1, J and J_1 yield the same homomorphisms of the invariants of $T(X,Y)$ into those of $T(X',Y')$.

The above reasoning yields the following conclusions. The invariants of $T(A,C)$ as defined above are independent of the choice of the resolutions of A and C; maps f: $A \to A'$, g: $C' \to C$ induce a map of the invariants of $T(A,C)$ into those of $T(A',C')$; homotopic maps $f \simeq f_1$, $g \simeq g_1$ induce the same maps of the invariants. Thus the invariants of $T(A,C)$ may be regarded as a functor covariant in A, contravariant in C and invariant with respect to homotopies.

We now proceed with the computation of the initial terms $\mathrm{I}_2^{p,q}$, $\mathrm{II}_2^{p,q}$ of the two spectral sequences associated with $T(A,C)$. By XV,6 this amounts to computing the doubly graded modules

$$H_{\mathrm{I}}H_{\mathrm{II}}(T(X,Y)), \qquad H_{\mathrm{II}}H_{\mathrm{I}}(T(X,Y))$$

where X is an injective resolution of A and Y is a projective resolution of C.

We begin by computing $H_{\mathrm{II}}(T(X,Y))$. Since in H_{II} only the second differentiation operators are used, we may concentrate our attention on a fixed row $X^{p_1,*}$ of X and a fixed row $Y_{p_2,*}$ of Y. Then

$$H_{\mathrm{II}}^{p,q}(T(X,Y)) = \sum_{p_1+p_2=p} H^q(T(X^{p_1,*},Y_{p_2,*})).$$

Since $X^{r,*}$ is an injective resolution of A^r and $Y_{s,*}$ is a projective resolution of C_s we find

$$H^q(T(X^{p_1,*},Y_{p_2,*})) = R^qT(A^{p_1},C_{p_2}).$$

We thus find

(1) $$H_{\mathrm{II}}^{p,q}(T(X,Y)) = \sum_{p_1+p_2=p} R^qT(A^{p_1},C_{p_2})$$

or equivalently

(1′) $$H_{\mathrm{II}}^{*,q} = R^qT(A,C).$$

The differentiation operators on both sides of (1′) coincide. Since $H^p(H_{\mathrm{II}}^{*,q})$ is precisely the module of degree (p,q) of $H_{\mathrm{I}}H_{\mathrm{II}}$ we obtain

(2) $$\mathrm{I}_2^{p,q} = H^p(R^qT(A,C)).$$

We now proceed with $H_{\mathrm{I}}(T(A,C))$. Since only the first differentiation operators are employed in computing H_{I} we may limit our attention to fixed columns X^{*,q_1}, Y_{*,q_2} of X and Y. Then

$$H_{\mathrm{I}}^{p,q}(T(X,Y)) = \sum_{q_1+q_2=q} H^p(T(X^{*,q_1},Y_{*,q_2})).$$

Since X^{*,q_1}, $Z(X^{*,q_1})$, $B(X^{*,q_1}), \ldots, H(X^{*,q_1})$ are all composed of injective modules, the complex X^{*,q_1} splits. Similarly Y_{*,q_2} splits. Therefore by IV,7.4 the maps α and α' are defined and are isomorphisms. This leads to the identification

$$H^p(T(X^{*,q_1}, Y_{*,q_2})) = \sum_{p_1+p_2=p} T(H^{p_1}(X^{*,q_1}),H_{p_2}(Y_{*,q_2})).$$

Combining the last two formulae we obtain

$$H_{\mathrm{I}}(T(X,Y)) = T(H_{\mathrm{I}}(X),H_{\mathrm{I}}(Y)). \tag{3}$$

The differentiation operators on both sides coincide and are both given by the second differentiation operators. Since $H_{\mathrm{I}}(X)$ is an injective resolution of $H(A)$ and $H_{\mathrm{I}}(Y)$ is a projective resolution of $H(C)$, we may apply (1) to (3). We find in this way the terms $H_{\mathrm{II}}^{p,q}H_{\mathrm{I}}$, i.e. the terms $\mathrm{II}_2^{p,q}$ as

$$\mathrm{II}_2^{p,q} = \sum_{p_1+p_2=p} R^qT(H^{p_1}(A),H_{p_2}(C)). \tag{4}$$

We recall here that in the notation for the second spectral sequence, the second index q indicates the degree of the filtration. In (2) the degree of the filtration is p.

The case of two variables considered above was only an example. The discussion applies to any number of variables provided all covariant variables are resolved injectively and contravariant variables are resolved projectively.

There is a dual discussion for homology invariants based on projective resolutions of covariant variables and injective resolutions of contravariant variables. The homology invariants of $T(A,C)$ consists of *hyperhomology* modules $\mathscr{L}_nT(A,C)$ possessing two filtrations, and of two spectral sequences beginning with the terms

$$\mathrm{I}_{p,q}^2 = H_p(L_qT(A,C)) \tag{2a}$$

$$\mathrm{II}_{p,q}^2 = \sum_{p_1+p_2=p} L_qT(H_{p_1}(A),H^{p_2}(C)) \tag{4a}$$

where the filtration degree in (2a) is p while in (4a) it is q.

3. REGULARITY CONDITIONS

As above let A and C be complexes, X an injective resolution of A and Y a projective resolution of C. Since $X^{p,q} = 0$ and $Y_{p,q} = 0$ for $q < 0$, it follows that

$$T^{p,q}(X,Y) = 0 \qquad \text{for } q < 0$$

where T is a functor covariant in the first variable and contravariant in the second. Therefore the first filtration is regular. Thus, by xv,6 (Case 1), we have the edge homomorphism

$$\mathrm{I}_2^{n,0} \to H^n(T(X,Y))$$

which becomes the homomorphism

$$(1) \qquad H^n(R^0T(A,C)) \to \mathscr{R}^nT(A,C).$$

As for the second filtration, it need not be regular, but nevertheless by xv,6, we have the edge homomorphism,

$$H^n(T(X,Y)) \to \mathrm{II}_2^{n,0}$$

which yields

$$(2) \qquad \mathscr{R}^nT(A,C) \to R^0T(H(A),H(C)).$$

PROPOSITION 3.1. *The composition of the homomorphisms* (1) *and* (2) *coincides with the homomorphism*

$$\alpha'\colon\ H(R^0T(A,C)) \to R^0T(H(A),H(C))$$

of IV, 6.1a, *applied to the left exact functor* R^0T.

PROOF. Let $\bar{\alpha}$ be the composition of (1) and (2). Clearly $\bar{\alpha}$ is natural. In view of IV,6.1a it therefore suffices to show that $\bar{\alpha}$ is the identity if A and C have differentiation zero. In this case the resolutions X and Y may be constructed simply by choosing resolutions for the modules A^p and C_q and letting the first differentiation operators be zero. Thus $T(X,Y)$ will have the first differentiation operator zero. The fact that $\bar{\alpha}$ is the identity follows from the final remark of xv,6.

In practice, we shall not be able to say much about the cohomology invariants of $T(A,C)$ unless we know that the second filtration also is regular. We have no general criteria for this, but the following two cases include all the situations actually encountered.

Case 1. The complex A is bounded from below (i.e. $A^p = 0$ for p sufficiently small) and the complex C is bounded from above (i.e. $C^p = 0$ for p sufficiently large). Then by 1.3, the resolution X of A may be chosen with $X^{p,q} = 0$ for p small and Y may be chosen with $Y^{p,q} = 0$ for p large. Therefore since $T^{p,q}(X,Y) = \Sigma T(X^{p_1,q_1}, Y_{p_2,q_2})$, $p_1 + p_2 = p$, $q_1 + q_2 = q$, it follows that $T^{p,q}(X,Y) = 0$ for p sufficiently small. In this case the second filtration of $T(X,Y)$ is regular.

Case 2. Suppose that the rings Λ_1 and Λ_2 over which the nodules A and C are given are such that $\mathrm{Ext}^q_{\Lambda_1} = 0$, $\mathrm{Ext}^q_{\Lambda_2} = 0$ for q sufficiently large. Then, by 1.4, the resolutions X and Y may be chosen so that $X^{p,q} = 0 = Y^{p,q}$ for $|q|$ sufficiently large. It follows that $T^{p,q}(X,Y) = 0$ for $|q|$ sufficiently large. In this case the second filtration is regular.

In the sequel, when considering the cohomology invariants we shall automatically assume that we are in one of these two cases. We thus have the sequences

(3) $$H^p(R^qT(A,C)) \underset{p}{\Rightarrow} \mathscr{R}^nT(A,C)$$

(4) $$\sum_{p_1+p_2=p} R^qT(H^{p_1}(A),H_{p_2}(C)) \underset{q}{\Rightarrow} \mathscr{R}^nT(A,C)$$

called, respectively, the first and the second cohomology spectral sequences of $T(A,C)$.

If we assume that $R^qT(A,C) = 0$ for $q > 0$, then the sequence (3) collapses and (4) becomes

(5) $$\sum_{p_1+p_2=p} R^qT(H^{p_1}(A),H_{p_2}(C)) \underset{q}{\Rightarrow} H^n(R^0T(A,C)).$$

If we assume that A is an acyclic right complex over a module M and C is an acyclic left complex over a module N, then (4) collapses and (3) becomes

(6) $$H^p(R^qT(A,C)) \underset{p}{\Rightarrow} R^nT(M,N).$$

If further $R^qT(A,C) = 0$ for $q > 0$ then (6) yields

(7) $$H^n(R^0T(A,C)) \approx R^nT(M,N).$$

This generalizes the rule for expressing the derived functors R^nT using resolutions of the variables.

We now briefly state the corresponding facts for homology invariants. If X is a projective resolution of A and Y an injective resolution of C, then $T_{p,q}(X,Y) = 0$ for $q < 0$ and the second filtration $T(X,Y)$ is regular. We have the edge homomorphisms

$$\mathrm{II}^2_{n,0} \to H_n(T(X,Y)) \to \mathrm{I}^2_{n,0}$$

which give

$$L_0T(H(A),H(C)) \to \mathscr{L}_0T(A,C) \to H(L_0T(A,C)).$$

PROPOSITION 3.1a. *The composition of the above two homomorphisms coincides with the homomorphism*

$$\alpha\colon\ L_0T(H(A),H(C)) \to H(L_0T(A,C))$$

of IV,6.1.

In the sequel, in order to assume the regularity of the first filtration we shall automatically assume that we are in one of the following two cases:

Case 1a. The complex A is bounded from below, while the complex C is bounded from above.

Case 2a. The same as case 2.

In either of these two cases we have the sequences

$$H_p(L_qT(A,C)) \underset{p}{\Rightarrow} \mathscr{L}_nT(A,C) \tag{3a}$$

$$\sum_{p_1+p_2=p} L_qT(H_{p_1}(A),H^{p_2}(C)) \underset{q}{\Rightarrow} \mathscr{L}_nT(A,C) \tag{4a}$$

called, respectively, the first and the second homology spectral sequences of $T(A,C)$.

4. MAPPING THEOREMS

PROPOSITION 4.1. *The natural transformation t: $R^0T \to T$ induces an isomorphism of all the cohomology invariants of $R^0T(A,C)$ onto those of $T(A,C)$.*

PROOF. In view of xv,3.2 it suffices to verify the conclusion for the initial terms of the spectral sequences (the regularity conditions of § 3 being tacitly assumed). This, however, is clear, since by v,5.3 t induces isomorphisms $R^qR^0T \approx R^qT$.

The above proposition shows that T and R^0T have the same cohomology invariants. Thus without any loss of generality, we may assume that T is left exact.

PROPOSITION 4.2. *Let f: $A \to A'$, g: $C' \to C$ be maps of complexes such that the induced mappings*

$$H(RT(A,C)) \to H(RT(A',C'))$$

$$RT(H(A),H(C)) \to RT(H(A'),H(C'))$$

are isomorphisms. Then f and g induce isomorphisms of all the cohomology invariants of $T(A,C)$ onto those of $T(A',C')$.

This is an immediate consequence of xv,3.2.

THEOREM 4.3. *Let f: $A \to A'$, g: $C' \to C$ be maps of complexes which induce isomorphisms $H(A) \to H(A')$, $H(C') \to H(C)$. Let T be a left exact functor such that $R^qT(A,C) = 0 = R^qT(A',C')$ for $q > 0$. Then, if the regularity conditions of § 3 are satisfied, f and g induce an isomorphism $H(T(A,C)) \to H(T(A',C'))$.*

PROOF. Since $R^qT(A,C) = 0$ for $q > 0$, the first spectral sequence of $T(A,C)$ collapses and reduces to an isomorphism $H^n(T(A,C)) \approx R^nT(A,C)$. Thus the second spectral sequence yields

$$\sum_{p_1+p_2=p} R^qT(H^{p_1}(A),H_{p_2}(C)) \underset{q}{\Rightarrow} H^n(T(A,C)).$$

The same holds with A,C replaced by A',C'. Since f and g induce isomorphisms of the terms on the left, the conclusion follows from xv,3.2.

We leave to the reader the statement of analogous propositions for homology invariants.

5. KÜNNETH RELATIONS

We shall suppose here that the functor T is *left exact* and satisfies

(1) $$R^nT = 0 \qquad \text{for } n > 1.$$

We are in Case 3 of xv,6. Thus the first spectral cohomology sequence yields the exact sequence

$$\cdots \to \mathrm{I}_2^{n,0} \to \mathscr{R}^nT \to \mathrm{I}_2^{n-1,1} \to \mathrm{I}_2^{n+1,0} \to \mathscr{R}^{n+1}T \to \mathrm{I}_2^{n,1} \to \cdots$$

which can be conveniently recorded as the exact triangle of graded modules

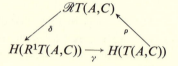

$$\begin{array}{ccc} & \mathscr{R}T(A,C) & \\ {}^{\delta}\swarrow & & \nwarrow^{\rho} \\ H(R^1T(A,C)) & \xrightarrow[\gamma]{} & H(T(A,C)) \end{array}$$

with the degrees of γ, ρ, δ being 2, 0, -1 respectively.

The second spectral cohomology sequence yields exact sequences

$$0 \to \mathrm{II}^{n-1,1} \to \mathscr{R}^nT \to \mathrm{II}^{n,0} \to 0$$

which can be recorded as the exact sequence

$$0 \longrightarrow R^1T(H(A),H(C)) \xrightarrow{\sigma} \mathscr{R}T(A,C) \xrightarrow{\tau} T(H(A),H(C)) \longrightarrow 0$$

with σ, τ having degrees 1, 0.

We already know from 3.1 that the composition $\tau\rho$ is the homomorphism $\alpha'\colon H(T(A,C)) \to T(H(A),H(C))$. It can be shown by a similar argument that the composition $\delta\sigma$ coincides with the homomorphism $\alpha\colon R^1T(H(A),H(C)) \to H(R^1T(A,C))$ which is defined since, by (1), the functor R^1T is right exact.

In sum all the information available can be recorded in a single diagram

(2) $$\begin{array}{ccccccc} 0 \longrightarrow & R^1T(H(A),H(C)) & \xrightarrow{\sigma} & \mathscr{R}T(A,C) & \xrightarrow{\tau} & T(H(A),H(C)) & \longrightarrow 0 \\ & \downarrow{\scriptstyle\alpha} & {}^{\delta}\swarrow & & \nwarrow^{\rho} & \uparrow{\scriptstyle\alpha'} & \\ & H(R^1T(A,C)) & & \xrightarrow[\gamma]{\qquad\qquad} & & H(T(A,C)) & \end{array}$$

in which the top row and center triangle are exact, and the remaining two triangles are commutative.

Suppose now that for some integer n we have

(3) $$H^k(R^1T(A,C)) = 0 \qquad \text{for } k = n-1, n-2$$

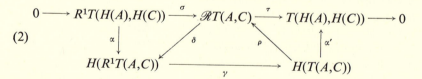

Then ρ yields an isomorphism

$$\rho\colon\ H^n(T(A,C)) \approx \mathscr{R}^n T(A,C)$$

so that the top row of the diagram yields the exact sequence

$$(4)\quad 0 \longrightarrow \sum_{p+q=n-1} R^1T(H^p(A),H_q(C)) \xrightarrow{\beta'} H^n(T(A,C)) \xrightarrow{\alpha'} \sum_{p+q=n} T(H^p(A),H_q(C)) \longrightarrow 0$$

where $\beta' = \rho^{-1}\sigma$.

THEOREM 5.1. *Let $n > 1$ be given. If T is a left exact functor (cov. in A, contrav. in C) such that $R^nT = 0$, and if A and C are complexes such that* (3) *holds, then we have the exact sequence* (4).

In particular the theorem may be applied to the functor $\operatorname{Hom}_\Lambda(A,C)$ where Λ is a hereditary ring. We obtain

COROLLARY 5.2. *If Λ is a hereditary ring and A and C are complexes such that*

$$H^k[\operatorname{Ext}^1_\Lambda(A,C)] = 0 \qquad \text{for } k = n-1, n-2,$$

then we have the exact sequence

$$\begin{aligned} 0 \longrightarrow \sum_{p+q=n-1} \operatorname{Ext}^1_\Lambda(H_p(A),H^q(C)) &\xrightarrow{\beta'} H^n(\operatorname{Hom}_\Lambda(A,C)) \\ &\xrightarrow{\alpha'} \sum_{p+q=n} \operatorname{Hom}_\Lambda(H_p(A),H^q(C)) \longrightarrow 0 \end{aligned}$$

To obtain analogous results for homology invariants we assume that T is right exact and satisfies $L_nT = 0$ for $n > 1$. In diagram (2) we then replace $\mathscr{R}$ by $\mathscr{L}$, R^1 by L_1, we interchange α and α' and reverse all arrows. We obtain:

THEOREM 5.1a. *Let $n > 1$ be given. If T is right exact (cov. in A, contrav. in C), and satisfies $L_nT = 0$, and if A and C are complexes such that*

$$H_k(L_1T(A,C)) = 0 \qquad \text{for } k = n-1, n-2$$

then we have the exact sequence

$$\begin{aligned} (4a)\qquad 0 \longrightarrow \sum_{p+q=n} T(H_p(A),H^q(C)) &\xrightarrow{\alpha} H_n(T(A,C)) \\ &\xrightarrow{\beta} \sum_{p+q=n-1} L_1T(H_p(A),H^q(C)) \longrightarrow 0. \end{aligned}$$

COROLLARY 5.2a. *If Λ is a hereditary ring and A and C are complexes such that*

$$H_k(\operatorname{Tor}^\Lambda_1(A,C)) = 0 \qquad \text{for } k = n-1, n-2$$

then we have the exact sequence

$$\begin{aligned} (5a)\qquad 0 \to \sum_{p+q=n} H_p(A) \otimes_\Lambda H_q(C) &\to H_n(A \otimes_\Lambda C) \\ &\to \sum_{p+q=n-1} \operatorname{Tor}_1(H_p(A),H_q(C)) \to 0. \end{aligned}$$

In stating 5.1 and 5.1a we tacitly assumed that the regularity conditions of § 4 are satisfied. This is no longer necessary in 5.2 and 5.2a since the ring Λ is supposed hereditary, so that we are in Case 2 of § 3.

The reader should compare these results with those of IV,8 and VI,3.

6. BALANCED FUNCTORS

THEOREM 6.1. *Let T be a right balanced functor (cov. in A, contrav. in C), X an injective resolution of a complex A, and Y a projective resolution of a complex C. Then the maps*

$$T(X,C) \xrightarrow{\xi} T(X,Y) \xleftarrow{\mu} T(A,Y)$$

induced by the augmentations, yield isomorphisms between the cohomology invariants of $T(A,C)$ and the invariants of the double complexes $T(X,C)$ and $T(A,Y)$.

PROOF. Since the regularity conditions of § 3 are tacitly assumed, it suffices, in view of XV,3.2, to establish the isomorphisms of the initial terms of the spectral sequences. We shall limit our attention to μ, the proof for ξ being quite analogous.

We begin by considering the operator H_{II}. Since only the second differentiation operator is involved, we may concentrate on a single row $X^{p_1,*}$ of X and a single row $Y_{p_2,*}$ of Y. Then

$$H_{\mathrm{II}}^{p,q}(T(X,Y)) = \sum_{p_1+p_2=p} H^q(T(X^{p_1,*},Y_{p_2,*})$$

$$H_{\mathrm{II}}^{p,q}(T(A,Y)) = \sum_{p_1+p_2=p} H^q(T(A^{p_1},Y_{p_2,*})).$$

Since $X^{p_1,*}$ is an injective resolution of A^{p_1}, while $Y_{p_2,*}$ is a projective resolution of C_{p_2} and the functor T is right balanced, the two terms of the right hand side coincide with $R^qT(A^{p_1},C_{p_2})$. We thus obtain

$$H_{\mathrm{II}}(T(X,Y)) = H_{\mathrm{II}}(T(A,Y)). \tag{1}$$

Applying H_{I} to both sides we obtain the equality of the initial terms in the first spectral sequences.

Before we proceed with the second spectral sequences we establish:

LEMMA 6.2. *If T is right balanced and C is a complex such that both $B^n(C)$ and $H^n(C)$ are projective for all n then for any complex A the map*

$$\alpha'\colon\ HT(A,C) \to T(H(A),H(C))$$

is an isomorphism.

To prove the lemma we first observe that all the modules C^n, $Z^n(C)$, $Z'^n(C)$, $B^n(C)$, $B'^n(C)$ and $H^n(C)$ are projective. If we therefore denote by $\mathcal{M}_C$ the category consisting of these modules and all their homomorphisms into one another, we find that all exact sequences $0 \to D' \to D \to D'' \to 0$ in $\mathcal{M}_C$ split. Thus if we regard T as a functor defined only if the second variable is in $\mathcal{M}_C$, we find that T is exact with respect to the second variable. Since all modules in $\mathcal{M}_C$ are projective and T is right balanced, it follows that T is also exact with respect to the first variable, provided the second variable is in $\mathcal{M}_C$. Thus it follows from IV,7.2 that α' is an isomorphism.

We now return to the proof of 6.1 and apply the operator H_{I}. We find a commutative diagram

$$\begin{array}{ccc} H_{\mathrm{I}}(T(X,Y)) & \xleftarrow{\mu'} & H_{\mathrm{I}}(T(A,Y)) \\ \downarrow{\scriptstyle\alpha'} & & \downarrow{\scriptstyle\alpha'} \\ T(H_{\mathrm{I}}(X),H_{\mathrm{I}}(Y)) & \xleftarrow[\mu'']{} & T(H(A),H_{\mathrm{I}}(Y)) \end{array}$$

We assert that the vertical maps are isomorphisms. Indeed, since only the first differentiation operator is involved we may replace Y by one of its columns $Y^{*,q}$. Since $B^p(Y^{*,q})$ and $H^p(Y^{*,q})$ are both projective modules, the conditions of 6.2 are satisfied and therefore the vertical maps in the diagram are isomorphisms. Applying H_{II} to the diagram we obtain the commutative diagram

$$\begin{array}{ccc} H_{\mathrm{II}}H_{\mathrm{I}}(T(X,Y)) & \leftarrow & H_{\mathrm{II}}H_{\mathrm{I}}(T(A,Y)) \\ \downarrow & & \downarrow \\ H_{\mathrm{II}}(T(H_{\mathrm{I}}(X),H_{\mathrm{I}}(Y))) & \leftarrow & H_{\mathrm{II}}(T(H(A),H_{\mathrm{I}}(Y))) \end{array}$$

in which the vertical maps are isomorphisms. Since $H_{\mathrm{I}}(X)$ is an injective resolution of $H(A)$ and $H_{\mathrm{I}}(Y)$ is a projective resolution of $H(C)$ it follows from (1) that the lower horizontal map also is an isomorphism. This concludes the proof.

THEOREM 6.1a. *Let T be a left balanced functor (cov. in A, contrav. in C), X a projective resolution of a complex A, and Y an injective resolution of a complex C. Then the maps*

$$T(X,C) \leftarrow T(X,Y) \rightarrow T(A,Y)$$

induced by the augmentations, yield isomorphisms between the cohomology invariants of $T(A,C)$ and the invariants of the double complexes $T(X,C)$ and $T(A,Y)$.

It should be noted that 6.1 justifies the notation $\mathscr{R}^nT(A,C)$ used in XVI,2; these modules are indeed the hypercohomology modules of $T(A,C)$. Similarly the module $\mathscr{H}^n(\Lambda,M)$ of XVI,8 is the hypercohomology of $\mathrm{Hom}_\Lambda(K,M)$.

7. COMPOSITE FUNCTORS

We apply the concepts developed in this chapter to study a composite functor $V = TU$ where for simplicity we assume that U is a covariant functor of one variable defined for Λ-modules whose values are Γ-modules, and T is a covariant functor of one variable defined for Γ-modules.

Given a Λ-module A, choose an injective resolution X of A and let $Y = U(X)$. We consider the cohomology invariants of $T(Y)$ and introduce the notation $W^q = (R^qT)U$. Then

$$\mathrm{I}_2^{p,q} = H^p(R^qT(Y)) = H^p(W^q(X)) = R^pW^q(A),$$

$$\mathrm{II}_2^{p,q} = R^qT(H^p(Y)) = R^qT(R^pU(A)).$$

We thus obtain

$$R^pW^q(A) \underset{p}{\Rightarrow} \mathscr{R}^nT(U(X)), \tag{1}$$

$$R^qT(R^pU(A)) \underset{q}{\Rightarrow} \mathscr{R}^nT(U(X)). \tag{2}$$

Further it is clear that since the functor R^nT is invariant under homotopies, the module $\mathscr{R}^nT(U(X))$ is independent of the choice of the injective resolution X of A.

Both spectral sequences are in the case 2 of xv,6 and therefore we have the edge homomorphisms

$$R^nW^0(A) \xrightarrow{\varphi} \mathscr{R}^nT(U(X)) \longrightarrow R^0W^n(A),$$

$$R^nT(R^0U(A)) \longrightarrow \mathscr{R}^nT(U(X)) \xrightarrow{\psi} R^0T(R^nU(A)).$$

If T is left exact, then $R^0T = T$ and $W^0 = (R^0T)U = V$. We thus obtain a homomorphism $\psi\varphi$

$$R^nV \to T(R^nU). \tag{3}$$

If T is exact, both spectral sequences collapse, and φ and ψ become isomorphisms. Thus (3) is then an isomorphism.

Assume now that U is exact. Then the second spectral sequence collapses to an isomorphism $R^nT(U(A)) \approx \mathscr{R}^nT(U(X))$ and the first spectral sequence then becomes

$$R^p((R^qT)U) \underset{p}{\Rightarrow} (R^nT)U$$

This yields edge homomorphisms

$$R^n((R^0T)U) \to (R^nT)U \to R^0((R^nT)U)$$

and the exact sequence of terms of low degree:

$$0 \to R^1((R^0T)U) \to (R^1T)U \to R^0((R^1T)U) \to R^2((R^0T)U) \to (R^2T)U$$

which for T left exact becomes

$$0 \to R^1V \to (R^1T)U \to R^0((R^1T)U) \to R^2V \to (R^2T)U.$$

Similar considerations apply to left derived functors and to functors with a larger number of variables of various variances.

Appendix: Exact Categories

by David A. Buchsbaum

Introduction. Throughout this book, the authors dealt with functors defined on categories of modules over certain rings and whose values again were modules over a ring. It will be shown here that the theory may be generalized to functors defined on abstract categories that will be described below, and whose values are again in such abstract categories. The advantages of such an abstract treatment are manifold. We list a few:
(1°) The dualities of the type

kernel	— cokernel
projective	— injective
$Z(A)$	— $Z'(A)$

that were observed throughout the book may now be formulated as explicit mathematical theorems.
(2°) In treating derived functors, it suffices to consider left derived functors of a covariant functor of several variables; all other types needed may then be obtained by a dualization process.
(3°) Further applications of the theory of derived functors are bound to show that the consideration of modules over a ring Λ will be insufficient. Rings with additional structure such as grading, differentiation, topology, etc. will have to be considered. With the theory developed abstractly, these generalizations are readily available.

The following treatment has some points in common with that of MacLane (*Bull. Amer. Math. Soc.* (1950), pp. 485–516). No proofs will be given here; they will be found in a separate publication.

1. Definition of exact categories. An *exact category* $\mathscr{A}$ is given by the following four data:

(i) a collection of objects A;

(ii) a distinguished object Φ, called the zero object;

(iii) an abelian group $H(A,B)$ given for any two objects $A, B \in \mathscr{A}$. The elements $\varphi \in H(A,B)$ will be called *maps*. We shall frequently write $\varphi\colon A \to B$ instead of $\varphi \in H(A,B)$. The zero element of any of the groups $H(A,B)$ will be denoted by 0;

(iv) a homomorphism $H(B,C) \otimes H(A,B) \to H(A,C)$ given for each triple of objects $A, B, C \in \mathscr{A}$. The image of $\psi \otimes \varphi$ in $H(A,C)$ will be denoted by $\psi\varphi$ and will be called the *composition* of ψ and φ. The primitive terms (i)–(iv) are subjected to four axioms:

Axiom I. If $\alpha\colon A \to B$, $\beta\colon B \to C$, $\gamma\colon C \to D$ then $\gamma(\beta\alpha) = (\gamma\beta)\alpha$.
Axiom II. $H(\Phi,\Phi) = 0$.
Axiom III. For each $A \in \mathscr{A}$ there is a map $e_A\colon A \to A$ such that $e_A\beta = \beta$ for each $\beta\colon B \to A$, and $\gamma e_A = \gamma$ for each $\gamma\colon A \to C$.

It is easy to verify that $H(A,\Phi) = 0 = H(\Phi,A)$ for all $A \in \mathscr{A}$ and that the *identity* map e_A of Axiom III is unique.

A map $\varphi\colon A \to B$ will be called an *equivalence* if there exists a map $\varphi'\colon B \to A$ such that $\varphi'\varphi = e_A$, $\varphi\varphi' = e_B$. It is easy to see that φ' is unique; we write $\varphi' = \varphi^{-1}$. Clearly φ^{-1} is also an equivalence and $(\varphi^{-1})^{-1} = \varphi$. If $\psi\colon B \to C$ is another equivalence, then $\psi\varphi$ also is an equivalence and $(\psi\varphi)^{-1} = \varphi^{-1}\psi^{-1}$.

Definition. *We shall say that the pair of maps*

$$A \xrightarrow{\alpha} B \xrightarrow{\beta} C$$

has property (*E*) *if the following three conditions hold:*

(1) $\beta\alpha = 0$.

(2) *If* $\alpha'\colon A' \to B$ *and* $\beta\alpha' = 0$, *then there exists a unique* $\gamma\colon A' \to A$ *with* $\alpha' = \alpha\gamma$.

(3) *If* $\beta'\colon B \to C'$ *and* $\beta'\alpha = 0$, *then there exists a unique* $\delta\colon C \to C'$ *with* $\beta' = \delta\beta$.

Axiom IV. For any map $\alpha\colon A \to B$ there exist objects K, I, I', F and maps

$$(*) \qquad K \xrightarrow{\delta} A \xrightarrow{\tau} I \xrightarrow{\theta} I' \xrightarrow{\kappa} B \xrightarrow{\pi} F$$

such that

(4) $\alpha = \kappa\theta\tau$

(5) θ is an equivalence

(6) $K \xrightarrow{\delta} A \xrightarrow{\tau} I$ has property (E)

(7) $I' \xrightarrow{\kappa} B \xrightarrow{\pi} F$ has property (E).

Theorem 1: *If*

$$K_1 \xrightarrow{\delta_1} A \xrightarrow{\tau_1} I_1 \xrightarrow{\theta_1} I_1' \xrightarrow{\kappa_1} B \xrightarrow{\pi_1} F_1$$

also satisfy (4)–(7) *then there exist unique maps* χ, μ, ζ, ω *such that the diagram*

$$\begin{array}{ccccccccccc} K & \xrightarrow{\delta} & A & \xrightarrow{\tau} & I & \xrightarrow{\theta} & I' & \xrightarrow{\kappa} & B & \xrightarrow{\pi} & F \\ \downarrow{\scriptstyle\chi} & & \downarrow{\scriptstyle e_A} & & \downarrow{\scriptstyle\mu} & & \downarrow{\scriptstyle\zeta} & & \downarrow{\scriptstyle e_B} & & \downarrow{\scriptstyle\omega} \\ K_1 & \xrightarrow{\delta_1} & A & \xrightarrow{\tau_1} & I_1 & \xrightarrow{\theta_1} & I_1' & \xrightarrow{\kappa_1} & B & \xrightarrow{\pi_1} & F_1 \end{array}$$

is commutative. The maps χ, μ, ζ, ω *are equivalences.*

2. Exact sequences. In view of Theorem 1, we shall call the pairs (K,δ), (I,τ), (I',κ) and (F,π) the *kernel*, *coimage*, *image* and *cokernel* of α. The sense in which these notions are uniquely associated with α is clear from Theorem 1.

A sequence

$$A_m \xrightarrow{\alpha_m} A_{m+1} \longrightarrow \cdots \xrightarrow{\alpha_{n-1}} A_n \qquad m+1<n$$

is now called exact if it satisfies the usual condition:

$$\operatorname{Ker} \alpha_q = \operatorname{Im} \alpha_{q-1} \qquad m<q<n.$$

The following two theorems are crucial.

THEOREM 2. *The maps* $A \xrightarrow{\alpha} B \xrightarrow{\beta} C$ *have the property* (E) *if and only if the sequence*

$$\Phi \longrightarrow A \xrightarrow{\alpha} B \xrightarrow{\beta} C \longrightarrow \Phi$$

is exact.

THEOREM 3. *A map* $\theta\colon A \to B$ *is an equivalence if and only if*

$$\Phi \longrightarrow A \xrightarrow{\theta} B \longrightarrow \Phi$$

is exact.

Monomorphisms and epimorphisms are now defined in the usual way.

With this done it is possible to establish the usual lemmas encountered when dealing with exact sequences. In particular, the "5 lemma", I,1.1 may now be proved.

3. Duality. For each exact category $\mathscr{A}$ we define the dual category $\mathscr{A}^*$ as follows. The objects of $\mathscr{A}^*$ are symbols A^* with $A \in \mathscr{A}$; the zero object of $\mathscr{A}^*$ is Φ^*; the group $H(A^*,B^*)$ is defined as $H(B,A)$; for each map $\varphi\colon B \to A$ in $\mathscr{A}$ we denote by $\varphi^*\colon A^* \to B^*$ the corresponding "dual" map in $\mathscr{A}^*$. The composition in $\mathscr{A}^*$ is given by $\psi^*\varphi^* = (\varphi\psi)^*$.

It is now a trivial matter to verify that $\mathscr{A}^*$ is an exact category. Clearly $(\mathscr{A}^*)^* \approx \mathscr{A}$.

Given any diagram of objects and maps in $\mathscr{A}$, we obtain a dual diagram in $\mathscr{A}^*$, with the maps reversed. It is clear that the dual of a commutative diagram is commutative, and that the dual of an exact sequence is exact. The dual of a monomorphism is an epimorphism and vice versa.

As an illustration as to how one can utilize the dual category, we discuss the "5 lemma" I,1.1. In the hypothesis, a certain commutative diagram with exact rows is given. Then there are two conclusions (1) and (2). It is easy to see that conclusion (2) is precisely conclusion (1) applied to the dual diagram. Thus (2) is actually a consequence of (1) and vice versa.

Suppose now that Λ is a ring. The totality of all left Λ-modules and Λ-homomorphisms (with the usual composition) forms an exact category $\mathscr{M}_\Lambda$. In this category, $H(A,B) = \operatorname{Hom}_\Lambda(A,B)$. However, the dual category $\mathscr{M}^*_\Lambda$ admits no such concrete interpretation. This explains the fact that the duality principle could not be efficiently used, as long as we were restricted to categories concretely defined, in which the objects were sets and the maps were maps of those sets.

Another use of the dual categories is the following. Let $T(A,C)$ be an additive functor defined on the exact categories $\mathscr{A}$ and $\mathscr{C}$ and with values in an exact category $\mathscr{D}$. Suppose that T is covariant in A and contravariant in C. Then replacing $\mathscr{C}$ by $\mathscr{C}^*$, the functor T is converted into a covariant functor in both variables. Another procedure consists in replacing $\mathscr{A}$ and $\mathscr{D}$ by $\mathscr{A}^*$ and $\mathscr{D}^*$.

A few remarks are needed concerning exact categories intended to represent graded groups, graded modules over a graded ring, etc. If only maps of degree zero are considered, then no change in the description of abstract categories is needed. If we wish to consider maps of all degrees, then it is necessary to assume that $H(A,B)$ is graded and that the composition of homogeneous maps adds the degrees. Axiom IV is assumed only for homogeneous maps, and they are the only ones for which the notions of kernel, image, exactness, etc. are defined. In defining the dual $\mathscr{A}^*$ of such a graded exact category, we set $H^n(A^*,B^*) = H^{-n}(B,A) = H_n(B,A)$. This is in keeping with the general principles of IV,5.

4. Homology. An *object with differentiation* in an exact category $\mathscr{A}$ is a pair (A,d) consisting of an object $A \in \mathscr{A}$ and a map $d\colon A \to A$ with $dd = 0$. The definition of $Z(A)$, $Z'(A)$, $B(A)$, $B'(A)$ and $H(A)$ then takes place essentially as in IV,1. The same holds for the definition of the connecting homomorphisms and the exact sequences of IV,1.1.

The self-duality of the definition of $H(A)$ may now be stated in terms of the dual object (A^*,d^*) in the category $\mathscr{A}^*$. We have

$$Z(A^*) = [Z'(A)]^*, \qquad B(A^*) = [B'(A)]^*$$

$$Z'(A^*) = [Z(A)]^*, \qquad B'(A^*) = [B(A)]^*$$

$$H(A^*) = [H(A)]^*$$

In discussing complexes, we avoid direct sum considerations and therefore define a *complex* in A as a sequence

$$\cdots \longrightarrow A^{n-1} \xrightarrow{d^{n-1}} A^n \xrightarrow{d^n} A^{n+1} \xrightarrow{d^{n+1}} A^{n+2} \longrightarrow \cdots$$

with $d^{n+1}d^n = 0$.

We may mention that for each exact category $\mathscr{A}$, the objects (A,d) with differentiation may themselves be converted into an exact category $\mathscr{A}_d$. This is the analogue of the construction of the ring of dual numbers $\Gamma = (\Lambda,d)$ of IV,2. Similarly the complexes in A may be treated as objects in an exact category $\mathscr{A}_c$.

5. Direct sums. So far we have carefully avoided any use of direct sums and products. We see no way of discussing infinite direct sums and products in an exact category $\mathscr{A}$. A finite direct sum (= direct product) may be defined as follows. A family of maps

$$A_\alpha \xrightarrow{i_\alpha} A \xrightarrow{p_\alpha} A_\alpha$$

where α belongs to a *finite* set of indices, is a *direct sum representation* of A if

$$p_\alpha i_\alpha = e_{A_\alpha}$$

$$p_\beta i_\alpha = 0 \qquad \text{for } \beta \neq \alpha$$

$$\textstyle\sum_\alpha i_\alpha p_\alpha = e_A$$

This of course does not guarantee the existence of a direct sum of given factors. For this purpose we introduce

AXIOM V (Existence of direct sums.) For any two objects $A_1, A_2 \in \mathscr{A}$ there is an object $A \in \mathscr{A}$ and maps

$$A_\alpha \xrightarrow{i_\alpha} A \xrightarrow{p_\alpha} A_\alpha \qquad \alpha = 1, 2$$

which yield a direct sum representation of A.

It can then be proved that the direct sum of any finite number of factors exists and is essentially unique (up to equivalences).

Using this axiom it is possible to discuss double (and multiple) complexes $A = \{A^{p,q},d_1,d_2\}$ provided that for each n only a finite number of the objects $A^{p,q}$ with $p + q = n$ is different from Φ. It is then possible to assign to each such double complex an essentially unique complex, and thus define $H^n(A)$.

It is now also possible to duplicate the discussion of IV,6 and IV,7 concerning the homomorphisms α and α' for functors of any number of variables. Note that for functors of one variable, the discussion does not utilize Axiom V.

6. Projective and injective objects. An object $P \in \mathscr{A}$ is called *projective* if any diagram

$$\begin{array}{ccccc} & & P & & \\ & & \downarrow & & \\ A & \to & A'' & \to & \Phi \end{array}$$

in which the row is exact, may be imbedded in a commutative diagram

$$\begin{array}{ccccc} & & P & & \\ & \swarrow & \downarrow & & \\ A & \to & A'' & \to & \Phi \end{array}$$

Similarly, $Q \in \mathscr{A}$ is called *injective* if any diagram

$$\begin{array}{ccccc} \Phi & \to & A' & \to & A \\ & & \downarrow & & \\ & & Q & & \end{array}$$

in which the row is exact, may be imbedded in a commutative diagram

$$\begin{array}{ccccc} \Phi & \to & A' & \to & A \\ & & \downarrow & \swarrow & \\ & & Q & & \end{array}$$

The two notions are dual in the sense that $P \in \mathscr{A}$ is projective if and only if $P^* \in \mathscr{A}^*$ is injective.

For further work we need the following axioms:

Axiom VI (Existence of projectives). Given $A \in \mathscr{A}$ there is an exact sequence $\Phi \to M \to P \to A \to \Phi$ with P projective.

Axiom VI* (Existence of injectives). Given $A \in \mathscr{A}$ there is an exact sequence $\Phi \to A \to Q \to N \to \Phi$ with Q injective.

The axioms are clearly dual to one another.

With Axiom VI assumed, propositions 2.1 (restricted to finite sums), 2.4 and 2.5 of Chapter I may be established. Similarly, if Axiom VI* is assumed, the dual propositions 3.1, 3.4 and 3.5 of Chapter I automatically follow. The same applies to the discussion of V,1, V,2, and XVII,1.

We are now ready to discuss satellite functors and derived functors.

Let T be a covariant functor defined on an exact category $\mathscr{A}$ satisfying Axioms V and VI with values in an exact category $\mathscr{B}$ (with no axioms beyond I–IV). We can then define the left satellite functor S_1T of T. To define the right satellite functor, we assume that $\mathscr{A}$ satisfies Axiom VI* instead of VI. We then define the covariant functor T^*: $\mathscr{A}^* \to \mathscr{B}^*$ and set

$$S^1T = (S_1T^*)^*.$$

All the main results of Chapter III, can be duplicated.

The derived functors L_nT and R^nT are handled similarly, except that now T may be a covariant functor of any number of variables.

The requirement that all variables be covariant is made entirely to simplify the notation; the contravariant variables may always be replaced by their duals.

7. The functors Ext^n. For each exact category $\mathscr{A}$, the functor $H(A,C)$ may be regarded as a functor contravariant in A, covariant in C and with values in the exact category $\mathscr{M}$ of abelian groups. This functor is left exact; for a fixed $A_0 \in \mathscr{A}$, $H(A_0,C)$ is exact if and only if A_0 is projective; for a fixed $C_0 \in \mathscr{A}$, $H(A,C_0)$ is exact if and only if C_0 is injective. Thus $H(A,C)$ is right balanced.

If $\mathscr{A}$ satisfies Axioms V and VI, then Ext^n (A,C) may be defined as the right derived functor with respect to the variable A (i.e. using a projective resolution of A). If A satisfies Axioms V and VI* then injective resolutions of C may be used to define Ext^n (A,C). If A satisfies Axioms V, VI and VI*, either or both may be used.

The discussion of dimension in VI,2 can be carried over mutatis mutandis. The global dimension of an exact category $\mathscr{A}$ is the highest integer n for which Ext^n $(A,C) \neq 0$. A category has global dimension zero if and only if $H(A,C)$ is exact, i.e. if all elements of $\mathscr{A}$ are projective (or injective). This takes the place of semi-simple rings.

8. Other applications. We should like to mention here some applications of exact categories which step outside the framework of this book.

The axiomatic homology and cohomology theories of Eilenberg-Steenrod (*Foundations of Algebraic Topology*, Princeton, 1952) may be defined using an arbitrary exact category $\mathscr{A}$ as the range of values of the theory. Thus, replacing $\mathscr{A}$ by $\mathscr{A}^*$ replaces a homology theory by a cohomology theory, and vice versa. This duality principle simplifies the exposition of the theory. Furthermore, the uniqueness proof (*loc. cit.*, Ch. IV) remains valid for such generalized homology and cohomology theories.

The Pontrjagin duality for discrete and compact abelian groups readily shows that the category $\mathscr{C}$ of compact abelian groups is the dual of the category $\mathscr{M}$ of discrete abelian groups. Thus we conclude that $\mathscr{C}$ satisfies Axioms V, VI and VI*. In fact, in $\mathscr{C}$, the injectives are the toroids (since the only discrete abelian projectives are the free groups); and the projectives in $\mathscr{C}$ are those compact groups whose character groups are divisible.

List of Symbols

Each symbol is listed with the number of the page where it is introduced and explained.

Index of Terminology